Building Regulations in Brief

Ray Tricker and Rozz Algar

Fifth edition

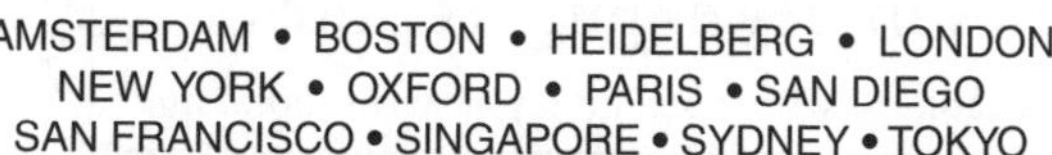

AMSTERDAM • BOSTON • HEIDELBERG • LONDON
NEW YORK • OXFORD • PARIS • SAN DIEGO
SAN FRANCISCO • SINGAPORE • SYDNEY • TOKYO

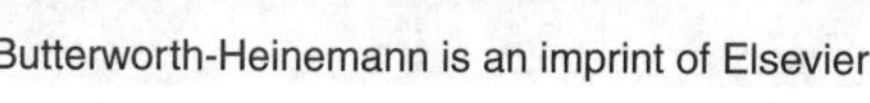

Butterworth-Heinemann is an imprint of Elsevier

Butterworth-Heinemann is an imprint of Elsevier
Linacre House, Jordan Hill, Oxford OX2 8DP
30 Corporate Drive, Suite 400, Burlington, MA 01803

First edition 2003
Second edition 2004
Third edition 2005
Fourth edition 2006

British Library Cataloguing in Publication Data
A catalogue record for this book is available from the British Library

Library of Congress Cataloging-in-Publication Data
A catalog record for this book is available from the Library of Congress

ISBN-13: 978-0-7506-8444-6

For information on all Butterworth-Heinemann publications
visit our web site at http://books.elsevier.com

Typeset by Charon Tec Ltd (A Macmillan Company), Chennai, India
www.charontec.com

Printed and bound in the UK
06 07 08 09 10 10 9 8 7 6 5 4 3 2 1

Contents

Appendices referred to in this book can be found at books.elsevier.com/companions/978 07506 84446

Foreword

Subject to specified exemptions, all building work in England and Wales (a separate system of building control applies to Scotland and Northern Ireland) is governed by Building Regulations. This is a statutory instrument, which sets out the minimum requirements and performance standards for the design and construction of buildings, and extensions to buildings.

The current regulations are the Building Regulations 2000. These take into consideration some major changes in technical requirement (such as conservation of fuel and power) and some procedural changes allowing local authorities to regularize unauthorized development.

Although the 2000 regulations are comparatively short, they rely on their technical detail being available in a series of Approved Documents and a vast number of British, European and international standards, codes of practice, drafts for development, published documents and other non-statuary guidance documents.

The main problem, from the point of view of the average builder and DIY enthusiast, is that the Building Regulations are too professional for their purposes. They cover every aspect of building, are far too detailed and contain too many options. All the builder or DIY person really requires is sufficient information to enable them to comply with the regulations in the simplest and most cost-effective manner possible.

Building inspectors, acting on behalf of local authorities, are primarily concerned with whether a building complies with the requirements of the Building Regulations and to do this, they need to 'see the calculations'. But how do the DIY enthusiast and/or builder obtain these calculations? Where can they find, for instance, the policy and requirements for load bearing elements of a structure?!

Builders, through experience, are normally aware of the overall requirements for foundations, drains, walls, central heating, air conditioning, safety, security, glazing, electricity, plumbing, roofing, floors, etc., but they still need a reminder when they come across a different situation for the first time (e.g. what if they are going to construct a building on soft soil, how deep should the foundations have to be?).

On the other hand, the DIY enthusiast, keen on building his own extension, conservatory, garage or workshop etc. usually has no past experience and needs the relevant information – but in a form that he can easily understand without having had the advantage of many years experience. In fact, what he really needs is a rule of thumb guide to the basic requirements.

From a number of surveys it has emerged that the majority of builders and virtually all DIY enthusiasts are self taught and most of their knowledge is gained through experience. When they hit a problem, it is usually discussed over a pint in the local pub with friends in the building trade as opposed to seeking professional help. What they really need is a reference book to enable them to understand (or remind themselves of) the official requirements.

The aim of my book, therefore, is to provide the reader with an in-brief guide that can act as an *aide-mémoire* to the current requirements of the Building Regulations. Intended readers are primarily builders and the DIY fraternity (who need to know the regulations but do not require the detail), but the book, with its ready reference and no-nonsense approach, will be equally useful to students, architects, designers, building surveyors and inspectors, etc.

This edition of the book includes the requirements of Part B Fire Safety for dwelling houses (volume 1) and non dwelling houses (volume 2). Also included are outline details of the new proposed Guidance Document for Electronic Communication Services.

Note: If any reader has any thoughts about the contents of this book (such as areas where perhaps they feel I have not given sufficient coverage, omissions and/or mistakes, etc.) then please let me know by e-mailing me at ray@herne.org.uk and I will make suitable amendments in the next edition of this book.

Preface

The Great Fire of London in 1666 was probably the single most significant event to shape today's legislation! The rapid growth of fire through co-joined timber buildings highlighted the need to consider the possible spread of fire between properties and this consideration resulted in the publication of the first building construction legislation in 1667 requiring all buildings to have some form of fire resistance.

Two hundred years later, the Industrial Revolution had meant poor living and working conditions in ever expanding, densely populated urban areas. Outbreaks of cholera and other serious diseases, through poor sanitation, damp conditions and lack of ventilation, forced the government to take action and building control took on the greater role of health and safety through the first Public Health Act of 1875. This Act had two major revisions in 1936 and 1961, leading to the first set of national building standards (i.e. the Building Regulations 1965). Over the years these regulations have been amended and updated and the current document is the Building Regulations 2000.

The Building Regulations are approved by the Secretary of State and are intended to provide guidance to some of the more common building situations as well as providing a practical guide to meeting the requirements of Regulation 7 of the Building Act 1984, which states:

Materials and workmanship

7. Building work shall be carried out –

(a) with adequate and proper materials which –
  - (i) are appropriate for the circumstances in which they are used,
  - (ii) are adequately mixed or prepared, and
  - (iii) are applied, used or fixed so as adequately to perform the functions for which they are designed; and

(b) in a workmanlike manner.

What are the current regulations?

The current legislation is the Building Regulations 2000 (Statutory Instrument No 2531) which is made by the Secretary of State for the Environment under powers delegated by parliament under the Building Act 1984. Since then, the Building Regulations have received a number of Building Amendment Regulations as shown below.

Table P.1 Statutory instruments currently in place

The Building Regulations 2000 (SI 2000 No 2531)
Made *13 September 2000*
Laid before Parliament *22 September 2000*
Came into force *1 January 2001*

Statutory Instrument	Made	Laid before Parliament	Coming into force
SI 2001 No 3335	4 Oct 2001	11 Oct 2001	1 Apr 2002
SI 2002 No 440	28 Feb 2002	5 Mar 2002	1 Apr 2002
SI 2002 No 2871	16 Nov 2002	25 Nov 2002	1 Jan 2004
SI 2003 No 2692	17 Oct 2003	27 Oct 2003	1 May 2003
SI 2004 No 1465	28 May 2004	8 Jun 2004	1 Dec 2004
SI 2004 No 3210	6 Dec 2004	10 Dec 2004	31 Dec 2004
SI 2006 No 652	9 Mar 2006	15 Mar 2006	6 April 2006
SI 2006 No 3318	13 Dec 2006	18 Dec 2006	6 April 2007

Note: Copies of the above documents are available from TSO (☎ 0870 600 5522) and through booksellers. They can also be viewed on the DCLG website at www.communities.gov.uk

The Building Act 1984

By Act of Parliament, the Secretary of State is responsible for ensuring that the health, welfare and convenience of persons living in or working in (or nearby) buildings is secured. This Act is called the Building Act 1984 and one of its prime purposes is to assist in the conservation of fuel and power, prevent waste, undue consumption, and the misuse and contamination of water.

It imposes on owners and occupiers of buildings a set of requirements concerning the design and construction of buildings and the provision of services, fittings and equipment used in (or in connection with) buildings.

The Building Act 1984 consists of five parts:

Part 1 The Building Regulations
Part 2 Supervision of Building Work etc. other than by a Local Authority
Part 3 Other provisions about buildings
Part 4 General
Part 5 Supplementary

Part 5 then contains seven schedules whose prime function is to list the principal areas requiring regulation and to show how the Building Regulations are to be controlled by local authorities. These schedules are:

Schedule 1 – Building Regulations;
Schedule 2 – Relaxation of building regulations;

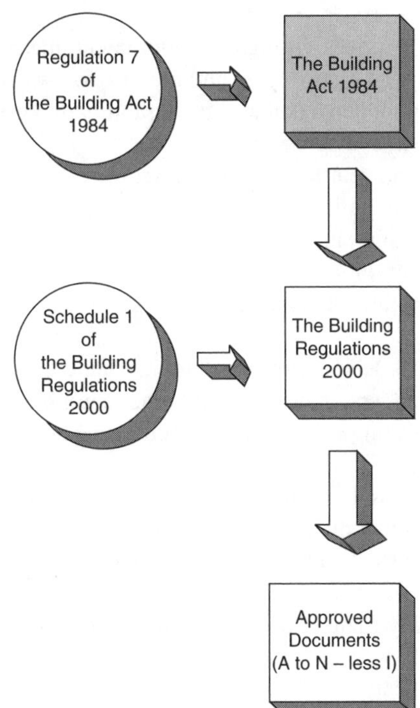

Figure P.1 Implementing the Building Act

Schedule 3 – Inner London;
Schedule 4 – Provisions consequential upon public body's notice;
Schedule 5 – Transitional provisions;
Schedule 6 – Consequential amendments;
Schedule 7 – Repeals.

Schedule 1 is the most important (from the point of view of builders) as it shows, in general terms, how the Building Regulations are to be administered by local authorities, the approved methods of construction and the approved types of materials that are to be used in (or in connection with) buildings.

The Building Act 1984 does not apply to Scotland or to Northern Ireland.

The Building Regulations describe the mandatory requirements for completing **all** building work including:

- accommodation for specific purposes (e.g. for disabled persons);
- air pressure plants;
- cesspools (and other methods for treating and disposing of foul matter);
- dimensions of rooms and other spaces (inside buildings);
- drainage (including waste disposal units);

- emission of smoke, gases, fumes, grit or dust (or other noxious or offensive substances);
- fire precautions (services, fittings and equipment, means of escape);
- lifts (escalators, hoists, conveyors and moving footways);
- materials and components (suitability, durability and use);
- means of access to and egress from;
- natural lighting and ventilation of buildings;
- open spaces around buildings;
- prevention of infestation;
- provision of power outlets;
- resistance to moisture and decay;
- site preparation;
- solid fuel, oil, gas, electricity installations (including appliances, storage tanks, heat exchangers, ducts, fans and other equipment);
- standards of heating, artificial lighting, mechanical ventilation and air-conditioning;
- structural strength and stability (overloading, impact and explosion, underpinning, safeguarding of adjacent buildings);
- telecommunications services (wiring installations for telephones, radio and television);
- third party liability (danger and obstruction to persons working or passing by building work);
- transmission of heat;
- transmission of sound;
- waste (storage, treatment and removal);
- water services, fittings and fixed equipment (including wells and bore-holes for supplying water); and
- matters connected with (or ancillary to) any of the foregoing matters.

The Building Regulations

Building Regulations 2000 (Statutory Instrument No 2531) has been made by the Secretary of State for the Environment under powers delegated by Parliament under the Building Act 1984. They are a set of minimum requirements and basic performance standards designed to secure the health, safety and welfare of people in and around buildings and to conserve fuel and energy in England and Wales.

They are legal requirements laid down by parliament and based on the Building Act 1984. The Building Regulations:

- are approved by parliament;
- deal with the minimum standards of design and building work for the construction of domestic, commercial and industrial buildings;
- set out the procedure for ensuring that building work meets the standards laid down;

- are designed to ensure structural stability;
- promote the use of suitable materials to provide adequate durability, fire and weather resistance, and the prevention of damp;
- stipulate the minimum amount of ventilation and natural light to be provided for habitable rooms;
- ensure the health and safety of people in and around buildings (by providing functional requirements for building design and construction);
- promote energy efficiency in buildings;
- contribute to meeting the needs of disabled people.

The level of safety and standards acceptable are set out as guidance in the approved documents. Compliance with the detailed guidance of the Approved Documents is usually considered as evidence that the Building Regulations themselves have been complied with.

Approved Documents

The Building Regulations are supported by separate documents which correspond to the different areas covered by the regulations. These are called 'Approved Documents' and they contain practical and technical guidance on ways in which the requirements of Schedule 1 and Regulation 7 of the Building Act 1984 can be met.

Each Approved Document reproduces the actual *requirements* contained in the Building Regulations relevant to the subject area. This is then followed by *practical and technical guidance* (together with examples) showing how the requirements can be met in some of the more common building situations. There may, however, be alternative ways of complying with the requirements to those shown in the Approved Documents and you are, therefore, under no obligation to adopt any particular solution in an Approved Document if you prefer to meet the requirement(s) in some other way.

The current set of approved documents are in 13 parts, A to N (less 'I') and consist of:

A Structural
B Fire safety
C Site preparation and resistance to moisture
D Toxic substances
E Resistance to the passage of sound
F Ventilation
G Hygiene
H Drainage and waste disposal
J Combustion appliances and fuel storage systems
K Protection from falling, collision and impact
L Conservation of fuel and power
M Access and facilities for disabled people

N Glazing – safety in relation to impact, opening and cleaning
P Electrical safety

Parts A to D, F to K (except for paragraphs H2 and J6), N and P of Schedule 1 do not require anything to be done except for the purpose of securing reasonable standards of health and safety for persons in (or about) buildings and for any others who may be affected by buildings, or matters connected with buildings.

Notes:

(1) Paragraphs H2 and J6 are excluded from Regulation 8 because they deal directly with the prevention of contamination of water.
(2) Parts E and M (which deal, respectively, with resistance to the passage of sound, and access to and use of buildings) are excluded from Regulation 8 because they address the welfare and convenience of building users.
(3) Part L is excluded from Regulation 8 because it addresses the conservation of fuel and power.

Planning permission

Planning permission is the single biggest hurdle for anyone who has acquired land on which to build a house, or wants to extend or carry out other building work on property. There is never a guarantee that permission will be given and without it no project can start. Yet the system is not at all user-friendly.

There is a bewildering array of formalities to go through and ever more stringent requirements to satisfy. Planning permission has never been more difficult to get, nor so sought after. Every year over half-a-million applications are made and the number is rising.

The purpose of the planning system is to protect the environment as well as public amenities and facilities. It is **not** designed to protect the interests of one person over another. Within the framework of legislation approved by parliament, councils are tasked to ensure that development is allowed where it is needed, while ensuring that the character and amenity of the area are not adversely affected by new buildings or changes in the use of existing buildings and/or land.

Provided, that the work you are completing does not affect the external appearance of the building, you are allowed to make certain changes to your home without having to apply to the local council for permission. These are called 'Permitted Development Rights', but the majority of building work, that you are likely to complete will, however, probably require you to have planning permission – so be warned!

The actual details of planning requirements are complex but for most domestic developments, the planning authority is only really concerned with construction work such as an extension to the house (e.g. a conservatory) or the provision of a new garage or new outbuildings. Structures like walls and fences also need to be considered because their height or siting might well infringe the rights of their neighbours and other members of the community. The planning authority will also want to approve any change of use, such as converting a house into flats or running a business from premises previously occupied as a dwelling only.

The voluntary Code for Sustainable Homes was launched as part of a package of measures towards zero carbon development on 13th December 2006. Full technical guidance on how to comply with the code was published in April 2007. Further details of the code can be found in Chapter 4.

Also in Chapter 4 you will find details of the new Homebuyers Information Pack (HIP). Whilst this does not necessarily fall within the natural bounds of this book (e.g. doesn't have a direct impact on building regulations etc.), it does make up part of the responsibilities of a home owner and much of the documentation created at the time of building or extending or renovating a home will be relevant to the new HIP.

Aim of this book

The prime aim of this book is to provide builders and DIY people with an aide memoire and a quick reference to the requirements of the Building Regulations. This book provides a user-friendly background to the Building Act 1984 and its associated Building Regulations. It explains the meaning of the Building Regulations, their current status, requirements, associated documentation and how local authorities and councils view their importance. It goes on to describe the content of the guidance documents (i.e. the 'Approved Documents') published by the Secretary of State and, in a series of 'what ifs', provides answers to the most common questions that DIY enthusiasts and builders might ask concerning building projects.

The book is structured as follows:

Chapter 1 – The Building Act 1984
Chapter 2 – The Building Regulations 2000
Chapter 3 – The requirements of the Building Regulations
Chapter 4 – Planning permission
Chapter 5 – How to comply with the requirements of the Building Regulations
Chapter 6 – Meeting the requirements of the Building Regulations

These chapters are then supported by the following appendices:

Appendix A Access and facilities for disabled people
Appendix B Conservation of fuel and power
Appendix C Sound insulation
Appendix D Electrical safety

and concludes with a bibliography, useful names and addresses, and a full index.

The following symbols will help you get the most out of this book:

an important requirement or point

a good idea or suggestion

further amplification or information.

Main changes in the 2006 edition

The new 2006 Edition of Part B (Fire Safety) came into force on 6 April 2007 and, similar to Part L (Conservation of Fuel and Power), the requirements have now been split into two Volumes. One is for dwelling houses and the second is for buildings other than dwelling houses. Quite a lot of the information, however, is duplicated between these two Volumes which probably account for Vol 2 being twice as thick as Vol 1!

These changes affect:

- all new building work in England and Wales;
- new extensions to existing buildings;
- material changes of use of a whole building;
- material alterations to existing buildings.

For Domestic Buildings the new Regulations now include fire safety guidance on:

- loft conversions;
- the use of door closing devices in dwellings;
- the use of sprinklers in tall buildings;

and for Non-Domestic Buildings, fire safety guidance is included on:

- the use of sprinklers in single storey warehouses;
- the use of sprinklers in care homes and other houses of multiple occupation.

Note: Copies of these documents are available from RIBA Bookshops (☎ 020 7374 2737 – sales@ribabookshops.com) or an on-line version can be downloaded from http://www.planningportal.gov.uk/england/professionals/en/4000000000084.html

In more detail, the changes made by the publication of Part B:2006 concern:

Adult placements

Introduction of a Code of Practice for fire safety in adult placements (e.g. registered group homes, sheltered housing etc.).

Car parks

Non-combustible materials now need to be used in the construction of a car park in order for it to be regarded as 'open sided'.

Cavity barriers

Window and door frames now only suitable for use as cavity barriers if they are constructed of steel or timber of an appropriate thickness.

Certification schemes

The use of Independent Certification Schemes (such as the Part P Competent Person Self-certification Scheme) now being used for evidence of compliance.

Compartment walls

Requirements for the predicted deflection of a floor (i.e. in the event of a fire) to be included in the design of compartment walls.

Compartmentation

Guidance on the junction between compartment walls and roofs clarified.

Concealed spaces

This section has been completely restructured.

Fire alarms

Guidance for buildings other than dwellings has been updated to take account of the 2002 edition of BS 5839-1.

Fire dampers

Guidance on the specification and installation of fire dampers.

Fire risk analysis

Part B now includes a requirement for the responsible person (i.e. the person carrying out work to a building) to make available to the owner (other than houses occupied as single private dwellings) 'fire safety information' concerning the design and construction of the building or extension plus details of the services, fittings and equipment that have been provided in order that they (when required under the new Regulatory Reform (Fire Safety) Order 2005 – Statutory Instrument 2005 No 1541) may complete a fire risk analysis.

The sort of information required must include basic advice on the proper use and maintenance of systems provided in the building such as:

- emergency egress windows;
- fire doors;
- smoke alarms;
- sprinklers etc.

For small buildings, basic information about the location and nature of fire protection measures might be all that is necessary. For larger buildings, a more detailed record of the fire safety strategy and procedures for operating and maintaining fire protection measures will probably be needed.

Note: Although this primarily concerns places of work, in some circumstances (such as domestic training) it can extend to dwelling houses.

Integral garages

The provision of a sloping floor now included as an alternative to the 100 mm step between dwelling houses and integral garages.

Management of premises

New guidance on the need to ensure that management systems are realistic.

Means of escape

- additional guidance on the protection of ventilation systems;
- additional options to provide sprinkler protection and/or a protected stairway instead of alternative escape routes has been included for flats with more than one storey;
- additional guidance with respect to work on existing houses;
- alternative approach for loft conversions to two storey buildings removed;
- guidance applicable to small premises updated;
- guidance on replacement windows included;
- guidance on means of escape for disabled people;
- guidance on means of escape from buildings that are subject to open special planning;
- guidance on the provision of cavity barriers associated with subdivided corridors clarified;
- Guidance on the use of air circulation systems in houses with protected stairways;
- locks and child resistant safety catches may be provided on escape windows;
- guidance to the interaction with fire fighters during phased evacuation of tall buildings;
- new guidance on the design of residential care homes and the inclusion of sprinklers and/or free swing door closing devices;

- new guidance on the provision of galleries and inner rooms;
- new method for calculating acceptable final exit widths;
- requirements for smoke control in the common areas of flats changed;
- new option to provide sprinkler protection instead of alternative escape routes for dwelling houses with a floor more than 7.5 m above ground level.

Residential sprinklers

The use of BS 9251:2005 sprinkler systems to provide automatic fire suppression where reasonably necessary.

Roof coverings

Guidance on roof coverings updated to incorporate new European system of classification set out in BS EN 13501-5:2005.

Smoke (i.e. fire) alarms

- smoke alarms now need to be installed in accordance with BS 5839-6:2004;
- all smoke alarms should have a standby power supply;
- if a dwelling house is extended, then smoke alarms should be provided in circulation spaces.

Self-closing devices

Fire doors now only need to not be provided with self closing devices, if they are between a dwelling house and an integral garage.

Sprinkler protection in flats

Sprinkler systems should be provided in blocks of flats exceeding 30 m in height.

Under floor voids

Extensive cavities in floor voids should be subdivided with cavity barriers.

Vehicle access

There should be access for a pump appliance to within 45 m of **all** points of a dwelling house.

Warehouses

A maximum compartment size has been introduced for unsprinkled single storey warehouse buildings.

1 The Building Act 1984

1.1 Aim of the Building Act 1984 (Building Act 1984 Section 1)

By Act of Parliament, the Secretary of State is responsible for ensuring that the health, welfare and convenience of persons living in or working in (or nearby) buildings is secured.

This Act is called the Building Act 1984 and one of its prime purposes is to assist in the conservation of fuel and power, to prevent waste, undue consumption, misuse and contamination of water. It imposes on owners and occupiers of buildings a set of requirements concerning the design and construction of buildings and the provision of services, fittings and equipment used in (or in connection with) buildings. These involve, and cover:

- a method of controlling (inspecting and reporting) buildings;
- how services, fittings and equipment may be used;
- the inspection and maintenance of any service, fitting or equipment used.

1.1.1 What about the rest of the United Kingdom?

As shown in Table 1.1, the Building Act 1984 does **not** apply to Scotland or Northern Ireland.

Scotland

Within Scotland, the requirements for buildings are controlled by the *Building (Scotland) Act 2003*. The *Building (Scotland) Regulations 2004* then set the

Table 1.1 Building legislation

	Act	Regulations	Implementation
England and Wales	Building Act 1984	Building Regulations 2000	Approved Documents
Scotland	Building (Scotland) Act 2003	Building (Scotland) Regulations 2004	Technical Handbooks
Northern Ireland	Building Regulations (Northern Ireland) Order 1979	Building Regulations (Northern Ireland) 2000	'Deemed to satisfy' by meeting supporting publications

functional standards under this Act. The methods for implementing these requirements are similar to England and Wales, except that the guidance documents (i.e. for achieving compliance) are contained in two *Technical Handbooks*, one for domestic work and one for non-domestic. Each handbook has a general section and 6 technical sections.

The main procedural difference between the Scottish system and the others is that a building warrant is **still** required before work can start in Scotland.

	England and Wales		Scotland		Northern Ireland
Part A	Structure	Section 1	Structure	Technical Booklet D	Structure
Part B	Fire safety	Section 2	Fire	Technical Booklet E	Fire safety
Part C	Site preparation and resistance to contaminants and water	Section 3	Environment	Technical Booklet C	Preparation of site and resistance to moisture
Part D	Toxic substances	Section 3	Environment	Technical Booklet B	Materials and workmanship
Part E	Resistance to the passage of sound	Section 5	Noise	Technical Booklet G	Sound insulation of dwellings
Part F	Ventilation	Section 3	Environment	Technical Booklet K	Ventilation
Part G	Hygiene	Section 3	Environment	Technical Booklet P	Sanitary appliances and unvented hot water storage systems
Part H	Drainage and waste disposal	Section 3	Environment	Technical Booklet J	Solid waste in buildings
				Technical Booklet N	Drainage
Part J	Combustion appliances and fuel storage systems	Section 3 Section 4	Environment Safety	Technical Booklet L	Heat-producing appliances and liquefied petroleum gas installations
Part K	Protection from falling, collision and impact	Section 4	Safety	Technical Booklet H	Stairs, ramps and protection from impact
Part L	Conservation of fuel and power	Section 6	Energy	Technical Booklet F	Conservation of fuel and power
Part M	Access and facilities for disabled people	Section 4	Safety	Technical Booklet R	Access for facilities and disabled people
Part N	Glazing	Section 6	Energy	Technical Booklet V	Glazing
Part P	Electrical safety	Section 4	Safety		

Northern Ireland

On the other hand, the *Building Regulations (Northern Ireland) Order 1979* (as amended by the *Planning and Building Regulations (Amendment) (NI) Order 1990*) is the main legislation for Northern Ireland and the *Building Regulations (Northern Ireland) 2000* then details the requirements for meeting this legislation.

Supporting publications (such as British Standards, BRE publications and/or Technical Booklets published by the Department) are then used to ensure that the requirements are implemented (*deemed to satisfy*).

1.2 What happens if I contravene any of these requirements? (Building Act 1984 Sections 2, 7, 35, 36 and 112)

If you contravene the Building Regulations or wilfully obstruct a person acting *in the execution of the Building Act 1984 or of its associated Building Regulations*, then on summary conviction, you could be liable to a fine or, in exceptional circumstances, even a short holiday in one of HM Prisons!

1.3 Who polices the Act?

Under the terms of the Building Act 1984, local authorities are responsible for ensuring that any building work being completed conforms to the requirements of the associated Building Regulations. They have the authority to:

- make you take down and remove or rebuild anything that contravenes a regulation;
- make you complete alterations so that your work complies with the Building Regulations;
- employ a third party (and then send you the bill!) to take down and rebuild non-conforming buildings or parts of buildings.

They can, in certain circumstances, even take you to court and have you fined – especially if you fail to complete the removal or rebuilding of the nonconforming work.

The above authority to prosecute and order remedial work to be completed applies equally whether you are the actual owner or merely the occupier – so be warned!

1.4 Are there any exemptions from Building Regulations? (Building Act 1984 Sections 3, 4 and 5)

The following are exempt from the Building Regulations:

- A '*public body*' (i.e. local authorities, county councils and any other body '*that acts under an enactment for public purposes and not for its own profit*').

This can be rather a grey area and it is best to seek advice if you think that you come under this category;

- Buildings belonging to '*statutory undertakers*' (e.g. a water board).

Note: From 1 April 2001, maintained schools ceased to have exemption from the Building Regulations and school-specific standards have now been incorporated into the latest editions of Approved Documents.

Purpose-built student living accommodation (including flats) should thus be treated as hotel/motel accommodation in respect of space requirements and internal facilities.

1.4.1 What about Crown buildings? (Building Act 1984 Sections 44a, d and 87)

Although the majority of the requirements of the Building Regulations are applicable to Crown buildings (i.e. a building in which there is a Crown or Duchy of Lancaster or Duchy of Cornwall interest) or government buildings (held in trust for Her Majesty) there are occasional deviations and before submitting plans for work on a Crown building you should seek the advice of the Treasury.

1.4.2 What about buildings in Inner London? (Building Act 1984 Sections 44, 46 and 88)

You will find that the majority of the requirements found in the Building Regulations are also applicable to buildings in Inner London boroughs (i.e. Inner Temple and Middle Temple). There are, however, some important deviations (see Section 1.7.3) and before submitting plans you should seek the advice of the local authority concerned.

1.4.3 What about the UK Atomic Energy Authority? (Building Act 1984 Section 45)

The Building Regulations do **not** apply to buildings belonging to or occupied by the United Kingdom Atomic Energy Authority (UKAEA) unless they are dwelling houses and offices.

1.4.4 What about the British Airports Authority? (Building Act 1984 Section 45)

The Building Regulations do **not** apply to buildings belonging to or occupied by the British Airports Authority, unless it is a house, hotel or building used as offices or showrooms.

1.4.5 What about the Civil Aviation Authority? (Building Act 1984 Section 45)

The Building Regulations do **not** apply to buildings belonging to or occupied by the Civil Aviation Authority, unless it is a house, hotel or building used as offices or showrooms.

1.5 What about civil liability? (Building Act 1984 Section 38)

It is an aim of the Building Act 1984 that all building work is completed safely and without risk to people employed on the site or visiting the site etc. Any contravention of the Building Regulations that causes injury (or death) to any person is liable to prosecution in the normal way.

1.6 What does the Building Act 1984 contain?

The Building Act 1984 consists of five parts:

Part 1 The Building Regulations
Part 2 Supervision of building work etc. other than by a local authority
Part 3 Other provisions about buildings
Part 4 General
Part 5 Supplementary

These parts are then broken down into a number of sections and subsections as shown in Appendix A to this chapter.

1.7 What are the Supplementary Regulations?

Part 5 of the Building Act contains seven schedules whose function is to list the principal areas requiring regulation and to show how the Building Regulations are to be controlled by the local authority. These schedules are:

- Schedule 1 – Building Regulations;
- Schedule 2 – Relaxation of Building Regulations;
- Schedule 3 – Inner London;
- Schedule 4 – Provisions consequential upon public body's notice;
- Schedule 5 – Transitional provisions;
- Schedule 6 – Consequential amendments;
- Schedule 7 – Repeals.

1.7.1 What is Schedule 1 of Part 5 of the Building Act 1984?

The Building Regulations are a statutory instrument, authorized by parliament, which details how the generic requirements of the Building Act are to be met. Compliance with Building Regulations is required for **all**:

- alterations and extensions of buildings (including services, fixtures and fittings);
- provision of new services, fittings or equipment;

unless (in most circumstances) **the increased area of the alteration or extension is less than 30 m^2** (35.9 y^2) in which case the Building Regulations provide the generic and specific requirements for this work.

Building Regulations **also** apply to alterations and extensions being completed on buildings erected **before** the date on which the regulations came into force.

Schedule 1 of the Building Act 1984 shows, in general terms, how the Building Regulations are to be administered by local authorities, the approved methods of construction and the approved types of materials that are to be used in (or in connection with) buildings.

How are the Building Regulations controlled?

To assist local authorities, Section 1 shows:

- how notices are given;
- how plans of proposed work (or work already executed) are deposited;
- how copies of deposited plans are administered and retained;
- how documents are to be controlled;
- how work is tested;
- how samples are taken;
- how local authorities can seek external expertise to assist them in their duties;
- how certificates signifying compliance with the Building Regulations are to be issued;
- how local authorities can accept certificates from a person (or persons) nominated to act on their behalf;
- how proposed work can be prohibited;
- when a dispute arises, how local authorities can refer the matter to the Secretary of State;
- what fees (and what level of fees) local authorities can charge.

What are the requirements of the Building Regulations?

Schedule 1 describes the mandatory requirements for completing **all** building work. These include:

- accommodation for specific purposes (e.g. for disabled persons);
- air pressure plants;
- cesspools (and other methods for treating and disposing of foul matter);
- emission of smoke, gases, fumes, grit or dust (or other noxious and/or offensive substances);
- dimensions of rooms and other spaces (inside buildings);
- drainage (including waste disposal units);
- electrical safety;
- fire precautions (services, fittings and equipment, means of escape);
- lifts (escalators, hoists, conveyors and moving footways);
- materials and components (suitability, durability and use);
- means of access to and egress from;
- natural lighting and ventilation of buildings;
- open spaces around buildings;
- prevention of infestation;
- provision of power outlets;

- resistance to moisture and decay;
- site preparation;
- solid fuel, oil, gas and electricity installations (including appliances, storage tanks, heat exchangers, ducts, fans and other equipment);
- standards of heating, artificial lighting, mechanical ventilation and air-conditioning;
- structural strength and stability (overloading, impact and explosion, under-pinning, safeguarding of adjacent buildings);
- third party liability (danger and obstruction to persons working or passing by building work);
- transmission of heat;
- transmission of sound;
- waste (storage, treatment and removal);
- water services, fittings and fixed equipment (including wells and bore-holes for supplying water);

and matters connected with (or ancillary to) any of the foregoing matters.

1.7.2 What is Schedule 2 of the Building Act 1984?

This Schedule provides guidance in connection with work that has been carried out prior to a local authority (under the Building Act 1984 Section 36) dispensing with or relaxing some of the requirements contained in the Building Regulations.

This Schedule is quite difficult to understand and if it affects you, then I would strongly advise that you discuss it with the local authority before proceeding any further.

1.7.3 What is Schedule 3 of the Building Act 1984?

Schedule 3 applies to how Building Regulations are to be used in Inner London and, as well as ruling which sections of the Act may be omitted, also details the requirements for drainage to Inner London buildings and shows how by-laws concerning the relation to the demolition of buildings (in Inner London) may be made.

What sections of the Building Act 1984 are not applied to Inner London?

In Inner London, because of its existing and changed circumstances (compared to other cities in England and Wales), certain sections of the Building Act are inappropriate (see Tables 1.2 and 1.3) and additional requirements – which are applicable to Inner London **only** – have been approved instead. These primarily cover drainage and demolition of buildings.

What about the buildings and drainage to buildings in Inner London?

Under the terms of the Building Act 1984, it is not lawful in an Inner London borough to erect a house/other building, or to rebuild a house/other building

Table 1.2 Sections inapplicable to Inner London

Section	Sub-section
Buildings	• Provision of food storage accommodation in house. • Entrances, exits etc. to be required in certain cases. • Means of escape from fire. • Raising of chimney. • Cellars and rooms below subsoil water level. • Consents under Section 74.
Defective premises, demolition etc.	• Dangerous building. • Dangerous building – emergency measures. • Ruinous and dilapidated buildings and neglected sites. • Notice to local authority of intended demolition. • Local authority's power to serve notice about demolition. • Notices under Section 81. • Appeal against notice under Section 81.

Table 1.3 Sections inapplicable to Temples

Section	Sub-section
Drainage	• Drainage of building. • Use and ventilation of soil pipes. • Repair etc. of drain.
Buildings	• Provision of food storage accommodation in house. • Entrances, exits etc. to be required in certain cases. • Means of escape from fire. • Raising of chimney. • Cellars and rooms below subsoil water level. • Consents under Section 74.
Defective premises, demolition etc.	• Dangerous building. • Dangerous building – emergency measures. • Ruinous and dilapidated buildings and neglected sites. • Notice to local authority of intended demolition. • Local authority's power to serve notice about demolition. • Notices under Section 81. • Appeal against notice under Section 81.

that has been pulled down to (or below) floor level, **unless** that house/building is provided with drains in conformance with the borough council's requirements. These drains must be suitable for the drainage of the whole building and all works, apparatus and materials used in connection with these drains must satisfy the council's requirements.

It is not lawful to occupy a house or other building in Inner London that has been erected or rebuilt in contravention of the above restriction.

The basic requirements of all Inner London borough councils are that:

- the drains must be connected into a sewer that is (or is intended to be constructed) nearby;
- if a suitable sewer is not available then a covered cesspool or other place should be used, provided that it is not under any house or other building;
- the drains must provide efficient gravitational drainage at all times and under all circumstances and conditions.

If it is impossible or unfeasible to provide gravitational drainage to all parts of the building, then (but depending on the circumstances) the council may allow pumping and/or some other form of lifting apparatus to be used.

In **all** circumstances the council have the authority (under this Schedule of the Act) to order the owner/occupier:

- to construct a covered drain from the house or building into the sewer;
- to provide proper paved or water-resistant sloping surfaces for carrying surface water into the drain;
- to provide proper sinks, inlets and outlets (siphoned or otherwise trapped), for preventing the emission of effluvia from the drain – or any connection to it;
- to provide a proper water supply and water-supplying pipes, cisterns and apparatus for scouring the drain;
- to provide proper sand traps, expanding inlets and other apparatus for preventing the entry of improper substances into the drain.

You are not allowed to commence any work on drains, dig out the foundations of a house or to rebuild a house in Inner London **unless**, at least seven days previously, you have provided a notice of intent to the borough council.

If a house or building in an Inner London borough (regardless of when it was first erected), has insufficient drainage and there is no proper sewer within 200 feet of any part of the house or building, the borough council may serve on the owner written notice requiring that person:

- to construct a covered watertight cesspool or tank or other suitable receptacle (provided that it is not under the house); and
- to construct and lay a covered drain leading from the house or building into that cesspool, tank or receptacle.

The Inner London borough councils have the authority to carry out irregular inspections of drains and cesspools constructed by the owner and, if they prove to be unsuitable, they have the authority to make the owner alter, repair or abandon them if they contravene council regulations.

What about Inner London's by-laws?

By authority of the Building Act 1984, the Greater London Authority (GLA) may make by-laws in relation to the demolition of buildings in the Inner London boroughs and regulate and (in certain circumstances) mandate, concerning:

- the fixing of floor level fans on buildings undergoing demolition;
- the hoarding up of windows in a building where all the sashes and glass have been removed;
- the demolition of internal parts of buildings before any external walls are taken down;
- using screens and mats as a precaution against dust;
- the hours during which ceilings may be broken down and mortar may be shot, or be allowed to fall, into any lower floor.

The GLA may also make by-laws with respect to closets, sanitary conveniences, ashpits, cesspools and receptacles for dung (and their accessories) for buildings being erected or altered in Inner London.

1.7.4 What is Schedule 4 of the Building Act 1984?

Schedule 4 of the Building Act 1984 concerns the authority and ruling of public bodies' notices and certificates.

What is a public body's plans certificate?

When a public body (i.e. local authorities, county councils and any other body '*that acts under an enactment for public purposes and not for its own profit*') is satisfied that the work specified in their (as well as another) public body's notices has been completed as detailed (and in full accordance with the Building Regulations) then that public body will give that local authority a certificate of completion.

This certificate is called a 'Public Bodies Plans Certificate' and can relate either to the whole, or to part of, the work specified in the public body's notice. Acceptance by the local authority signifies satisfactory completion of the planned work and the public body's notice ceases to apply to that work.

What is a public body's final certificate?

When a public body is satisfied that all work specified in their (or another's) public body's notice has been completed in compliance with the Building Regulations, then that public body will give the local authority a certificate of completion. This is referred to as a 'Final Certificate'.

How long is the duration of a public body's notice?

A public body's notice comes into force when it is accepted by the local authority and continues in force until the expiry of an agreed period of time.

Local authorities are authorized by the Building Regulations to extend the notice in certain circumstances.

1.7.5 What is Schedule 5 of the Building Act 1984?

Schedule 5 lists the transitional effect of the Building Act 1984 concerning:

- The Public Health Act 1936;
- The Clean Air Act 1956;
- The Housing Act 1957;
- The Public Health Act 1961;
- The London Government Act 1963;
- The Local Government Act 1972;
- The Health and Safety at Work etc. Act 1974;
- The Local Government (Miscellaneous Provisions) Act 1982.

1.7.6 What is Schedule 6 of the Building Act 1984?

Schedule 6 lists the consequential amendments that will have to be made to existing Acts of Parliament owing to the acceptance of the Building Act 1984. These amendments concern:

- The Restriction of Ribbon Development Act 1935;
- The Public Health Act 1936;
- The Atomic Energy Authority Act 1954;
- The Clean Air Act 1956;
- The Housing Act 1957;
- The Radioactive Substances Act 1960;
- The Public Health Act 1961;
- The London Government Act 1963;
- The Offices, Shops and Railway Premises Act 1963;
- The Faculty Jurisdiction Measure 1964;
- The Fire Precautions Act 1971;
- The Local Government Act 1972;
- The Safety of Sports Grounds Act 1975;
- The Local Land Charges Act 1975;
- The Development of Rural Wales Act 1976;
- The Local Government (Miscellaneous Provisions) 1976;
- The Interpretation Act 1978;
- The Highways Act 1980;
- New Towns Act 1981;
- The Local Government (Miscellaneous Provisions) 1982;
- The Public Health (Control of Disease) Act 1984.

1.7.7 What is Schedule 7 of the Building Act 1984?

Schedule 7 lists the cancellation (repeal) of some sections of existing Acts of Parliament, owing to acceptance of the Building Act 1984. These cancellations concern:

- The Public Health Act 1936;
- The Education Act 1944;
- The Water Act 1945;
- The Town and Country Planning Act 1947;
- The Atomic Energy Authority Act 1954;
- The Radioactive Substances Act 1960;
- The Public Health Act 1961;
- The London Government Act 1963;
- The Greater London Council (General Powers) Act 1967;
- The Fire Precautions Act 1971;
- The Local Government Act 1972;
- The Water Act 1973;
- The Health and Safety at Work etc. Act 1974;
- The Control of Pollution Act 1974;
- The Airports Authority Act 1975;
- The Local Government (Miscellaneous Provisions) Act 1976;
- The Criminal Law Act 1977;
- The City of London (Various Powers) Act 1977;
- The Education Act 1980;
- The Highways Act 1980;
- The Water Act 1981;
- The Civil Aviation Act 1982;
- The Local Government (Miscellaneous Provisions) Act 1982.

1.8 What are 'Approved Documents'? (Building Act 1984 Section 6)

The Secretary of State makes available a series of documents (called 'Approved Documents') which are intended to provide practical guidance with respect to the requirements of the Building Regulations (for details see Chapter 3).

1.9 What is the 'Building Regulations Advisory Committee'? (Building Act 1984 Section 14)

The Building Act allows the Secretary of State to appoint a committee (known as the Building Regulations Advisory Committee) to review, amend, improve and produce new Building Regulations and associated documentation (e.g. such as Approved Documents – see above).

1.10 What is 'type approval'? (Building Act 1984 Sections 12 and 13)

Type approval is where the Secretary of State is empowered to approve a particular type of building matter as complying, either generally or specifically, with a particular requirement of the Building Regulations. This power of approval is normally delegated by the Secretary of State to the local council or other nominated public body.

1.11 Does the Fire Authority have any say in Building Regulations? (Building Act 1984 Section 15)

When a requirement 'encroaches' on something that is normally handled by the Fire Authority under the Fire Precautions Act 1971 (e.g. provision of means of escape, structural fire precautions etc.) then the local authority **must** consult the fire authority before making any decision.

1.12 How are buildings classified? (Building Act 1984 Section 35)

For the purpose of the Building Act, buildings are normally classified:

- by reference to size;
- by description;
- by design;
- by purpose;
- by location.

or (to quote the Building Act of 1984) '*any other suitable characteristic*'!

1.13 What are the duties of the local authority? (Building Act 1984 Section 91)

It is the duty of local authorities to ensure that requirements of the Building Act 1984 are carried out (and that the appropriate associated Building Regulations are enforced) subject to:

- the provisions of Part I of the Public Health Act 1936 (relating to united districts and joint boards);

- Section 151 of the Local Government, Planning and Land Act 1980 (relating to urban development areas);
- Section 1(3) of the Public Health (Control of Disease) Act 1984 (relating to port health authorities).

1.13.1 What document controls must local authorities have in place? (Building Act 1984 Sections 92 and 93)

All notices, applications, orders, consents, demands and other documents, authorized, required by or given to, that are required by this Act or by a local authority (or an officer of a local authority), need to be in writing and in the format laid down by the Secretary of State.

All documents that a local authority is required to provide under the Building Act 1984 shall be signed by:

- the proper officer for this authority;
- the district surveyor (for documents relating to matters within his province);
- an officer authorized by the authority to sign documents (of a particular kind).

A document bearing the signature (including a facsimile of a signature by whatever process chosen) of an officer is deemed (for the purposes of the Building Act 1984 and any of its associated Building Regulations and orders made under it) to have been given, made or issued by the local authority, unless otherwise proved.

1.13.2 How do local authorities 'serve' notices and documents? (Building Act 1984 Section 94)

Any notice, order, consent, demand or other document that is authorized or required by the Building Act 1984 can be given or served to a person:

- by delivering it to the person concerned;
- by leaving it, or sending it in a prepaid letter addressed to him, at his usual or last known residence.

Or if it is not possible to ascertain the name and address of the person to or on whom it should be given or served (or if the premises are unoccupied) then the notice, order, consent, demand or other document can be addressed to the 'owner' or 'occupier' of the premises (naming them) and delivering it to '*some person on the premises*' or, if there isn't anyone at the premises to whom it can be delivered, then a copy of the document can be fixed to a conspicuous part of the premises.

1.14 What are the powers of the local authority? (Building Act 1984 Sections 97–101)

The powers of the local authority, as given by the Building Act 1984 and its associated Building Regulations, include:

- overall responsibility for the construction and maintenance of sewers and drains and the laying and maintenance of water mains and pipes;
- the authority to make the owner or occupier of any premises complete essential and remedial work in connection with the Building Act 1984 (particularly with respect to the construction, laying, alteration or repair of a sewer or drain);
- the authority to complete remedial and essential work themselves (on repayment of expenses) if the owner or occupier refuses to do this work himself;
- the ability to sell any materials that have been removed, by them, from any premises when executing works under this Act (paying all proceeds, less expenses, from this sale to the owner or occupier).

This does not apply to any refuse that is, or has been, removed by the local authority.

1.14.1 Have the local authority any power to enter premises? (Building Act 1984 Section 95)

An authorized officer of a local authority has a right to enter any premises, at all 'reasonable hours' (except for a factory or workplace in which 24 hours' notice has to be given) for the purpose of:

- ascertaining whether there is (or has been) a contravention of this Act (or of any Building Regulations) that it is the duty of the local authority to enforce;
- ascertaining whether or not any circumstances exist that would require local authority action or for them having to complete any work;
- taking any action, or executing any work, authorized or required by this Act, or by Building Regulations;
- carrying out their functions as a local authority.

If the local authority is refused admission to any premises (or the premises are unoccupied) then the local authority can apply to a Justice of the Peace for a warrant authorizing entry.

1.15 Who are approved inspectors? (Building Act 1984 Section 49)

An approved inspector is a person who is approved by the Secretary of State (or a body such as a local authority or county council designated by the Secretary of State) to inspect, supervise and to authorize building work. Lists of approved inspectors are available from all local authorities.

Building Act 1984 Section 57
If an approved inspector gives a notice or certificate that falsely claims to comply with the Building Regulations and/or the Building Act of 1984, then he is liable to prosecution.

1.15.1 What is an initial notice? (Building Act 1984 Section 47)

An approved inspector will have to present an initial notice and plan of work to the local authority. Once accepted, the approved inspector is authorized to inspect and supervise all work being completed and to provide certificates and notices. Acceptance of an initial notice by a local authority is treated (for the purposes of conformance with Section 13 of the Fire Precautions Act 1971 regarding suitable means of escape) as '*depositing plans of work*'.

Under Section 47 of the Building Act, the local authority is required to accept all certificates and notices, unless the initial notice and plans contravene a local ruling. Whilst the initial notice continues in force, the local authority are not allowed to give a notice in relation to any of the work being carried out or take any action for a contravention of Building Regulations.

If the local authority rejects the initial notice for any reason, then the approved inspector can appeal to a magistrates' court for a ruling. If still dissatisfied, he can appeal to the crown court.

Cancellation of initial notice (Building Act 1984 Sections 52 and 53)

If work has not commenced within three years (beginning the date on which the certificate was accepted), the local authority can cancel the initial notice.

If an approved inspector is unable to carry out or complete his functions, or is of the opinion that there is a contravention of the Building Regulations, then he can cancel the initial notice lodged with the local authority.

Equally, if the person carrying out the work has good reason to consider that the approved inspector is unable (or unwilling) to carry out his functions, then that person can cancel the initial notice given to the local authority.

The fact that the initial notice has ceased to be in force does not affect the right of an approved inspector, however, to give a new initial notice relating to any of the work that was previously specified in the original notice.

1.15.2 What are plans certificates? (Building Act 1984 Section 50)

When an approved inspector has inspected and is satisfied himself that the plans of work specified in the initial notice do not contravene the Building Regulations in any way, he will provide a certificate (referred to as a 'plans certificate') to the local authority. This plans certificate:

- can relate to the whole or part of the work specified in the initial notice;
- does not have any effect unless the local authority accepts it;
- may only be rejected by the local authority '*on prescribed grounds*'.

If, however, work has not commenced within three years (beginning the date on which the certificate was accepted), the local authority may rescind their acceptance, by notice to the approved inspector or the person shown on the initial notice, giving their grounds for cancellation.

1.15.3 What are final certificates? (Building Act 1984 Section 51)

Once the approved inspector is satisfied that all work has been completed in accordance with the work specified in the initial notice, he will provide a certificate (referred to as a 'Final Certificate') to the local authority and the person who carried out the work. This certificate will detail his acceptance of the work and, once acknowledged by the local authority, the approved inspector's job will have been completed and (from the point of view of local authority) he will have been considered '*to have discharged his duties*'.

1.15.4 Who retains all these records? (Building Act 1984 Section 56)

Local authorities are required to keep a register of all initial notices and certificates given by approved inspectors and to retain all relevant and associated documents concerning those notices and certificates. The local authority is further required to make this register available for public inspection during normal working hours.

1.15.5 Can public bodies supervise their own work? (Building Act 1984 Sections 54 and 55)

If a public body (e.g. local authority or county council) is of the opinion that building work that is to be completed on one of its own buildings can be adequately supervised by one of its employees and/or agents, then they can provide the local authority with a notice (referred to as a 'public bodies notice') together with their plan of work.

Once accepted by the local authority, the public body is authorized to inspect and supervise all work being completed and to provide certificates and notices. Acceptance by a local authority of public bodies notice is treated (for the purposes of conformance with Section 13 of the Fire Precautions Act 1971 regarding suitable means of escape) as 'depositing plans of work'.

If the local authority rejects the public bodies notice for any reason, then they can appeal to a magistrates' court for a ruling. If still dissatisfied, they can appeal to the crown court.

1.16 What causes some plans for building work to be rejected? (Building Act 1984 Sections 16 and 17)

The local authority will reject all plans for building work that:

- are defective;
- contravene any of the Building Regulations.

In all cases the local authority will advise the person putting forward the plans why they have been rejected (giving details of the relevant regulation or section) and, where possible, indicate what amendments and/or modifications will have to be made in order to get them approved. The person who initially put forward the plans is then responsible for making amendments/alterations and resubmitting them for approval.

If a plan for proposed building work is accompanied by a certificate from a person or persons approved by the Secretary of State (or someone designated by him), then only in extreme circumstances can the local authority reject the plans.

1.17 Can I apply for a relaxation in certain circumstances? (Building Act 1984 Sections 7–11, 30 and 39)

The Building Act allows the local authority to dispense with, or relax, a Building Regulation if they believe that that requirement is unreasonable in relation to a particular type of work being carried out.

Schedule 2 of the Building Act 1984 provides guidance and rules for the application of Building Regulations to work that has been carried out **prior to** the local authority (under the Building Act 1984 Section 36) dispensing with, or relaxing, some of the requirements contained in the Building Regulations. This schedule is quite difficult to understand and if it affects you, then I would strongly advise that you discuss it with the local authority before proceeding any further.

For the majority of cases, applications for dispensing with or relaxing Building Regulations can be settled locally. In more complicated cases, however, the local authority can seek guidance from the Secretary of State who will give a direction as to whether the requirement may be relaxed or dispensed with (unconditionally or subject to certain conditions).

If a question arises between the local authority and the person who has executed (or has proposed to execute any) work regarding:

- the application of Building Regulations;
- whether the plans are in conformity with the Building Regulations;
- whether the work has been executed in conformance with these plans;

then the question can be referred to the Secretary of State for determination. In these cases, the Secretary of State's decision will be deemed final.

The Building Act allows the local authority to charge a fee for reviewing and deciding on these matters with different fees for different cases.

1.18 Can I change a plan of work once it has been approved? (Building Act 1984 Section 31)

If the person intending to carry out building work has had their plan (or plans) passed by the local authority, but then wants to change them, that person will have to submit (to the local authority) a set of revised plans showing precisely how they want to deviate from the approved plan and ask for their approval. If the deviation or change is a small one this can usually be achieved by talking to the local planning officer, but if it is a major change, then it could result in the resubmission of a complete plan of the revised building work.

1.19 Must I complete the approved work in a certain time? (Building Act 1984 Section 32)

Once a building plan has been passed by the local authority, then '**work must commence**' within three years from the date that it was approved. Failure to do so could result in the local authority cancelling the approved plans and you will have to resubmit them if you want to carry on with your project.

The phrase 'work must commence' can vary from local authority to local authority. Normally this will mean physically laying the foundations of the building but in other cases it could mean that far more work has to be completed in the three year time span. It is always best to check with the local authority and ask for clarification about this restriction when your plans are first approved.

1.20 How is my building work evaluated for conformance with the Building Regulations? (Building Act 1984 Section 32)

Part of the local authority's duty is to make regular checks that all building work being completed is in conformance with the approved plan and the Building

Regulations. These checks would normally be completed at certain stages of the work (e.g. the excavation of foundations) and tests will include:

- tests of the soil or subsoil of the site of the building;
- tests of any material, component or combination of components that has been, is being, or is proposed to be used in the construction of a building;
- tests of any service, fitting or equipment that has been, is being, or is proposed to be provided in or in connection with a building.

The cost of carrying out these tests will normally be charged to the owner or occupier of the building.

The local authority has the power to ask the person responsible for the building work to complete some of these tests on their behalf.

1.20.1 Can I build on a site that contains offensive material? (Building Act 1984 Section 29)

If the site you are intending to erect a building or extension on is:

- ground that has been filled up with material impregnated with faecal or offensive animal or offensive vegetable mater;
- ground upon which any such material has been deposited;

then that material must be removed or rendered innocuous before work can commence.

This requirement normally rests with the current owner/occupier of the building, but in certain circumstances (for example, if the site was previously used as a chicken farm or similar) then the previous owner might be held responsible. If the requirements of this particular section are applicable to you, then it is recommended that you seek guidance from the local authority before committing yourself!

1.21 What about dangerous buildings? (Building Act 1984 Sections 77 and 78)

If a building, or part of a building or structure, is in such a dangerous condition (or is used to carry loads which would make it dangerous) then the local authority may apply to a magistrates' court to make an order requiring the owner:

- to carry out work to avert the danger;
- to demolish the building or structure, or any dangerous part of it, and remove any rubbish resulting from the demolition.

The local authority can also make an order restricting its use until such time as a magistrates' court is satisfied that all necessary works have been completed.

1.21.1 Emergency measures

In emergencies, the local authority can make the owner take immediate action to remove the danger or they can complete the necessary action themselves. In these cases, the local authority is entitled to recover from the owner such expenses reasonably incurred by them. For example:

- fencing off the building or structure;
- arranging for the building/structure to be monitored.

1.21.2 Can I demolish a dangerous building? (Building Act 1984 Section 80)

You must have good reasons for knocking down a building, such as making way for rebuilding or improvement (which in most cases would be incorporated in the same planning application).

Be careful, penalties can be very severe for demolishing something illegally!

You are not allowed to begin any demolition work (even on a dangerous building) unless you have given the local authority notice of your intention and this has either been acknowledged by the local authority or the relevant notification period has expired. In this notice you will have to:

- specify the building to be demolished;
- state the reason(s) for wanting to demolish it;
- show how you intend to demolish it.

Copies of this notice will have to be sent to:

- the local authority;
- the occupier of any building adjacent to the building in question;
- British Gas;
- the area electricity board in whose area the building is situated.

This regulation does not apply to the demolition of an internal part of an occupied building, or a greenhouse, conservatory, shed or prefabricated garage (that forms part of that building) or an agricultural building defined in Section 26 of the General Rate Act 1967.

1.21.3 Can I be made to demolish a dangerous building? (Building Act 1984 Sections 81, 82 and 83)

If the local authority considers that a building is so dangerous that it should be demolished, they are also entitled to issue a notice to the owner requiring him:

- to shore up any building adjacent to the building to which the notice relates;
- to weatherproof any surfaces of an adjacent building that are exposed by the demolition;

- to repair and make good any damage to an adjacent building caused by the demolition or by the negligent act or omission of any person engaged in it;
- to remove material or rubbish resulting from the demolition and clear the site;
- to disconnect, seal and remove any sewer or drain in or under the building;
- to make good the surface of the ground that has been disturbed in connection with this removal of drains etc.;
- (in accordance with the Water Act 1945 (interference with valves and other apparatus) and the Gas Act 1972 (public safety)), to arrange with the relevant statutory undertakers (e.g. the water authority, British Gas and the electricity supplier) for the disconnection of gas, electricity and water supplies to the building;
- to leave the site in a satisfactory condition following completion of all demolition work.

Before complying with this notice, the owner must give the local authority 48 hours' notice of commencement.

In certain circumstances, the owner of an adjacent building may be liable to assist in the cost of shoring up their part of the building and waterproofing the surfaces. It could be worthwhile checking this point with the local authority!

1.22 What about defective buildings? (Building Act 1984 Sections 76, 79 and 80)

If a building or structure is, because of its ruinous or dilapidated condition, liable to cause damage to (or be a nuisance to) the amenities of the neighbourhood, then the local authority can require the owner:

- to carry out necessary repairs and/or restoration; or
- to demolish the building or structure (or any part of it) and to remove all of the rubbish or other material resulting from this demolition.

If, however, the building or structure is in a defective state and remedial action (envisaged under Sections 93 to 96 of the Public Health Act) would cause an unreasonable delay, then the local authority can serve an abatement notice stating that within nine days **they** intend to complete such works as they deem necessary to remedy the defective state and recover the '*expenses reasonably incurred in so doing*' from the person on whom the notice was sent.

If appropriate, the owner can (within seven days) after the local authority's notice has been served, serve a counter-notice stating that he intends to remedy the defects specified in the first-mentioned notice himself.

A local authority is **not** entitled to serve a notice, or commence any work in accordance with a notice that they have served, if the execution of the works would (to their knowledge) be in contravention of a building preservation order that has been made under Section 29 of the Town and Country Planning Act.

1.23 What are the rights of the owner or occupier of the premises? (Building Act 1984 Sections 102–107)

When a person has been given a notice by a local authority to complete work, he has the right to appeal to a magistrates' court on any of the following grounds:

- that the notice or requirement is not justified by the terms of the provision under which it purports to have been given;
- that there has been some informality, defect or error in (or in connection with) the notice;
- that the authority have refused (unreasonably) to approve completion of alternative works, or that the works required by the notice to be executed are unreasonable or unnecessary;
- that the time limit set to complete the work is insufficient;
- that the notice should lawfully have been served on the occupier of the premises in question instead of on the owner (or vice versa);
- that some other person (who is likely to benefit from completion of the work) should share in the expense of the works.

1.24 Can I appeal against a local authority's ruling? (Building Act 1984 Sections 40 and 41)

If you have grounds for disagreeing with the local authority's ruling to remove or renew 'offending work', then you are entitled to appeal to the local magistrates' court and they will rule whether the local authority were correct and entitled to give you this ruling, or whether they should withdraw the notice.

If you then disagree with the magistrates' ruling, you have the right to appeal to the crown court.

Where the Secretary of State has given a ruling, however, this ruling shall be considered as being final.

1.24.1 What about compensation? (Building Act 1984 Sections 103–110)

If an owner or occupier considers that a ruling obtained from the local authority is incorrect, he can appeal (in the first case) to the local magistrates' court. If, on appeal, the magistrates rule against the local authority, then the owner/occupier of the building concerned is entitled to compensation from the local authority. If, on the other hand, the magistrates rule in favour of the local authority, then the local authority is entitled to recover any expenses that they have incurred.

Be sure of your facts before you ask a magistrates' court for a ruling!

1.24.2 What happens if the plans mean building over an existing sewer etc.? (Building Act 1984 Section 18)

Before the local authority can approve a plan for building work which means having to first erect a building or extension over an existing sewer or drain, they must notify and seek the advice of the water authority.

As part of the Public Health Act 1936 and the Control of Pollution Act 1974, local authorities are required to keep maps of all sewers etc.

Appendix 1A Contents of the Building Act 1984

Part 1 The Building Regulations

Section	Sub-section
Power to make building regulations	• Power to make building regulations. • Continuing requirements.
Exemption from building regulations	• Exemption of particular classes of buildings etc. • Exemption of educational buildings and buildings of statutory undertakers. • Exemption of public bodies from procedural requirements of building regulations.
Approved Documents	• Approval of documents for purposes of building regulations. • Compliance or non-compliance with Approved Documents.
Relaxation of building regulations	• Relaxation of building regulations. • Application for relaxation. • Advertisement of proposal for relaxation of building regulations. • Type relaxation of building regulations.
Type approval of building matter	• Power of Secretary of State to approve type of building matter. • Delegation of power to approve.
Consultation	• Consultation with Building Regulations Advisory Committee and other bodies. • Consultation with fire authority.
Passing of plans	• Passing or rejection of plans. • Approval of persons to give certificates etc. • Building over sewer etc. • Use of short-lived materials. • Use of materials unsuitable for permanent building. • Provision of drainage. • Drainage of buildings in combination. • Provision of facilities for refuse. • Provision of exits etc. • Provision of water supply. • Provision of closets. • Provision of bathrooms. • Provision for food storage. • Site containing offensive material.
Determination of questions	

Section	Sub-section
Proposed departure from plans	
Lapse of deposit of plans	
Tests for conformity with building regulations	
Classification of buildings	
Breach of building regulations	• Penalty for contravening building regulations. • Removal or alteration of offending work. • Obtaining of report where Section 36 notice given. • Civil liability.
Appeals in certain cases	• Appeal against refusal etc. to relax building regulations. • Appeal against Section 36 notice. • Appeal to Crown Court. • Appeal and statement of case to High Court in certain cases. • Procedure on appeal to Secretary of State on certain matters.
Application of building regulations to Crown etc.	• Application to Crown. • Application to United Kingdom Atomic Energy Authority.
Inner London	

Part 2 Supervision of Building Work etc. otherwise than by a local authority

Section	Sub-section
Supervision of plans and work by approved inspectors	• Giving and acceptance of initial notice. • Effect of initial notice. • Approved inspectors. • Plans certificates. • Final certificates. • Cancellation of initial notice. • Effect of initial notice ceasing to be in force.
Supervision of their own work by public bodies	
Supplementary	• Appeals. • Recording and furnishing of information. • Offences. • Construction of Part 11.

Part 3 Other provisions about buildings

Section	Sub-section
Drainage	• Drainage of building. • Use and ventilation of soil pipes.

(*Continued*)

Appendix A *(Continued)*

Section	Sub-section
	• Repair etc. of drain. • Disconnection of drain. • Improper construction or repair of water closet or drain.
Provision of sanitary conveniences	• Provision of closets in building. • Provision of sanitary conveniences in workplace. • Replacement of earth closets etc. • Loan of temporary sanitary conveniences. • Erection of public conveniences.
Buildings	• Provision of water supply in occupied house. • Provision of food storage accommodation in house. • Entrances, exits etc. to be required in certain cases. • Means of escape from fire. • Raising of chimney. • Cellars and rooms below subsoil water level. • Consents under Section 74.
Defective premises, demolition etc.	• Defective premises. • Dangerous building. • Dangerous building – emergency measures. • Ruinous and dilapidated buildings and neglected sites. • Notice to local authority of intended demolition. • Local authority's power to serve notice about demolition. • Notices under Section 81. • Appeal against notice under Section 81.
Yards and passages	• Paving and drainage of yards and passages. • Maintenance of entrances to courtyards.
Appeal to Crown Court	
Application of provisions to Crown property	
Inner London	
Miscellaneous	• References in Acts to building byelaws. • Facilities for inspecting local Acts.

Part 4 General

Section	Sub-section
Duties of local authorities	
Documents	• Form of documents. • Authentication of documents. • Service of documents.
Entry on premises	• Power to enter premises. • Supplementary provisions as to entry.
Execution of works	• Power to require occupier to permit work. • Content and enforcement of notice requiring works. • Sale of materials. • Breaking open of streets.
Appeal against notice requiring works	

Section	Sub-section
General provisions about appeals and applications	• Procedure on appeal or application to magistrates' court. • Local authority to give effect to appeal. • Judge not disqualified by liability to rates.
Compensation and recovery of sums	• Compensation for damage. • Recovery of expenses etc. • Payments by instalments. • Inclusion of several sums in one complaint. • Liability of agent or trustee. • Arbitration.
Obstruction	
Prosecutions	• Prosecution of offences. • Continuing offences.
Protection of members etc. of authorities	
Default powers	• Default powers of Secretary of State. • Expenses of Secretary of State. • Variation or revocation of order transferring powers.
Local inquiries	
Orders	
Interpretation	• Meaning of 'building'. • Meaning of 'building regulations'. • Meaning of 'construct' and 'erect'. • Meaning of deposit of plans. • Construction and availability of sewers. • General interpretation. • Construction of certain references concerning temples.
Savings	• Protection for dock and railway undertakings. • Saving for Local Land Charges Act 1975. • Saving for other laws. • Restriction of application of Part IV to Schedule 3.

Part 5 Supplementary

Section	Sub-section
Supplementary	• Transitional provisions. • Consequential amendments and repeals. • Commencement. • Short title and extent.
Schedule 1 – Building regulations.	
Schedule 2 – Relaxation of building regulations.	
Schedule 3 – Inner London.	
Schedule 4 – Provisions consequential upon public body's notice.	
Schedule 5 – Transitional provisions.	
Schedule 6 – Consequential amendments.	
Schedule 7 – Repeals.	

2

The Building Regulations 2000

Even when planning permission is not required, most building works, including alterations to existing structures, are subject to minimum standards of construction to safeguard public health and safety.

2.1 What is the purpose of the Building Regulations?

The Building Regulations are legal requirements laid down by parliament, based on the Building Act 1984. They are approved by parliament and deal with the **minimum** standards of design and building work for the construction of domestic, commercial and industrial buildings.

Building Regulations ensure that new developments or alterations and/or extensions to buildings are all carried out to an agreed standard that protects the health and safety of people in and around the building.

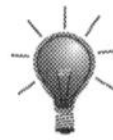

Building standards are enforced by your local building control officer, but for matters concerning drainage or sanitary installations, you will need to consult their technical services department.

Builders and developers are required by law to obtain building control approval, which is an independent check that the Building Regulations have been complied with. There are two types of building control providers – the local authority and approved inspectors.

2.2 Why do we need the Building Regulations?

As mentioned in the Preface, the Great Fire of London in 1666 was the single most significant event to have shaped today's legislation. The rapid growth of the fire through timber buildings built next to each other highlighted the need for builders to consider the possible spread of fire between properties when rebuilding work commenced. This resulted in the first building construction legislation that required all buildings to have some form of fire resistance.

During the Industrial Revolution (200 years later) poor living and working conditions in ever expanding, densely populated urban areas caused outbreaks of cholera and other serious diseases. Poor sanitation, damp conditions and lack of ventilation forced the government to take action and building control took on the greater role of health and safety through the first Public Health Act of 1875. This Act had two major revisions in 1936 and 1961 and led to the first set of national building standards – the Building Regulations 1965.

The current legislation is the Building Regulations 2000 (Statutory Instrument No 2531) which is made by the Secretary of State for the Environment under powers delegated by parliament under the Building Act of 1984.

The Building Regulations are a set of minimum requirements designed to secure the health, safety and welfare of people in and around buildings and to conserve fuel and energy in England and Wales. They are basic performance standards and the level of safety and acceptable standards are set out as guidance in the Approved Documents (which are quite frequently referred to as 'Parts' of the Building Regulations). Compliance with the detailed guidance of the Approved Documents is usually considered as evidence that the Regulations themselves have been complied with.

Alternate ways of achieving the same level of safety, or accessibility, are also acceptable.

2.3 What building work is covered by the Building Regulations?

The Building Regulations cover all new building work. This means that if you want to put up a new building, extend or alter an existing one, or provide new and/or additional fittings in a building such as drains or heat-producing appliances, washing and sanitary facilities and hot water storage (particularly unvented hot water systems), the Building Regulations will probably apply. Statutory Instrument 2006 no. 652 (SI 652) has amended the Building Regulations to cover situations where a building becomes a building to which energy efficiency requirements would now apply.

In addition, under SI 652 the definition of exempt buildings has been substantionally altered (e.g. it now means that such things as greenhouses when receiving electricity from a source shared with, or located inside a dwelling, will have to comply with Part P of Schedule 1). They may also apply to certain changes of use of an existing building (even though construction work may not be intended) as the 'change of use' could involve the building having to meet different requirements of the Regulations.

It should be remembered, however, that although it may appear that the Regulations do not apply to some of the work you wish to undertake, the end result of doing that work could well lead to you contravening some of the Regulations. You should also recognize that some work – whether or not controlled – could have implications for an adjacent property. In such cases it would be advisable to take professional advice and consult the local authority or an approved inspector. Some examples are:

- removing a buttressed support to a party wall;
- underpinning part of a building;
- removing a tree close to a wall of an adjoining property;
- adding floor screed to a balcony which may reduce the height of a safety barrier;
- building parapets which may increase snow accumulation and lead to an excessive increase in loading on roofs.

2.4 What are the requirements associated with the Building Regulations?

The Building Regulations contain a list of requirements (referred to as 'Schedule 1') that are designed to ensure the health and safety of people in and around buildings; to promote energy conservation; and to provide access and facilities for disabled people. In total there are 14 parts (A–P less I) to these requirements and these cover subjects such as structure, fire and electrical safety, ventilation, drainage etc.

The requirements are expressed in broad, functional terms in order to give designers and builders the maximum flexibility in preparing their plans.

2.5 What are the Approved Documents?

Approved Documents contain practical and technical guidance on ways in which the requirements of each part of the Building Regulations can be met.

Each Approved Document reproduces the *requirements* contained in the Building Regulations relevant to the subject area. This is then followed by *practical and technical guidance*, with examples, on how the requirements can be met in some of the more common building situations. There may, however, be alternative ways of complying with the requirements to those shown in the Approved Documents and you are, therefore, under no obligation to adopt any particular solution in an Approved Document if you prefer to meet the relevant requirement(s) in some other way.

Just because an Approved Document has not been complied with, however, does not necessarily mean that the work is wrong. The circumstances of each particular case should be considered when an application is made to make sure that adequate levels of safety will be achieved.

Note: The Building Regulations are constantly reviewed to meet the growing demand for better, safer and more accessible buildings as well as the need to reflect emerging harmonized European Standards. Building Regulations were last consolidated in SI 2000:2531, since then several of the Approved Documents have been republished as new editions and others are now under active review.

Where there are any issues common to one or more parts (such as the guidance on airtightness in Part L corresponding to the requirements for ventilation in Part F) these have been taken into consideration.

Any changes necessary are brought into operation after consultation with all interested parties. This has meant several amendments since the publication of the Building Regulations in 2000 with the emphasis in more recent years being on:

- increased thermal insulation to conserve energy and reduce global warming;
- providing better access and facilities for disabled people;
- a more comprehensive, one stop approach to fire safety requirements;
- the need for protection against sound from within a dwelling-house, other parts of the building and/or adjoining buildings;
- improvement of acoustic conditions in schools.

The Approved Documents are in 14 parts (A to P less I) and consist of:

A Structure
B Fire safety
C Site preparation and resistance to contaminants and water moisture
D Toxic substances
E Resistance to the passage of sound
F Ventilation
G Hygiene
H Drainage and waste disposal
J Combustion appliances and fuel storage systems
K Protection from falling, collision and impact
L Conservation of fuel and power
M Access and facilities for disabled people
N Glazing – safety in relation to impact, opening and cleaning
P Electrical safety

Parts A to D, F to K, N and P (except for paragraphs H2 and J6 which are excluded from regulation 8 because they deal directly with preventing contamination of water) of Schedule 1 do not require anything to be done except for the purpose of securing reasonable standards of health and safety for persons in or about buildings (and any others who may be affected by buildings) or matters connected with buildings.

Parts E and M (which deal, respectively, with resistance to the passage of sound, and access and facilities for disabled people) are excluded from regulation 8 because they address the welfare and convenience of building users. Part L is excluded from regulation 8 because it addresses the conservation of fuel and power.

You can buy a copy of the Approved Documents (and the Building Act 1984 if you wish) from the Stationery Office (TSO), PO Box 29, Duke Street, Norwich, NR3 1GN (Tel: 0870 600 5522, Fax: 0870 600 5533, www.tso. co.uk), or some book shops. Occasionally they are available from libraries. Alternatively, you can download pdf copies of the Approved Documents from www.planningportal.gov.uk (then england/professionals/en/1115314110382.html).

2.5.1 Part A Structure

So that buildings do not collapse, requirements ensure that:

- all structural elements of a building can safely carry the loads expected to be placed on them;
- foundations are adequate for any movement of the ground (for example, caused by landslip or subsidence);
- large buildings are strong enough to withstand (for example) an explosion without collapsing.

2.5.2 Part B Fire Safety

The Regulations consider 7 aspects of fire safety in the construction of buildings both dwelling houses and other buildings. These are:

(1) the building shall be designed and constructed so that there are appropriate provisions for the early warning of fire, and appropriate means of escape in case of fire;
(2) internal spread of fire should be inhibited within the building by ensuring linings adequately resist the spread of flame over their surface and have a rate of heat release, or fire growth which is reasonable;
(3) the building shall be designed and constructed so that, in the event of fire, its stability will be maintained for a reasonable period;
(4) that walls common to two or more dwellings are designed and constructed to adequately resist the spread of fire between those buildings;
(5) that where a building is sub-divided fire spread shall be inhibited through the use of fire resisting materials or fire suppression systems;
(6) that external walls of a building are able to resist the spread of fire from one building to another and that roofs are able to resist the spread of fire from one room to another;
(7) that buildings are designed and constructed so as to provide reasonable assistance to fire-fighters in the protection of life and to enable fire appliances to gain access to the building.

2.5.3 Part C Site preparation and resistance to contaminants and moisture

There are four requirements to this part:

(1) that before any building works commence, all vegetation and topsoil are removed;
(2) that any contaminated ground is either treated, neutralized or removed before a building is erected;

(3) that subsoil drainage is provided to waterlogged sites;
(4) that **all** floors, walls and roof of a building should not be adversely affected by interstitial condensation.

2.5.4 Part D Toxic substances

This part requires walls to be constructed in such a way that any fumes filling a cavity are prevented from penetrating the building.

2.5.5 Part E Resistance to passage of sound

This part has four main requirements:

(1) that dwellings shall provide reasonable resistance to sound from other parts of the same building and/or from adjoining buildings;
(2) that internal walls and floors of dwellings shall provide reasonable resistance to sound;
(3) that the common internal parts of buildings (containing flats or rooms for residential purposes) shall prevent unreasonable reverberation;
(4) that school rooms shall be acoustically insulated against noise.

2.5.6 Part F Ventilation

There are two aspects considered by this part:

(1) adequate ventilation must be provided to kitchens, bath and shower rooms, sanitary accommodation and other habitable rooms (both domestic and non-domestic);
(2) roofs need to be well vented (or designed) to prevent moist air causing condensation damage.

2.5.7 Part G Hygiene

There are three aspects included in this part:

(1) buildings are required to have satisfactory sanitary conveniences and washing facilities;
(2) all dwellings are required to have a fixed bath or shower with hot and cold water;
(3) unvented hot water systems over a certain size are required to have safety provisions to prevent explosion.

2.5.8 Part H Drainage and waste disposal

There are four aspects of this part:

(1) new drains taking foul water from buildings are required to discharge into a foul water sewer (or other suitable outfall), be watertight and be accessible for cleaning;

(2) where no public sewer is available, holding tanks or sewage treatment plants should be made available;
(3) new drains taking rainwater from roofs of buildings need to be watertight, accessible for cleaning and (if there is no sewer available) discharge to a suitable surface water sewer or ditch, soakaway, or watercourse;
(4) storage facilities, reasonably close to the building, need to be provided for refuse collection.

2.5.9 Part J Combustion appliances and fuel storage systems

There are three main aspects to this part:

(1) heat producing appliances must be provided with a supply of fresh air to prevent carbon monoxide poisoning to the building's occupants;
(2) chimneys and flues need to be adequately designed so that smoke and other products of combustion are safely discharged to the outside air;
(3) fireplaces and heat producing appliances should be designed and positioned so as to avoid the building's structure from igniting.

2.5.10 Part K Protection from falling, collision and impact

There are five main aspects to this part:

(1) to avoid accidents on stairs, ladders and ramps; the physical dimensions need to be suitable for the use of the building;
(2) to avoid persons falling off stairwells, balconies, floors, some roofs; light wells and basement areas (or similar sunken areas) connected to a building need to be suitably guarded according to the building's use;
(3) to avoid vehicles falling off buildings; car park floors, ramps and other raised areas need to be provided with vehicle barriers;
(4) to avoid danger to people from colliding with an open window, skylight, or ventilator; some form of guarding may be needed;
(5) measures need to be taken to avoid the opening and closing of powered sliding or open-upwards doors and gates falling onto any person and/or trapping them.

Possible future amendment	Approved Document K contains general guidance on stair and ramp design. The guidance in Approved Document M (2004) reflects more recent ergonomic research conducted to support BS 8300 and takes precedence over Approved Document K in conflicting areas. Further research on stairs is currently being undertaken and will be reflected in future revisions of Approved Document K.

2.5.11 Part L Conservation of fuel and power

There are three main aspects to this part:

(1) limiting heat gains and losses;
(2) providing and commissioning energy-efficient fixed building services with effective controls; and
(3) providing the owner with sufficient information about the building, the fixed building services and their maintenance requirements, so that the building can be operated in such a manner as to use no more fuel and power than is reasonable in the circumstances.

These apply to:

- new dwellings;
- existing dwellings;
- new buildings (other than dwellings);
- existing buildings (other than dwellings).

2.5.12 Part M Access and facilities for disabled people

There are six main aspects to this part:

(1) dwellings – people, including disabled people, should be able:
  - to reach the principal, or suitable alternative, entrance to the dwelling from the point of access;
  - to gain access into and within the principal storey of the dwelling;
  - to gain access to sanitary conveniences at no higher storey than the principal storey.

(2) regardless of disability, age or gender it should be possible for people in buildings (other than dwellings):
  - to reach the principal entrance to the building from the site boundary, from car parking (within the site) and from other buildings on the same site (e.g. such as a university campus, school or hospital);
  - to have access into and within, any storey of the building;
  - to have access and use of the building's facilities.

(3) the structure and amenities of a building should not constitute a hazard to users (especially people with impaired sight);
(4) suitable accommodation should be made available for people in wheelchairs (or people with other disabilities) in audience or spectator seating;
(5) people with a hearing or sight impairment should be provided with some form of aid to communication in auditoria, meeting rooms, reception areas, ticket offices and at information points;
(6) sanitary accommodation should be available for **all** users of the building.

2.5.13 Part N Glazing – safety in relation to impact, opening and cleaning

There are four main aspects in this part:

(1) glazing in locations where people might collide with the glass should either be robust enough not to break, or be constructed of safety glass, or be provided with suitable guarding;
(2) large sheets of glazing need to be made obvious so that people do not collide with that glazing;
(3) where non-dwelling windows, skylights and ventilators are openable by people, controls and/or limiters need to be provided to ensure safe operation and prevent persons falling through a window;
(4) safe access for cleaning both sides of non-dwelling windows, skylights etc. over 2.0 m above ground needs to be available.

Possible future amendment	Approved Document N contains guidance on the use of symbols and markings on glazed doors and screens. The guidance now given in Approved Document M (2003) is as a result of more recent experience of 'door manifestation' and takes precedence over the guidance currently provided in Approved Document N in conflicting areas until such time as Approved Document N is revised.

2.5.14 Part P – Electrical safety

This part requires that the design and installation of electrical installations should be planned so as to protect persons operating, maintaining or altering the installations from fire or injury.

It is understood that the Government is also intending to introduce a scheme whereby domestic installations are checked at regular intervals (as well as when they are sold and/or purchased) to make sure that they comply with the Building Regulations. This would mean, of course, that if you had an installation that was **not** correctly certified, then your house insurance may well **not** be valid.

2.5.15 Future Approved Documents and Guidance

For a while now there has been a Consultation Draft under consideration for Electronic Communication Services. However, as we all know the world of electronic communications is rapidly changing and therefore the decision has been made to deal with this area under a Guidance Document rather than under the restrictions of an Approved Document, as this will mean amendments and updates are quicker and easier to manage and therefore will come to the public more frequently. The Guidance Documents currently reflect the details that have been laid out in the Draft Document for some time now, however, we can no doubt expect changes shortly!

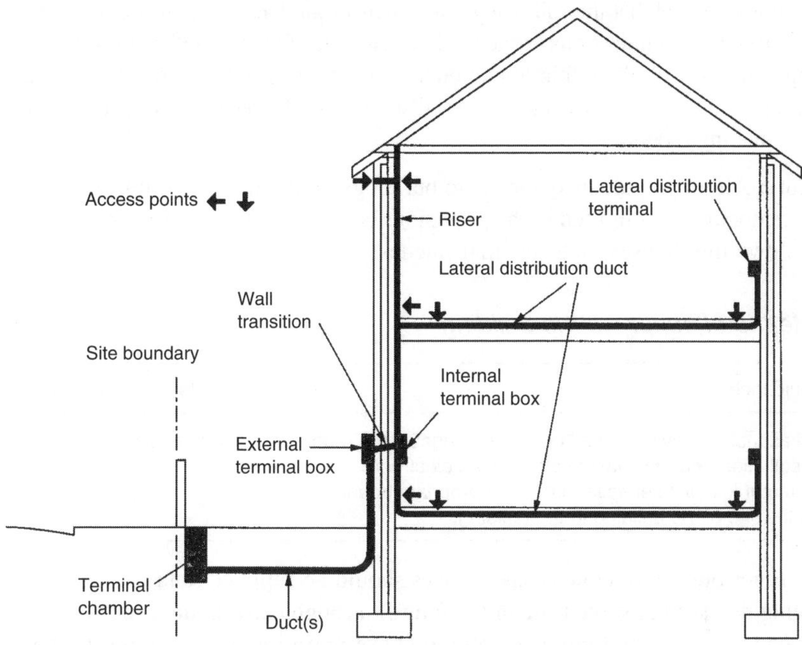

Figure 2.1 Outline guidance for the distribution of electronic communication services in a building

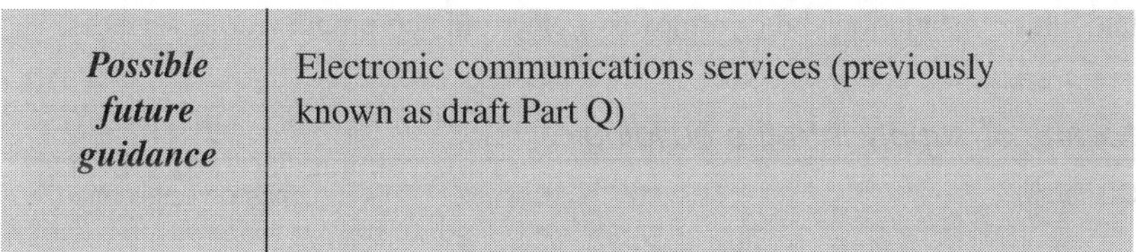

Possible future guidance	Electronic communications services (previously known as draft Part Q)

Electronic communications services, in their widest sense, convey a range of communications by a range of media. These can include speech, music (and other sounds), visual images, broadband and/or signals communicated by electric, magnetic, electromagnetic, collector-chemical or electromechanical media.

Note: The generic term 'broadband' is used to describe the technology that delivers higher capacity two-way communication to private homes or business premises. Broadband can be delivered by several technologies, including copper and fibre optic cables, radio or satellite services.

The increased use of these electronic communications services in buildings has resulted in them being supplied/routed in quite a number of different ways. The problems of trying to route cables carrying these services into and around existing buildings has become quite a difficulty to both owners and occupiers of these buildings and may, as a result, have caused a certain amount of damage and disruption to the building fabric and surrounding ground.

It is intended that the following three areas of guidance will ensure that future electronic communications services will be capable of being installed into an existing building with the minimum amount of inconvenience to the building owner/occupier and without any unnecessary disruption of the building fabric and/or the surrounding ground.

Note: Although guidance applies to buildings of all purpose groups, the guidance document is limited to the provisions considered necessary for dwellings, i.e. dwelling-houses, flats and maisonettes.

Means of supply to the building

Guidance	Limits on application
Reasonable provision shall be made to enable the ready installation and removal of cable-based electronic communications services from an appropriate boundary of the site of the building to the building.	Not yet fully established

Electronic communications services should be supplied (and be capable of being removed at some time in the future) through existing ducts between the boundary of the building site and a point of entry into the building, **without** having to excavate the ground within the curtilage of the site of the building.

Sufficient terminal chambers and associated ducts capable of serving all of accommodation units (e.g. flats) in the building should be provided so as to facilitate future installation and/or removal of electronic communications services.

Means of supply into the building

Guidance	Limits on application
Reasonable provision shall be made to enable electronic communications services from cable or wireless networks to be readily supplied from the exterior of the building to the interior (of the building).	Not yet fully established

Electronic communications services should be supplied to the inside of a dwelling (without having to unnecessarily disturb the fabric of the building) via an external terminal box, through a suitable wall transition, to an internal terminal box.

Means of supply around the building

Guidance	Limits on application
There shall be reasonable provision of [cable ducts] within the building to facilitate the future installation and supply of electronic communications services to each floor of the building and to at least one suitable room on each floor of the building.	Not yet fully established

Electronic communications services should be capable of being distributed around the building, currently or at some time in the future, inside ducts to at least:

- each floor of each dwelling (including the loft space and basement where these exist);
- one habitable room at each level of each dwelling.

The guidance **does not** require anything to be done except for the purposes of '*securing reasonable levels of convenience for building users*'.

2.6 Are there any exemptions?

The Building Regulations do not apply to:

- a building belonging to statutory undertakers;
- a building belonging to the United Kingdom Atomic Energy Authority;
- a building belonging to the British Airports Authority;
- a building belonging to the Civil Aviation Authority;

unless it is a house, hotel or a building used as offices or showrooms not forming part of any of the above premises.

Note: From 1 April 2001, maintained schools ceased to have exemption from the Building Regulations and school-specific standards have now been incorporated into the latest editions of Approved Documents.

Purpose-built student living accommodation (including flats) should thus be treated as hotel/motel accommodation in respect of space requirements and internal facilities.

2.7 What happens if I do not comply with an Approved Document?

Not actually complying with an Approved Document (which is after all only meant as a guidance document) doesn't mean that you are liable to any civil or criminal prosecution. If, however, you have contravened a Building Regulation then not having complied with the recommendations contained in the Approved Documents may be held against you.

2.8 Do I need Building Regulations approval?

If you are considering carrying out building work to your property then you may need to apply to your local authority for *Building Regulations approval.*

For most types of building work (e.g. extensions, alterations, conversions and drainage works), you will be required to submit a Building Regulations application prior to commencing any work.

Certain types of extensions and small detached buildings are exempt from Building Regulations control – particularly if the increased area of the alteration or extension is less than 30 m^2. However, you may **still** be required to apply for planning permission.

If you are in any doubt about whether you need to apply for permission, you should contact your local authority building control department before commencing any work to your property (in all cases, you may require planning permission).

2.8.1 Building work needing formal approval

The Building Regulations apply to any building that involves:

- the erection of a new building or re-erection of an existing building;
- the extension of a building;
- the '*material alteration*' of a building;
- the '*material change of use*' of a building;
- the installation, alteration or extension of a controlled service or fitting to a building;
- work relating to a change in the energy status;
- work required by Regulation 17D (consequential improvements in energy performance).

2.8.2 Typical examples of work needing approval

- Altered openings for new windows in roofs or walls;
- Cellars (particularly in London);
- Electrical installations;
- Erection of new buildings that are not exempt;
- Home extensions such as for a kitchen, bedroom, lounge, etc.;
- Installation of baths, showers, WCs which involve new drainage or waste plumbing;
- Installation of cavity insulation;
- Installation of new heating appliances;
- Internal structural alterations, such as the removal of a load-bearing wall or partition;
- Loft conversions;
- New chimneys or flues;

- Replacing roof coverings (unless exactly like for like repair);
- Underpinning of foundations.

2.8.3 Exempt buildings

There are certain buildings and work that are exempt from control. This is generally because they are buildings controlled by other legislation or because it would not be reasonable to control.

The following list, although not extensive, provides an indication of the main exemptions. These Regulations do not apply to:

- local authorities;
- county councils;
- public bodies;
- the Metropolitan Police Authority;
- any building constructed in accordance with the Explosives Acts 1875 and 1923;
- any building erected under the Nuclear Installations Act 1965;
- any building included in the schedule of monuments maintained under Section 1 of the Ancient Monuments and Archaeological Areas Act 1979;
- buildings not frequented by people;
- greenhouses and agricultural buildings – unless they are being used for retailing, packing or exhibiting (however, should that greenhouse/small detached building receive its electricity from a source shared with/or located inside a dwelling then Part P would apply);
- temporary buildings, i.e. a building which is not intended to remain erected for more than 28 days;
- ancillary buildings, e.g. an office on a building site;
- a detached building with a floor area less than 15 m^2 and containing no sleeping accommodation;
- a small detached building with a floor area less than 30 m^2, which contains no sleeping accommodation, is less than 1 m from the boundary and is constructed substantially of non-combustible material;
- a detached building designed and intended to shelter people from the effects of nuclear, chemical or conventional weapons;
- a conservatory whose floor area is less than 30 m^2 provided that it is wholly or partly glazed;
- a porch whose floor area is less than 3 m^2;
- a covered yard or covered way whose floor area is less than 30 m^2;
- a carport open on at least two sides whose floor area is less than 30 m^2.

The power to dispense with or relax any requirement contained in these Regulations rests with the local authority. It is therefore advisable to contact your local authority building control officer with details of your particular exemption claim so that you obtain a written reply agreeing the exemption. This will aid any future sale of the property!

Are there any other exemptions from the requirement to give building notice or deposit full plans?

The installations listed in Table 2.1 are exempt from having to give building notice or deposit full plans, **provided** that the person carrying out the work is as indicated in the second column.

In addition, provided any associated building work required to ensure that the appliance (service or fitting detailed above) complies with the applicable requirements contained in Schedule 1 – unless it is a heat producing gas appliance) which

(a) has a net rated heat input of 70 kilowatts or less; and
(b) is installed in a building with no more than three storeys (excluding any basement).

'*appliance*' includes any fittings or services, other than a hot water storage vessel that does not incorporate a vent pipe to the atmosphere, which form part of the space heating or hot water system served by the combustion appliance; and

'*building work*' does **not** include the provision of a masonry chimney.

2.8.4 Where can I obtain assistance in understanding the requirements?

Local councils can provide assistance with:

- advice about how to incorporate the most efficient energy safety measures into your scheme;
- advice about the use of materials;
- advice on electrical safety;
- advice on fire safety measures (including safe evacuation of buildings in the event of an emergency);
- at what stages local councils need to inspect your work;
- deciding what type of application is most appropriate for your proposal;
- how to apply for Building Regulations approval;
- how to prepare your application (and what information is required);
- how to provide adequate access for disabled people;
- what your Building Regulation Completion Certificate means to you.

2.9 How do I obtain Building Regulations approval?

You, as the owner or builder, are required to fill in an application form and return it, along with basic drawings and relevant information, to the building control office at least two days before work commences. Alternatively, you may submit full detailed plans for approval. Whatever method you adopt, it

Table 2.1 Exemptions from giving building notice or depositing full plans

Type of work	Person carrying out work
Installation of a **heat-producing gas appliance**.	A person, or an employee of a person, who is a member of a class of persons approved in accordance with regulation 3 of the Gas Safety (Installation and Use) Regulations 1998.
Installation of **heating or hot water service system connected to a heat-producing gas appliance**, or associated controls.	A person registered by CORGI Services Limited in respect of that type of work.
Installation of: (a) an **oil-fired combustion appliance** which has a rated heat output of 100 kilowatts or less and which is installed in a building with no more than 3 storeys (excluding any basement) or in a dwelling; (b) **oil storage tanks** and the pipes connecting them to combustion appliances; or (c) **heating and hot water service systems** connected to an oil-fired combustion appliance.	An individual registered by Oil Firing Technical Association Limited, NAPIT Certification Limited or Building Engineering Services Competence Accreditation Limited in respect of that type of work.
Installation of: (a) a **solid fuel burning combustion appliance** which has a rated heat output of 50 kilowatts or less which is installed in a building with no more than 3 storeys (excluding any basement); or (b) **heating and hot water service systems** connected to a solid fuel burning combustion appliance.	A person registered by HETAS Limited, NAPIT Certification Limited or Building Engineering Services Competence Accreditation Limited in respect of that type of work.
Installation of a **heating or hot water service system**, or associated controls, in a dwelling.	A person registered by Building Engineering Services Competence Accreditation Limited in respect of that type of work.
Installation of a **heating, hot water service, mechanical ventilation** or air conditioning system, or associated controls, in a building other than a dwelling.	A person registered by Building Engineering Services Competence Accreditation Limited in respect of that type of work.
Installation of an **air conditioning or ventilation system** in an existing dwelling, which does not involve work on systems shared with other dwellings.	A person registered by CORGI Services Limited or NAPIT Certification Limited in respect of that type of work.

(*Continued*)

Table 2.1 (*Continued*)

Type of work	Person carrying out work
Installation of a **commercial kitchen ventilation system** which does not involve work on systems shared with parts of the building occupied separately.	A person registered by CORGI Services Limited in respect of that type of work.
Installation of a **lighting system or electric heating system**, or associated electrical controls.	A person registered by the Electrical Contractors Association Limited in respect of that type of work.
Installation of **fixed low or extra-low voltage electrical installations**.	A person registered by BRE Certification Limited, British Standards Institution, ELECSA Limited, NICEIC Group Limited or NAPIT Certification Limited in respect of that type of work.
Installation of **fixed low or extra-low voltage electrical installations** as a necessary adjunct to or arising out of other work being carried out by the registered person.	A person registered by CORGI Services Limited, ELECSA Limited, NAPIT Certification Limited, NICEIC Group Limited or Oil Firing Technical Association Limited in respect of that type of electrical work.
Installation, **as a replacement, of a window, rooflight, roof window or door** (being a door which together with its frame has more than 50 per cent of its internal face area glazed) in an existing building.	A person registered under the Fenestration Self-Assessment Scheme by Fensa Ltd, or by CERTASS Limited or the British Standards Institution in respect of that type of work.
Installation of a **sanitary convenience, washing facility or bathroom** in a dwelling, which does not involve work on shared or underground drainage.	A person registered by CORGI Services Limited or NAPIT Certification Limited in respect of that type of work.
(1) Subject to paragraph (2), any building work, other than the **provision of a masonry chimney**, which is necessary to ensure that any appliance, service or fitting which is installed and which is described in the preceding entries in column 1 above, complies with the applicable requirements contained in Schedule 1. (2) Paragraph (1) does not apply to: (a) building work which is necessary to ensure that a heat-producing gas appliance complies with the applicable requirements contained in Schedule 1 unless the appliance (i) has a rated heat output of 100 kilowatts or less; and (ii) is installed in a building with no more than 3 storeys (excluding any basement), or in a dwelling; (b) the provision of a masonry chimney.	

may save time and trouble if you make an appointment to discuss your scheme with the building control officer well before you intend carrying out any work.

The building control officer will be happy to discuss your intentions, including proposed structural details and dimensions together with any lists of the materials you intend to use, so that he can point out any obvious contravention of the Building Regulations before you make an official application for approval. At the same time he can suggest whether it is necessary to approach other authorities to discuss planning, sanitation, fire escapes and so on.

The building control officer will ask you to inform the office when crucial stages of the work are ready for inspection (by a surveyor) in order to make sure the work is carried out according to your original specification. Should the surveyor be dissatisfied with any aspect of the work, he may suggest ways to remedy the situation.

When the building is finished you must notify the council.

It would be to your advantage to ask for written confirmation that the work was satisfactory as this will help to reassure a prospective buyer when you come to sell the property.

2.10 What are building control bodies?

Your local authority has a general duty to see that all building work complies with the Building Regulations. To ensure that your particular building work complies with the Building Regulations you must use one of the two services available to check and approve plans and to inspect your work as appropriate. The two services are the local authority building control service or the service provided by the private sector in the form of an approved inspector. Both building control bodies will charge for their services. Both may offer advice before work is started.

2.10.1 What will the local authority do?

This rather depends on whether you are submitting:

(1) full plans application submission; or
(2) building notice application.

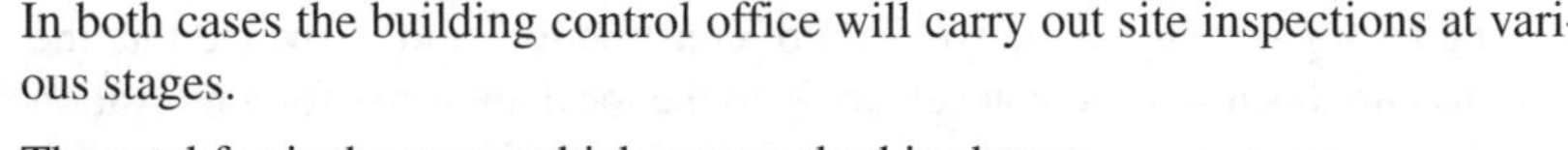

In both cases the building control office will carry out site inspections at various stages.

The total fee is the same whichever method is chosen.

Full plans

If you use the full plans procedure, the local authority will check your plans and consult any appropriate authorities (such as fire and water authorities).

If your plans comply with the Building Regulations you will receive a notice that they have been approved. If the local authority are not satisfied, then you may be asked to make amendments or provide more details. Alternatively, a conditional approval may be issued which will either specify modifications that must be made to the plans, or will specify further plans that must be deposited. A local authority may only apply conditions if you have either requested them to do so or have consented to them doing so. A request or consent must be made in writing. If your plans are rejected the reasons will be stated in the notice.

Building notice

If you use the building notice procedure, as with full plans applications, the work will normally be inspected as it proceeds; but you will not receive any notice indicating whether your proposal has been passed or rejected. Instead, you will be advised where the work itself is found (by the building control officer) not to comply with the Regulations.

Where a building notice has been given, the person carrying out building work or making a material change of use is required to provide plans showing how they intend conforming with the requirements of the Building Regulations. The local authority may also require further information such as structural design calculations of plans.

2.10.2 What will the approved inspector do?

If you use an approved inspector they will give you advice, check plans, issue a plans certificate, inspect the work etc. as agreed between you both. You and the inspector will jointly notify the local authority on what is termed an initial notice. Once that has been accepted by the local authority, the approved inspector will then be responsible for the supervision of building work. Although the local authority will have no further involvement, you may still have to supply them with limited information to enable them to be satisfied about certain aspects linked to Building Regulations (e.g. about the point of connection to an existing sewer).

If the approved inspector is not satisfied with your proposals you may alter your plans according to his advice; or you may seek a ruling from the Secretary of State regarding any disagreement between you. The approved inspector might also suggest an alternative form of construction, and, provided that the work has not been started, you can apply to the local authority for a relaxation or a dispensation from one (or more) of the Regulations' requirements and, in the event of a refusal by the local authority, appeal to the Secretary of State.

If, however, you do not exercise these options and you do not do what the approved inspector has advised to achieve compliance, the inspector will not be able to issue a final certificate. The inspector will also be obliged to notify the local authority so that they can consider whether to use their powers of enforcement.

2.10.3 What is the difference between a full plans application and the building notice procedure?

A person who intends carrying out any building work or making a material change of use to a building, shall:

- either provide the local authority with a building notice or
- deposit full plans with the local authority

subject to the following exclusions listed in Section 2.11.1 below.

For a full plans application, plans need to be produced showing all constructional details, preferably well in advance of your intended commencement on site. For the building notice procedure less detailed plans are required. In both cases, your application or notice should be submitted to the local authority and should be accompanied by any relevant calculations, to demonstrate compliance with safety requirements concerning the structure of the building.

If the use of the building is a 'designated use' under the Fire Precautions Act 1971, the application method **must** be a 'full plans' submission. This is to allow the local building control office to consult the fire brigade to see if they have any comments on the adequacy of the building's proposed means of escape in the event of fire.

Approved plans are valid for at least three years.

2.11 How do I apply for building control?

If your prospective work will involve any form of structure, you could need building control approval.

Some types of work may need both planning permission and building control approval; others may need only one or the other. The process of assessing a proposed building project is carried out through an evaluation of submitted information and plans and the inspection of work as the building progresses.

Take advantage of the free advice that local authorities offer, and discuss your ideas well in advance.

2.11.1 What applications do not require submission plans?

The following building works do **not** require the submission of plans:

- in respect of any work specified in an initial notice, an amendment notice or a public body's notice, which is in force;

- where a person intends to have electrical installation work completed by a competent firm registered under the NICIEC Approved Contractor scheme;
- where a person intends to have installed (by a person, or an employee of a person approved in accordance with Regulation 3 of the Gas Safety (Installation and Use) Regulations 1998) a heat-producing gas appliance;
- where Regulation 20 of the Building (Approved Inspectors etc.) Regulations 2000 (local authority powers in relation to partly completed work) applies.

2.11.2 Other considerations

Depending on the type of work involved, you may need to get approval from several sources before starting. The list below provides a few examples:

- There may be legal objections to alterations being made to your property.
- A solicitor might need to be consulted to see if any covenants or other forms of restriction are listed in the title deeds to your property and if any other person or party needs to be consulted before you carry out your work.
- You may need planning permission for a particular type of development work.
- If a building is listed or is within a Conservation Area or an Area of Outstanding Natural Beauty, special rules apply.

2.12 Full plans application

This type of application can be used for any type of building work, but it **must** be used where the proposed premises are to be used as a factory, office, shop, hotel, boarding house or railway premises.

A full plans application requires the submission of fully detailed plans, specifications, calculations and other supporting details to enable the building control officer to ascertain compliance with the Building Regulations. The amount of detail depends on the size and type of building works proposed, but as a minimum will have to consist of:

- a description of the proposed building work or material change of use;
- plan(s) showing what work will be completed; plus
- a location plan showing where the building is located relative to neighbouring streets.

The full plans application may be accompanied by a request (from the person carrying out such building work) that on completion of the work, he wishes the local authority to issue a completion certificate.

Two copies of the full plans application need to be sent to the local authority except in cases where the proposed building work relates to the erection, extension or material alteration of a building (other than a dwelling-house or flat) and where fire safety imposes an additional requirement, in which case five copies are required.

A full plans application will be thoroughly checked by the local authority who are required to pass or reject your plans within a certain time limit (usually eight weeks); or they may add conditions to an approval (with your written agreement). If they are satisfied that the work shown on the plans complies with the Regulations, you will be issued with an approval notice (within a period of five weeks or up to two months) showing that your plans were approved as complying with the Building Regulations.

If your plans are rejected, and you do not consider it is necessary to alter them, you will have two options available to you:

- you may seek a 'determination' from the Secretary of State if you believe your work complies with the Regulations (but you must apply before work starts);
- if you acknowledge that your proposals do not necessarily comply with a particular requirement in the Regulations and feel that it is too onerous in your particular circumstances, you may apply for a relaxation or dispensation of that particular requirement from the local authority. You can make this sort of application at any time you like but it is obviously sensible to do so as soon as possible and preferably before work starts. If the local authority refuses your application, you may then appeal to the Secretary of State within a month of the date of receipt of the rejection notice.

2.12.1 Consultation with sewerage undertaker

Where applicable, the local authority shall consult the sewerage undertaker as soon as practicable after the plans have been deposited, and before issuing any completion certificate in relation to the building work.

2.12.2 Advantages of submitting full plans application

The advantages of the full plans method are that:

- a (free) completion certificate will be issued on satisfactory completion of the work;
- a formal notice of approval or rejection will be issued within five weeks (unless the applicant agrees to extend this to two months);
- only when work starts on site (and the building control officer has completed his initial visit) is the remaining part of the fee invoiced;
- the plans can be examined and approved in advance (for an advance payment of (typically) 25% of the total fee).

2.13 Building notice procedure

Under the building notice procedure no approval notice is given. There is also no procedure to seek a determination from the Secretary of State if there is a disagreement between you and the local authority – unless plans are subsequently deposited. However, the advantage of the building notice procedure is that it will allow you to carry out **minor works** without the need to prepare full plans. You must, however, feel confident that the work will comply with the Regulations or you risk having to correct any work you carry out at the request of the local authority.

A building notice is particularly suited to minor works (for example, a householder wishing to install another WC). For such building work, detailed plans are unnecessary and most matters can be agreed when the building control officer visits your property. You do not need to have detailed plans prepared, but in some cases you may be asked to supply extra information.

As no formal approval is given, good liaison between the builder and the building control officer is essential to ensure that work does not have to be re-done.

The submission of a marked-up sketch showing the location of the building, although not mandatory, is recommended.

This type of application may be used for all types of building work, so long as no part of the premises is used for any of the purposes mentioned above under the full plans application.

2.13.1 What do I have to include in a building notice?

A building notice shall:

- state the name and address of the person intending to carry out the work;
- be signed by that person or on that person's behalf;
- contain, or be accompanied by:
  - a description of the proposed building work or material change of use;
  - particulars of the location of the building;
  - the use or intended use of that building.

Extension of a building

When planning a building extension, the building notice needs to be accompanied by:

- a plan to a scale of not less than 1:1250 showing:
  - the size and position of the building, or the building as extended and its relationship to adjoining boundaries;

  - the boundaries of the curtilage of the building, or the building as extended, and the size, position and use of every other building or proposed building within that curtilage;
  - the width and position of any street on or within the boundaries of the curtilage of the building or the building as extended;
- a statement specifying the number of storeys (each basement level being counted as one storey), in the building to which the proposal relates;
- particulars of:
  - the provisions to be made for the drainage of the building or extension;
  - the steps to be taken to comply with any local enactment which applies.

Insertion of insulating material into the cavity walls of a building

For cavity wall insulations, the building notice needs to be accompanied by a statement which specifies:

- the name and type of insulating material to be used;
- the name of any European Technical Approval issuing body that has approved the insulating material;
- the requirements of Schedule 1 in relation to which the issuing body has approved the insulating material;
- any European Economic Area (EEA) national standard with which the insulating material conforms;
- the name of any body that has issued any current approval to the installer of the insulating material.

Provision of a hot water storage system

A building notice in respect of a proposed hot water system shall be accompanied by a statement which specifies:

- the name, make, model and type of hot water storage system to be installed;
- the name of the body (if any) that has approved or certified the system;
- the name of the body (if any) that has issued any current registered operative identity card to the installer or proposed installer of the system.

Electrical installations

All proposals to carry out electrical installation work **must** be notified to the local authority's building control body before work begins, **unless** the proposed installation work is undertaken by a person who is a competent person registered with an electrical self-certification scheme and does not include the provision of a new circuit.

Non-notifiable work (such as replacing a socket outlet or other fixed electrical equipment) can be completed by a DIY enthusiast (family member or friends) but needs to be installed in accordance with manufacturers' instructions and done in such a way that they do not present a safety hazard. This work does **not** need to be notified to a local authority building control body (unless it is installed in an area of high risk such as a kitchen or a bathroom etc.) **but** all DIY electrical work (unless completed by a qualified professional – who is responsible for issuing a Minor Electrical Installation Certificate) will still need to be checked, certified and tested by a competent electrician.

Any work that involves adding a new circuit to a dwelling will need to be either notified to the building control body (who will then inspect the work) or needs to be carried out by a competent person who is registered under a Government Approved Part P Self-Certification Scheme.

Work involving any of the following will also have to be notified:

- consumer unit replacements;
- electric floor or ceiling heating systems;
- extra-low-voltage lighting installations, other than pre-assembled, CE-marked lighting sets;
- garden lighting or power installations;
- installation of a socket outlet on an external wall;
- installation of outdoor lighting and/or power installations in the garden or that involves crossing the garden;
- installation of new central heating control wiring;
- outdoor lighting and power installations;
- solar photovoltaic (PV) power supply systems;
- small-scale generators such as microCHP units.

Note: Where a person who is **not** registered to self-certify, intends to carry out the electrical installation, then a Building Regulation (i.e. a building notice or full plans) application will need to be submitted together with the appropriate fee, based on the estimated cost of the electrical installation. The building control body will then arrange to have the electrical installation inspected at first fix stage and tested upon completion.

2.14 How long is a building notice valid?

A building notice shall cease to have effect three years from the date when that notice was given to the local authority, unless, before the expiry of that period:

- the building work to which the notice related has commenced; or
- the material change of use described in the notice was made.

The approved plans may be used (i.e. built to) for at least three years, **even if the Building Regulations change during this time**.

2.15 What can I do if my plans are rejected?

If your plans were initially rejected, you can start work **provided** you give the necessary notice of commencement required under Regulation 14 of the Building Regulations and are satisfied that the building work itself now complies with the Regulations. However, it would **not** be advisable to follow this course if you are in any doubt and have not taken professional advice. Instead:

- you should resubmit your full plans application with amendments to ensure that they comply with Building Regulations; or
- if you think your plans comply (and that the decision to reject is, therefore, unjustified) you can refer the matter to the Secretary of State for the Environment, Transport and the Regions, or the Secretary of State for Wales (as appropriate) for their determination, but usually only before the work has started; or
- you could (in particular cases) ask the local authority to relax or dispense with their rejection. If the local authority refuse your application you could then appeal to the appropriate Secretary of State within one month of the refusal.

In the first two cases, the address to write to is the Department for Communities and Local Government (DCLG), Eland House, Bressenden Place, London SW1E 5DU. In Wales, you should refer the matter to the Secretary of State for Wales, Welsh Office, Crown Buildings, Cathays Park, Cardiff CF1 3NQ.

A fee is payable for determinations but not for appeals. The fee is half the plan fee (excluding VAT) subject to a minimum of £50 and a maximum of £500. The DETR or the Welsh Office will then seek comments from the local authority on your application (or appeal) which will be copied to you. You will then have a further opportunity to comment before a decision is issued by the Secretary of State.

2.15.1 Do my neighbours have the right to object to what is proposed in my Building Regulations application?

Basically – ***no***! But whilst there is no requirement in the Building Regulations to consult neighbours, it would be prudent to do so. In any event, you should be careful that the work does not encroach on their property since this could well lead to bad feeling and possibly an application for an injunction for the removal of the work.

Objections may be raised under other legislation, particularly if your proposal is subject to approval under the Town and Country Planning legislation or the Party Wall etc. Act of 1996.

The Party Wall Act 1996 came into force on 1 July 1997 and is largely based on Part VI of the London Building Acts (Amendment) Act 1939 – which started life as a Private Members Bill sponsored by the Earl of Lytton.

In a nutshell, this Act says that if you intend to carry out building work which involves:

- work on an existing wall shared with another property;
- building on the boundary with a neighbouring property;
- excavating near an adjoining building;

you **must** find out whether that work falls within the scope of the Act. If it does, then you must serve the statutory notice on all those defined by the Act as 'adjoining owners'. You should, however, remember that reaching agreement with adjoining owners on a project that falls within the scope of the Act does **not** remove the possible need for planning permission or Building Regulations approval.

Note: If you are not sure whether the Act applies to the work that you are planning, you should seek professional advice (see 'Useful contact names and addresses' at the end of the book).

A free explanatory booklet on the Party Wall etc. Act 1996 (*The Party Wall Act 1996 – Explanatory Booklet* (product code 02 BR 00862)) is available from the DCLG, www.communities.gov.uk (Tel: 0870 1226 236, Fax: 0870 1226 237).

2.16 What happens if I wish to seek a determination but the work in question has started?

You will only need to seek a determination if you believe the proposals in your full plans application comply with the Regulations but the local authority disagrees. You may apply for a determination either before or after the local authority has formally rejected your full plans application. The legal procedure is intended to deal with compliance of '*proposed*' work only and, in general, applications relating to work which is substantially completed cannot be accepted. Exceptionally, however, applications for '*late*' determinations may be accepted – but it is in your best interest to always ensure that you apply for a determination well before you start work.

2.17 When can I start work?

Again, it depends on whether you are using the local authority or the approved inspector.

2.17.1 Using the local authority

Once you have given a building notice or submitted a full plans application, you can start work at any time. However, you must give the local authority a commencement notice at least two clear days (not including the day on which you give notice and any Saturday, Sunday, bank or public holiday) before you start.

If you start work before you receive a decision on your full plans application, you will prejudice your ability to seek a determination from the Secretary of State if there is a dispute.

2.17.2 Using an approved inspector

If you use an approved inspector you may, subject to any arrangements you may have agreed with the inspector, start work as soon as the initial notice is accepted by the local authority (or is deemed to have been accepted if nothing is heard from the local authority within five working days of the notice being given). Work may not start if the initial notice is rejected, however.

2.18 Planning officers

Before construction begins, planning officers determine whether the plans for the building or other structure comply with the Building Regulations and if they are suited to the engineering and environmental demands of the building site. Building inspectors are then responsible for inspecting the structural quality and general safety of buildings.

2.19 Building inspectors

Building inspectors examine the construction, alteration, or repair of buildings, highways and streets, sewer and water systems, dams, bridges, and other structures to ensure compliance with building codes and ordinances, zoning regulations, and contract specifications.

Building codes and standards are the primary means by which building construction is regulated in the UK to assure the health and safety of the general public. Inspectors make an initial inspection during the first phase of construction and then complete follow-up inspections throughout the construction project in order to monitor compliance with regulations.

The inspectors will visit the worksite before the foundation is poured to inspect the soil condition and positioning and depth of the footings. Later, they return to the site to inspect the foundation after it has been completed. The size and type of structure, as well as the rate of completion, determine the number of other site visits they must make. Upon completion of the project, they make a final comprehensive inspection.

2.20 Notice of commencement and completion of certain stages of work

A person who proposes carrying out building work shall not start work unless:

- he has given the local authority notice that he intends to commence work; and
- at least two days have elapsed since the end of the day on which he gave the notice.

2.20.1 Notice of completion of certain stages of work

The person responsible for completing the building work is also responsible for notifying the local authority a minimum of five days **prior** to commencing any work involving excavations for foundations, foundations themselves, any damp-proof course any concrete or other material to be laid over a site and drains or sewers.

Upon completion of this work (especially work that will eventually be covered up by later work) the person responsible for the building work shall give five days' notice of intention to backfill.

A person who has laid, haunched or covered any drain or sewer shall (not more than five days after that work has been completed) give the local authority notice to that effect.

Where a building is being erected and that building (or any part of it) is to be occupied before completion, the person carrying out that work shall give the local authority at least five days' notice before the building, or any part of it is, occupied.

The person carrying out the building work shall **not**:

- cover up any foundation (or excavation for a foundation), any damp-proof course or any concrete or other material laid over a site; or
- cover up (in any way) any drains or sewers **unless** he has given the local authority notice that he intends to commence that work and at least one day has elapsed since the end of the day on which he gave the notice.

Where a person fails to comply with the above, then the local authority can insist that he shall cut into, lay open or pull down '*so much of the work as to enable the authority to ascertain whether these Regulations have been complied with, or not*'. If the local authority then notifies the owner/builder that certain work contravenes the requirements in these Regulations, then the owner/builder shall, after completing the remedial work, notify the local authority of its completion.

This requirement does not apply in respect of any work specified in an initial notice, an amendment notice or a public body's notice that is in force.

2.20.2 What kind of tests are the local authorities likely to make?

To establish whether building work has been carried out in conformance with the Building Regulations, local authorities will test to ensure that all work has been carried out:

- in a workmanlike manner;
- with adequate and proper materials which:
  - are appropriate for the circumstances in which they are used,
  - are adequately mixed or prepared and
  - are applied, used or fixed so as to adequately perform the functions for which they are designed;
- complies with the requirements of Part H of Schedule 1 (drainage and waste disposal);
- so as to enable them to ascertain whether the materials used comply with the provisions of these Regulations.

2.20.3 Energy rating

Where a new dwelling is being created, the person carrying out the building work shall calculate (and inform the local authority of) the dwelling's energy rating not later than five days after the work has been completed and, where a new dwelling is created, at least five days before intended occupation of the dwelling.

If the building is not to be immediately occupied, then the person carrying out the building work shall affix (not later than five days after the work has been completed) in a conspicuous place in the dwelling, a notice stating the energy rating of the dwelling.

Details of the correct procedures for calculating the energy rating are available from local authorities.

2.21 What are the requirements relating to building work?

In all cases, building work shall be carried out so that it:

(*a*) *it complies with the applicable requirements contained in Schedule 1; and*
(*b*) *in complying with any such requirement there is no failure to comply with any other such requirement.*

Building work shall be carried out so that, ***after*** *it has been completed:*

(*a*) *any building which is extended or to which a material alteration is made; or*

(*b*) *any building in, or in connection with which, a controlled service or fitting is provided, extended or materially altered; or*
(*c*) *any controlled service or fitting, complies with the applicable requirements of Schedule 1 or, where it did not comply with any such requirement, is no more unsatisfactory in relation to that requirement than before the work was carried out.*

2.22 Do I need to employ a professional builder?

Unless you have a reasonable working knowledge of building construction it would be advisable before you start work to get some professional advice (e.g. from an architect, or a structural engineer, or a building surveyor) and/or choose a recognized builder to carry out the work. It is also advisable to consult the local authority building control officer or an approved inspector in advance.

2.23 Unauthorized building work

If, for any reason, building work has been done without a building notice or full plans of the work being deposited with the local authority; or a notice of commencement of work being given, then the applicant may apply, in writing, to the local authority for a regularization certificate.

This application will need to include:

- a description of the unauthorized work;
- a plan of the unauthorized work; and
- a plan showing any additional work that is required for compliance with the requirements relating to building work in the Building Regulations.

Local authorities may then '*require the applicant to take such reasonable steps, including laying open the unauthorized work for inspection by the authority, making tests and taking samples, as the authority think appropriate to ascertain what work, if any, is required to secure that the relevant requirements are met*'.

When the applicant has taken any such steps required by the local authority, the local authority will notify the applicant:

- if no work is required to secure compliance with the relevant requirements;
- of the work which is required to comply with the relevant requirements;
- of the requirements which can be dispensed with or relaxed.

2.23.1 What happens if I do work without approval?

The local authority has a general duty to see that all building work complies with the Regulations – except where it is formally under the control of an approved inspector. Where a local authority is controlling the work and finds after its completion that it does not comply, then the local authority may require you to alter or remove it. If you fail to do this the local authority may serve a notice requiring you to do so and you will be liable for the costs.

2.23.2 What are the penalties for contravening the Building Regulations?

If you contravene the Building Regulations by building without notifying the local authority or by carrying out work which does not comply, the local authority can prosecute. If you are convicted, you are liable to a penalty not exceeding £5000 (at the date of publication of this book) plus £50 for each day on which each individual contravention is not put right after you have been convicted. If you do not put the work right when asked to do so, the local authority have power to do it themselves and recover costs from you.

2.24 Why do I need a completion certificate?

A completion certificate certifies that the local authority are satisfied that the work complies with the relevant requirements of Schedule 1 of the Building Regulations, '*in so far as they have been able to ascertain after taking all reasonable steps*'.

A completion certificate is a valuable document that should be kept in a safe place! Full rates for the property would also apply from that time onward.

2.25 How do I get a completion certificate when the work is finished?

The local authority shall give a completion certificate only when they have received the completion notice and have been able to ascertain that the relevant requirements of Schedule 1 (specified in the certificate) have been satisfied.

Where full plans are submitted for work that is also subject to the Fire Precautions Act 1971, the local authority must issue you with a completion certificate concerning compliance with the fire safety requirements of the Building Regulations once work has finished. In other circumstances, you may ask to be given one when the work is finished, but you must make your request **when you first submit your plans**.

If you use an approved inspector, they must issue a final certificate to the local authority when the work is completed.

2.26 Where can I find out more?

You can find out more from:

- the local authority's building control department;
- an approved inspector; or
- other sources.

2.26.1 Local authority

Each local authority in England and Wales (i.e. unitary, district and London boroughs in England and county and county borough councils in Wales) has a building control section whose general duty is to see that work complies with the Building Regulations – except where it is formally under the control of an approved inspector. Most local authorities have their own website and these usually contain a wealth of useful information, the majority of which is downloadable as read-only pdf files.

Individual local authorities co-ordinate their services regionally and nationally (and provide a range of national approval schemes) via LABC (Local Authority Building Control) Services. You can find out more about LABC Services through its website at www.labc-services.co.uk but your local authority building control department will be pleased to give you information and advice. They may offer to let you see their copies of the Building Act 1984, the Building Regulations 2000 and their associated Approved Documents that provide additional guidance.

The *Fire and Building Regulations Procedural Guide* which deals with procedures for building work to which the Fire Precautions Act 1971 applies, and the Department of the Environment, Transport and the Regions (DETR) leaflet on safety of garden walls, are amongst the documentation and advice that is available, free of charge, from your local authority.

The DETR's (and the Welsh Office's) separate booklets on planning permission for small businesses and householders are also available free of charge from your local authorities.

2.26.2 Approved inspectors

Approved inspectors are companies or individuals authorized under the Building Act 1984 to carry out building control work in England and Wales.

The Construction Industry Council (CIC) is responsible for deciding all applications for approved inspector status. You can find out more about the

CIC's role (including how to apply to become an approved inspector) through its website at www.cic.org.uk.

A list of approved inspectors can be viewed at the Association of Corporate Approved Inspectors (ACAI) website at www.acai.org.uk.

2.26.3 Other sources

Most of the documents can be purchased from The Stationery Office, 29 Duke Street, Norwich, NR3 1GN or from any main bookshop. Orders to TSO can be telephoned to 0870 600 5522 or faxed to 0870 600 5533 and their website is www.tso.co.uk. Copies should also be available in public reference libraries.

Appendix 2A Example application form

Building Control Service

Building Act 1984 – Building Regulations 2000

DEPOSIT OF BUILDING NOTICE

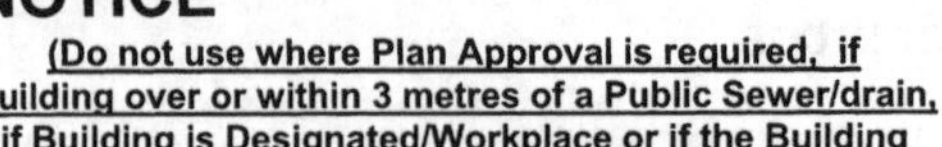
(Do not use where Plan Approval is required, if Building over or within 3 metres of a Public Sewer/drain, if Building is Designated/Workplace or if the Building will front onto a Private street.)

Submit **Two** Copies of **Forms** To:	Complete in BLOCK Capitals and BLACK Ink.
Director of Planning and Technical Services **Riddiford District Council** **Riddiford House** **Riddiford, Devon** **EX19 8DW**	Application No: .... FEE: .... Cheque/PO/Cash: .... VAT: .... Total Fee: .... Accepted: .... *For Office Use Only*

Please Read The Notes Before Completing This Form

1. **Applicants Details (See Note 1):** Owners Details (if different)
   Name:
   Address:
   Post Code: Tel. No:
   Fax No: E-Mail:

2. **Agents Details** (if any) to whom correspondence should be sent
   Name:
   Address:
   Post Code: Tel. No:
   Fax No: E-Mail:

3. **Location** of building to which work relates
   Address:
   Post Code: Tel. No:

4. **Proposed Work**
   Description:
   Date of commencement (See Note 6):

5. **Use of building**
   a. If new building or extension state proposed use:
   b. If existing building state present use:

6. **Method of Drainage**
   a. Foul water:
   b. Surface water:

7. **Means of water supply:**

8. Have you received Planning Consent for this work? YES/NO
   If yes, please give Consent No. and Date of Decision:

9. **Conservation of Fuel & Power** (See Note 3.2) Please indicate the method used to show compliance with Part L

For Dwellings	For Other Buildings
☐ Elemental Method	☐ Elemental Method
☐ Target U-Value Method	☐ Whole Building Method
☐ Carbon Index Method (Calculations to be enclosed)	☐ Carbon Emission Calculation Method
☐ Material Alteration / Change of Use	☐ Material Alteration / Change of Use

10. **Fees** (See Guidance Note on Fees for information)
    a. If Table A work - please state number of small domestic buildings:
    - please state number of different types of buildings:
    b. If Table B work - please state floor area m^2:
    c. If Table C work - please state estimated cost of work excluding VAT £:

11. **Statement**
    This notice is given in relation to the building work as described, and is submitted in accordance with Regulation 11(1)(a) and is accompanied by the appropriate fee

    Name: Signature: Date:

PLEASE TURN OVER **JANUARY 2005**

Notes:

1 The applicant is the person proposing to carry out building work, e.g. the building's owner.

2 Two copies of this notice should be completed and submitted. If a sewer connection is proposed, please submit an additional set of plans.

3 Where the proposed work includes the erection of a new building, material change of use or extension this notice shall be accompanied by the following:

3.1 a block plan to a scale of not less than 1:1250 showing:

3.1.1 the size and the position of the building, or the building as extended, and its relationship to adjoining boundaries;

3.1.2 the boundaries of the curtilage of the building, or the building as extended, and the size, position and use of every other building or proposed building within that curtilage;

3.1.3 the width and position of any street or within the boundaries of the curtilage of the building or the building as extended;

3.1.4 the provision to be made for the drainage of the building or extension.

3.2 the requirements of Part L (Conservation of Fuel and Power) must be considered for all dwellings and all other buildings. The requirements also apply to material alterations and changes of use.

3.2.1 extensions to dwellings less than 6 square metres would only need to be as energy efficient as the existing dwelling. Therefore question 9 is not applicable.

3.2.2 for other buildings with a floor area exceeding 100 m^2 floor area, details of artificial lighting systems will be required with the application.

3.2.3 Regulation 16 requires the provision of Energy Ratings calculated by the Government's Standard Assessment Procedure (SAP) for new dwellings and dwellings created as the result of material changes of use. The SAP calculation must be provided to the Authority within a minimum of five days prior to the occupation of the dwelling or five days after the completion, whichever comes first. Dwellings includes Flats.
The person carrying out the building work must display in the dwelling, as soon as practicable after the energy rating has been calculated, a notice of that rating. The notice must be displayed in a conspicuous place in the dwelling. Should occupation of the dwelling take place before physical completion, then the energy rating notice should be given to the occupier.
The requirement to display an energy rating notice in the dwelling, or give a notice to the occupier, is not necessary where the person carrying out the building work intends to occupy, or occupies, the dwelling as his/her residence.

4 Where the proposed work involves the insertion of insulating material into the cavity walls of a building this building notice shall be accompanied by a statement as to:

4.1 the name and type of insulating material to be used;

4.2 the name and type of insulating material is approved by the British Board of Agrément or conforms to a British Standard specification;

4.3 whether or not the installer is a person who is subject of a British Standards Institution Certificate of Registration or has been approved by the British Board of Agrément for the insertion of that material.

5 Where the proposed work involves the provision of an un-vented hot water storage system, this building notice shall be accompanied by a statement as to:

5.1 the name and type of system to be provided;

5.2 whether or not the system is approved by the British Board of Agrément

5.3 whether or not the installer has been approved by the British Board of Agrément for the provision of that system.

6 Persons carrying out building work must give written notice of the commencement of the work at least 48 hours beforehand.

7 A fee is usually payable to contribute towards the cost of site inspections, being a single payment which covers all necessary site visits until satisfactory completion of the work in accordance with the Building Regulations.

8 The building notice fee is calculated in accordance with current fees regulations and paid at the time of deposit of application. A Guidance Note of Fees is available on request.

Table A prescribes the fees payable for small domestic buildings. Table B prescribes the fees payable for small alterations and extensions to a dwelling house, and the addition of a small garage or carport.
Table C prescribes the fees payable for all other cases.

9 Subject to certain provisions of the Public Health Act 1936 owners and occupiers of premises are entitled to have their private foul and surface water drains and sewers connected to the public sewers, where available. Special arrangements apply to trade effluent discharge. Persons wishing to make such connections must give not less than 21 days' notice to the appropriate authority.

10 These notes are for general guidance only, particulars regarding the submission of Building Notices are contained in Regulation 12 of the Building Regulations 2000 and, in respect of fees, in the Building (Prescribed Fees etc.) Regulations 1991 - as amended.

11 Persons proposing to carry out building work or make a material change of use are reminded that permission may be required under the Town and Country Planning Acts.

12 Further information and advice concerning the Building Regulations and planning matters may be obtained from your local authority.

13 A building notice shall cease to have effect on the expiry of 3 years from the date on which that notice was given to the local authority, unless the work has commenced.

14 A Building Notice cannot be used where the building is proposed to be put, or is currently put, to a use as a Workplace to which the Fire Precautions (Workplace) Regulations 1997 applies (and/or is designated as defined in the Fire Precautions Act 1971).

15 All Electrical work required to meet the requirements of Part P (Electrical Safety) must be designed, installed, inspected and tested by a person competent to do so.

15.1 Prior to completion the Council should be satisfied that Part P has been complied with. This may require an appropriate BS 7671:2001 electrical installation certificate to be issued for the work by a person competent to do so.

15.2 Compliance with Part P2

- **(a) The electrical installation certificate will normally include all the details necessary to identify and give details of each circuit.**
- **(b) The person signing the certificate will also ensure that all appropriate notices required by BS 7671:2001 (for example those warning of differing wiring colours) are correctly provided and fixed.**
- **(c) Provided the information in (a) and (b) is provided electrical installation diagrams should not normally be necessary. In cases where the works are of unusual complexity (perhaps in mixed use developments for example), the right to require such diagrams will remain at the discretion of the Building Control Surveyor.**

Appendix 2B Example planning permission form

Building Control Service

Building Act 1984 – Building Regulations 2000

DEPOSIT OF FULL PLANS

Submit **Two** Copies of **Forms** To:	Complete in BLOCK Capitals and BLACK Ink.
Director of Planning and Technical Services **Riddiford District Council** **Riddiford House** **Riddiford, Devon** **EX19 8DW**	APPLICATION NUMBER:.................... EXPIRY DATE: .... **PLAN FEE** **INSPECTION FEE** Cheque/PO/Cash: .... VAT: .... Net Amount .... Total Fee: .... VAT .... Accepted: .... Total Fee .... *For Office Use Only*

Please Read The Notes Before Completing This Form

1. **Applicants Details (See Note 1):** Owners Details (if different)
   Name:
   Address:

   Post Code: Tel. No:
   Fax No: E-Mail:

2. **Agents Details** (if any) to whom correspondence should be sent
   Name:
   Address:

   Post Code: Tel. No:
   Fax No: E-Mail:

3. **Location** of building to which work relates
   Address:

   Post Code: Tel. No:

4. **Proposed Work**
   Description:
   Date of commencement:

5. **Use of building**
   a. If new building or extension state proposed use:
   b. If existing building state present use:

6. **Method of Drainage**
   a. Foul water:
   b. Surface water:

7. **Means of water supply:**

8. Have you received Planning Consent for this work? YES/NO
   If yes, please give Consent No. and Date of Decision:

9. **Conservation of Fuel & Power** (See Note 5.3) Please indicate the method used to show compliance with Part L

   For Dwellings
   ☐ Elemental Method
   ☐ Target U-Value Method
   ☐ Carbon Index Method (Calculations to be enclosed)
   ☐ Material Alteration / Change of Use

   For Other Buildings
   ☐ Elemental Method
   ☐ Whole Building Method
   ☐ Carbon Emission Calculation Method
   ☐ Material Alteration / Change of Use

10. **Fire Safety Requirements**
    Is the building proposed to be put, or is currently put, to a use as a Workplace to which the Fire Precautions (Workplace) Regulations 1997 applies and/or is designated as defined in the Fire Precautions Act 1971) ? Yes ☐ / No ☐

    If Yes, please indicate the design method you have used to show compliance with the Requirements of Part B of Schedule 1 to the Building Regulations.
    An additional copy of all plans/specifications are to be submitted if you have answered Yes to the above.

PLEASE TURN OVER **JANUARY 2005**

11. **Conditions**
Do you consent to the plans being passed subject to conditions where appropriate Yes ☐ / No ☐

12. **Extension of time**
Do you consent to the prescribed period being extended to two months Yes ☐ / No ☐

13. **Fees** (See Guidance Note on Fees for information)
a. If Table A work - please state number of small domestic buildings:
- please state number of different types of buildings:
b. If Table B work - please state floor area: m^2
If Table C work - please state estimated cost of work excluding VAT £:

14. **Statement**
This notice is given in relation to the building work as described, and is submitted in accordance with Regulation 11(1)(b) and is accompanied by the appropriate fee. I understand that further fees will be payable following the first inspection by the Local Authority.

Name: Signature: Date:

Should you have difficulty in filling out these forms please contact this office.

NOTES

1 The applicant is the person proposing to carry out the building works, e.g. the building's owner.

2 Two copies of this notice should be completed and submitted with two copies of plans and particulars in accordance with the provisions of Building Regulation 13. If a sewer connection is proposed an additional set of plans will be required.

3 Subject to certain exceptions a Full Plans Submission attracts fees payable by the person by whom or on whose behalf the work is to be carried out. Fees are payable in two stages. The first must accompany the deposit of plans and the second fee is payable after the first site inspection of work in progress. This second fee is a single payment in respect of each individual building, to cover all site visits and consultations, which may be necessary, until the work is satisfactorily completed.

3.1 Table A prescribes the plan and inspection fees payable for small domestic buildings. Table B prescribes the fees payable for small alterations and extensions to a dwelling home, and the addition of a small garage or carport. Table C prescribes the fees payable for all other cases.

3.2 The appropriate fee is dependent upon the type of work proposed. Fee scales and methods of calculation are set out in the Guidance Notes on Fees, which is available on request.

4 Subject to certain provisions of the Public Health Act 1936 owners and occupiers of premises are entitled to have their private foul and surface water drains and sewers connected to the public sewers, where available. Special arrangements apply to trade effluent discharge. Persons wishing to make such connections must give not less than 21 days' notice to the appropriate authority.

5 The requirements of Part L (Conservation of Fuel and Power) must be considered for all dwellings and other buildings. The requirements also apply to material alterations and change of use.

5.1 Extensions to dwellings less than 6 square metres would only need to be as energy efficient as the existing dwelling. Therefore question 9 is not applicable.

5.2 For other buildings with a floor area exceeding 100 m^2 floor area, details of artificial lighting systems will be required with the application.

5.3 Regulation 16 requires the provision of Energy Ratings calculated by the Government's Standard Assessment Procedure (SAP) for new dwellings and dwellings created as a result of material changes of use.

The SAP calculation must be provided to the Authority within a minimum of five days prior to the occupation of the dwelling or five days after the completion, whichever comes first. Dwellings include Flats. The person carrying out the building work must display in the dwelling, as soon as practicable after the energy rating has been calculated, a notice of that rating. The notice must be displayed in a conspicuous place in the dwelling. Should occupation of the dwelling take place before physical completion, then the energy rating notice should be given to the occupier.
The requirement to display an energy rating notice in the dwelling, or give a notice to the occupier, is not necessary where the person carrying out the building work intends to occupy, or occupies, the dwelling as his/her residence.

6 Section 16 of the Building Act 1984 provides for the passing of plans subject to conditions. The conditions may specify modifications to the deposited plans and/or that further plans shall be deposited.

7 These notes are for general guidance only, particulars regarding the deposit of plans are contained in Regulation 13 of the Building Regulations 2000 and, in respect of fees, in the Building (Prescribed Fees, etc.) Regulations 1991 as amended.

8 Persons proposing to carry out building work or make a material change of use of a building are reminded that permission may be required under the Town and Country Planning Acts.

9 Further information concerning the Building Regulations and Planning matters may be obtained from your Local Authority.

10 All Electrical work required to meet the requirements of Part P (Electrical Safety) must be designed, installed, inspected and tested by a person competent to do so.

10.1 Prior to completion the Council should be satisfied that Part P has been complied with. This may require an appropriate BS 7671:2001 electrical installation certificate to be issued for the work by a person competent to do so.

10.2 Compliance with Part P2

(a) The electrical installation certificate will normally include all the details necessary to identify and give details of each circuit.

(b) The person signing the certificate will also ensure that all appropriate notices required by BS 7671:2001 (for example those warning of differing wiring colours) are correctly provided and fixed.

(c) Provided the information in (a) and (b) is provided electrical installation diagrams should not normally be necessary. In cases where the works are of unusual complexity (perhaps in mixed use developments for example), the right to require such diagrams will remain at the discretion of the Building Control Surveyor.

Appendix 2C Example of an application for listed building consent

Planning and Technical Services Department

Application For Listed Building Consent

Planning (Listed Buildings and Conservation Areas) Act 1990

Submit Three Copies of Plans and Forms To: (See Note 1.)	Complete in BLOCK Capitals and BLACK Ink.
Director of Planning and Technical Services Riddiford District Council Riddiford House Riddiford, Devon EX19 8DW	APPLICATION No: .... DATE RECEIVED: .... *For Office Use Only*

Please Read The Notes Before Completing This Form

1. **Applicants Details**
   Name:
   Address:
   Post Code: Tel. No.:
   Are you, or your partner related to or connected with an elected member or officer of the Riddiford District Council? If yes give details:

2. **Agents Details** (if any) to whom correspondence should be sent
   Name:
   Address:
   Post Code: Tel. No.:

3. Full address or location of the building to which this application relates:
   Address:
   Post Code: Tel. No.:

4. Applicants interest in building (e.g. owner, lease, prospective purchaser, etc.)

5. Proposed work e.g. demolition, alteration, extension. Description:

6. Justification why the works are considered desirable or necessary.

7. Drawings and plans submitted with application.
   Details:

Note: The plans should be sufficient to identify the buildings and all the alterations and extensions should be shown in detail; the works should also be shown in relation to any adjacent buildings.

PLEASE TURN OVER

8. * I/We hereby apply for listed building consent to execute the works described in the application and all the accompanying plans and drawings and in accordance therewith.

DATE: SIGNED: ....

On behalf of:

* Delete where appropriate (insert applicants name if signed by agent)

Planning (Listed Buildings and Conservation Areas) Act 1990 provide that an application for listed building consent shall not be entertained unless it is accompanied by one of four certificates. If you are the sole owner (see(a) below) of all the land, Certificate A, which is printed below, is appropriate: only one copy need be completed. If you can not complete certificate A you will have to give notice to the other owners and complete certificate B. Certificate C and D are appropriate only if you have made efforts to trace the other owners and have failed.
The forms of these notices and certificates are prescribed in the regulations.

9. Planning (Listed Buildings and Conservation Areas) Act 1990.

CERTIFICATE A

I hereby certify that no person other than *myself / the applicant was an owner (a) of the building to which the application relates at the beginning of the period of 20 days before the date of the accompanying application.

SIGNED: ....
*On behalf of: Date:
- Delete where appropriate (insert applicants name if signed by agent)

NOTE: (a) "owner" means a person having freehold interest or a leasehold interest the unexpired term of which was not less than 7 years.

NOTES:

1. Planning Policy Guidance Note 15 (para. 3.4) requires that the application must make clear the justification for the works (see question 6).

2. In support of the application, please provide 3no. plans / drawings showing, in red, the location of the property and the work to be carried out. The drawings should preferably be of a scale of no less than 1:100 and should clearly indicate existing and new work.

3. Any object or structure fixed to the listed building or forming part of the land and comprised within the curtilage of the building is treated as part of the listed building.

4. If an appeal is made to the Secretary of State concerning this application, the regulations require that a copy of the following documents shall be furnished to the Secretary of State by the applicant:

(a) the application made to the Local Authority Planning Department together with all the relevant plans, drawings, particulars and documents (including a copy of the certificate) submitted with it.
(b) the notice of decision (if any) and all other relevant correspondence with the Local Planning Authority.

5. If consent is granted for the demolition of a listed building, the effect of section 8(2)(b) of the 1990 Act is that demolition may not be undertaken until notice of the proposal has been given to the Royal Commission and the Commission subsequently have either been given reasonable access to the building for at least one month following the grant of consent and before works commenced or have stated that they have completed their record of the building or that they do not wish to record it.

Appendix 2D Typical application for agricultural/forestry determination

Planning and Technical Services Department

Application For
Agricultural/Forestry Determination

Town and Country Planning (General Permitted Development) Order 1995 Part 6 & 7, Class A, Schedule 2

Submit Two Copies of Plans and Forms To:	Complete in BLOCK Capitals and BLACK Ink.
Director of Planning and Technical Services Riddiford District Council Riddiford House Riddiford, Devon EX19 8DW	APPLICATION No: .... DATE RECEIVED: .... *For Office Use Only*

Please Read The Notes Before Completing This Form

1. **Applicants Details**
Name:
Address:
Post Code: Tel. No.:

2. **Agents Details** (if any) to whom correspondence should be sent
Name:
Address:
Post Code: Tel. No.:

3. **Full Postal address & Location of Works / Building(s)**

4. **Description of proposed works / building(s) [see note 4]**
Details:

5. Please state:
(a) Does this holding consist of more that one separate block of land? Yes ☐ / No ☐
If yes, please give the size of each block of land and indicate which block the proposal relates to :-

(b) Size of holding [see note 4]

(c) Size of building(s) in metres
(d) Height (e.g. to ridge) in metres
(e) Materials to be used
Purpose to which building(s) works will be put

PLEASE TURN OVER **JANUARY 2005**

6. Please give details on any Agricultural Building(s)/extensions erected within the previous two years

* I / We enclose the appropriate fee of £35.00

DATE: **SIGNED:**

* Delete where appropriate

On behalf of:
(insert applicants name if signed by agent)

In accordance with the scale of charges, I enclose a remittance of £

NOTES:

1. An application for Agricultural Determination is required for development consisting of the erection of a building or the significant alteration of a private way. (Note: "Significant extension and significant alteration" mean any extension or alteration of the building where the cubic content of the original building will be exceeded by more than 10%, or the height of the building as extended or altered would exceed the height of the original building.)
2. The accompanying plans must show the siting, design and external appearance of the building, or as the case may be, siting and means of construction of the private way.
3. Please ensure that any dimensions shown on any drawings/plans accompanying this application are given in metric measurement.
4. If you intend to erect a new building as opposed to altering or extending one, it will be necessary to provide a plan demonstrating that the size of the holding exceeds 5 hectares.
5. If the holding is less than 5 hectares, planning permission will always be required. In these cases this application form should NOT be used.
6. The application must be accompanied by a fee of £35.00

Appendix 2E Example of an application for consent to display advertisements

Planning and Technical Services Department

Application For Consent To Display Advertisements

Town and Country Planning (Control of Advertisements) Regulations
Town and Country Planning Act 1990

Submit Three Copies of Plans and Forms To:	Complete in BLOCK Capitals and BLACK Ink.
Director of Planning and Technical Services Riddiford District Council Riddiford House Riddiford, Devon EX19 8DW	APPLICATION No: .... DATE RECEIVED: .... *For Office Use Only*

Please Read The Notes Before Completing This Form

1. **Applicants Details**
   Name:
   Address:
   Post Code: Tel. No. :

2. **Agents Details** (if any) to whom correspondence should be sent
   Name:
   Address:
   Post Code: Tel. No. :

3. **Full Postal Address & Location of Land or Building** on which the advertisement is to be displayed

4. **State the purpose for which the land or building is now being used**
   Details:

5. (A) Has the applicant an interest in the land or building?

   (B) If not, has the permission of the owner or of any other persons entitled to give permission for the display of advertisement been obtained? (see note 4).

6. (A) State the nature of the advertisement (e.g. hoarding, shop sign, projecting sign etc.)

   (B) Is the advertisement already being displayed? Yes ☐ / No ☐

PLEASE TURN OVER **JANUARY 2005**

7. (A) Will the advertisement be illuminated? Yes ☐ / No ☐

(B) If so state the type of illumination (e.g. internally, externally, floodlighting etc.)

(C) Will the illumination be static or intermittent?

8. Period for which consent is sought (see note 2).

9. Any additional information that the applicant may wish to supply.

* I / We hereby apply for consent to display the advertisement as described in the application and all the accompanying plans and drawings and in accordance therewith.

DATE: **SIGNED:**

* Delete where appropriate **On behalf of:**

(insert applicants name if signed by agent)

In accordance with the scale of charges, I enclose a remittance of £

Notes:

1. General. Under the Town and Country Planning (Control of Advertisements) Regulations many outdoor advertisements require express consent before they can be lawfully displayed. Applicants should refer to the regulations for details or call the Local Planning Authority.
2. Period of Consent. Normally the maximum period for which consent may be granted is 5 years. The Council may not grant for a longer period without special approval from the Secretary of State. If consent is required for a specific period of less than 5 years this should be stated in reply to question 8.
3. Drawings Required. The drawing can be in black and white on paper. It should show the size of the advertisement and its position on the land or the building in question. In the case of a sign it should also give the materials to be used, fixings, colours, height above the ground and where it would project from a building, the amount of the projection. The drawing should include a site location plan which should be of sufficient detail to enable the site to be identified.
4. Owners Consent. It is a condition of every consent by or under the regulations that before the advertisement to which the consent relates is displayed, the permission of the owner of the land or other persons entitled to grant permission shall be obtained.
5. Other Consents. Consent under the Town and Country Planning (Control of Advertisements) Regulations does not relieve the applicant from obtaining any other consents which may be necessary, under Listed Building & Conservation Areas Act etc.
6. Fees for advertisement applications.

Adverts Relating to the Business on the Premises	£60.00
Advance Signs Directing the Public to a Business	£60.00
(Unless Business can be seen from the signs position)	£220.00
Other Advertisements (e.g. Hoardings)	£220.00

Appendix 2F Regularization (Example)

Building Control Service

Building Act 1984 – Building Regulations 2000

REGULARIZATION

Submit **Two** Copies of **Forms** To:	Complete in BLOCK Capitals and BLACK Ink.
Director of Planning and Technical Services **Riddiford District Council** **Riddiford House** **Riddiford, Devon** **EX19 8DW**	Application No: .... FEE: .... Cheque/PO/Cash: .... VAT: .... Total Fee: .... Accepted: .... *For Office Use Only*

Please Read The Notes Before Completing This Form

1. **Applicants Details (See Note 1):** **Owners Details (if different)**
   Name:
   Address:
   Post Code: Tel. No.:
   Fax No: E-Mail:

2. **Agents Details** (if any) to whom correspondence should be sent
   Name:
   Address:
   Post Code: Tel. No.:
   Fax No: E-Mail:

3. **Location** of building to which work relates
   Address:
   Post Code: Tel. No.:

4. **Work requiring Regularization certificate (See Note 3):**
   Description:

5. **Use of building**
   a. Previous use:
   b. Present use:

6. **Services**
   a. Foul water:
   Surface water:
   c. Means of water supply:

7. Have you received Planning Consent for this work? Yes / No
   If yes, please give Consent No. & Date of Decision:

8. **Conservation of Fuel & Power** (See Note 3.3) Please indicate the method used to show compliance with Part L

For Dwellings	For Other Buildings
☐ Elemental Method	☐ Elemental Method
☐ Target U-Value Method	☐ Whole Building Method
☐ Carbon Index Method (Calculations to be enclosed)	☐ Carbon Emission Calculation Method
☐ Material Alteration / Change of Use	☐ Material Alteration / Change of Use

9. **Particulars of building work**
   Date work commenced: Date work completed:
   Name of Builder: Address:
   Postcode: Tel. No.:

PLEASE TURN OVER **JANUARY 2005**

10. **Fire Safety Requirements**
Is the Building proposed to be put, or is currently put, to a use as a Workplace to which the Fire Precautions (Workplace) Regulations 1997 applies and/or is designated as defined in the Fire Precautions Act 1971? Yes ☐ / No ☐

If Yes, please indicate the method you have used to show compliance with the Requirements of Part B of Schedule 1 to the Building Regulations.

An additional copy of all plans/specifications are to be submitted if you have answered Yes to the above.

11. **Fees** (See Guidance Note on Fees for information)
a. If Table A work - please state number of small domestic buildings:
- please state number of different types of buildings:
b. If Table B work - please state floor area m^2:
c. If Table C work - please state estimated cost of work excluding VAT £:

12. **Statement**
This notice is given in relation to the building work as described, and is submitted in accordance with Regulation 13(a) and is accompanied by the appropriate fee.

Name: Signature: Date:

Should you have difficulty in filling out these forms please contact this office.

NOTES

1 The applicant is the building's owner.

2 Two copies of this notice should be completed and submitted.

3 Where the work includes the erection of a new building, material alteration, material change of use or extension, this notice shall be accompanied, so far as is reasonably practicable, by the following:

3.1 a block plan to a scale of not less than 1:1250 showing:

3.1.1 a plan of the unauthorized work;

3.1.2 the provision made for the drainage of the building or extension;

3.1.3 a plan showing any additional work required to be carried out to secure the unauthorized work complies with the requirements relating to building work in the building regulations which were applicable to that work when it was carried out.

3.2 where the building or extension has been erected over a sewer or drain shown on the relative map of public sewers, the precautions taken in building over a sewer or drain.

3.3 the requirements of Part L (Conservation of Fuel and Power) must be considered for all dwellings and all other buildings. The requirements also apply to material alterations and changes of use.

3.3.1 for some small extensions to dwellings, these would only need to be as energy efficient as the existing dwelling. Therefore question 8 is not applicable.

3.3.2 for other buildings with a floor area exceeding 100 m^2 floor area, details of artificial lighting systems will be required with the application.

3.3.3 Regulation 16 requires the provision of Energy Ratings calculated by the Government's Standard Assessment Procedure (SAP) for new dwellings and dwellings created as the result of material changes of use. The SAP calculations must be provided to the Authority as soon as possible. Dwellings include Flats. The Person who carried out the building work must display in the dwelling, as soon as practicable after the energy rating has been calculated, a notice of that rating. The notice must be displayed in a conspicuous place in the dwelling. Should occupation take place before physical completion, then the energy rating notice should be given to the occupier.
The requirement to display an energy rating notice in the dwelling, or give a notice to the occupier, is not necessary where the person who carried out the building work intends to occupy, or occupies, the dwelling as his/her residence.

4 Where the work involved the insertion of insulating material into the cavity walls of a building this application shall be accompanied by a statement as to:

4.1 the name and type of insulating material used;

4.2 whether or not the insulating material is approved by the British Board of Agrément or conforms to a British Standard specification;

4.3 whether or not the installer was a person who is subject of a British Standards Institution Certification of Registration or has been approved by the British Board of Agrément for the insertion of that material.

5 Where the work involved the provision of an unvented hot water storage system, this application shall be accompanied by a statement as to:

5.1 the name and type of system to be provided;

5.2 whether or not the system is approved by the British Board of Agrément;

5.3 whether or not the British Board of Agrément has approved the installer for the provision of that system.

6 In accordance with Regulation 13(a), the Council may require an applicant to take such reasonable steps, including laying open the unauthorized work for inspection, making tests and taking samples as the Authority think appropriate to ascertain what work, if any, is required to secure compliance with the relevant Regulations.

7 The **Regularization Fee** is payable at the time the submission is made. A Guidance Note on Fees is available on request.

8 These notes are for general guidance only, particulars regarding the submission of Regularization applications are contained in Regulation 13(a) of the Building Regulations 2000 and, in respect of fees, in the Building (Prescribed Fees etc.) Regulations 1991 – as amended.

9 Further information and advice concerning the Building Regulations and planning matters may be obtained from your Local Authority.

10 The responsibility for demonstrating that the electrical work complies with Part P lies squarely with the person requesting a Regularization certificate, and the Authority is under no obligation to issue a certificate until satisfied on that point.

10.1 The applicant should provide any electrical certification issued at the time the work was carried out. If not, a periodic Inspection Report will be necessary to show compliance together with approximate exposure/tracing of the cables as is considered necessary.

3

The requirements of the Building Regulations

Introduction

The current Statutory Instruments concerning Building Regulations can be found in the table below. This chapter provides a breakdown of the key elements of each Approved Document, showing not only the regulation, but also an overview of the actual requirement 'in a nutshell'. This acts as a good starting point to understand the regulations which is then expanded upon in much more detail by specific building elements in Chapter 6.

Statutory Instrument	Made	Laid before Parliament	Coming into force
SI 2001 No 3335	4 Oct 2001	11 Oct 2001	1 Apr 2002
SI 2002 No 440	28 Feb 2002	5 Mar 2002	1 Apr 2002
SI 2002 No 2871	16 Nov 2002	25 Nov 2002	1 Jan 2004
SI 2003 No 2692	17 Oct 2003	27 Oct 2003	1 May 2003
SI 2004 No 1465	28 May 2004	8 Jun 2004	1 Dec 2004
SI 2004 No 3210	6 Dec 2004	10 Dec 2004	31 Dec 2004
SI 2006 No 652	9 Mar 2006	15 Mar 2006	6 April 2006
SI 2006 No 3318	13 Dec 2006	18 Dec 2006	6 April 2007

Note: Copies of the above documents are available from TSO (☎ 0870 600 5522) and through booksellers. They can also be viewed on the DCLG website at www.communities.gov.uk.

3.1 Part A – Structure

Number	Title	Regulation	Requirement (in a nutshell)
A1	**Loading**	*(1) The building shall be constructed so that the combined dead, imposed and wind loads are sustained and transmitted by it to the ground –* *(a) safely; and* *(b) without causing such deflection or deformation of any part of the building, or such movement of the ground, as will impair the stability of any part of another building.* *(2) In assessing whether a building complies with sub-paragraph* *(a) regard shall be had to the imposed and wind loads to which it is likely to be subjected in the ordinary course of its use for the purpose for which it is intended.*	The safety of a structure depends on: • the loading (see BS 6399, Parts 1 and 3); • properties of materials; • design analysis; • details of construction; • safety factors; • workmanship.
A2	**Ground movement**	*The building shall be constructed so that ground movement caused by –* *(a) swelling, shrinkage or freezing of the subsoil; or* *(b) land-slip or subsidence (other than subsidence arising from shrinkage), in so far as the risk can be reasonably foreseen, will not impair the stability of any part of the building.*	• Horizontal and vertical ties should be provided.

3.2 Part B – Fire safety

Number	Title	Regulation	Requirement (in a nutshell)
B1	**Means of warning and escape**	*The building shall be designed and constructed so that there are appropriate provisions for the early warning of fire, and appropriate means of escape in case of fire from the building to a place of safety outside the building capable of being safely and effectively used at all material times.* Requirement B1 does not apply to any prison provided under Section 33 of the Prisons Act 1952 (power to provide prisons etc.).	• There shall be an early warning fire alarm system for persons in the building. • There shall be sufficient escape routes that are suitably located to enable persons to evacuate the building in the event of a fire. • Safety routes shall be protected from the effects of fire. • In an emergency, the occupants of any part of the building shall be able to escape without any external assistance.
B2	**Internal fire spread (linings)**	*(1) To inhibit the spread of fire within the building the internal linings shall:* *(a) adequately resist the spread of flame over their surfaces; and* *(b) have, if ignited, a rate of heat release or a rate of fire growth which is reasonable in the circumstances.* 'Internal linings' means the materials or products used in lining any partition, wall, ceiling or other internal structure.	• The spread of flame over the internal linings of the building shall be restricted. • The heat released from the internal linings shall be restricted.
B3	**Internal fire spread (structure)**	*(1) The building shall be designed and constructed so that, in the event of fire, its stability will be maintained for a reasonable period.* *(2) A wall common to two or more buildings shall be designed and constructed so that it adequately*	Dependent on the use of the building, its size and the location of the element of construction: • load bearing elements of a building structure shall be capable of withstanding the effects of fire for an appropriate period without loss of stability;

		resists the spread of fire between those buildings. For the purposes of this sub-paragraph a house in a terrace and a semi-detached house are each to be treated as a separate building. (3) *Where reasonably necessary to inhibit the spread of fire within the building. measure shall be taken, to an extent appropriate to the size and intended use of the building, comprising either or both of the following:* (a) *sub-division of the building with fire-resisting construction;* (b) Installation of suitable automatic fire suppression systems. (4) *The building shall be designed and constructed so that the unseen spread of fire and smoke within concealed spaces in its structure and fabric is inhibited.* Requirement B3(3) does not apply to material alterations to any prison provided under Section 33 of the Prisons Act 1952.	• the building shall be sub-divided by elements of fire-resisting construction into compartments; • all openings in fire-separating elements shall be suitably protected in order to maintain the integrity of the element (i.e. the continuity of the fire separation); • any hidden voids in the construction shall be sealed and sub-divided to inhibit the unseen spread of fire and products of combustion.
B4	**External fire spread**	(1) *The external walls of the building shall adequately resist the spread of fire over the walls and from one building to another, having regard to the height, use and position of the building.* (2) *The roof of the building shall adequately resist the spread of fire over the roof and from one building to another, having regard to the use and position of the building.*	• External walls shall be constructed so as to have a low rate of heat release and thereby be capable of reducing the risk of ignition from an external source and the spread of fire over their surfaces. • The amount of unprotected area in the side of the building shall be restricted so as to limit the amount of thermal radiation that can pass through the wall.

(*Continued*)

Part B (*Continued*)

Number	Title	Regulation	Requirement (in a nutshell)
			• The roof shall be constructed so that the risk of spread of flame and/or fire penetration from an external fire source is restricted.
B5	**Access and facilities for the fire service**	(1) *The building shall be designed and constructed so as to provide reasonable facilities to assist fire fighters in the protection of life.* (2) *Reasonable provision shall be made within the site of the building to enable fire appliances to gain access to the building.*	• There shall be sufficient means of external access to enable fire appliances to be brought near to the building for effective use. • There shall be sufficient means of access into and within the building for firefighting personnel to affect search and rescue and fight fire. • The building shall be provided with sufficient internal fire mains and other facilities to assist firefighters in their tasks. • The building shall be provided with adequate means for venting heat and smoke from a fire in a basement.

3.3 Part C – Site preparation and resistance to contaminants and moisture

Number	Title	Regulation	Requirement (in a nutshell)
C1	**Preparation of site and resistance to moisture**	(1) *The ground to be covered by the building shall be reasonably free from any material that might damage the building or affect its stability, including vegetable matter, topsoil and pre-existing foundations.* (2) *Reasonable precautions shall be taken to avoid danger to health and safety caused by contaminants on or in the ground covered, or to be covered by the building and any land associated with the building.* (3) *Adequate subsoil drainage shall be provided if it is needed to avoid* (a) *the passage of the ground moisture to the interior of the building;* (b) *damage to the building, including damage through the transport of water-borne contaminants to the foundations of the building.* **Note:** For the purpose of this requirement, 'contaminant' means any substance which is or may become harmful to persons or buildings including substances, which are corrosive, explosive, flammable, radioactive or toxic.	Buildings should be safeguareded from the adverse effects of: • vegetable matter; • contaminants on or in the ground to be covered by the building; • ground water.

(*Continued*)

Part C (*Continued*)

Number	Title	Regulation	Requirement (in a nutshell)
C2	**Resistance to moisture**	*The floors, walls and roof of the building shall adequately protect the building and people who use the building from harmful effects caused by:* *(a) ground moisture;* *(b) precipitation and wind-driven spray;* *(c) interstitial and surface condensation; and* *(d) spillage of water from or associated with sanitary fittings or fixed appliances.*	• A solid or suspended floor shall be built next to the ground to prevent undue moisture from reaching the upper surface of the floor. • A wall shall be erected to prevent undue moisture from the ground reaching the inside of the building, and (if it is an outside wall) adequately resisting the penetration of rain and snow to the inside of the building. • The roof of the building shall be resistant to the penetration of moisture from rain or snow to the inside of the building. • All floors next to the ground, walls and roof shall not be damaged by moisture from the ground, rain or snow and shall not carry that moisture to any part of the building which it would damage.

3.4 Part D – Toxic substances

Number	Title	Regulation	Requirement (in a nutshell)
D1	**Cavity insulation**	*If insulating material is inserted into a cavity in a cavity wall reasonable precautions shall be taken to prevent the subsequent permeation of any toxic fumes from that material into any part of the building occupied by people.*	Fumes given off by insulating materials such as by urea formaldehyde (UF) foams should not be allowed to penetrate occupied parts of buildings to an extent where they could become a health risk to persons in the building by reaching an irritant concentration.

3.5 Part E – Resistance to the passage of sound

Number	Title	Regulation	Requirement (in a nutshell)
E1	**Protection against sound from other parts of the building and adjoining buildings**	*Dwelling-houses, flats and rooms for residential purposes shall be designed and constructed in such a way that they provide reasonable resistance to sound from other parts of the same building and from adjoining buildings.*	Dwellings shall be designed so that the noise from domestic activity in an adjoining dwelling (or other parts of the building) is kept to a level that: • does not affect the health of the occupants of the dwelling; • will allow them to sleep, rest and engage in their normal activities in satisfactory conditions. Dwellings shall be designed so that any domestic noise that is generated internally does not interfere with the occupants' ability to sleep, rest and engage in their normal activities in satisfactory conditions.
E2	**Protection against sound within a dwelling-house etc.**	*Dwelling-houses, flats and rooms for residential purposes shall be designed and constructed in such a way that:* *(a) internal walls between a bedroom or a room containing a water closet, and other rooms and* *(b) internal floors* *provide reasonable resistance to sound.*	**Note:** Requirement E2 does not apply to: (a) an internal wall which contains a door; (b) an internal wall which separates an en suite toilet from the associated bedroom; (c) existing walls and floors in a building which is subject to a material change of use.

(Continued)

Part E (*Continued*)

Number	Title	Regulation	Requirement (in a nutshell)
E3	**Reverberation in the common internal parts of buildings containing flats or rooms for residential purposes**	*The common internal parts of buildings which contain flats or rooms for residential purposes shall be designed and constructed in such a way as to prevent more reverberation around the common parts than is reasonable.*	Suitable sound absorbing material shall be used in domestic buildings so as to restrict the transmission of echoes. Requirement E3 only applies to corridors, stairwells, hallways and entrance halls which give access to the flat or rooms for residential purposes.
E4	**Acoustic conditions in schools**	*(1) Each room or other space in a school building shall be designed and constructed in such a way that it has the acoustic conditions and the insulation against disturbance by noise appropriate to its intended use.* *(2) For the purposes of this part – 'school' has the same meaning as in section 4 of the Education Act 1996; and 'school building' means any building forming a school or part of a school.*	Suitable sound insulation materials shall be used within a school building so as to reduce the level of ambient noise (particularly echoing in corridors etc.).

3.6 Part F – Ventilation

Number	Title	Regulation	Requirement (in a nutshell)
F1	**Means of ventilation**	*There shall be adequate means of ventilation provided for people in the building.* Requirement F1 does not apply to a building or space within a building – (a) into which people do not normally go; or (b) which is used solely for storage; or (c) which is a garage used solely in connection with a single dwelling.	• Ventilation (mechanical and/or air-conditioning systems designed for domestic buildings) shall be capable of restricting the accumulation of moisture and pollutants originating within a building.

3.7 Part G – Hygiene

Number	Title	Regulation	Requirement (in a nutshell)
G1	**Sanitary conveniences and washing facilities**	*(1) Adequate sanitary conveniences shall be provided in rooms provided for that purpose, or in bathrooms. Any such room or bathroom shall be separated from places where food is prepared.* *(2) Adequate washbasins shall be provided in –* *(a) rooms containing water closets; or* *(b) rooms or spaces adjacent to rooms containing water closets.* *Any such room or space shall be separated from places where food is prepared.* *(3) There shall be a suitable installation for the provision of hot and cold water to washbasins provided in accordance with paragraph (2).* *(4) Sanitary conveniences and washbasins to which this paragraph applies shall be designed and installed so as to allow effective cleaning.*	**All** dwellings (house, flat or maisonette should have at least one closet and one washbasin: • closets (and/or urinals) should be separated by a door from any space used for food preparation or where washing-up is done; • washbasins should, ideally, be located in the room containing the closet; • the surfaces of a closet, urinal or washbasin should be smooth, non-absorbent and capable of being easily cleaned; • closets (and/or urinals) should be capable of being flushed effectively; • closets (and/or urinals) should only be connected to a flush pipe or discharge pipe; • washbasins should have a supply of hot and cold water; • closets fitted with flushing apparatus should discharge through a trap and discharge pipe into a discharge stack or a drain.
G2	**Bathrooms**	*A bathroom shall be provided containing either a fixed bath or shower bath, and there shall be a suitable installation for the provision of hot and cold water to the bath or shower bath.*	All dwellings (house, flat or maisonette) should have at least one bathroom with a fixed bath or shower and the bath or shower should: • have a supply of hot and cold water;

		Requirement G2 applies only to dwellings.	• discharge through a grating, a trap and branch discharge pipe to a discharge stack or (if on a ground floor); • discharge into a gully or directly to a foul drain; • be connected to a macerator and pump (of an approved type) if there is no suitable water supply or means of disposing of foul water.
G3	**Hot water storage**	*A hot water storage system that has a hot water storage vessel which does not incorporate a vent pipe to the atmosphere shall be installed by a person competent to do so, and there shall be precautions:* *(a) to prevent the temperature of stored water at any time exceeding 100°C; and* *(b) to ensure that the hot water discharged from safety devices is safely conveyed to where it is visible but will not cause danger* **to** *persons in or about the building.*  Requirement G3 does not apply to: (a) a hot water storage system that has a storage vessel with a capacity of 15 litres or less; (b) a system providing space heating only; (c) a system that heats or stores water for the purposes only of an industrial process.	A hot water storage system shall: • be installed by a competent person; • not exceed 100°C; • discharge safely; • not cause danger to persons in or about the building.

3.8 Part H – Drainage and waste disposal

Number	Title	Regulation	Requirement (in a nutshell)
H1	**Foul water drainage**	*(1) An adequate system of drainage shall be provided to carry foul water from appliances within the building to one of the following, listed in order of priority –* *(a) a public sewer; or, where that is not reasonably practicable,* *(b) a private sewer communicating with a public sewer; or, where that is not reasonably practicable,* *(c) either a septic tank which has an appropriate form of secondary treatment or another wastewater treatment system; or, where that is not reasonably practicable,* *(d) a cesspool.* *(2) In this Part 'foul water' means wastewater which comprises or includes* *(a) waste from a sanitary convenience, bidet or appliance used for washing receptacles for foul waste; or* *(b) water which has been used for food preparation, cooking or washing.* 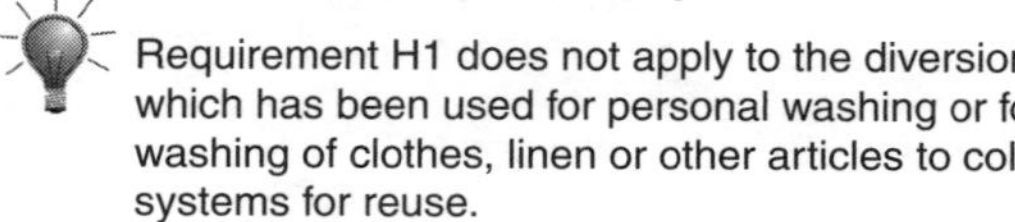 Requirement H1 does not apply to the diversion of water which has been used for personal washing or for the washing of clothes, linen or other articles to collection systems for reuse.	The foul water drainage system shall: • convey the flow of foul water to a foul water outfall (i.e. sewer, cesspool, septic tank or settlement (i.e. holding) tank); • minimize the risk of blockage or leakage; • prevent foul air from the drainage system from entering the building under working conditions; • be ventilated; • be accessible for clearing blockages; • not increase the vulnerability of the building to flooding. H1 is applicable to domestic buildings and small non-domestic buildings. Further guidance on larger buildings is provided in Appendix A to Approved Document H. Complex systems in larger buildings should be designed in accordance with BS EN 12056.

H2	**Wastewater treatment systems and cesspools**	*(1) Any septic tank and its form of secondary treatment, other wastewater treatment system or cesspool, shall be so sited and constructed that –* *(a) it is not prejudicial to the health of any person;* *(b) it will not contaminate any watercourse, underground water or water supply;* *(c) there are adequate means of access for emptying and maintenance; and* *(d) where relevant, it will function to a sufficient standard for the protection of health in the event of a power failure.* *(2) Any septic tank, holding tank which is part of a wastewater treatment system or cesspool shall be –* *(a) of adequate capacity;* *(b) so constructed that it is impermeable to liquids; and* *(c) adequately ventilated.* *(3) Where a foul water drainage system from a building discharges to a septic tank, wastewater treatment system or cesspool, a durable notice shall be affixed in a suitable place in the building containing information on any continuing maintenance required to avoid risks to health.*	Wastewater treatment systems shall: • have sufficient capacity to enable breakdown and settlement of solid matter in the wastewater from the buildings; • be sited and constructed so as to prevent overloading of the receiving water. Cesspools shall have sufficient capacity to store the foul water from the building until they are emptied. Wastewater treatment systems and cesspools shall be sited and constructed so as not to: • be prejudicial to health or a nuisance; • adversely affect water sources or resources; • pollute controlled waters; • be in an area where there is a risk of flooding. Septic tanks and wastewater treatment systems and cesspools shall be constructed and sited so as to: • have adequate ventilation; • prevent leakage of the contents and ingress of subsoil water; • having regard to water table levels at any time of the year and rising groundwater levels. Drainage fields shall be sited and constructed so as to: • avoid overloading of the soakage capacity; and • provide adequately for the availability of an aerated layer in the soil at all times.

(*Continued*)

Part H (*Continued*)

Number	Title	Regulation	Requirement (in a nutshell)
H3	**Rainwater drainage**	*(1) Adequate provision shall be made for rainwater to be carried from the roof of the building.* *(2) Paved areas around the building shall be so constructed as to be adequately drained.* *(3) Rainwater from a system provided pursuant to sub-paragraphs (1) or (2) shall discharge to one of the following, listed in order of priority –* *(a) an adequate soakaway or some other adequate infiltration system; or, where that is not reasonably practicable,* *(b) a watercourse; or, where that is not reasonably practicable,* *(c) a sewer.* Requirement H3(2) applies only to paved areas – (a) which provide access to the building pursuant to paragraph M2 of Schedule 1 (access for disabled people); (b) which provide access to or from a place of storage pursuant to paragraph H6(2) of Schedule 1 (solid waste storage); or (c) in any passage giving access to the building, where this is intended to be used in common by the occupiers of one or more other buildings. Requirement H3(3) does not apply to the gathering of rainwater for reuse.	Rainwater drainage systems shall: • minimize the risk of blockage or leakage; • be accessible for clearing blockages; • ensure that rainwater soaking into the ground is distributed sufficiently so that it does not damage foundations of the proposed building or any adjacent structure; • ensure that rainwater from roofs and paved areas is carried away from the surface either by a drainage system or by other means; • ensure that the rainwater drainage system carries the flow of rainwater from the roof to an outfall (e.g. a soakaway, a watercourse, a surface water or a combined sewer).

H4	**Building over sewers**	*(1) The erection or extension of a building or work involving the underpinning of a building shall be carried out in a way that is not detrimental to the building or building extension or to the continued maintenance of the drain, sewer or disposal main.* *(2) In this paragraph 'disposal main' means any pipe, tunnel or conduit used for the conveyance of effluent to or from a sewage disposal works, which is not a public sewer.* *(3) In this paragraph and paragraph H5 'map of sewers' means any records kept by a sewerage undertaker under Section 199 of the Water Industry Act 1991.* Requirement H4 applies only to work carried out – (a) over a drain, sewer or disposal main which is shown on any map of sewers; or (b) on any site or in such a manner as may result in interference with the use of, or obstruction of the access of any person to, any drain, sewer or disposal main which is shown on any map of sewers.	Building or extension or work involving underpinning shall: • be constructed or carried out in a manner which will not overload or otherwise cause damage to the drain, sewer or disposal main either during or after the construction; • not obstruct reasonable access to any manhole or inspection chamber on the drain, sewer or disposal main; • in the event of the drain, sewer or disposal main requiring replacement, not unduly obstruct work to replace the drain, sewer or disposal main, on its present alignment; • reduce the risk of damage to the building as a result of failure of the drain, sewer or disposal main.
H5	**Separate systems of drainage**	*Any system for discharging water to a sewer which is provided pursuant to paragraph H3 shall be separate from that provided for the conveyance of foul water from the building.* Requirement H5 applies only to a system provided in connection with the erection or extension of a building	Separate systems of drains and sewers shall be provided for foul water and rainwater where: (a) the rainwater is not contaminated; and (b) the drainage is to be connected either directly or indirectly to the public sewer system and either –

(*Continued*)

Part H (*Continued*)

Number	Title	Regulation	Requirement (in a nutshell)
		where it is reasonably practicable for the system to discharge directly or indirectly to a sewer for the separate conveyance of surface water which is – (a) shown on a map of sewers; or (b) under construction either by the sewerage undertaker or by some other person (where the sewer is the subject of an agreement to make a declaration of vesting pursuant to Section 104 of the Water Industry Act 1991).	(i) the public sewer system in the area comprises separate systems for foul water and surface water; or (ii) a system of sewers which provides for the separate conveyance of surface water is under construction either by the sewerage undertaker or by some other person (where the sewer is the subject of an agreement to make a declaration of vesting pursuant to Section 104 of the Water Industry Act 1991).
H6	**Solid waste storage**	*(1) Adequate provision shall be made for storage of solid waste.* *(2) Adequate means of access shall be provided –* *(a) for people in the building to the place of storage; and* *(b) from the place of storage to a collection point (where one has been specified by the waste collection authority under Section 46 (household waste) or Section 47 (commercial waste) of the Environmental Protection Act 1990 or to a street (where no collection point has been specified)).*	Solid waste storage shall be: • designed and sited so as not to be prejudicial to health; • of sufficient capacity having regard to the quantity of solid waste to be removed and the frequency of removal; • sited so as to be accessible for use by people in the building and of ready access from a street for emptying and removal.

3.9 Part J – Combustion appliances and fuel storage systems

Number	Title	Regulation	Requirement (in a nutshell)
J1	**Air supply**	*Combustion appliances shall be so installed that there is an adequate supply of air to them for combustion, to prevent over-heating and for the efficient working of any flue.* Requirement J1 only applies to fixed combustion appliances (including incinerators).	The building shall: • enable the admission of sufficient air for: – the proper combustion of fuel and the operation of flues; and – the cooling of appliances where necessary; • enable normal operation of appliances without the products of combustion becoming a hazard to health; • enable normal operation of appliances without their causing danger through damage by heat or fire to the fabric of the building; • have been inspected and tested to establish suitability for the purpose intended; • have been labelled to indicate performance capabilities.
J2	**Discharge of products of combustion**	*Combustion appliances shall have adequate provision for the discharge of products of combustion to the outside air.* Requirement J2 only applies to fixed combustion appliances (including incinerators).	
J3 J4	**Protection of building** **Provision of information**	*Combustion appliances and flue-pipes shall be so installed, and fireplaces and chimneys shall be so constructed and installed, as to reduce to a reasonable level the risk of people suffering burns or the building catching fire in consequence of their use.* Requirement J3 only applies to fixed combustion appliances (including incinerators).	Oil and LPG fuel storage installations shall be located and constructed so that they are reasonably protected from fires that may occur in buildings or beyond boundaries.
J5	**Protection of liquid fuel storage systems**	*Where a hearth, fireplace, flue or chimney is provided or extended, a durable notice containing information on the performance capabilities of the hearth, fireplace, flue or chimney shall be affixed in a suitable place in the building for the purpose of enabling combustion appliances to be safely installed.*	Oil storage tanks used wholly or mainly for private dwellings shall be: • reasonably resistant to physical damage and corrosion; • designed and installed so as to minimize the risk of oil escaping during the filling or maintenance of the tank; • incorporate secondary containment when there is a significant risk of pollution; • be labelled with information on how to respond to a leak.

(*Continued*)

Part J (*Continued*)

Number	Title	Regulation	Requirement (in a nutshell)
J6	**Protection against pollution**	*Liquid fuel storage systems and the pipes connecting them to combustion appliances shall be so constructed and separated from buildings and the boundary of the premises as to reduce to a reasonable level the risk of the fuel igniting in the event of fire in adjacent buildings or premises.* Requirement J5 applies only to – (a) fixed oil storage tanks with capacities greater than 90 litres and connecting pipes; and (b) fixed liquefied petroleum gas storage installations with capacities which are located outside the building and which serve fixed combustion appliances (including incinerators) in the building. *Oil storage tanks and the pipes connecting them to combustion appliances shall –* *(a) be so constructed and protected as to reduce to a reasonable level the risk of the oil escaping and causing pollution; and* *(b) have affixed in a prominent position a durable notice containing information on how to respond to an oil escape so as to reduce to a reasonable level the risk of pollution.* Requirement J6 applies only to fixed oil storage tanks with capacities of 3500 litres or less, and connecting pipes, which are – (a) located outside the building; and (b) serve fixed combustion appliances (including incinerators) in a building used wholly or mainly as a private dwelling but does not apply to buried systems.	

3.10 Part K – Protection from falling, collision and impact

Number	Title	Regulation	Requirement (in a nutshell)
K1	**Stairs, ladders and ramps**	*Stairs, ladders and ramps shall be so designed, constructed and installed as to be safe for people moving between different levels in or about the building.* Requirement K1 applies only to stairs, ladders and ramps which form part of the building.	All stairs, steps and ladders shall provide reasonable safety between levels in a building. In a public building the standard of stair, ladder or ramp may be higher than in a dwelling, to reflect the lesser familiarity and greater number of users.
K2	**Protection from falling**	(a) *Any stairs, ramps, floors and balconies and any roof to which people have access, and* (b) *any light well, basement area or similar sunken area connected to a building,* *shall be provided with barriers where it is necessary to protect people in or about the building from falling.* Requirement K2 (a) applies only to stairs and ramps which form part of the building.	Pedestrian guarding should be provided for any part of a floor, gallery, balcony, roof, or any other place to which people have access and any light well, basement area or similar sunken area next to a building.

(*Continued*)

Part K (*Continued*)

Number	Title	Regulation	Requirement (in a nutshell)
K3	**Vehicle barriers and loading bays**	(1) *Vehicle ramps and any levels in a building to which vehicles have access, shall be provided with barriers where it is necessary to protect people in or about the building.* (2) *Vehicle loading bays shall be constructed in such a way, or be provided with such features, as may be necessary to protect people in them from collision with vehicles.*	Vehicle barriers should be provided that are capable of resisting or deflecting the impact of vehicles. Loading bays shall be provided with an adequate number of exits (or refuges) to enable people to avoid being crushed by vehicles.
K4	**Protection from collision with open windows etc.**	*Provision shall be made to prevent people moving in or about the building from colliding with open windows, skylights or ventilators.* Requirement K4 does not apply to dwellings.	All windows, skylights, and ventilators shall be capable of being left open without danger of people colliding with them.
K5	**Protection against impact from and trapping by doors**	(1) *Provision shall be made to prevent any door or gate –* (a) *which slides or opens upwards, from falling onto any person; and* (b) *which is powered, from trapping any person.* (2) *Provision shall be made for powered doors and gates to be opened in the event of a power failure.* (3) *Provision shall be made to ensure a clear view of the space on either side of a swing door or gate.*	Requirement K5 does not apply to – (a) dwellings, or (b) any door or gate that is part of a lift.

3.11 Part L – Conservation of fuel and power

Number	Title	Regulation	Requirement (in a nutshell)
L	**Conservation of fuel and power** (in new and existing dwellings and buildings other than dwellings)	*Reasonable provision shall be made for the conservation of fuel and power in buildings by:* *(a) limiting heat gains and losses* *(i) through thermal elements and other parts of the building fabric; and* *(ii) from pipes, ducts and vessels used for space heating, space cooling and hot water services;* *(b) providing and commissioning energy-efficient fixed building services with effective controls; and* *(c) providing to the owner sufficient information about the building, the fixed building services and their maintenance requirements so that the building can be operated in such a manner as to use no more fuel and power than is reasonable in the circumstances.* **Note:** In addition to Part L, some of the other Approved Documents also have requirements concerning the conservation of fuel and power. In particular these include: • Part E (Resistance to the passage of sound); • Part F (Ventilation); • Part C (Site preparation and resistance to moisture); and • Part J (Combustion appliances and fuel storage systems). And where relevant, these requirements have been included within this part of the book.	Limiting heat gains and losses.

3.12 Part M – Access to and use of buildings

Number	Title	Regulation	Requirement (in a nutshell)
M1	**Access and use**	*Reasonable provision shall be made for people to:* *(a) gain access to and* *(b) use* *the buildings and its facilities.*  The requirements of this part do not apply to: (a) an extension of, or material alteration of, a dwelling; or (b) any part of a building which is used solely to enable the building or service or fitting in the building to be inspected, repaired or maintained.	(1) In addition to the requirements of the Disability Discrimination Act 1995, Approved Document M also requires that: precautions need to be taken to ensure that: 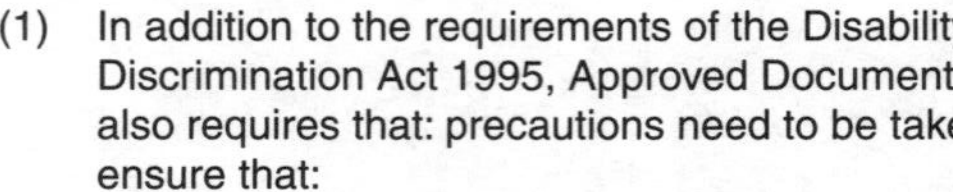 • new non-domestic buildings and/or dwellings (e.g. houses and flats used for student living accommodation etc.); • extensions to existing non-domestic buildings; • non-domestic buildings that have been subject to a material change of use (e.g. so that they become a hotel, boarding house, institution, public building or shop);
M2	**Access to extensions of buildings other than dwellings**	*Suitable independent access shall be provided in any building that is to be extended. Reasonable provision shall be made within the extension for sanitary convenience.* 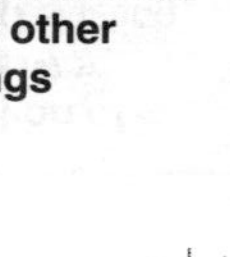 	(2) Are capable of allowing people, regardless of their disability, age or gender to: (a) gain access to buildings; (b) gain access within buildings; (c) be able to use the facilities of the buildings (both as visitors and as people who live or work in them).

M3 **Sanitary conveniences in extensions to buildings other than dwellings**

Requirement M2 does not apply where suitable access to the extension is provided through the building that is extended.

If sanitary conveniences are provided in any building that is to be extended, reasonable provision shall be made within the extension for sanitary conveniences.

Requirement M3 does not apply where there is reasonable provision for sanitary conveniences elsewhere in the building, such that people occupied in, or otherwise having occasion to enter the extension, can gain access to and use those sanitary conveniences.

M4 **Sanitary conveniences in dwellings**

Reasonable provision shall be made in the entrance storey for sanitary conveniences, or where the entrance contains no habitable rooms, reasonable provision for sanitary convenience shall be made in either the entrance storey or principal storey.

In requirement M4:

- 'entrance storey' means the storey which contains the principal entrance
- 'principal storey' means the storey nearest to the entrance storey which contains a habitable room, or if there are two stories equally near, either such storey.

(3) Use sanitary conveniences in the principal storey of any new dwelling.

3.13 Part N – Glazing – safety in relation to impact, opening and cleaning

Number	Title	Regulation	Requirement (in a nutshell)
N1	**Protection against impact**	*Glazing with which people are likely to come into contact whilst moving in or about the building shall –* *(a) if broken on impact, break in a way which is unlikely to cause injury; or* *(b) resist impact without breaking; or* *(c) be shielded or protected from impact.*	All glazing installed in buildings shall be: • sufficiently robust to withstand impact from a falling or passing person; or • protected from a falling or passing person.
N2	**Manifestation of glazing**	*Transparent glazing with which people are likely to come into contact while moving in or about the building, shall incorporate features which make it apparent.*	Requirement N2 does not apply to dwellings.
N3	**Safe opening and closing of windows etc.**	*Windows, skylights and ventilators which can be opened by people in or about the building shall be so constructed or equipped that they may be opened, closed or adjusted safely.*	Requirement N3 does not apply to dwellings.
N4	**Safe access for cleaning windows etc.**	*Provision shall be made for any windows, skylights, or any transparent or translucent walls, ceilings or roofs to be safely accessible for cleaning.*	Requirement N4 does not apply to – (a) dwellings; or (b) any transparent or translucent elements whose surfaces are not intended to be cleaned.

3.14 Part P – Electrical safety

Number	Title	Regulation	Requirement (in a nutshell)
P1	**Design and installation**	Reasonable provision shall be made in the design and installation of electrical installations in order to protect persons operating, maintaining or altering the installations from fire or injury.	These requirements only apply to electrical installations that are intended to operate at low or extra-low voltage: (a) in (or attached to) a dwelling; (b) in common parts of a building serving one or more dwellings (but excluding power supplies to lifts); (c) in a building that receives its electricity from a source located within (or shared with) a dwelling; (d) in a garden associated with a building where the electricity is from a source located within or shared with a dwelling; (e) on land associated with a building where the electricity is from a source located within or shared with a dwelling.

4

Planning permission

Before undertaking any building project, you must first obtain the approval of local government authorities. Many people (particularly householders) are initially reluctant to approach local authorities because, according to local gossip, they are 'likely to be obstructive'. In fact the reality of it is quite the reverse as their purpose is to protect all of us from irresponsible builders and developers and they are normally most sympathetic and helpful to any builder and/or DIY person who wants to comply with the statutory requirements and has asked for their advice.

There are two main controls that districts rely on to ensure that adherence to the local plan is ensured, namely planning permission and Building Regulation approval. Quite a lot of people are confused as to their exact use and whilst both of these controls are associated with gaining planning permission, actually receiving planning permission does not automatically confer Building Regulation approval and vice versa. You **may** require **both** before you can proceed. Indeed, there may be a variation in the planning requirements (and to some extent the Building Regulations) from one area of the country to another. Consequently, the information given on the following pages should be considered as a guide only and not as an authoritative statement of the law.

You are allowed to make certain changes to your home without having to apply to the local council for permission **provided** that it does not affect the external appearance of the building. These are called permitted development rights. The majority of building work that you are likely to complete will, however, probably require you to have planning permission and it is the nation's planning system that plays an important role in today's society by helping to protect the environment in our towns, cities and the countryside.

For example, if you are thinking about carrying out work on a listed building or work that requires the pruning or felling of a tree protected by a Tree Preservation Order, then you will need to contact your local authority planning department before carrying out any work. You never know, you might even require listed building consent or be required to follow certain procedures if carrying out work to trees.

You do **not** require planning permission to carry out any internal alterations to your home, house, flat or maisonette, provided that it does not affect the external appearance of the building.

4.1 Planning controls

Planning controls exist primarily to regulate the use and siting of buildings and other constructions – as well as their appearance. What might seem to be a minor development in itself, could have far-reaching implications that you had not previously considered (for example, erecting a structure that would ultimately obscure vision at a busy junction and thereby constitute a danger to traffic). Equally, the local authority might refuse permission on the grounds that the planned scheme would not blend sympathetically with its surroundings. Your property could also be affected by legal restrictions such as a right of way, which could prejudice planning permission.

The actual details of planning requirements are complex but in respect of domestic developments, the planning authority is concerned primarily with the construction work such as an extension to the house or the provision of a new garage or new outbuildings that is being carried out. Structures like walls and fences also need to be considered because their height or siting might well infringe the rights of neighbours and other members of the community. The planning authority will also want to approve any change of use, such as converting a house into flats or running a business from premises previously occupied as a dwelling only.

4.1.1 Why are planning controls needed?

The purpose of the planning system is to protect the environment as well as public amenities and facilities. It is not designed to protect the interests of one person over another. Within the framework of legislation approved by Parliament, councils are tasked to ensure that development is allowed where it is needed, while ensuring that the character and amenity of the area are not adversely affected by new buildings or changes in the use of existing buildings or land.

Some people think the planning system should be used to prevent any change in their local environment, while others may think that planning controls are an unnecessary interference on their individual rights. The present position is that **all** major works need planning permission from the council but many minor works do not. Parliament thinks this is the right balance as it enables councils to protect the character and amenity of their area, while individuals have a reasonable degree of freedom to alter their property.

If you live in a listed building of historical or architectural interest or your house is in a Conservation Area, you should seek advice before considering **any** alterations.

4.2 Who requires planning permission?

Although the rules and requirements vary according to whether you actually own a house or a flat/maisonette, generally speaking, the principles and procedures for making planning applications are exactly the same for owners of houses and for freeholders (or leaseholders) of flats and maisonettes. Planning regulations, however, have to cover many different situations and so even the provisions that affect the average householder are quite detailed.

You will not need to apply for planning permission to paint your flat or maisonette but, if you are a leaseholder, you may first need to get permission from your landlord or management company.

4.3 Who controls planning permission?

The planning system is made up of a cascade of documents. Currently, under the provisions of the Building Act 1984, national policy is mainly set out in Planning Policy Guidance notes (PPGs). Regions set out regional policy through Regional Planning Guidance notes (RPGs). Structure Plans establish broad planning policies at County Council level, and finally Local Plans set out detailed planning policy at District Council level (where Unitary Councils exist these two documents are generally combined into a Unitary Development Plan). Each layer has to be in conformity with the policies above it in the hierarchy.

4.3.1 Planning and Compulsory Purchase Act 2004

This system has changed significantly thanks to the passing of the Planning and Compulsory Purchase Act 2004. This creates a new hierarchy of policies, and includes complex guidance on how local authorities are supposed to move from the old system to the new.

Under the 2004 Act County Structure Plans are abolished entirely, and the regional tier of policies is reclassified as a Regional Spatial Strategy (RSS). Local Plans are changed radically, and renamed Local Development Frameworks (LDFs).

These Frameworks are made up of Local Development Documents (LDDs) which set out specific policies for the whole area or which give detailed guidance for a particular site. LDDs can include Development Plan Documents (DPDs), Supplementary Planning Documents (SPDs), a Statement of Community Involvement (SCI), an Annual Monitoring Report and a Local Development Scheme.

The transitional arrangements allow Councils to redesignate their current Local Plans as part of their Local Development Framework, but there may well be many cases where Council's will have to shelve Local Plans that are currently being developed and revert to older documents. The potential for confusion is substantial and readers are advised to talk to their local council planning officials.

4.3.2 District Local Plan

Local Plans are prepared by district councils for their areas (except local plans concerning waste and minerals, which are prepared by the county council), and they set out the planning policies for the whole of the district and are used as the basis for assessing **all** planning applications. The district council is responsible

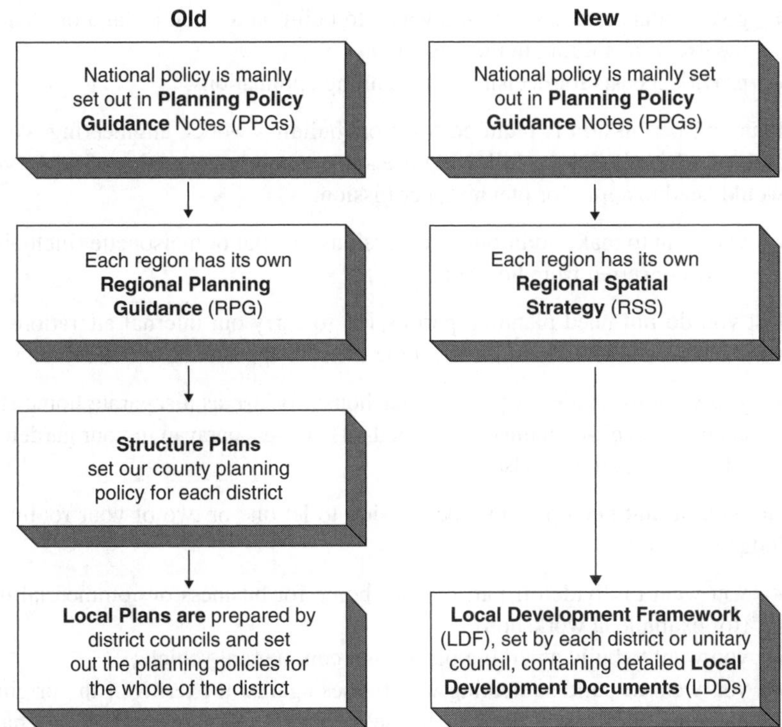

Figure 4.1 Planning responsibilities (old and new)

for keeping their plan under constant review and for making it available to everybody (usually via their council website).

The published Local Plan is used as a guide to the location of development over a ten-year period. For example, they:

- will identify where new homes, jobs and other types of development may be built;
- may require related development to be provided, such as children's play areas, parking facilities and road improvements;
- will outline restrictions where certain types of development are unacceptable.

In preparing Local Plans, districts are responsible for consulting local people and for ensuring that their views are taken into account, thereby giving them a chance to influence the way in which their area is affected.

At all times, Local Plans must also take account of national, regional and county planning policy.

A Local Plan consists of a written statement (which sets out and explains the policies and proposals) and the proposals map, which shows where they apply. Together, these elements of the plan:

- allow local people to clearly see if their homes, businesses or other property would be affected by what is proposed;

- give guidance to anyone who wants to build on a piece of land or change the use of a building in the area; and
- provide a basis for decisions on planning applications.

Planning permission is required for most building works, engineering works and use of land, and the following are some common examples of when you would need to apply for planning permission:

- you want to make additions or extensions to a flat or maisonette (including those converted from houses);

But you do **not** need planning permission to carry out internal alterations or work, provided that it does not affect the external appearance of the building.

- you want to divide off part of your house for use as a separate home (for example, a self-contained flat or bed-sit) or use a caravan in your garden as a home for someone else;

But you do **not** need planning permission to let one or two of your rooms to lodgers.

- you want to divide off part of your home for business or commercial use (for example, a workshop);
- you want to build a parking place for a commercial vehicle;
- you want to build something which goes against the terms of the original planning permission for your house – for example, your house may have been built with a restriction to stop people putting up fences in front gardens because it is on an 'open plan' estate;
- the work you want to complete might obstruct the view of road users;
- the work would involve a new or wider access to a trunk or classified road.

If you have any queries about a particular case, the first thing to do is to ask the planning department of your local council who will have records of all planning permissions in its area. You may also be able to find out more about planning law in your local library. If you are concerned about a legal problem involving planning, you may need to get professional advice or ask your local Citizens Advice Bureau.

A DETR booklet (*Planning Permission, A Guide for Business*) giving advice about working from home and whether planning permission is likely to be required is available from councils.

4.4 What is planning permission?

The planning control process is administered by your local authority and the system '*exists to control the development and use of land and buildings for the best interests of the community*'.

The process is intended to make the environment better for everyone and acts as a service to manage the types of constructions, modifications of premises, uses of land, and ensures the right mix of premises in any one vicinity (that individuals may plan to make) is maintained. The key feature of the process is to allow a party to propose a plan and for other parties to object if they wish to, or are qualified to.

4.5 What types of planning permission are available?

There are three types of planning permission available: outline, reserved and full.

4.5.1 Outline

This is an application for a development 'in principle' without giving too much detail on the actual building or construction. It basically lets you know, in advance, whether the development is likely to be approved. Assuming permission is granted under these circumstances you will then have to submit a further application in greater detail. In the main, this applies to large-scale developments only and you will probably be better off making a full application in the first place.

4.5.2 Reserved matters

This is the follow-up stage to an outline application to give more substance and more detail.

4.5.3 Full planning permission

Is the most widely used and is for erection or alteration of buildings or changes of use. There are no preliminary or outline stages and when consent is granted it is for a specific period of time. If this is due to lapse, a renewal of limited permission can be applied for.

4.6 How do I apply for planning permission?

There are fees to pay for each application for planning permission and your local planning office can provide you with the relevant details. You must make sure you have paid the correct fee – as permission can be refused if there is a discrepancy on fees paid.

When your forms and plans are ready, they need to be submitted to the planning office. The planning office will arrange for them to be listed in the local newspaper under 'latest planning applications' and will write to each neighbouring property and (normally) give them 21 days in which to raise any objections.

At the planning office, officials will produce a file after the 21 days have expired, with any objections or supporting information, and will make a recommendation on the application, ready for presenting it at the next planning committee or subcommittee meeting. At this meeting, they will discuss the case, reject it, ask for modifications or accept it. Whichever the decision the planning officer will feed back the decision to the applicant.

There is an appeal procedure, which your local authority planning officer can advise you about.

The Planning Portal is the UK government's planning resource. It provides extensive details about the planning system, applying for planning permission, finding out about development near you, appeals against a planning decision and researches the latest government policy.

The Portal is split into three sections:

- **general public** – a guide to applying for planning permission and accessing local information;
- **planning professionals** – the complete resource for researching and submitting planning applications;
- **government users** – a dedicated knowledge base for all levels of government.

For more details go to www.planningportal.gov.uk

4.7 Do I really need planning permission?

Most alterations and extensions to property and changes of use of land need to have some form of planning permission, which is achieved by submitting a planning application to the local authority. The purpose of this control is to protect and enhance our surroundings, to preserve important buildings and natural areas and strengthen the local economy.

However, not all extensions and alterations to dwelling houses require planning permission. Certain types of development are permitted without the need to make an official request, and it is always wise to contact the local authority before commencing any work.

Whether or not planning permission is required, good design is always important. Extensions and alterations should be in scale and in harmony with the remainder of the house. The builder should ensure that details such as window openings and matching materials are taken into account.

Householders are encouraged (by councils) to employ a skilled designer when preparing plans for extensions and alterations. Alternatively, the authority's planning officers are able to offer general design guidance prior to the submission of your scheme.

Table 4.1 provides an indication of the basic requirements for planning permission and building regulation approval.

Table 4.1 is only meant as guidance. A more complete description of the above synopsis is contained in Chapter 5. In all circumstances it is recommended that you talk to your local planning officer before contemplating any work. The cost of a local phone call could save you a lot of money (and stress) in the long term!

Table 4.1 Basic requirements for planning permission and building regulation approval

Type of work	Planning permission		Building Regulation approval	
Advertising	No	If the advertisement is less than 0.3 m^2 and not illuminated.	Possibly	Consult your local planning officer.
Building a conservatory	Possibly	You can extend your house by building a conservatory, provided that the total of both previous and new extensions does not exceed the permitted volume.	Yes	If area exceeds 30 m^2 (35.9 y^2). **Note:** Schedule 2 of the regulation is currently under consultation as there is a view that porches, conservatories, covered ways and carports under 30 m^2 should be subject to some regulations.
Building a garage	Possibly	You can build a garage up to 10 m^3 (13.08 y^3) in volume without planning permission, if it is within 5 m (16 ft 3″) of the house or an existing extension. Further away than this, it can be up to half the area of the garden, but the height must not exceed 4 m (13 ft).	Yes	
Building a garden wall or fence	Yes	If it is more than 1 m (3 ft 3″) high and is a boundary enclosure adjoining a highway.	No	
	Yes	If it is more than 2 m (6 ft 6″) high elsewhere.	No	
Building a hard standing for a car	No	Provided that it is within your boundary and is not used for a commercial vehicle.	No	
Building a new house	Yes		Yes	
Building an extension	Possibly	You can extend your house by building an extension, provided that the total of both previous and new extensions does not exceed the permitted volume.	Yes	If area exceeds 30 m^2 (35.9 y^2).

(*Continued*)

Table 4.1 (*Continued*)

Type of work	Planning permission		Building Regulation approval	
		'Building extensions' can be a potential minefield and it is best to consult the local planning officer before contemplating any work.		**Note:** Schedule 2 of the regulation is currently under consultation as there is a view that porches, conservatories, covered ways and carports under 30 m^2 should be subject to some regulations. However this currently **does not** stipulate that any amendments will also apply to extensions.
Building a porch	No	Unless: • the floor area exceeds 3 m^2 (3.6 y^2) • any part is more than 3 m (9 ft 9″) high • any part is less than 2 m (6 ft 6″) from a boundary adjoining a highway or public footpath	Yes	If area exceeds 30 m^2 (35.9 y^2). **Note:** Schedule 2 of the regulation is currently under consultation as there is a view that porches, conservatories, covered ways and carports under 30 m^2 should be subject to some regulations.
Central heating	No		No Yes	If electric. If gas, solid fuel or oil.
Constructing a small outbuilding	Possibly	Provided the building is less than 10 m^3 (13.08 y^3) in volume, not within 5 m (16 ft 3″) of the house or an existing extension. Erecting 'outbuildings' can be a potential minefield and it is best to consult the local planning officer before commencing work.	Yes	If area exceeds 30 m^2 (35.9 y^2). If it is within 1 m (3 ft 3″) of a boundary, it must be built from incombustible materials.
Converting a house to business premises (including bedsitters)	Yes	Even where construction work may not be intended.	Yes	Unless you are not proposing any building work to make the change.
Converting an old building	Yes		Yes	

(*Continued*)

Table 4.1 (*Continued*)

Type of work	Planning permission		Building Regulation approval	
Decoration and repair inside and outside a building	No	Unless it is of a listed building or within a Conservation Area. Consult your local authority.	No	Unless it is a listed building or within a Conservation Area. Consult your local authority.
Demolition	Yes	If it is a listed building or in a Conservation Area. If the whole house is to be demolished.	No	For a complete detached house.
	Possibly	For partial demolition (seek advice from your local planning officer before proceeding).	Yes	For a partial demolition to ensure that the remaining part of the house (or adjoining buildings/extensions) is structurally sound.
Electrical work	No		Probably	All proposals to carry out electrical installation work **must** be notified to the local authority's building control body before work begins, **unless** the proposed installation work is undertaken by a person who is a competent person registered with an electrical self-certification scheme and does not include the provision of a new circuit.
Erecting aerials, satellite dishes wind turbines, and flagpoles	No	Unless it is a stand-alone antenna or mast greater than 3 m in height.	No	
	Possibly	If erecting a satellite dish, especially in a Conservation Area or if it is a listed building (consult your local planning officer).	No	

(*Continued*)

Table 4.1 (*Continued*)

Type of work	Planning permission		Building Regulation approval	
	Possibly	If erecting a wind turbine either stand alone or on the side of a dwelling (consult your local planning officer).	No	
Felling or lopping trees	No	Unless the trees are protected by a Tree Preservation Order or you live in a Conservation Area.	No	
Infilling	Possibly	Consult your local planning officer.	Yes	If a new development.
Installing a swimming pool	Possibly	Consult your local planning officer.	Yes	For an indoor pool.
Laying a path or a driveway	No	Unless it provides access to a main road.	No	
Loft conversions and roof extensions	No	Provided the volume of the house is unchanged and the highest part of the roof is not raised.	Yes	
	Yes	For front elevation dormer windows or rear ones over a certain size.	Yes	
Material change of use	Possibly	Even if no building or engineering work is proposed	Yes	
Oil-storage tank	No	Provided that it is in the garden and has a capacity of not more than 3500 litres (778 gallons) and **no** point is more than 3 m (9 ft 9″) high and **no** part projects beyond the foremost wall of the house facing the highway.	No	

(*Continued*)

Table 4.1 (*Continued*)

Type of work	Planning permission		Building Regulation approval	
Planting a hedge	No	Unless it obscures view of traffic at a junction or access to a main road.	No	
Plumbing	No		No	For replacements (but you will need to consult the Technical Services Department for any installation which alters present internal or external drainage).
			Yes	For an unvented hot water system.
Replacing windows and doors	No	Unless: • they project beyond the foremost wall of the house facing the highway • the building is a listed building • the building is in a Conservation Area	Possibly	Installation must be carried out by a registered approved person.
	Yes	To replace shop windows.	Yes	And due consideration to new Ventilation regulations (Part F).
Structural alterations inside	No	As long as the use of the house is not altered.	Possibly	Consult your local authority.
	Yes	If the alterations are major such as removing or part removing of a load bearing wall or altering the drainage system.	Yes	
	Yes	If they are to an office or shop.	Yes	

The above table is only meant as guidance. A more complete description of the above synopsis is contained in Chapter 5. In all circumstances it is recommended that you talk to your local planning officer before contemplating any work. The cost of a local phone call could save you a lot of money (and stress) in the long term!

4.8 How should I set about gaining planning permission?

If you are in the planning stages for your work and you know planning permission will be required, it is wise to get the plans passed before you go to any expense or make any decisions that you may find hard to reverse – such as signing a contract for work. If your plans are rejected, you will still have to pay your architect or whoever prepared your plans for submission but you won't have to pay any penalty clauses to the building contractor.

It is always best to submit an application in the early stages – if you try to be clever by submitting plans at the last minute (in the hope that neighbours will not have time to react) then you could be in for an expensive mistake! It's much better to do things properly and up-front.

An architect (surveyor or general contractor) can be asked to prepare and submit your plans on your behalf if you like, but as the owner and person requiring the development, it will be your name that goes on the application, even if all the correspondence goes between your architect and the planning department.

You don't have to own the land to make a planning application, but you will need to disclose your interest in the property. This might happen if you plan to buy land, with the intention of developing it, subject to planning approval. It would, therefore, be in your best interest to obtain the consent before the purchase proceeds.

To submit your application you will need to use the official forms, available from the local authority planning department. It's a good idea to collect these personally, as you may get the opportunity to talk through your ideas with a planning officer and in doing so probably get some useful feedback. You will also need to include detailed plans of the present and proposed layout as well as the property's position in relation to other properties and roads or other features.

New work requires details of materials used, dimensions and all related installations, similar to that required for Building Regulations.

4.9 What sort of plans will I have to submit?

There are three types of plans (namely site, block and building) that can accompany your application and, as indicated above, the choice will depend on the work proposed.

4.9.1 Site plan

A site plan indicates the development location and relationship to neighbouring property and roads etc. Minimum scale is 1:2500 (or 1:1250 in a built-up area). The land to which the application refers is outlined in red ink. Adjacent land, if owned by the applicant, is outlined in blue ink.

Block plan

A block plan is a detailed plan of a construction or structural alteration that shows the existing and proposed building, all trees, waterways, ways of access, pipes and drainage and any other important features. Minimum scales are 1:1500.

Building plans

Building plans are the detailed drawings of the proposed building works and would show plans, elevations and cross-sections that accurately describe every feature of the proposal. These plans are normally very thorough and include types of material, colour and texture, the layers of foundations, floor constructions, and roof constructions etc.

4.10 What is meant by 'building works'?

In the context of the Building Regulations, 'building works' means:

(a) the erection or extension of a building;
(b) the provision or extension of a controlled service or fitting;
(c) the material alteration of a building, or a controlled service or fitting;
(d) work required by Regulation 6 (requirements relating to material change of use);
(e) the insertion of insulating material into the cavity wall of a building;
(f) work involving the underpinning of a building.

4.11 What important areas should I take into consideration?

The following are some of the most important areas that should be considered before you submit a planning application.

4.11.1 Advertisement applications

If your proposal is to display an advertisement, you will need to make a separate application on a special set of forms. Three copies of the forms and the relevant drawings must be supplied. These must include a location plan and sufficient detail to show the size, materials and colour of the sign and its position. No certificate of ownership is needed, but it is illegal to display signs on the property without the consent of the owner.

4.11.2 Conservation Area consent

If you live in a Conservation Area, you will need Conservation Area consent to do the following:

- demolish a building with a volume of more than 115 m^3 (there are a few exceptions and further information will be available from your council);

- demolish a gate, fence, wall or railing over 1 m high if it is next to a highway (including a public footpath or bridleway) or public open space; or over 2 m high elsewhere.

4.11.3 Listed building consent

You will need to apply for listed building consent if either of the following cases apply:

- you want to demolish a listed building;
- you want to alter or extend a listed building in a manner which would affect its character as a building of special architectural or historic interest.

You may also need listed building consent for any works to separate buildings within the grounds of a listed building. Check the position carefully with the council – it is a criminal offence to carry out work which needs listed building consent without obtaining it beforehand.

4.11.4 Trees

Many trees are protected by Tree Preservation Orders (TPOs), which mean that, in general, you need the council's consent to prune or fell them. In addition, there are controls over many other trees in Conservation Areas.

Ask the council for a copy of the department's free leaflet *Protected Trees: a guide to tree preservation procedures.*

4.12 What are the government's restrictions on planning applications?

All applications for planning permission will have to take into account the following Acts and regulations.

Planning (Listed Buildings and Conservation Areas) Act 1990

Under the terms of the Planning (Listed Buildings and Conservation Areas) Act 1990, local councils must maintain a list of buildings within their boroughs, which have been classified as being of special architectural or historic interest. Councils are also required to keep maps showing which properties are within Conservation Areas.

Town and Country Planning (Control of Advertisements) Regulations 1992

In accordance with the Town and Country Planning (Control of Advertisements) Regulations 1992, councils need to maintain a publicly available register of applications and decisions for consent to display advertisements.

The Local Government (Access to Information) (Variation) Order 1992

The Local Government (Access to Information) (Variation) Order 1992 ensures that information relating to proposed development by councils cannot be treated as exempt when the planning decision is made.

Town and Country Planning (General Development Procedure) Order 1995

Every council must keep the following registers available for public inspection in accordance with the Town and Country Planning (General Development Procedure) Order 1995:

- planning applications, including accompanying plans and drawings;
- applications for a certificate of lawfulness of existing or proposed use or development;
- Enforcement Notices and any related stop notices.

All applications for planning permission must receive publicity.

Other areas

As well as the legal requirement to make the planning register available for public inspection, councils will also allow the public to have access to all other relevant information such as letters of objection/support for an application or correspondence about considerations. Three clear days before any committee meeting, the file will normally be made available for public inspection and this file will remain available (i.e. for further public inspection) after the committee meeting. Although commercial confidentiality could well be a valid consideration, the council will not use it so as to prevent important information about materials and facilities also being available.

4.13 How do I apply for planning permission?

Once you have established that planning permission is required, you will need to submit a planning application. Remember, it may take up to eight weeks, or even longer, to get planning permission, so apply early.

You will have to prepare a plan showing the position of the site in question (i.e. the site plan) so that the authority can determine exactly where the building is located. You must also submit another, larger-scale, plan to show the relationship of the building to other premises and highways (i.e. the block plan). In addition, it would help the council if you also supplied drawings to give a clear idea of what the new proposal will look like, together with details of both the colour and the kind of materials you intend using. You may prepare the drawings yourself, provided you are able to make them accurate.

Under normal circumstances you will have to pay a fee in order to seek planning permission, but there are exceptions. The planning department will advise you.

4.13.1 Application forms and plans

It is important to make sure that you make your planning application correctly. The following checklist may help:

- Obtain the application forms from the planning department or from the local council's website.
- Read the 'Notes for Applicants' carefully – again available from the planning department, or the local council's website.
- Fill in the relevant parts of the forms and remember to sign and date them.
- Submit the correct number and type of supporting plans. Each application should be accompanied by a site plan of not less than 1:2500 scale and detailed plans, sections and elevations, where relevant.
- Fill in and sign the relevant certificate relating to land ownership.

It is in your own interest to provide plans of good quality and clarity and so it is probably advisable to get help from an architect, surveyor, or similarly qualified person to prepare the plans and carry out the necessary technical work for you. You can obtain the necessary application form from the planning department of your local council and you will find that this is laid out simply, with guidance notes to help you fill it in. Alternatively, you can ask a builder or architect to make the application on your behalf. This is sensible if the development you are planning is in any way complicated, because you will have to include measured drawings with the application form.

Applying online via the Planning Portal?

You can apply for planning permission online via the Planning Portal (www.planningportal.gov.uk).

The Portal's service will also let you:

- create a site location plan (compulsory for all applications);
- attach supporting documents (such as photographs);
- pay the application fee online (where enabled).

Local authorities working with the Planning Portal offer two different ways of applying for permission.

- If your local authority has integrated its systems with the Portal you can complete the whole process online and pay for the application electronically (where enabled).
- If the council systems have not been integrated you can still use the Portal's service to complete the application forms electronically then print and post them to your local authority. *Some local authorities request up to five copies of the forms, so completing them electronically can save time and make sure there are no discrepancies.*

The portal provides a detailed map and list of all authorities who currently allow applications on line.

4.14 What is the planning permission process?

If you think you might need to apply for planning permission, then this is the process to follow:

Step 1

Contact the planning department of your council. Tell the planning staff what you want to do and ask for their advice.

Step 2

If they think you need to apply for planning permission, ask them for an application form. They will tell you how many copies of the form you will need to send back and how much the application fee will be. Ask if they foresee any difficulties which could be overcome by amending your proposal. It can save time or trouble later if the proposals you want to carry out also reflect what the council would like to see. The planning department will also be able to tell you if Building Regulations approval will also be required.

Step 3

Decide what type of application you need to make. In most cases this will be a full application but there are a few circumstances when you may want to make an outline application – for example, if you want to see what the council thinks of the building work you intend to carry out before you go to the trouble of making detailed drawings (but you will still need to submit details at a later stage).

Step 4

Send the completed application forms and supporting documents to your council, together with the correct fee. Each form must be accompanied by a plan of the site and a copy of the drawings showing the work you propose to carry out. (The council will advise you on what drawings are needed.)

Extracts from Ordnance Survey maps can be supplied for planning applications submitted by private individuals and for school/college use. There is usually a charge for this service.

Step 5

The planning department will acknowledge receipt of your application, and publicly announce it – via letters to the neighbourhood parish council and anyone directly affected by the proposal, by publishing details of the application in the local press, notifying your neighbours and/or putting up a notice on or near the site. The council may also consult other organizations, such as the highway authority or the parish council (or community council in Wales).

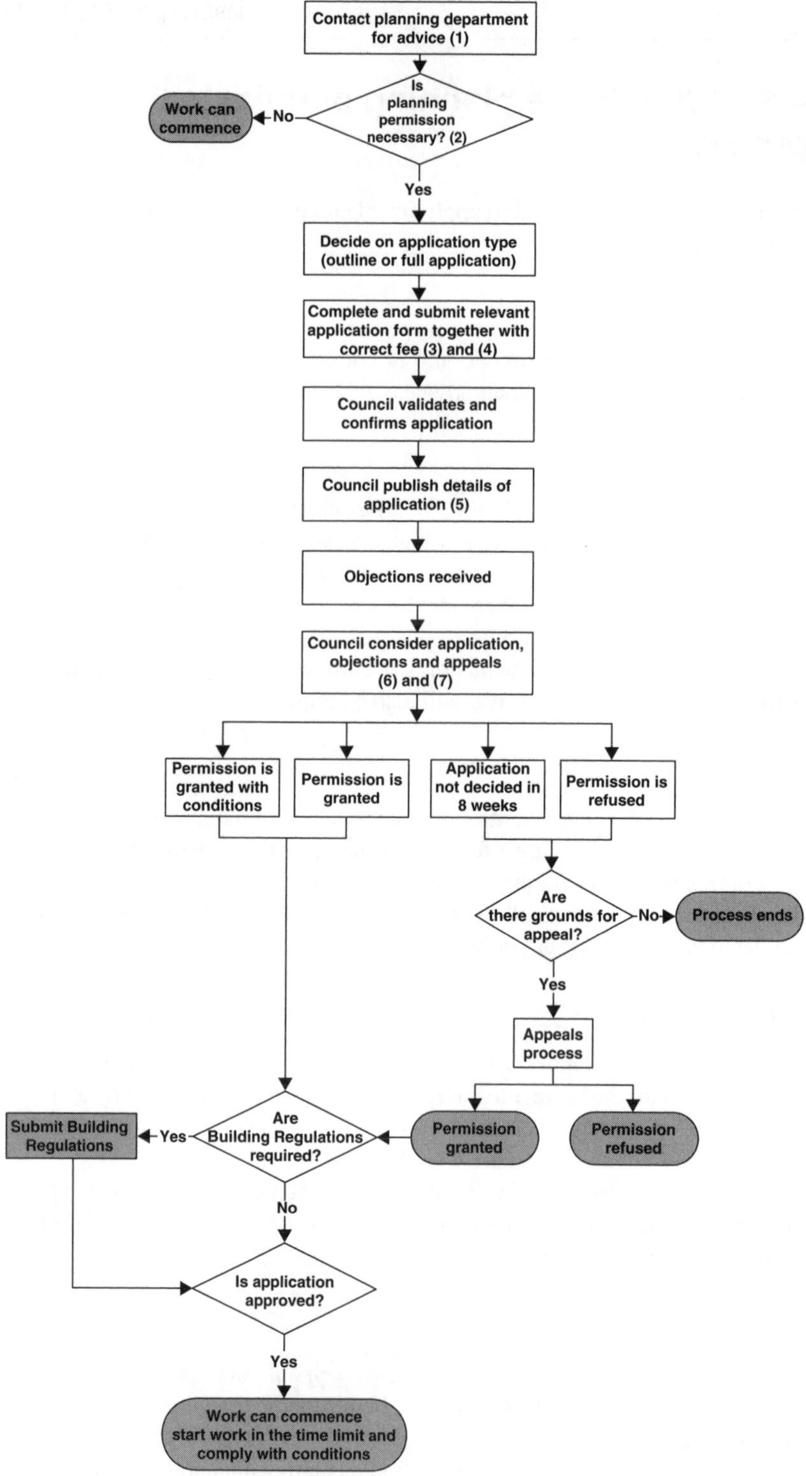

Figure 4.2 Planning permission

A copy of the application will also be placed on the planning register at the council offices so that it can be inspected by any interested member of the public. Anyone can object to the proposal, but there is a limited period of time in which to do this and they must specify the grounds for objection.

Under the Local Government Act 1972 (as amended), the public have the right to inspect and copy the following documents:

- the agenda for a council committee or sub-committee meeting; reports for the public part of the meeting;
- the minutes of such meetings and any background papers, including planning applications, used in preparing reports.

These documents can be inspected and copied from three clear days before a meeting. There is no charge to inspect a document but councils will charge for making photocopies.

Step 6

The planning department may prepare a report for the planning committee, which is made up of elected councillors. Or the council may give a senior officer in the planning department the responsibility for deciding your application on its behalf. If a report has been made, then this will be presented to a meeting of the council committee, with recommendations on the decision to be made, based on the implications and objections received.

You are entitled to see and have a copy of any report submitted to a local government committee. You are also entitled to see certain background papers used in the preparation of reports. The background papers will generally include the comments of consultees, objectors and supporters which are relevant to the determination of your application. Such material should normally be made available at least three working days before the committee meeting.

Step 7

The councillors or council officers who decide your application must consider whether there are any good planning reasons for refusing planning permission or for granting permission subject to conditions. The council cannot reject a proposal simply because many people oppose it. It will also look at whether your proposal is consistent with the development plan for the area.

The committee will consider the merits of a proposal; ensure the proposed work meets all the conditions of any local plan or requirements for a district and that the process has been followed properly. The kinds of planning issue it can also consider include potential traffic problems; the effect on amenity and the impact the proposal may have on the appearance of the surrounding area. Moral issues, the personal circumstances of the applicant or the effect the development might have on nearby property prices are not relevant to planning and will not normally be taken into account by the council. The committee will arrive at its decision and the result will be communicated back to the applicants via the planning department.

4.14.1 How long will the council take?

You can expect to receive a decision from the planning department within eight weeks and, **once granted, planning permission is valid for five years**. If the work is not begun within that time, you will have to apply for planning permission again.

If the council cannot make a decision within eight weeks then it must obtain your written consent to extend the period. If it has not done so, you can appeal to the Secretary of State for the Environment, Transport and the Regions, or, in Wales, to the National Assembly for Wales (see below). But appeals can take several months to decide and it may be quicker to reach agreement with the council.

Do not be afraid to discuss the proposal with a representative of the planning department **before** you submit your application. They will do their best to help you meet the requirements.

4.14.2 What can I do if my application is refused?

If the council refuses permission or imposes conditions, it must give reasons. If you are unhappy or unclear about the reasons for refusal or the conditions imposed, talk to staff at the planning department. Ask them if changing your plans might make a difference. If your application has been refused, you may be able to submit another application with modified plans free of charge within 12 months of the decision on your first application.

The planning department will always grant planning permission unless there are very sound reasons for refusal, in which case the department must explain the decision to you so that you can amend your plans accordingly and resubmit them for further consideration.

A second application is normally exempt from a fee.

The following are some of the main objection areas that your application may meet.

The property is a listed building

Listed buildings are protected for their special architectural or historical value. A *Listed Building Consent* may be needed for alterations but grants could be available towards repair and restoration!

If it's a listed building, it probably has some historic importance and will have been listed by the Department of the Environment. This could apply to houses, factories, warehouses and even walls or gateways. Most alterations which affect the external appearance or design will require listed building consent in addition to other planning consents.

The property is in a Conservation Area

This is an area defined by the local authority, which is subject to special restrictions in order to maintain the character and appearance of that area. Again, other planning consents may be needed for areas designated as green belt, areas of outstanding natural beauty, national parks or sites of specific scientific interest.

The application does not comply with the local development plan

Local authorities often publish a development plan, which sets out policies and aims for future development in certain areas. These are to maintain specific environmental standards and can include very detailed requirements such as minimum or maximum dimensions of plot sizes, number of dwellings per acre, height and style of dwellings etc. It is important to check if a plan exists for your area, as proposals can meet with some fierce objections from residents protecting their environment.

The property is subject to a covenant

This is an agreement between the original owners of the land and the persons who acquired it for development. They were implemented to safeguard residential standards and can include things like the size of outbuildings, banning use of front gardens for parking cars, or even just the colours of exterior paintwork.

There is existing planning permission

A previous resident or owner may have applied for planning permission, which may not have expired yet. This could save time and expense if a new application can be avoided. If you are considering a planning application, you should consider the above questions. Normally your retained expert – architect, surveyor or builder – can advise and help you to get an application passed. Most information can be collected from your local planning department, or if you need to find out about covenants, look for the appropriate land registry entry.

The proposal infringes a right of way

If your proposed development would obstruct a public path that crosses your property, you should discuss the proposals with the council at an early stage. The granting of planning permission will **not** give you the right to interfere with, obstruct or move the path. A path cannot be legally diverted or closed unless the council has made an order to divert or close it to allow the development to go ahead. The order must be advertised and anyone may object. You must not obstruct the path until any objections have been considered and the order has been confirmed. You should bear in mind that confirmation is not automatic; for example, an alternative line for the path may be proposed, but not accepted.

4.14.3 What matters cannot be taken into account?

- Competition
- Disturbance from construction work
- Loss of property value
- Loss of view

- Matters controlled under other legislation such as Building Regulations (e.g. structural stability, drainage, fire precautions etc.)
- Moral issues
- Need for development
- Private issues between neighbours (e.g. land and boundary disputes, damage to property, private rights of way, deeds, covenants etc.)
- Sunday trading
- The identity or personal characteristics of the applicant.

4.14.4 What are the most common stumbling blocks?

In no particular order of priority, these are:

- Adequacy of parking
- Archaeology
- Design, appearance and materials
- Effect on listed building or Conservation Area
- Government advice
- Ground contamination
- Hazardous materials
- Landscaping
- Light pollution
- Local planning policies
- Nature conservation
- Noise and disturbance from the use (but not from construction work)
- Overlooking and loss of privacy
- Previous planning decisions
- Previous appeal decisions
- Road access
- Size, layout and density of buildings
- The effect on the street or area (but not loss of private view)
- Traffic generation and overall highway safety.

4.15 Can I appeal if my application is refused?

If you think the council's decision is unreasonable, you can appeal to the Secretary of State or (in Wales) to the National Assembly for Wales. Appeals must be made within six months of the date of the council's notice of decision. You can also appeal if the council does not issue a decision within eight weeks.

A free booklet *Planning Appeals – A Guide* is available from the Planning Inspectorate, Tollgate House, Houlton Street, Bristol BS2 9DJ or Crown Buildings, Cathays Park, Cardiff CF10 3NQ.

Appeals are intended as a last resort and they can take several months to decide. It is often quicker to discuss with the council whether changes to your proposal would make it more acceptable. The planning authority will supply you with the necessary appeal forms.

Be careful not to proceed without approval, as you might find yourself obliged to restore the property to its original condition.

4.16 Before you start work

There are many kinds of alterations and additions to houses and other buildings which do not require planning permission. Whether or not you need to apply, you should think about the following before you start work.

4.16.1 What about neighbours?

Have the neighbours any rights to complain?
Many of us live in close proximity to others and your neighbours should be the first individuals you talk to.

What if your alteration infringes their access to light, or a view? Such disputes are notorious for causing bad feeling but with a little consideration, at an early stage, you can avoid a good deal of unpleasantness later.

Plans for the local area can normally be viewed at the local town hall, but most planning applications will involve consultation with neighbours and statutory consultees such as the Highways Authority and the drainage authorities. The extent of consultation will, quite naturally, reflect on the nature and scale of the proposed development – together with its location. Applications to make an alteration to your property can also be refused because you live in an area of outstanding natural beauty, a national park, or a Conservation Area or your property is listed. Any alterations to public utilities such as drains or sewers, or changes to public access such as footpaths will require consultation with the local council. They will have to approve your plans. Even a sign on or above your property may need to be of a certain size or shape.

Some properties may also be the home of a range of protected species such as bats or owls. These animals are protected by the Wildlife and Countryside Act 1981 and the Nature Conservancy Council must give approval to any work that may potentially disturb them. Likewise many members of the public are extremely defensive of trees that grow where they live. Tree Preservation Orders may control the extent to which you can fell or even prune a tree, even if it is on your property. Trees in Conservation Areas are particularity protected, and you will need to supply at least six weeks' notice before working upon them.

New street names and house numbers and names need approval from the council.

Let your neighbours know about the work you intend to carry out to your property. They are likely to be as concerned about work which might affect them as you would be about changes which might affect your enjoyment of your own property. For example, your building work could take away some of their light or spoil a view from their windows. If the work you carry out seriously overshadows a neighbour's window and that window has been there for 20 years or more, you may be affecting his or her 'right to light' and you could be open to legal action. It is best to consult a lawyer if you think you need advice about this.

You may be able to meet some of your neighbour's worries by modifying your proposals. Even if you decide not to change what you want to do, it is usually better to have told your neighbours what you are proposing before you apply for planning permission and before any building work starts.

If you do need to make a planning application for the work you want to carry out, the council will ask your neighbours for their views. If you or any of the people you are employing to do the work need to go on to a neighbour's property, you will, of course, need to obtain their consent before doing so.

4.16.2 What about design?

Everybody's taste varies and different styles will suit different types of property. Nevertheless, a well-designed building or extension is likely to be much more attractive to you and to your neighbours. It is also likely to add more value to your house when you sell it. It is therefore worth thinking carefully about how your property will look after the work is finished.

Extensions often look better if they use the same materials and are in a similar style to the buildings that are there already – but good design is impossible to define and there may be many ways of producing a good result. In some areas, the council's planning department issues design guides or other advisory leaflets that may help you.

4.16.3 What about crime prevention?

You may feel that your home is secure against burglary and you may already have taken some precautions such as installing security locks to windows. However, alterations and additions to your house may make you more vulnerable to crime than you realize. For example, an extension with a flat roof, or a new porch, could give access to upstairs windows which previously did not require a lock. Similarly, a new window next to a drainpipe could give access. Ensure that all windows are secure. Also, your alarm may need to be extended to cover any extra rooms or a new garage. The crime prevention officer at your local police station can provide helpful advice on ways of reducing the risk.

4.16.4 What about lighting?

If you are planning to install external lighting for security or other purposes, you should ensure that the intensity and direction of light does not disturb others. Many people suffer extreme disturbance due to excessive or poorly designed lighting. Ensure that beams are **not** pointed directly at windows of other houses. Security lights fitted with passive infra-red detectors (PIRs) and/or timing devices should be adjusted so that they minimize nuisance to neighbours and are set so that they are not triggered by traffic or pedestrians passing outside your property.

4.16.5 What about covenants?

Covenants or other restrictions in the title to your property or conditions in the lease may require you to get someone else's agreement before carrying out some kinds of work to your property. This may be the case even if you do not need to apply for planning permission. You can check this yourself or consult a lawyer.

You will probably need to use the professional services of an architect or surveyor when planning a loft conversion. Their service should include considerations of planning control rules.

4.16.6 What about listed buildings?

Buildings are listed because they are considered to be of special architectural or historic interest and as a result require special protection. Listing protects the **whole** building, both inside and out and possibly also adjacent buildings if they were erected before 1 July 1948.

The prime purpose of having a building listed is to protect the building and its surroundings from changes that will materially alter the special historic or architectural importance of the building or its setting.

The list of buildings is prepared by the Department of Culture, Media and Sport and properties are scheduled into one of three grades, Grade I, Grade II* and Grade II, with Grade I being the highest grade. Over 90% of all listed properties fall within Grade II. (In Scotland the grades are A, B and C.)

All buildings erected prior to 1700 and substantially intact are listed, as are most buildings constructed between 1700 and 1840, although some selection does take place. The selection process is more discriminating for buildings erected since 1840 because so many more properties remain today. Buildings less than 30 years old are generally only listed if they are of particular architectural or historic value and are potentially under threat. Your district council holds a copy of the statutory list for public inspection and this provides details on each of the listed properties.

See *Planning Policy Guidance Note 15 (PPG.15) – Planning and the Historic Environment*, which provides a practical understanding of the Planning (Listed Buildings and Conservation Areas) Act 1990 which can be viewed at your planning office or in main libraries, or purchased from The Stationery Office, 29 Duke Street, Norwich, NR3 1GN (Tel: 0870 600 5522; Fax: 0870 600 5533, www.tso.co.uk).

Owner's responsibilities?

If you are the owner of a listed building or come into possession of one, you are tasked with ensuing that the property is maintained in a reasonable state of repair. The council may take legal action against you if they have cause to believe that you are deliberately neglecting the property, or have carried out works without consent. Enforcement action may be instigated.

There is no statutory duty to effect improvements, but you must not cause the building to fall into any worse state than it was in when you became its owner.

This may necessitate some works, even if they are just to keep the building wind and watertight. However, you may need listed building consent in order to carry these works out!

A photographic record of the property when it came into your possession may be a useful asset, although you may also have inherited incomplete or unimplemented works from your predecessor, which you will become liable for.

If you are selling a listed building you may wish to indemnify yourself against future claims: speak to your solicitor.

4.16.7 What about Conservation Areas?

Tighter regulations apply to developments in Conservation Areas and to developments affecting listed buildings. Separate Conservation Area consent and/or listed building consent may be needed in addition to planning consent and Building Regulation consent.

Conservation Areas are '*areas of special architectural or historic interest the character and appearance of which it is desirable to preserve or enhance*'. (Civic Amenities Act 1967)

As the title indicates these designations cover more than just a building or property curtilage and most local authorities have designated Conservation Areas within their boundary. Although councils are not required to keep any statutory lists, you can usually identify Conservation Areas from a local plan's 'proposals maps' and appendices. Some councils may keep separate records or even produce leaflets for individual areas.

The purpose of designating a Conservation Area is to provide the council with an additional measure of control over an area that they consider being of special historic or architectural value. This does not mean that development proposals cannot take place, or that works to your property will be automatically refused. It means however that the council will have regard to the effect of your proposals on the designation in addition to their normal assessment. The council may also apply this additional tier of assessment to proposals that are outside the designated Conservation Area boundary, but which may potentially affect the character and appearance of the area.

As a result, local planning authorities may ask for more information to accompany your normal planning application concerning proposals within (or adjoining) a Conservation Area. This may include:

- a site plan to 1:1250 or 1:2500 scale showing the property in relation to the Conservation Area;
- a description of the works and the effect (if any) you think they may have on the character and appearance of the Conservation Area;
- a set of scale drawings showing the present and proposed situation, including building elevations, internal floor plans and other details as necessary.

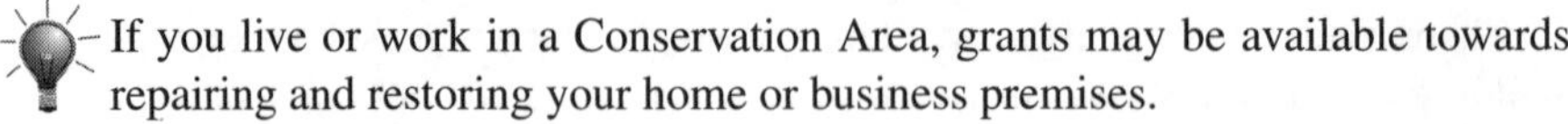

If you live or work in a Conservation Area, grants may be available towards repairing and restoring your home or business premises.

For major works you may need to involve an architect with experience of works affecting Conservation Areas.

4.16.8 What is Conservation Area consent?

Development within Conservation Areas is dealt with under the normal planning application process, except where the proposal involves demolition.

In this case you will need to apply for Conservation Area consent on the appropriate form obtainable from the planning department.

Here again the council will assess the proposal against its effect upon the special character and appearance of the designated area.

More details can be obtained by reference to *Planning Policy Guidance Note 15 (PPG.15) – Planning and the Historic Environment*, which provides a practical understanding of the Planning (Listed Buildings and Conservation Areas) Act 1990. These can be viewed at your planning office or in main libraries, or purchased from The Stationery Office, 29 Duke Street, Norwich, NR3 1GN (Tel: 0870 600 5522 Fax: 0870 600 5533, www.tso.co.uk).

4.16.9 What about trees in Conservation Areas?

Nearly all trees in Conservation Areas are automatically protected.

Trees in Conservation Areas are generally treated in the same way as if they were protected by a Tree Preservation Order, i.e. it is necessary to obtain the council's approval for works to trees in Conservation Areas before they are carried out. There are certain exceptions (where a tree is dead or in a dangerous condition) but it is always advisable to seek the opinion of your council's tree officer to ensure your proposed works are acceptable. Even if you are certain that you do not need permission, notifying the council may save the embarrassment of an official visit if a neighbour contacts them to tell them what you are doing.

If you wish to lop, top or fell a tree within a Conservation Area you must give six weeks' notice, in writing, to the local authority. This is required in order that they can check to see if the tree is already covered by a Tree Preservation Order (TPO), or consider whether it is necessary to issue a TPO to control future works on that tree.

Contact your council's landscape or tree officer for further information.

4.16.10 What are Tree Preservation Orders?

Trees are possibly the biggest cause of upset in town and country planning and many neighbours fall out over tree related issues. They may be too tall, may block out natural light, have overhanging branches, shed leaves on other property or the roots may cause damage to property. When purchasing a property the official searches carried out by your solicitor should reveal the presence of a TPO on the property or whether your property is within a Conservation Area within which trees are automatically protected.

However not all trees are protected by the planning regulations system – but trees that have protection orders on them must not be touched unless specific approval is granted. Don't overlook the fact that a preservation order could have been put on a tree on your land before you bought it and is still enforceable.

Planning authorities have powers to protect trees by issuing a TPO and this makes it an offence to cut down, top, lop, uproot, wilfully damage or destroy any protected tree(s) without first having obtained permission from the local authority. All types of tree can be protected in this way, whether as single trees or as part of a woodland, copse or other grouping of trees. Protection does not however extend to hedges, bushes or shrubs.

TPOs are recorded in the local land charges register which can be inspected at your council offices. The local authority regularly checks to see if trees on their list still exist and are in good condition. Civic societies and conservation groups also keep a close eye on trees. Before carrying out work affecting trees, you should check if the tree is subject to a TPO. If it is, you will need permission to carry out the work.

All trees in a Conservation Area are protected, even if they are not individually registered. If you intend to prune or alter a tree in any way you must give the local authority plenty of notice so they can make any necessary checks.

Even with a preservation order it is possible to have a tree removed, if it is too decayed or dangerous, or if it stands in the way of a development, the local authority may consider its removal, but will normally want a similar tree put in or near its place.

A TPO will not prevent planning permission being granted for development. However, the council will take the presence of TPO trees into account when reaching their decision.

If you have a tree on your property that is particularly desirable – either an uncommon species or a mature specimen – then you can request a preservation order for it. However, this will mean that in years to come you, and others, will be unable to lop it, remove branches or fell it unless you apply for permission.

What are my responsibilities?

Trees covered by TPOs remain the responsibility of the landowner, both in terms of any maintenance that may be required from time to time and for any damage they may cause. The council must formally approve any works to a TPO tree. If you cut down, uproot or wilfully damage a protected tree or carry out works such as lopping or topping which could be likely to seriously damage or destroy the tree then there are fines on summary conviction of up to £20 000, or, on indictment, the fines are unlimited. Other offences concerning protected trees could incur fines of up to £2500.

What should I do if a protected tree needs lopping or topping?

Although there are certain circumstances in which permission to carry out works to a protected tree are not required, it is generally safe to say that you should always write to your council seeking their permission before undertaking any works.

You should provide details of the trees on which you intend to do work, the nature of that work – such as lopping or topping – and the reasons why you think this is necessary. The advice of a qualified tree surgeon may also be helpful, see *Yellow Pages*.

You may be required to plant a replacement tree if the protected tree is to be removed.

4.16.11 What about nature conservation issues?

Many traditional buildings, particularly farm buildings, provide valuable wildlife habitats for protected species such as barn owls and bats.

Planning permission will not normally be granted for conversion and reuse of buildings if protected species would be harmed. However, in many cases, careful attention to the timing and detail of building work can safeguard or re-create the habitat value of a particular building. Guidance notes prepared by English Nature are available from councils or from their website www.english-nature.org.uk.

4.16.12 What about bats and their roosts?

Bats make up nearly one-quarter of the mammal species throughout the world. Some houses may hold roosts of bats or provide a refuge for other protected species. The Wildlife and Countryside Act 1981 gives special protection to **all** British bats because of their roosting requirements. English Nature (EN) or the Countryside Council for Wales (CCW) must be notified of any proposed action (e.g. remedial timber treatment, renovation, demolition and extensions) which is likely to disturb bats or their roosts. EN or CCW must then be allowed time to advise on how best to prevent inconvenience to both bats and householders.

Information on bats and the law is included in the booklet *Focus on Bats* which can be obtained free of charge from your local EN office. Similar booklets can be obtained from CCW local offices.

The type of stone barns and traditional buildings found in the UK have lots of potential bat roosting sites; the most likely places being gaps in stone rubble walls, under slates or within beam joints. These sites can be used throughout the year by varying numbers of bats, but could be particularly important for winter hibernation. As a result, the following points should be followed when considering or undertaking any work on a stone barn or similar building, particularly where bats are known to be in the area.

A survey for the presence of bats should be carried out by a member of the local bat group (contact via English Nature) before any work is done to a suitable barn during the summer bat breeding period.

The pipistrelle, the smallest of the European bats, has been found lurking in many strange places including vases, under floorboards, and between the panes of double glass.

Any pointing of walls should not be carried out between mid-November and mid-April to avoid potentially entombing any bats. When walls are to be pointed, areas of the walls high up on all sides of the building should be left unpointed to preserve some potential roosting sites. If any bats are found whilst work is in progress, work should be stopped and English Nature contacted for advice on how to proceed.

If any timber treatment is carried out, only chemicals safe for use in bat roosts should be used. A list of suitable chemicals is available from English Nature on request. Any pre-treated timber used should have been treated using the CCA method (copper chrome arsenic) which is safe for bats.

Work should not be commenced during the winter hibernation period (mid-November to mid-April). Any bats present during the winter are likely to be torpid, i.e. unable to wake up and fly away and are therefore particularly vulnerable.

If these guidelines are followed, then the accidental loss of bat roosts and death or injury to bats will be reduced.

Although vampire bats feed primarily on domestic animals, they have been known to feed on sleeping humans on rare occasions! Vampire bats have chemicals in their saliva that prevent the blood they are drinking from clotting. They consume five teaspoons of blood each day. The vampire bat has been known to transmit rabies to livestock and to man.

4.16.13 What about barn owls?

Barn owls also use barns and similar buildings as roosting sites in some areas. These are more obvious than bats and, therefore perhaps easier to take into account. Barn owls are also fully protected by law and should not be disturbed during their breeding season. Special owl boxes can be incorporated into walls during building work, details of which can be obtained from English Nature.

4.17 What could happen if you don't bother to obtain planning permission?

If you build something which needs planning permission without obtaining permission first, you may be forced to put things right later, which could prove troublesome and extremely costly. You might even have to remove an unauthorized building.

4.17.1 Enforcement

If you think that works are being carried out without planning permission, or not in accord with approved plans and/or conditions of consent, then seek the advice of the local planning officer who will then investigate, and if necessary take appropriate steps to deal with the problem. Conversely, if you are carrying out development works, it is important that you stick to the approved plans and condition. If changes become necessary please contact the development control staff before they are made.

4.18 How much does it cost?

A fee is required for the majority of planning applications and the council cannot deal with your application until the correct fee is paid. The fee is not refundable if your application is withdrawn or refused.

In most cases you will also be required to pay a fee when the work is commenced. These fees are dependent on the type of work that you intend to carry out. The fees outlined in Sections 14.18.1–10 are typical of the charges made by Local Authorities during 2004, when submitting an application and are an extract from '*Town and Country Planning* (*Fees for Applications and Amendments*) *Regulations 2003* (*Amended 2006*)'.

Work to provide access and/or facilities for disabled people to existing dwellings are exempt from these fees.

4.18.1 Householder applications

Outline applications (most types)	Site *not exceeding* 2.5 ha = **£265.00** per 0.1 ha (or part thereof) of site area, maximum **£6625** (2.5 ha) Site *exceeding* 2.5 ha = £80 per each additional 0.1 ha, maximum **£25 000**

Full applications and reserved matters

Dwellings – erection of new	*Up to* 50 dwellings, **£265.00** per dwelling house, maximum **£13 250** Over 50 dwellings, **£80** each additional unit, maximum **£50 000**
Dwellings – alteration (including outline)	**£135.00** per dwelling house, maximum **£265.00**
Approval of reserved matters where flat rate does **NOT** apply	A fee based upon the amount of floorspace and/or number of dwelling houses involved
Flat rate (only when maximum fee has been paid)	**£265.00**

4.18.2 Industrial/retail and other buildings applications

Industrial/retail buildings	Where no additional floorspace is created **£135.00** Works not creating more than **40 m^3** of additional floorspace **£135.00**
Outline application (see above)	More than **40 m^3** but not more than **75 m^3** of additional floor-space **£265.00** Each additional **75 m^3** (or part thereof) **£265.00**, maximum **£13 250** (=3750 m^3) *Over* 3750 m^3 **£80** each additional 75 m^3 up to maximum fee **£50 000**
Plant & machinery (erection, alteration, replacement)	**£265.00** per 0.1 ha (or part thereof) of the site area, maximum **£13 250** (5 ha) Over 5 ha **£80** each additional 0.1 ha up to maximum fee **£50 000**

4.18.3 Prior notice applications

Approvals for agricultural/forestry buildings/operations and demolition of buildings and telecommunications works	**£50.00**

4.18.4 Agricultural applications

Agricultural buildings	Buildings not exceeding **465 m^3** – **£50.00** Buildings exceeding **465 m^3** but more than **540 m^3** – **£265.00** More than **540 m^3** – **£265.00** for each additional **75 m^3** (or part thereof), maximum **£13 250** (4215 m^3) Over 4215 m^3 **£80** each additional 75 m^3 up to maximum fee **£50 000**
Erection of glasshouses/polytunnels	Works not creating more than **465 m^3** – **£50.00** (on land used for agriculture) Works creating more than **465 m^3** – **£1495.00**

4.18.5 Concessionary fees and exemptions

Works to improve the disabled person's access to a public building, or to improve their access, safety, health or comfort at their dwelling house	**No fee**
Applications by Parish Councils (all types)	**Half the normal fee**
Applications required by an Article 4 direction or removal of permitted development rights	**No fee**
Playing fields (for sports clubs etc)	**£265.00**
Revised or fresh applications of the **same character** or **description** by the **same applicant** within **12 months** of **refusal,** or the expiry of the statutory 8 week period where the applicant has appealed to the Secretary of State on grounds of **non-determination. Withdrawn applications of the same character or description must be made within 12 months of making the earlier one.**	**No fee**
Revised or fresh application of the **same character** or **description** within **12 months** of receiving permission	**No fee**
Duplicate applications made by the **same applicant for each application** submitted within 28 days of each other	**Full fee**
Alternative applications for one site submitted at the **same time**	**Highest of the fees applicable for each alternative and a sum equal to half the rest.**
Development crossing local authority boundaries	**Only one fee paid to the authority having the larger site but calculated for the whole scheme and subject to a special ceiling.**

4.18.6 Hazardous substances applications

Application for new consent	**£200.00**
New consent where maximum quantity specified exceeds twice the controlled quantity	**£400.00**

All other types of application	**£250.00**
Continuation of hazardous consent Under Section 17(1) of the 1992 Regulations	**£200.00**

4.18.7 Legal applications

Application for a certificate of lawfulness an existing use or operation	**Same fee payable** as if making a planning application
Application for a certificate of lawfulness for an existing activity in breach of planning condition(s)	**£135.00**
Application for a certificate of lawfulness for a proposed use or operation	**Half the fee payable** as if making a planning application

4.18.8 Advertisement applications

Adverts relating to the business on the premises	**£75.00**
Advance signs directing the public to a business (unless business can be seen from the signs position)	**£265.00**
Other advertisements (e.g. hoardings)	**£265.00**

4.18.9 Other applications

Exploratory drilling for oil or natural gas	**£265.00** per 0.1 ha (or part thereof) of site area, maximum **£19 875** (=7.5 ha) *Over* 7.5 ha **£80** each additional 0.1 ha, maximum fee **£50 000**
Winning, working, storage of minerals etc. and waste disposal	**£135.00** per 0.1 ha (or part thereof) of site area, maximum **£20 250** (=15 ha) *Over* 15 ha **£80** each additional 0.1 ha, maximum fee **£50 000**
Car parks, service roads or other accesses (existing uses only)	**£135.00**
Other operations on land	**£135.00** per 0.1 ha (or part thereof) of site area, maximum **£1350** (=1 ha)

Non-compliance with conditions	**£135.00**
Renewal of temporary permissions	**£135.00**
Removal or variation of conditions including renewal of unimplemented consents which have **not lapsed**	**£135.00** (full fee if consent has **lapsed**)
Change of use to sub-division of dwellings	**£265.00** per additional dwelling created, maximum **£13 250** (50 units) *Over* 50 units **£80** each additional unit, maximum fee **£50 000**
Other changes of use **except** waste or minerals	**£265.00**

Planning Portal – fee calculator

The Planning Portal is the UK Government's planning source. Within this site is a fee calculator. The fee calculator can help you by working out the cost of any particular planning application.

The calculator asks a series of questions to help determine the total cost of an application, ranging from a simple householder development to large-scale schemes such as housing schemes or industrial estates.

Go to www.planningportal.gov.uk/england/genpub/en and follow links through Useful Tools/Fee Calculator.

Extracted from *The Town & Country Planning (Fees for Applications & Deemed Applications) (Amendment) Regulations 2003. (Amended 2006.)*

4.19 Sustainable homes

On 13th December 2006, the Code for Sustainable Homes (a national standard for sustainable design and construction of new homes) was launched as part of a package of measures aimed at zero carbon development and which included an overarching consultation on the shift to zero carbon (i.e. Building a Greener Future) and a consultation on the draft of a new planning policy statement (i.e. Planning and Climate Change).

Full Technical Guidance on how to comply with the Code was published in April 2007 and consultation documents (e.g. summary of responses to the consultation and the Regulatory Impact Assessment and the Code for Sustainable Homes) can be viewed on the Communities and Local Government website (see www.communities.gov.uk/index.asp?id=1506120).

4.19.1 Who has drawn up the Code?

The initial idea for sustainable homes originated from the Government's Energy White paper ('Our energy future – creating a low carbon economy'), which was published in February 2003. Since then the Government has been working closely with Building Research Establishment (BRE) and the Construction Industry Research and Information Association (CIRIA) and others to develop the Code.

This new Code builds upon BRE's EcoHomes system of credits and weightings in a number of ways, for example;

- the introduction of minimum standards for energy and water efficiency at every level of the Code which require high levels of sustainability performance in these areas in order to achieve a high Code rating;
- a simpler system for awarding points, with more complex weightings removed;
- new areas of sustainability design, such as Lifetime Homes (i.e. homes that make life as easy as possible for as long as possible because they are thoughtfully designed) and inclusion of composting facilities;
- consideration of existing and ongoing European standardization work.

Note: BRE will continue to maintain and operate the EcoHomes scheme during the transition to the Code – see www.ecohomes.org.

4.19.2 Why is there a need for 'sustainable homes'?

Whilst great progress has been made during the last few years with homes that have been built under the revised standards (e.g. homes are now 40% more energy-efficient than 5 years ago and domestic appliances are far more efficient than 10 years ago) it is still true to say, however, that more than a quarter of all carbon emissions in the UK **still** come from energy used to heat, light and run our homes.

Demand for homes is only going to increase over the next few decades and, if we are to build the homes we need, then by 2050 as much as one-third of the total housing stock will have been built between now and then. Clearly then it is important that we do everything we can to ensure that all these new homes are built in such as way as to minimize their impact on the UKs carbon emissions.

Taking into consideration the burden that new, additional regulations will place on the building trade and building inspectors, the fact that some aspects of sustainability are not entirely suited to regulation (e.g. dictating to someone how to run their life and the fact that sustainability is not an exact science but more of a trade-off), the Government has taken the view that instead of direct regulation, the Code for sustainable homes will be one of 'voluntary compliance'. However, homebuilders will be 'encouraged' to follow Code principles because the Government is considering making assessment under Code standards mandatory in the future.

4.19.3 What will be the benefits of the new Code?

There will be benefits for the environment, the home building industry and individuals. In brief these could be seen as the following:

Benefits for the environment

- **Reduced greenhouse gas emissions** – with the minimum standard for energy efficiency at each level of the Code there will be a reduction in greenhouse gas emissions to the environment.
- **Better adaptation to climate change** – with minimum standards for water efficiency and other measures including better management of surface water run-off – will ensure that houses will be better adapted to cope with the impacts of climate change.
- **Reduced impact on the environment overall** – the inclusion of measures to promote the use of less polluting materials etc., will ensure houses have fewer negative impacts overall on the environment.

Benefits for home builders

- **A mark of quality** – the Code can be used by home builders to demonstrate the sustainability performance of their homes and differentiate themselves from the competitors.
- **Regulatory certainty** – the levels of performance of energy efficiency indicate the future direction of the Building Regulations, bringing regulatory certainty to homebuilders and acting as a guide to support effective business and investment planning.
- **Flexibility** – the Code is based on performance which means it sets levels of performance as opposed to stipulating how to achieve these levels, which will enable home builders to use the flexibility of this code to find cost-effective solutions to meet and exceed requirements.

Benefits for social housing providers

- **Lower running costs** – homes built to Code standards will have lower running costs through greater energy and water efficiency.
- **Improved comfort and satisfaction** – homes built to the Code will enhance comfort for tenants.
- **Raised sustainability credentials** – the Code will enable Social Housing Providers to demonstrate their sustainability credentials to the public, tenants and public bodies.

Benefits for consumers

- **Assisting choice** – the Code will provide valuable information to homebuyers on sustainability performance, helping them to make informed choices.

- **Reducing environmental footprint** – by asking what houses meet the Code, homebuyers will encourage industry to build more sustainable homes and help them reduce their own environmental footprint.
- **Lower running costs** – homes built to the Code will have lower running costs through greater efficiencies.
- **Improved well-being** – homes built to the Code will provide more pleasant and healthy places to live – e.g. more natural light.

4.19.4 How does the Code work?

The Code measures the sustainability of a home against design categories (see below) which rate the whole home as a complete package using stars (one star for entry level (i.e. above Building Regulations) and six stars the highest level) depending on the extent to which it has achieved Code standards.

Although the Code is closely linked to the Building Regulations, minimum standards for Code compliance have been set above the requirements of Building Regulations and it is intended that the Code will signal the future direction of Building Regulations in relation to carbon emissions from (and energy use in) homes, providing greater regulatory certainty for the home building industry.

Achieving a sustainability rating

The sustainability rating a home achieves represents its overall performance across nine Code design categories:

- Energy/CO_2
- Water
- Materials
- Surface water run-off
- Waste
- Pollution
- Health & well-being
- Management
- Ecology.

The Code sets minimum standards for these categories, which **must** be achieved if the house is to gain a one star rating. Energy efficiency and water efficiency categories also have a minimum standard that must be achieved at every level of the Code, recognizing their importance to the sustainability of any home.

Apart from these minimum standards, the Code is completely flexible; developers can choose which and how many standards they implement to obtain 'points' under the Code in order to achieve a higher sustainability rating.

Table 4.2 shows the nine design categories and the degree of flexibility afforded by each.

4.19.5 Is there any connection between 'sustainable homes' and 'sustainable locations'?

It should be noted that the Code will principally deal with sustainable homes as opposed to the *sustainability* of locations. This role will still be covered by the

Table 4.2 Code flexibility

Categories	Flexibility
Energy/CO_2	Minimum standards at each level of the Code
Water	
Materials	Minimum standard at Code entry level
Surface water run-off	
Waste	
Pollution	No minimum standards
Health & well-being	
Management	
Ecology	

existing planning system, which is recognized as a means of ensuring that developments are located on sustainable sites and that developments are such that they assist in the reduction of the need to travel. As a result of the planning system, the rate of development of brownfield sites in England has increased from 57% in 1997 to 70% in 2004.

4.19.6 How much will building a sustainable home cost?

At this early stage the Department for Communities and Local Government (DCLG) have calculated that in order to meet the minimum level, the average additional cost would be just over £600 per home. Obviously buildings that are designed to meet the higher levels of the Code will be more expensive and as a consequence purchase prices of such houses will also have a premium attached.

Note: For more details concerning the Code for sustainable homes as well as a summary of the responses to the consultation, see www.communities.gov.uk.

4.2 Home Information Pack (HIP)

Currently, if you plan to put a residential property containing four or more bedrooms up for sale on or after 1st August 2007 in England or Wales you will need to acquire a Home Information Pack (HIP) before your property goes on the market. This will be extended to smaller properties as rapidly as possible (although no specific date has been set) and will depend on the recruitment of sufficient energy assessors to provide energy ratings for properties. The responsibility for acquiring an HIP rests with the person marketing the property, usually the estate agent, developer or auctioneer, or indeed the seller themselves if the property is being sold privately.

4.20.1 Why are HIPs being introduced?

Currently much of the essential information needed by home buyers only comes to light when an offer has been made and as a consequence of that they are negotiating in the dark and often wasting money on legal fees, searches and surveys.

4.20.2 What are the benefits of the HIP?

By providing the information up front it is intended that this will make the home buying and selling process more efficient, more transparent and less stressful for all concerned.

The main benefits are seen as:

- **consumer satisfaction** – nearly 9 out of 10 consumers are dissatisfied by present processes;
- **consumer confidence** – estate agents marketing homes with an HIP will be required to belong to an approved scheme. This will mean greater consumer confidence as well as providing them with an option to an independent industry body if they have concerns or issues;
- **transparency** – currently key information only becomes available after the terms of sale are agreed;
- **sales success rate** – currently 28% of sales fail after terms are agreed;
- **cost** – failed transactions cost the consumer over £350 m a year;
- **speed** – transactions currently run at half the speed of the European average;
- **reliability** – 60% of property sales are in chains which means that any delay or failure from one sale will have a knock-on effect through the chain;
- **first time buyers** – as the HIP is free to the buyer, first time buyers will find it easier and cheaper to get on to the property ladder.

4.20.3 What is in the HIP?

HIP sets out that there are a range of required and authorized documents to be incorporated in the pack. Required documents must be included in the pack where appropriate and authorized documents may be included at the sellers discretion (the seller can use these to top up the pack voluntarily to include information they judge to be of interest to the prospective buyer).

The required documents are:

- an index;
- a sale statement summarizing terms of sale;
- evidence of title;
- standard searches – or at least proof that they have been requested (e.g. local authority enquiries and a drainage and water search);
- an Energy Performance Certificate;

and where appropriate:

- commonhold information (including a copy of the commonhold community statement) – commonhold is an alternative to the conventional method of owning flats and other independent properties under a lease;
- new homes warranty (provided by the builder);
- a report on a home that is not physically complete (i.e. not yet signed off by a building inspector).

Authorized documents include:

- guarantees and warranties;
- other searches;
- home condition report – whilst this is a voluntary inclusion, the Government strongly believes that these are likely to prove valuable to both sellers and buyers, and is working with industry to try and ensure there is take up of these.

4.20.4 Who provides the Home Information Pack?

At the time of writing, the DCLG are working in partnership with industry stakeholders (such as the National Association of Estate Agents, the Law Society, the Council of Mortgage Lenders, the Royal Institution of Chartered Surveyors etc.) to develop the programme in time for the June 2007 launch date.

Sellers are able to compile their own packs should they choose to, or through working with The Association of Home Information Pack Providers (AHIPP) who were set up so as to ensure that there would be proper representation for those companies who will be at the heart of a new industry. Full membership is open to those involved in the production and collation of HIPs, with Associate and Affiliate membership available to other interested organizations.

Since its formation in June 2005, AHIPP has grown to 91 members with a number of other important industry players expected to join in the near future.

4.20.5 The HIP Code of Practice

The AHIPP has decided to help regulate the new HIP industry on behalf of the consumer. It has achieved this by compiling a Code of Practice for HIP providers to follow, which ensures there are redress mechanisms in case of misleading content in the Home Information Pack.

To raise awareness, the following Kite Mark must be displayed on all packs that are Code Compliant.

To ensure that individual HIP providers are code compliant, an inspection regime has been put into place by an independently run scheme called the Property Codes Compliance Board (PCCB). This will overlook and control the CoPSO (Council of Property Search Organisations) search code and the AHIPP HIP Code.

4.20.6 Will the industry be ready in time?

Research carried out among the leading companies who are training and/or employing Home Inspectors and Domestic Energy Assessors (DEAs) firmly believe there will be over 2,500 people qualified to deliver Energy Performance Certificates (EPCs) by June 2007. Unfortunately at the time of writing less than 1000 people have been fully trained as inspectors and the whole HIP process has now been watered down from being applicable to ***all*** houses to being applicable to homes of more than four bedrooms.

It was estimated that around 1.7 million energy reports will be required each year and that in June (i.e. when Home Information Packs (HIPs) become mandatory) 91,000 reports would be needed. It was considered that this figure would grow to 153,000 reports in October if current levels of activity continue.

For further information on Home Information Pack Providers, see **www.hipassociation.co.uk**

4.20.7 Results from the HIP trial

In November 2006, HIP trial locations were set up in Bath, Cambridge, Huddersfield, Newcastle, Northampton and Southampton. Feedback from these initial six locations has shown that consumers like these packs and have found them a useful selling tool. Media reports, however, paint a completely different picture, but time will tell!

5

Requirements for planning permission and Building Regulations approval

Before undertaking any building project, you must first obtain the approval of local-government authorities. There are two main controls that districts rely on to ensure that adherence to the local plan is ensured, namely planning permission and Building Regulation approval.

Whilst both of these controls are associated with gaining planning permission, actually receiving planning permission does not automatically confer Building Regulation approval and vice versa. You **may** require **both** before you can proceed. Indeed, there may be variations in the planning requirements, and to some extent the Building Regulations, from one area of the country to another.

Provided, however, that the work you are completing does not affect the external appearance of the building, you are allowed to make certain changes to your home without having to apply to the local council for permission. These are called permitted development rights, but the majority of building work that you are likely to complete will still require you to have planning permission – so be warned!

The actual details of planning requirements are complex but for most domestic developments, the planning authority is only really concerned with construction work such as an extension to the house or the provision of a new garage or new outbuildings that is being carried out. Structures like walls and fences also need to be considered because their height or siting might well infringe the rights of your neighbours and other members of the community. The planning authority will also want to approve any change of use, such as converting a house into flats or running a business from premises previously occupied as a dwelling only.

Planning consent **may** be needed for minor works such as television satellite dishes, dormer windows, construction of a new access, fences, walls, and garden extensions. You are advised to consult with Development Control staff before going ahead with such minor works.

5.1 Decoration and repairs inside and outside a building

Planning permission		Building Regulation approval	
No	Unless it is of a listed building or within a Conservation Area Consult your local authority	No	Unless it is a listed building or within a Conservation Area Consult your local authority
No	As long as the use of the house is not altered	Possibly	Consult your local authority
Yes	If the alterations are major such as removing or part removing of a load bearing wall or altering the drainage system	Yes	

Generally speaking, you do not need to apply for planning permission:

- for repairs or maintenance;
- for minor improvements, such as painting your house or replacing windows;
- for internal alterations;
- for the insertion of windows, skylights or roof lights (but, if you want to create a new bay window, this will be treated as an extension of the house);
- for the installation of solar panels which do not project significantly beyond the roof slope (rules for listed buildings and houses in Conservation Areas are different however);
- to re-roof your house (but additions to the roof are treated as extensions to the house).

Occasionally, you may need to apply for planning permission for some of these works because your council has made an Article 4 direction withdrawing permitted development rights.

Do I need approval to carry out repairs to my house, shop or office?

No – if the repairs are of a minor nature – e.g. replacing the felt to a flat roof, repointing brickwork, or replacing floorboards.
Yes – if the repair work is major in nature – e.g. removing a substantial part of a wall and rebuilding it, or underpinning a building.

Do I need to apply for planning permission for internal decoration, repair and maintenance?

No.

Do I need to apply for planning permission for external decoration, repair and maintenance?

No – external work in most cases doesn't need permission, provided it does not make the building any larger.

Do I need approval to or alter the position of a WC, bath, etc. within my house, shop or flat?

No – unless the work involves new or an extension of drainage or plumbing.

Do I need approval to alter in any way the construction of fireplaces, hearths or flues within my house, shop or flat?

Yes.

Do I need to apply for planning permission if my property is a listed building?

Yes – if your property is a listed building consent will probably be needed for **any** external work, especially if it will alter the visual appearance, or use alternative materials. You also may need planning permission to alter, repair or maintain a gate, fence, wall or other means of enclosure.

Do I need to apply for planning permission if my property is in a Conservation Area?

Yes – if the building undergoing repair or decoration is in a Conservation Area, or comes under any type of covenant restricting changes you will probably need planning permission. You may also be restricted to replacing items such as roof tiles with the approved material, colour and texture, and have to use cast iron guttering rather than plastic etc.

5.2 Structural alterations inside

Planning permission		Building Regulation approval	
No	As long as the use of the house is not altered	Possibly	Consult your local authority
Yes	If the alterations are major such as removing or part removing of a load bearing wall or altering the drainage system	Yes	
Yes	If they are to an office or shop	Yes	

Do I need approval to make internal alterations within my house?

Yes – if the alterations are to the structure such as the removal or part removal of a load bearing wall, joist, beam or chimney breast, or would affect fire precautions of a structural nature either inside or outside your house. You also need approval if, in altering a house, work is necessary to the drainage system or to maintain the means of escape in case of fire.

Do I need approval to make internal alterations within my shop or office?

Yes.

Do I need approval to insert cavity wall insulation?

Yes.

Do I need approval to apply cladding?

Yes – if you live in a Conservation Area, a national park, an area of outstanding natural beauty or the Norfolk Broads. You will need to apply for planning permission before cladding the outside of your house with stone, tiles, artificial stone, plastic or timber.

If you are in any doubt about whether you need to apply for permission, you should contact your local authority planning department before commencing any work to your property. They will usually give you advice but if you want to obtain a formal ruling you can apply, on payment of a fee, for a lawful development certificate. You may also require Building Regulation approval.

5.3 Replacing windows and doors

Planning permission		Building Regulation approval	
Yes	If they are to an office or shop	Yes	
No	Unless: • they project beyond the foremost wall of the house facing the highway • the building is a listed building • the building is in a Conservation Area	Possibly	They will need to be installed by an approved person (see Table 2.1)
Yes	To replace shop windows	Yes	And comply with the requirements of Part F

Do I need approval to install replacement windows in my house, shop or office?

No – provided:

- the window opening is not enlarged. (If a larger opening is required, or if the existing frames are load-bearing, then a structural alteration will take place and approval is required).
- you do not remove those opening windows which are necessary as a means of escape in case of fire.
- the replacement of a window in an existing building is carried out by a person who is registered under the Fenestration Self-Assessment Scheme by FENSA Ltd (www.fensa.org.uk/faq/html).

Note: As FENSA does not apply to commercial premises or new build properties, replacement of windows in offices and other commercial premises (including the replacement of shop fronts) will, therefore, **ALL** require Local Authority Building Control approval.

Do I need approval to replace my shop front?

Yes.

Anyone who installs replacement windows or doors has to comply with strict thermal performance standards and when a property is sold, the purchaser's surveyors will normally ask for evidence that '*any replacement glazing installed after April 2002 complies with the new Building Regulations*'.

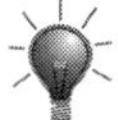

There will be two ways to prove compliance:

1. A certificate showing that the new work has been done by an installer who is registered under the FENSA Scheme (a scheme which allows installation companies to self-certify that their work complies with the Building Regulations), or
2. A certificate from the local authority saying that the installation has approval under the Building Regulations.

Note: Further information is available from your local building control or from the Glass and Glazing Federation (GGF) website www.ggf.org.uk.

5.4 Electrical work

Planning permission	Building Regulation approval	
No	Probably	But it must comply with Part P and other relevant Building Regulations Approved Documents

Do I need approval to replace electric wiring?

No – but:

- you must comply with Part P (and other relevant Building Regulations Approved Documents);
- you should meet the recommendations of the IET Wiring Regulations (i.e. BS 7671);
- your contract with the electricity supply company will have conditions about electrical safety which must not be broken. In particular, you should **not** interfere with the company's equipment which includes the cables to your consumer unit up to and including the separate isolator switch if provided.

Do I need approval to replace an existing electrical fitting?

No – non-notifiable work (such as replacing an electrical fitting) can be completed by a DIY enthusiast (family member or friend) but **still** needs to be installed in accordance with the manufacturer's instructions and done in such a way that they do not present a safety hazard.

This work does **not** need to be notified to a local authority building control body (unless it is installed in an area of high risk such as a kitchen or a bathroom etc.) **but** all DIY electrical work (unless completed by a qualified professional) will still need to be checked, certified and tested by a competent electrician.

Do I need approval to install a new electrical circuit?

Probably – any work that involves adding a new circuit to, in or around a dwelling will need to be either notified to the building control body (who will then inspect the work) or needs to be carried out by a competent person who is registered under a Government approved Part P self-certification scheme.

Work involving any of the following will also have to be notified:

- consumer unit replacements;
- electric floor or ceiling heating systems;
- extra-low voltage lighting installations (other than pre-assembled, CE-marked lighting sets);
- garden lighting and/or power installations;
- installation of a socket outlet on an external wall;
- installation of outdoor lighting and/or power installations in the garden or that involves crossing the garden;
- installation of new central heating control wiring;
- solar photovoltaic (pv) power supply systems;
- small-scale generators (such as microCHP units).

Note: Where a person who is **not** registered to self-certify, intends to carry out the electrical installation, then a Building Regulation (i.e. a building notice or full plans) application will need to be submitted together with the appropriate fee, based on the estimated cost of the electrical installation. The building control body will then arrange to have the electrical installation inspected at first fix stage and tested upon completion.

5.5 Plumbing

Planning permission	Building Regulation approval	
No	No	For replacements (but you will need to consult the technical services department for any installation which alters present internal or external drainage)
	Yes	For an unvented hot water system

Do I need approval to install hot water storage within my house, shop or flat?

Yes – if the water heater is unvented (i.e. supplied directly from the mains without an open expansion tank and with no vent pipe to atmosphere) and has storage capacity greater than 15 litres.

5.6 Central heating

Planning permission	Building Regulation approval	
No	No	If electric
	Yes	If gas, solid fuel or oil

Do I need approval to alter the position of a heating appliance within my house, shop or flat?

- **Gas**: Yes, unless the work is supervised by an approved installer under the Gas Safety (Installation and Use) Regulations 1984.
- **Solid fuel**: Yes.
- **Oil**: Yes.
- **Electric**: Yes, unless the work is carried out by a competent person who is registered under a Government Approved Part P Self-Certification Scheme.

5.7 Oil-storage tank

Planning permission		Building Regulation approval
No	Provided that it is in the garden and has a capacity of not more than 3500 litres (778 gallons) and **no** point is more than 3 m (9 ft 9″) high and **no** part projects beyond the foremost wall of the house facing the highway	No

Oil storage tanks, and the pipes connecting them to combustion appliances, should be constructed and protected so as to reduce the risk of the oil escaping and causing pollution.

5.8 Planting a hedge

Planning permission		Building Regulation approval
No	Unless it obscures view of traffic at a junction or access to a main road	No

You do not need planning permission for hedges or trees. However, if there is a condition attached to the planning permission for your property which restricts the planting of hedges or trees (for example, on an 'open plan' estate or where a sight line might be blocked), you will need to obtain the council's consent to relax or remove the condition before planting a hedge or tree screen. If you are unsure about this, you can check with the planning department of your council.

Hedges should not be allowed to block out natural light, and the positioning of fast growing hedges should be checked with your local authority. Recent incidents regarding hedging of the fast growing Leylandii trees have led to changes in the planning rules, where hedges previously had no restrictive laws.

5.9 Building a garden wall or fence

Planning permission		Building Regulation approval
Yes	If it is more than 1 m (3 ft 3″) high and is a boundary enclosure adjoining a highway	No
Yes	If it is more than 2 m (6 ft 6″) high elsewhere	No

Do I need approval to build or alter a garden wall or boundary wall?

No – subject to size.

You will need to apply for planning permission if:

- your house is a listed building or in the curtilage of a listed building; or
- the fence, wall or gate would be over 1 metre high and next to a highway used for vehicles; or over 2 metres high elsewhere.

In normal circumstances, the only restriction on walls and fences is the height allowed. This is 2 metres or no more than 1 metre if the walls or fence is near a highway or road junction, where its height might obscure a driver's view of other traffic, pedestrians or road users.

If there is a valid reason for a wall or fence higher than the prescribed dimensions, then it is possible to get planning consent. There may be security issues that would support an application for a high fence. If it has no affect on other people's valid interests and does not impair any amenity qualities in an area, there is no reason why a request should be refused.

Some walls have historic value and they, as well as arches and gateways, can be listed. Modifications, extensions and removal of these must have planning consent.

5.10 Felling or lopping trees

Planning permission		Building Regulation approval
No	Unless the trees are protected by a Tree Preservation Order or you live in a Conservation Area	No

Many trees are protected by Tree Preservation Orders (TPOs), which mean that, in general, you need the council's consent to prune or fell them. Nearly all trees in Conservation Areas are automatically protected.

Ask the council for a copy of the free leaflet *Protected Trees: a guide to tree preservation procedures*.

5.11 Laying a path or a driveway

Planning permission		Building Regulation approval
No	Unless it provides access to a main road	No

Do I need to apply for planning permission to install a pathway?

Generally no – but you may need approval from the highways department if the pathway crosses a pavement.

Do I need to apply for planning permission to lay a driveway?

No – unless it adjoins the main road.

Driveways

Provided a pathway or drive does not meet a public thoroughfare you will not need planning consent. There are no restrictions on the area of land around your house that you can cover with hard surfaces.

You will need to apply for planning permission only if the hard surface is not to be used for domestic purposes and is to be used instead, for example, for parking a commercial vehicle or for storing goods in connection with a business.

In the case of hardstanding you do not need permission to gain access to it within the confines of your land, but you would need permission for a hardstanding leading on to a public highway.

You must obtain the separate approval of the highways department of your council if you want to make access to a roadway or if a new driveway would cross a pavement or verge. The exception is if the roadway is unclassified and the drive or footway is related to a development that does not require planning permission. Your local authority highways department will be able to tell you if a road is classified or unclassified. If the road is classified then, depending on the volumes of traffic, it is harder to get permission. The busier the road the less likely a new driveway or footway will be allowed to meet it.

If a driveway crosses a pedestrian access, pavement or roadside verge, then the planning department will gain approval from the highways department. If this is the case, highways approval is required in addition to planning consent. The basic principle is to maintain safety and eliminate hazards.

You will also need to apply for planning permission if you want to make a new or wider access for your driveway onto a trunk or other classified road. The highways department of your council can tell you if the road falls into this category.

Pathways

Pathways do not normally need planning permission and you can lay paths however you like in the confines of your own property. The exception is for any path making access to a highway or public thoroughfare, in which case certain safety aspects arise. You may also need permission if your building is listed or is in a Conservation Area, so the style and size is suitable for the area.

If a pathway crosses a pedestrian access, pavement or roadside verge, then the planning department will gain approval from the highways department. If this is the case, highways approval is required in addition to planning consent. The basic principle is to maintain safety and eliminate hazards.

5.12 Building a hardstanding for a car, caravan or boat

Planning permission		Building Regulation approval
No	Provided that it is within your boundary and is not used for a commercial vehicle	No

Do I need to apply for planning permission to build a hardstanding for a car?

No – provided that it is within your boundary and is not used for a commercial vehicle.

Check local council rules.

Access from a new hardstanding to a highway requires planning consent. The exception is if the roadway is unclassified and the access to the hardstanding is related to a development that does not require planning permission. Your local authority highways department will be able to tell you if a road is classified or unclassified. If the road is classified then, depending on the volumes of traffic, it is harder to get permission. The busier the road the less likely a new driveway or footway will be allowed to meet it.

If the access crosses a pedestrian thoroughfare, pavement or roadside verge, then the planning department will gain approval from the highways department. If this is the case, highways approval is required in addition to planning consent. The basic principle is to maintain safety and eliminate hazards.

For a hardstanding on your own land, you do not need permission to gain access to it within the confines of your land, but you would need permission for a hardstanding leading on to a public highway.

There are different rules depending on what you use a hardstanding for. Planning permission is generally not needed provided there are no covenants limiting the installation of hardstanding for parking of cars, caravans or boats. There are still rules for commercial parking, however (e.g. taxis or commercial delivery vans) and a 'change of use' as a trade premises would probably need to be granted for this to be allowed.

You should check if there are any local covenants limiting changes in access to your premises or for hardstanding and parking of vehicles on it. If in doubt, contact the relevant local authority planning department for specific advice.

Do I need to apply for planning permission to build a hardstanding for a caravan and/or boat?

Some local authorities do not allow the parking of caravans or boats on driveways or hardstandings in front of houses. Check what the local rules are with your planning department, and if there's no restriction then you don't need to apply for permission.

There are no laws to prevent you, or your family from making use of a parked caravan while it's on your land or drive, but you cannot actually live in it as this would be classed as an additional dwelling. In addition, you cannot use a parked caravan for business use as this would constitute a change of use of the property.

If you want to put a caravan on your land to lease out as holiday accommodation or for friends or family to stay in while they visit you, then this would require planning permission. Rules on siting of static caravans or mobile homes are quite stringent.

5.13 Installing a swimming pool

Planning permission		Building Regulation approval	
Possibly	Consult your local planning officer	Yes	For an indoor pool

Swimming pools and saunas are subject to special requirements specified in Part 6 of BS 7671:2001.

5.14 Erecting aerials, satellite dishes television and radio aerials, wind turbines and flagpoles

Planning permission		Building Regulation approval
No	Unless it is a stand-alone antenna or mast greater than 3 m in height	No
Possibly	If erecting a satellite dish, especially in a Conservation Area or if it is a listed building (consult your local planning officer)	No
Possibly	If erecting a wind turbine, either stand-alone or attached to a dwelling (consult your local planning officer)	No

Do I need to apply for planning permission to erect satellite dishes, television and radio aerials, wind turbines and flagpoles?

No – unless it is a stand-alone antenna, flagpole or mast greater than 3 m in height.

Note: Flagpoles etc. erected in your garden are treated under the same rules as outbuildings, and cannot exceed 3 metres in height.

Normally there is no need for planning permission for attaching an aerial or satellite dish to your house or its chimneys. However, if it rises significantly higher than the roof's highest point then it may contravene local regulations or covenants. Planning permission may be required to install a wind turbine, either free standing or attached to a dwelling. It is recommended that you seek the advice of your local planning authority in relation to wind turbines as permissions vary depending on the region of the UK.

You should get specific advice if you plan to install a large satellite dish or aerial, such as a short wave mast, as the rules differ between authorities.

In certain circumstances, you will need to apply for planning permission to install a satellite dish on your house (see DTE's free booklet *A Householder's Planning Guide for the Installation of Satellite Television Dishes*, which can be obtained from your local council).

Conservation Areas have specific local rules on aerials and satellite dishes, so you need to approach your local planning department to find out the particular rules for your area. Certainly, if your house is a listed building, you may need listed building consent to install a satellite dish on your house.

Remember, if you are a leaseholder, you may need to obtain permission from the landlord.

Satellite dish locator

The Planning Portal website www.planningportal.gov.uk contains a very useful section called satellite dish locator. This provides a user-friendly way to check which parts of your house offer a suitable location for a satellite dish.

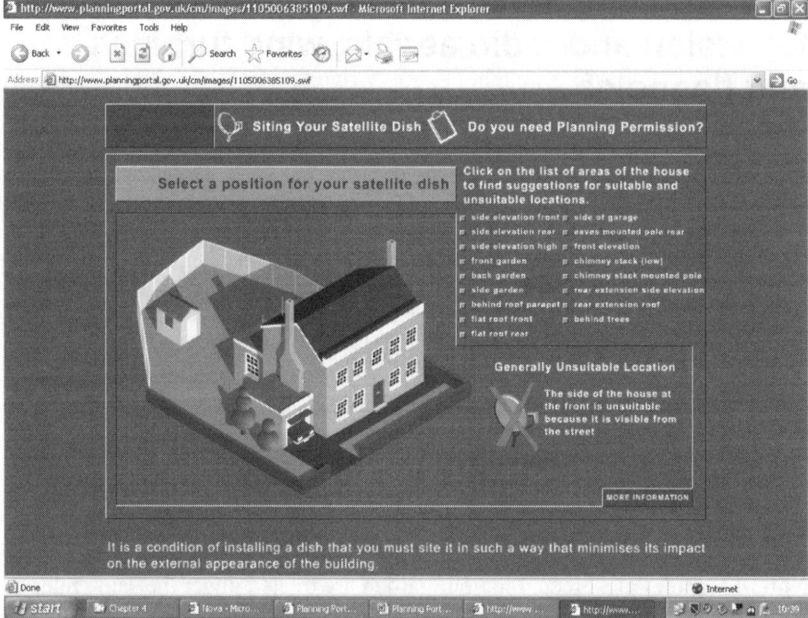

Figure 5.1 Satellite dish locator on the Planning Portal

5.15 Advertising

Planning permission		Building Regulation approval	
No	If the advertisement is less than 0.3 m^2 and not illuminated	Possibly	Consult your local planning officer

Do I need to apply for planning permission to erect an advertising sign?

Advertisement signs on buildings and on land often need planning consent. Some smaller signs and non-illuminated signs may not need consent, but it is always advisable to check with development control staff.

You are allowed to display certain small signs at the front of residential premises such as election posters, notices of meetings, jumble sales, car for sale etc. but business types of display and permanent signs may need to have planning permission granted. They may come under the category of 'advertising control' for which planning consent is required.

You may need to apply for advertisement consent to display an advertisement bigger than 0.3 m^2 on the front of, or outside, your property. This includes your house name or number or even a sign saying 'Beware of the dog'. Temporary notices up to 0.6 m^2 relating to local events, such as fêtes and concerts, may be displayed for a short period. There are different rules for estate agents' boards, but, in general, these should not be bigger than 0.5 m^2 on each side.

It is illegal to post notices on empty shops' windows, doors, and buildings, and also on trees. This is commonly known as 'fly posting' and can carry heavy fines under the Town and Country Planning Act.

Illuminated signs and all advertising signs outside commercial premises need to be approved. Most local authorities can give advice, by way of booklets or leaflets on what kinds of sign are allowed, not allowed or need approval.

You can get advice from the planning department of your local council; ask for a copy of the free booklet *Outdoor advertisements and signs*.

5.16 Building a porch

Planning permission		Building Regulation approval	
No	Unless: • the floor area exceeds 3 m^2 (3.6 y^2) • any part is more than 3 m (9 ft 9″) high • any part is less than 2 m (6 ft 6″) from a boundary adjoining a highway or public footpath	Yes	If area exceeds 30 m^2 (35.9 y^2) **Note:** Schedule 2 of the regulation is currently under consultation as there is a view that porches, conservatories, covered ways and carports under 30 m^2 should be subject to some regulations

Do I need planning permission for a porch?

Yes – depending on its size and position.

You will need to apply for planning permission:

- if your house is listed or is in a Conservation Area, national park, area of outstanding natural beauty;
- if the porch would have a ground area (measured externally) of more than 3 m^2;
- if the porch would be higher than 3 m above ground level;
- if the porch would be less than 2 m away from the boundary of a dwelling house with a highway (which includes all public roads, footpaths, bridleways and byways).

All measurements are taken externally.

However, a porch or conservatory built at ground level and under 30 m^2 in floor area is exempt provided that the glazing complies with the safety glazing requirements of the Building Regulations (Part N). Your local authority building control department or an approved inspector can supply further information on safety glazing. It is advisable to ensure that a conservatory is not constructed so that it restricts ladder access to windows serving a room in the roof or a loft conversion, particularly if that window is needed as an emergency means of escape in the case of fire.

The regulations are quite complicated and depend on previous works on the site, if any, so you should always check with development control staff.

5.17 Outbuildings

Planning permission		Building Regulation approval	
Possibly	Provided the building is less than 10 m^3 (13.08 y^3) in volume, not within 5 m (16 ft 3″) of the house or an existing extension	Yes	If area exceeds 30 m^2 (35.9 y^2)
	Erecting outbuildings can be a potential minefield and it is best to consult the local planning officer before commencing work		If it is within 1 m (3 ft 3″) of a boundary, it must be built from incombustible materials

Many kinds of buildings and structures can be built in your garden or on the land around your house without the need to apply for planning permission. These can include sheds, garages, greenhouses, accommodation for pets and domestic animals (e.g. chicken houses), summer houses, swimming pools, ponds, sauna cabins, enclosures (including tennis courts) and many other kinds of structure.

Outbuildings intended to go in the garden of a house do not normally require any planning permission, so long as they are associated with the residential amenities of the house and a few requirements are adhered to such as position and size.

You can build an outbuilding up to $10\,m^3$ ($13.08\,y^3$) in volume without planning permission if it is within 5 m (16 ft 3″) of the house or an extension. Further away than this, it can be up to half the area of the garden, but the height must not exceed 4 m (13 ft).

If your new building exceeds $10\,m^3$ (and/or comes within 5 m of the house) it would be treated as an extension and would count against your overall volume entitlement.

There are a few conditions to follow in order to avoid the need for planning consent:

- The structure should not result in more than half the original garden space being covered by the building.
- No part of the structure should extend beyond the original house limits on any side facing a public highway or footpath or service road.
- The height should not exceed 3 m (or 4 m if it has a ridged roof).

If your house is listed or is in a Conservation Area, national park, or area of outstanding natural beauty, then you will more than likely need to obtain planning consent. If in doubt, contact the relevant local authority planning department for specific advice.

Permission is required, however, for:

- any building/structure nearer to a highway than the nearest part of the original house, unless more than 20 m away from a highway;
- structures not required for domestic use;
- structures over 3 m high (or 4 m if it has a ridged roof);
- propane gas (LPG) tank;
- storage tank holding more than 3500 litres;
- a building or structure which would result in more than half of the grounds of your house being covered by buildings/structures.

You will also need to apply for planning permission if any of the following cases apply:

- You want to put up a building or structure which would be nearer to any highway than the nearest part of the original house, unless there would be at least 20 m between the new building and any highway. The term 'highway' includes public roads, footpaths, bridleways and byways.
- More than half the area of land around the original house would be covered by additions or other buildings.
- The building or structure is not to be used for domestic purposes and is to be used instead, for example, for parking a commercial vehicle, running a business or for storing goods in connection with a business.
- You want to put up a building or structure which is more than 3 m high, or more than 4 m high if it has a ridged roof (measured from the highest ground next to it).
- If your house is a listed building and you want to put up a building or structure with a volume of more than $10\,m^3$.

External water storage tanks

Many years ago the demand for external tanks for capturing rainwater made their installation quite commonplace. But it is rare today to need extra storage tanks, unless you are in a rural position.

If you are considering installing an external water tank you should seek guidance from your local authority, especially if the tank is to be mounted on a roof.

Fuel storage tanks

Storage of oil, or any other liquids, especially petrol, diesel and chemicals is strictly controlled and would not be allowed on residential premises. If you are considering installing an external oil storage tank for central heating use, then no planning permission is required, provided its capacity is no more than 3500 litres, it is no more than 3 m from the ground and it does not project beyond any part of a building facing a public thoroughfare.

You will need to apply for planning permission in the following circumstances:

- You want to install a storage tank for domestic heating oil with a capacity of more than 3500 litres or a height of more than 3 m above ground level.
- You want to install a storage tank, which would be nearer to any highway than the nearest part of the 'original house', unless there would be at least 20 m between the new storage tank and any highway. The term 'highway' includes public roads, footpaths, bridleways and byways.
- You want to install a tank to store Liquefied Petroleum Gas (LPG) or any liquid fuel other than oil.

Erecting any type of outbuilding can be a potential minefield and it is best to consult with the local planning officer before commencing work.

5.18 Garages

Planning permission		Building Regulation approval
Possibly	You can build a garage up to 10 m^3 (13.08 y^3) in volume without planning permission, if it is within 5 m (16 ft 3″) of the house or an existing extension. Further away than this, it can be up to half the area of the garden, but the height must not exceed 4 m (13 ft)	Yes

Do I need approval to build a garage extension to my house, shop or office?

Yes – but a carport extension built at ground level, open on at least two sides and under 30 m^2 in floor area, is exempt.

Note: Schedule 2 of the regulation is currently under consultation as there is a view that porches, conservatories, covered ways and carports under 30 m^2 should be subject to some regulations.

Do I need approval for a detached garage?

Yes – but a single storey garage at ground level, under 30 m^2 in floor area and with no sleeping accommodation, is exempt provided it is either built mainly using non-combustible material or, when built, it has a clear space of at least 1 m from the boundary of the property.

Garages planned to go in the garden of a house do not normally require any planning permission, so long as they are associated with the residential amenities of the house and a few requirements are adhered to such as position and size.

There are a few conditions to follow in order to avoid the need for planning consent, such as:

- the structure should not result in more than half the original garden space being covered by the building;
- no part of the structure should extend beyond the original house limits, on any side facing a public highway or footpath or service road;
- the height should not exceed 3 m (or 4 m if it has a ridged roof).

Integral garages (that is, those directly attached on the side or under existing rooms in your house) will nearly always require planning consent.

5.19 Building a conservatory

Planning permission		Building Regulation approval	
Possibly	You can extend your house by building a conservatory, provided that the total of both previous and new extensions does not exceed the permitted volume	Yes	If area exceeds 30 m^2 (35.9 y^2) **Note:** Schedule 2 of the regulation is currently under consultation as there is a view that porches, conservatories, covered ways and carports under 30 m^2 should be subject to some regulations

Do I need permission to erect a conservatory?

Possibly – see below.

Conservatories and sun lounges attached to a house are classed as extensions. If you want a conservatory or sun lounge separated from the house, this needs planning consent under similar rules for outbuildings.

If the answer to all the following questions is **no** then it is quite likely that planning permission will not be required:

- Is the conservatory going to be used as a separate dwelling, i.e. self-contained accommodation?
- Is your property listed, in a Conservation Area, national park or area of outstanding natural beauty?
- Will the conservatory cover more than half of the original garden space?
- Will any part of the conservatory within 2 m of the plot boundary be more than 4 m above ground level?
- Will any part of the conservatory be higher than the original roof of the main building?
- Will any part of the conservatory be nearer to a service road, public road or footpath than any part of the original building?
- Will the volume of the original house be increased so that any part of the conservatory within 2 metres of the plot boundary is more than 4 metres above ground level?
- Will the conservatory be behind the building line?
- Will your conservatory be 1 m away from the boundary (although most buildings tend to be nearer than this)?
- Will the conservatory be more than 50 ft away from the nearest road?

A conservatory has to be separated from the rest of the house to be exempt (i.e. patio doors).

Another thing to keep in mind is your neighbours' reaction – always keep them informed of what's happening and be prepared to alter the plans you had for locating the building if they object – it's better in the long run, believe me.

Will you need planning permission therefore? Generally no, as the building is classed as a 'portable building', nevertheless it is ***your*** responsibility to check with your local planning office.

However, a porch or conservatory built at ground level and under 30 m^2 in floor area is exempt provided that the glazing complies with the safety glazing requirements of the Building Regulations (Part N). Your local authority building control department or an approved inspector can supply further information on safety glazing. It is advisable to ensure that a conservatory is not constructed so that it restricts ladder access to windows serving a room in the roof or a loft conversion, particularly if that window is needed as an emergency means of escape in the case of fire.

The regulations are quite complicated and depend on previous works on the site, if any, so you should always check with development control staff.

Note: Conservatories will also need to have:

- heating systems equipped with independent temperature and on/off controls;
- thermal elements that have the appropriate U-values (see Annex B)
- glazed elements that comply with the standards (see Annex B).

5.20 Loft conversions, roof extensions and dormer windows

Planning permission		Building Regulation approval
No	Provided the volume of the house is unchanged and the highest part of the roof is not raised	Yes
Yes	For front elevation dormer windows or rear ones over a certain size	Yes

Do I need approval for a loft conversion?

Yes – see below.

Do I need to apply for planning permission to re-roof my house?

No – unless you live in a Conservation Area, a national park, an area of outstanding natural beauty or the Norfolk Broads.

Do I need to apply for permission to insert roof lights or skylights?

No.

Do I need to apply for planning permission to extend or add to my house?

Yes – in the following circumstances:

- If you want to build an addition or extension to any roof slope which faces a highway.
- If the roof extension would add more than 40 m^3 to the volume of a terraced house or more than 50 m^3 to any other kind of house.

These volume limits count as part of the allowance for extending the property (see extensions above).

- If the work would increase the height of the roof.
- If the intention is to create a separate dwelling, such as self-contained living accommodation or a granny flat.

If the answer to all the following questions is **no**, therefore, then it is quite likely that planning permission will not be required:

- Is the loft conversion going to be used as a separate dwelling, i.e. a self-contained flat?
- Is your property listed, in a Conservation Area, national park or area of outstanding natural beauty?
- Will any part of the loft conversion within 2 m of the plot boundary be more than 4 m above ground level?
- Will any part of the loft conversion be higher than the original roof of the main building?
- Will any part of the loft conversion be nearer to a service road, public road or footpath than any part of the original building?
- Will the roof be extended where it faces a public highway?
- Will the volume of the original house be increased beyond the following limits?
  - if the house is in a terrace, a Conservation Area, a national park, or an area of outstanding natural beauty 40 m^3 or 10% whichever is the greater, up to a maximum of 115 m^3.
  - for any other kind of house 50 m^3 or 15% whichever is the greater, up to a maximum of 115 m^3.

The volume is calculated from external measurement.

Do I need to apply for planning permission to alter a roof?

You will need to apply for planning permission if you live in a Conservation Area, a national park, an area of outstanding natural beauty or the Norfolk Broads and you want to build an extension to the roof of your house or any kind of addition that would materially alter the shape of the roof.

Roofs are expected to match those of the surrounding area, so consider this if you live in a protected area. Some areas require that the colour and style of the roof covering matches the original, and the pitch and construction should be the same. If you plan to save the expense of matching the roof, by opting for a flat roof, be sure that your local authority will accept this. Often high flat roofs are not desirable, due to the appearance of the house elevation. Provided the alterations to your roof do not make a noticeable change or don't increase its height you would normally not need to obtain planning permission.

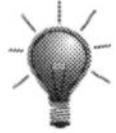

You do not normally need to apply for planning permission to re-roof your house or for the insertion of roof lights or skylights.

In the case of re-roofing, if the tiles are the same type then no approval is needed. If the new tiling or roofing material is substantially heavier or lighter than the existing material, or if the roof is thatched or is to be thatched where previously it was not, then an approval under Building Regulations is probably required.

5.21 Building an extension

Planning permission		Building Regulation approval	
Possibly	You can extend your house by building an extension, provided that the total of both previous and new extensions does not exceed the permitted volume	Yes	If area exceeds 30 m^2 (35.9 y^2) **Note:** Schedule 2 of the regulation is currently under consultation as there is a view that porches, conservatories, covered ways and carports under 30 m^2 should be subject to some regulations. However this currently **does not** stipulate that any amendments will also apply to extensions

Do I need approval to build an extension to my house?

Yes – if it would '*materially alter the appearance of the building*'.

Major alteration and extension nearly always need approval. However some small extensions such as porches, garages and conservatories may be 'Permitted Development' and, therefore, do not need planning consent.

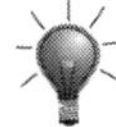

Building extensions can be a potential minefield and it is best to consult the local planning officer before contemplating any work.

If the answer to all the following questions is **no**, then it is quite likely that planning permission will not be required to build an extension to your house:

- Is the extension going to be used as a separate dwelling, i.e. a self-contained flat?
- Is the property listed?
- Will the extension cover more than half of the original garden space?
- Will any part of the extension within 2 m of the plot boundary be more than 4 m above ground level?
- Will any part of the extension be higher than the original roof of the main building?
- Will any part of the extension be nearer to a service road, public road or footpath than any part of the original building?
- Will the volume of the original house be increased beyond the following limits?

- If the house is in a terrace, a Conservation Area, a national park, or an area of outstanding natural beauty – 50 m^3 (or 10% whichever is the greater) up to maximum of 115 m^3.
- For any other kind of house, 70 m^3 (or 15% whichever is the greater) up to a maximum of 115 m^3.

The volume is calculated from external measurements.

Note: If a building is extended, or undergoes a material alteration, the completed building must comply with the relevant requirements of the Approved Documents or, where this is not feasible, be '*no more unsatisfactory than before*'.

Note: the area of windows, roof windows and doors in extensions should **not** be greater than 25% of the floor area of the extension **plus** the area of any windows or doors, which, as a result of the extension works, no longer exist or are no longer exposed.

You **will** need to apply for planning permission to extend or add to your house. You may also require planning permission if your house has previously been added to or extended. You may also require planning permission if the original planning permission for your house imposed restrictions on future development. (i.e. permitted development rights may have been removed by an 'Article 4 direction'. This is often the case with more recently constructed houses.)

You **will** also require planning permission if you want to make additions or extensions to a flat or maisonette.

Check with your local authority planning department if you are not sure.

You will, therefore, need to apply for planning permission:

- if an extension to your house comes within 5 m of another building belonging to your house (i.e. a garage or shed). The volume of that building counts against the allowance given above.
- for all additional buildings which are more than 10 m^3 in volume, if you live in a Conservation Area, a national park, an area of outstanding natural beauty or the Norfolk Broads. Wherever they are in relation to the house, these buildings will be treated as extensions of the house and reduce the allowance for further extensions.
- for a terraced house, end-of-terrace house, or any house in a Conservation Area, national park, an area of outstanding natural beauty or the Broads – where the volume of the original house would be increased by more than 10% or 50 m^3 (whichever is the greater).
- for any other type of house (i.e. detached or semi-detached) the volume of the original house would be increased by more than 15% or 70 m^3 (whichever is the greater). In any case the volume of the original house would be increased by more than 115 m^3.
- for alterations to the roof, including dormer windows (but permission is not normally required for skylights).

- for extensions nearer to a highway than the nearest part of the original house (unless the house, as extended, would be at least 20 m away from the highway).
- to extend or add to your house so as to create a separate dwelling, such as self-contained living accommodation or a granny flat.
- if an extension to your house comes within 5 m of another building belonging to your house.
- to build an addition, which would be nearer to any highway than the nearest part of the original house, unless there would be at least 20 m between your house (as extended) and the highway. The term 'highway' includes all public roads, footpaths, bridleways and byways.
- if more than half the area of land around the original house would be covered by additions or other buildings – although you may not have built an extension to the house, a previous owner may have done so.
- if the extension or addition exceeds the certain limits on height and volume.
- if the extension is higher than the highest part of the roof of the original house or any part of the extension is more than 4 m high and is within 2 m of the boundary of your property.

Any building which has been added to your property and which is more than 10 m^3 in volume and which is within 5 m of your house is treated as an extension of the house and so reduces the allowance for further extensions without planning permission.

Where the word 'original' is used above, in planning regulations terms this means the house as it was first built or as it was on 1 July 1948. Any extensions added since that date are counted towards the allowances.

Limitations

Planning permission is required if:

- **Height** – Any part is higher than the highest part of the house roof.
- **Projections** – Any part projects beyond the foremost wall of the house facing a highway.
- **Boundary** – Any part within 2 m (6 ft 6″) of a boundary is more than 4 m (13 ft) high.
- **Area** – It will cover more than half the original area of the garden.
- **Dwelling** – It is to be an independent dwelling.

You should measure the height of buildings from the ground level immediately next to it. If the ground is uneven, you should measure from the highest part of the surface, unless you are calculating volume.

- **Volume** – Planning permission is required if the extension results in an increase in volume of the original house by whichever is the greater of the following amounts:
  - for terraced houses 50 m^3 (65.5 y^3) or 10% up to a maximum of 115 m^3 (150.4 y^3);

- other houses 70 m^3 (91.5 y^3) or 15% up to a maximum of 115 m^3 (150.4 y^3);
- in Scotland, general category 24 m^2 (28.7 y^2) or 20%.

The volume of other buildings which belong to your house (such as a garage or shed) will count against the volume allowances. In some cases, this can include buildings that were built at the same time as the house or that existed on 1 July 1948.

5.21.1 Extensions to non-domestic buildings

With the new revision of Part M, an extension to a non-domestic building should now be treated in the same manner as a new building for compliance, which means that:

- there must be '*suitable independent access to the extension where reasonably practicable*';
- if a building is to be extended, '*reasonable provision must be made within the extension for sanitary conveniences*'.

Note: This requirement does not apply if it is possible for people using the extension to gain access to and be able to use sanitary conveniences in the existing building.

If a building has a total useful floor area greater than 1000 m^2 and the proposed building work includes:

- an extension; or
- the initial provision of any fixed building services; or
- an increase to the installed capacity of any fixed building services;

then 'consequential improvements' should be made to improve the energy efficiency of the whole building. These will include:

- upgrading all thermal units which have a high U-value;
- replacing all existing windows (less display windows), roof windows, rooflights or doors (excluding high-usage entrance doors) within the area served by the fixed building service with an increased capacity;
- replacing any heating system that is more than fifteen years old;
- replacing any cooling system that is more than fifteen years old;
- replacing any air handling system that is more than fifteen years old;
- upgrading any general lighting system that serves an area greater than 100 m^2 which has an average lamp efficacy of less than 40 lamp-lumens per circuit watt;
- installing energy metering;
- upgrading existing LZC energy systems if they provide less than 10% of the building's energy demand.

5.22 Conversions

Planning permission		Building Regulation approval	
Yes	For flats – even where construction works may not be intended	Yes	Unless you are not proposing any building work to make the change
Yes	For shops and offices unless no building work is envisaged		

Do I need approval to convert my house into flats?

Yes – even where construction works may not be intended.

Do I need approval to convert my house to a shop or office?

No – if you are not proposing any building work to make the change.

Do I need approval to convert part or all of my shop or office to a flat or house?

Yes.

Where building work is proposed you will probably need approval if it affects the structure or means of escape in case of fire. But you should check with the local fire authority and the county council, to see whether a fire certificate is actually required.

You will probably also need planning permission whether or not building work is proposed.

5.22.1 Converting an old building

Planning permission	Building Regulation approval
Yes	Yes

Do I need planning permission to convert an old building?

Yes.

Throughout the UK there are many under-used or redundant buildings, particularly farm buildings which may no longer be required, or suitable, for agricultural use. Such buildings of weathered stone and slate contribute substantially to the character and appearance of the landscape and the built environment. Their interest and charm stems from an appreciation of the functional requirements of the buildings, their layout and proportions, the type of building materials used and their display of local building methods and skills.

In most cases traditional buildings are best safeguarded if their original use can be maintained. However with changing patterns of land use and farming methods, changes of use or conversion may have to be considered.

The conversion or re-use of traditional buildings may, in the right locations, assist in providing employment opportunities, housing for local people, or holiday accommodation. Applicants and developers are encouraged to refer to the local plan for comprehensive guidance and to seek advice from a Planning Officer if further assistance is necessary.

All councils place the highest priority to good design and proposals. Those that fail to respect the character and appearance of traditional buildings, will not be permitted. Sensitive conversion proposals should ensure that existing ridge and eaves lines are preserved; new openings avoided as far as possible; traditional matching materials are used; and the impact of parking and garden areas is minimized. Buildings that are listed as being of 'special architectural or historic interest' require skilled treatment to conserve internal and external features.

In many instances, traditional buildings that are of simple, robust form with few openings may only be suitable for use as storage or workshops. Other uses, such as residential, may be inappropriate.

5.23 Change of use

Planning permission		Building Regulation approval
Possibly	Even if no building or engineering work is proposed	Yes

The use of buildings or land for a different purpose may need consent even if no building or engineering works are proposed. Again, it is always advisable to check with development control staff.

What is meant by material change of use?

A material change of use is where there is a change in the purposes for which or the circumstances in which a building is used, so that after that change:

- *(a) the building is used as a dwelling, where previously it was not;*
- *(b) the building contains a flat, where previously it did not;*
- *(c) the building is used as an hotel or a boarding house, where previously it was not;*
- *(d) the building is used as an institution, where previously it was not;*
- *(e) the building is used as a public building, where previously it was not;*
- *(f) the building is not a building described in Classes I to VI in Schedule 2, where previously it was;*
- *(g) the building, which contains at least one dwelling, contains a greater or lesser number of dwellings than it did previously;*
- *(h) the building contains a room for residential purposes, where previously it did not;*
- *(i) the building, which contains at least one room for residential purposes, contains a greater or lesser number of such rooms than it did previously; or*
- *(j) the building is used as a shop, where previously it was not.*

'Public building' means a building consisting of or containing:

- a theatre, public library, hall or other place of public resort;
- a school or other educational establishment;
- a place of public worship.

Material changes of use

Where there is a material change of use of a **whole** building to a hotel, boarding house, institution, public building or a shop (restaurant, bar or public house) the building must be upgraded, if necessary, so as to comply with Approved Document M1 (Access and use).

If an existing building undergoes a change of use so that **part** of it can be used as a hotel, boarding house, institution, public building or a shop, the work being carried out must ensure that:

- people can gain access from the site boundary and any on-site car parking space;
- sanitary conveniences are provided in that part of the building or it is possible for people (no matter their disability) to use sanitary conveniences elsewhere in the building.

When a building is subject to a material change of use, then:

- any thermal element that is being retained should be upgraded;
- any existing window (including roof window or rooflight) or door which separates a conditioned space from an unconditioned space (or the external environment) and which has a U-value that is worse than 3.3 W/m·K, should be replaced.

Material alterations

Material alterations (i.e. where work, or any part of it, would result in a building or controlled service or fitting not complying with a relevant requirement where previously it did, or making previous compliance more unsatisfactory) should, in order to comply with the requirements for conservation of heat and energy as follows:

Material alterations (domestic buildings)

If a building is subject to a material alteration by:

- substantially replacing a thermal element;
- renovating a thermal element;
- making an existing element part of the thermal envelope of the building (where previously it was not);

- providing a controlled fitting;
- providing (or extending) a controlled service; then:

in addition to the requirements of Part L, all applicable requirements from the following Approved Documents must be taken into account:

- Part A (structure);
- Paragraph B1 (means of warning and escape);
- Part B2 (internal fire spread-linings);
- Paragraph B3 (internal fire spread – structure);
- Paragraph B4 (external fire spread);
- Paragraph B5 (access and facilities for the fire service);
- Part M (access to and use of buildings).

Material alterations (buildings other than dwellings)

When an existing element becomes part of the thermal element of a building (where previously it did not) and it has a U-value worse than $3.3\,W/m^2{\cdot}K$ it should be replaced (unless they are display windows or high usage doors).

Extensions, material alterations or a material change of use

Where any electrical installation work is classified as an extension, a material alteration or a material change of use, the work must consider and include:

- confirmation that the mains supply equipment is suitable and can carry the additional loads envisaged;
- the rating and the condition of existing equipment (belonging to both the consumer and the electricity distributor) are sufficient;
- the amount of additions and alterations that will be required to the existing fixed electrical installation in the building;
- the necessary additions and alterations to the circuits which feed them;
- the protective measures required to meet the requirements;
- the earthing and bonding systems are satisfactory and meet the requirements.

Note: Appendix C to Part P of the Building Regulations offers guidance on some of the older types of installations that might be encountered during alteration work and Appendix D provides guidance on the application of the now harmonized European cable identification system.

What are the requirements relating to material change of use?

Where there is a material change of use of the whole of a building, any work carried out shall ensure that the building complies with the applicable requirements of the following paragraphs of Schedule 1:

(a) in all cases:

- means of warning and escape (B1)
- internal fire spread – linings (B2)
- internal fire spread – structure (B3)

- external fire spread – roofs (B4)(2)
- access and facilities for the fire service (B5)
- resistance to moisture (C1)(2)
- dwelling-houses and flats formed by material change of use (E4)
- ventilation (F1)
- sanitary conveniences and washing facilities (G1)
- bathrooms (G2)
- foul water drainage (H1)
- solid waste storage (H6)
- combustion appliances (J1, J2 & J3)
- conservation of fuel and power – dwellings (L1)
- conservation of fuel and power – buildings other than dwellings (L2)
- electrical safety (P1, P2).

In the case of a building exceeding 15 metres in height:

- external fire spread – walls (B4–(1)).

(b) in other cases:

Material change of use	Requirement	Approved Document
The building is used as a dwelling, where previously it was not	Resistance to moisture	C2 E1, E2, E3
The public building consists of a new school	Acoustic conditions in schools	E4
The building contains a flat, where previously it did not	Resistance to the passage of sound	E1, E2 & E3
The building is used as a hotel or a boarding house, where previously it was not	Structure	A1, A2 & A3 E1, E2, E3
The building is used as an institution, where previously it was not	Structure	A1, A2 & A3
The building is used as a public building, where previously it was not		A1, A2 & A3 E1, E2, E3
The building is not a building described in Classes I to VI in Schedule 2, where previously it was	Structure	A1, A2 & A3
The building, which contains at least one room for residential purposes, contains a greater or lesser number of dwellings than it did previously	Structure	A1, A2 & A3 E1, E2, E3
The building, which contains at least one dwelling, contains a greater or lesser number of dwellings than it did previously	Resistance to the passage of sound	E1, E2 & E3

In some circumstances (particularly when a historic building is undergoing a material change of use and where the special characteristics of the building need to be recognized) it may **not** be practical to improve sound insulation to the standards set out in Part E1 or resistance to contaminants and water as set out in Part C. In these cases, the aim should be to improve the insulation and resistance where it is practically possible – always provided that the work does not prejudice the character of the historic building, or increase the risk of long-term deterioration to the building fabric and/or fittings.

Note: BS 7913:1998 *The principles of the conservation of historic buildings* provides guidance on the principles that should be applied when proposing work on historic buildings.

Mixed use development

In mixed use developments the requirements of the Regulations may differ depending on whether it is part of a building used as a dwelling or part of a building which has a non-domestic use. In these cases the requirements for non-domestic use shall apply in any shared parts of the building.

5.23.1 Buildings suitable for conversion

Most local plans stipulate that conversion proposals '*should relate to buildings of traditional design and construction which enhance the natural beauty of the landscape' as opposed to 'non-traditional buildings, buildings of inappropriate design, or buildings constructed of materials which are of a temporary nature*'.

Isolated buildings

Planning permission will not normally be granted for the conversion or re-use of isolated buildings. Exceptionally, permission may be given for such buildings to be used for small-scale storage or workshop uses or for camping purposes.

An isolated building is normally:

- *a building, or part of a building, standing alone in the open countryside; or*
- *a building, or part of a building, comprised within a group which otherwise occupies a remote location having regard to the disposition of other buildings within the locality, to the character of the surroundings, and to the nature and availability of access and essential services.*

Assessing whether or not a particular building should be regarded as isolated may not always be straightforward and, in such instances early discussion with a planning officer at the national park authority is advised.

Structural condition

Buildings proposed for conversion should be large enough to accommodate the proposed use without the necessity for major alterations, extension or re-construction. In cases of doubt regarding the structural condition of any particular building, the authority will require the submission of a full structural survey to accompany a planning application. The authority can advise on this requirement and, if necessary, on persons who are suitably qualified to undertake such work and who practise locally.

Planning permission will not normally be granted for re-construction if substantial collapse occurs during work on the conversion of a building.

A list of local consulting engineers can be found in *Yellow Pages*.

Workshop conversions

Redundant farm buildings and buildings of historic interest are often well suited to workshop use and such conversions normally require minimal alterations. Potential problems of traffic generation and unneighbourliness can usually be addressed by the imposition of appropriate conditions.

The local authority will generally favourably consider proposals that make good use of traditional buildings by promoting local employment opportunities. In some instances grants may be available from other agencies to assist the conversion of buildings to workshop use.

Residential conversions

When reviewing proposals for converting a traditional building, the local authority will pay particular attention to the overall objectives of the housing policies of the local plan. If land that can be used for a new housing development is limited, residential conversions can make a valuable contribution to the local housing stock and support the social and economic well being of rural communities.

The local plan will require that residential conversions should, in most instances, contribute to the housing needs of the locality. Permission for such conversions are, in some districts, only granted subject to a condition restricting occupancy to local persons.

'Local persons' are normally defined as persons working, about to work, or having last worked in the locality or who have resided for a period of three years within the locality.

Renovation

Districts dedicate some areas as Environmentally Sensitive Areas (ESAs) and grants may be available towards the cost of renovating historical and important local buildings that have fallen into disrepair or towards the cost of renovation works to retain agricultural buildings in farming use, so as to retain their importance as landscape features. Further advice on the workings of the scheme may be obtained from the ESA project officers or the authority's building conservation officer.

Applicants are strongly advised to employ qualified architects or designers in preparing conversion proposals. Informal discussions with a planning officer at an early stage in considering design solutions are also encouraged.

5.24 Building a new house

Planning permission	Building Regulation approval
Yes	Yes

Do I need planning permission to erect a new house?

Yes.

All new houses or premises of any kind require planning permission.

Private individuals will normally only encounter this if they intend to buy a plot of land to build on, or buy land with existing buildings that they want to demolish to make way for a new property to be built.

In all cases like this, unless you are an architect or a builder, you **must** seek professional advice. If you are using a solicitor to act on your behalf in purchasing a plot on which to build, he will include the planning questions within all the other legal work, as well as investigating the presence of covenants, existing planning consent together with other constraints or conditions.

The architect, surveyor or contractor you hire will then need to take into account the planning requirements as part of their planning and design procedures. They will normally handle planning applications for any type of new development.

If you are hiring a professional (or more than one – say a building contractor to do the work and a surveyor or architect to plan and design) be sure to find out exactly who does what and that approval is obtained before going to too much expense, should a refusal arise.

5.25 Infilling

Planning permission		Building Regulation approval	
Possibly	Consult your local planning officer	Yes	If a new development

Can I use an unused, but adjoining, piece of land to build a house (e.g. build a new house on land that used to be a large garden)?

Often there may be no official grounds for denying consent, but residents and individuals can impose quite some delay. It is worth testing the likelihood of a successful application by talking to the neighbours and judging opinions.

Planning consent is often quite difficult to obtain in these cases as this sort of development normally causes a lot of opposition as it is in a settled residential area and people do not like change.

New developments will undoubtedly also need to follow building regulations. This, and all site visits from inspectors, is normally arranged by your building contractor.

There are plenty of substantial building projects that don't require any planning permission. However, it is undoubtedly a good idea to consult a range of people before you consider any work.

5.26 Demolition

Planning permission		Building Regulation approval	
Yes	If it is a listed building or in a Conservation Area	No	For a complete detached house
	If the whole house is to be demolished	Yes	For a partial demolition to ensure that the remaining part of the house (or adjoining buildings/extensions) are structurally sound

You must have good reasons for knocking a building down, such as making way for rebuilding or improvement (which in most cases would be incorporated in the same planning application). Penalties are severe for demolishing something illegally.

You do not need to make a planning application to demolish a listed building or to demolish a building in a Conservation Area. However, you may need listed building or Conservation Area consent.

Elsewhere, you will **not** need to apply for planning permission:

- to demolish a building such as a garage or shed of less than 50 m^3; or
- if the demolition is urgently necessary for health and safety reasons; or
- if the demolition is required under other legislation; or
- where the demolition is on land that has been given planning permission for redevelopment; or
- to demolish a gate, fence, wall or other means of enclosure.

In all other cases, such as demolishing a house or block of flats, the council may wish to agree the details of how you intend to carry out the demolition and how you propose to restore the site afterwards. You will need to apply for a formal decision on whether the council wishes to approve these details. This is called a 'prior approval application' and your council will be able to explain what it involves.

You are not allowed to begin any demolition work (even on a dangerous building) unless you have given the local authority notice of your intention and this has either been acknowledged by the local authority or the relevant notification period has expired. In this notice you will have to:

- specify the building to be demolished;
- state the reason(s) for wanting to demolish it;
- show how you intend to demolish it.

Copies of this notice will have to be sent to:

- the local authority;
- the occupier of any building adjacent to the building;
- British Gas;
- the area electricity board in whose area the building is situated.

This regulation does not apply to the demolition of an internal part of an occupied building, or a greenhouse, conservatory, shed or prefabricated garage (that forms part of that building) or an agricultural building defined in Section 26 of the General Rate Act 1967.

5.26.1 What about dangerous buildings? (Building Act 1984 Sections 77 and 78)

If a building, or part of a building or structure, is in such a dangerous condition (or is used to carry loads that would make it dangerous) then the local authority may apply to a magistrates' court to make an order requiring the owner:

- to carry out work to avert the danger;
- to demolish the building or structure, or any dangerous part of it, and remove any rubbish resulting from the demolition.

The local authority can also make an order restricting its use until such time as a magistrates' court is satisfied that all necessary works have been completed.

These works are controllable by the local authority under Sections 77 and 78 of the Building Act 1984. In inner London the legislation is under the London Building (Amendment) Act 1939.

This involves responding to all reported instances of dangerous walls, structures and buildings within each local authority's area on a 24 hour 365 days a year basis.

Refer to the relevant local authority building control office during office hours or their local authority emergency switchboard, out of hours.

If the building or structure poses a potential danger to the safety of people, the local authority will take the appropriate action to remove the danger. The local authority has powers to require the owners of buildings or structures to remedy the defects or they can direct their own contractors to carry out works to make the building or structure safe. In addition, the local authority may provide advice on the structural condition of buildings during fire fighting to the fire brigade.

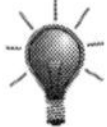

If you are concerned that a building or other structure may be in a dangerous condition, then you should report it to the local council.

Emergency measures

In emergencies the local authority can make the owner take immediate action to remove the danger, or they can complete the necessary action themselves. In these cases, the local authority is entitled to recover from the owner such expenses reasonably incurred by them. For example:

- fencing off the building or structure;
- arranging for the building/structure to be watched.

5.26.2 Can I be made to demolish a dangerous building? (Building Act 1984 Sections 81, 82 and 83)

If the local authority considers that a building is so dangerous that it should be demolished, they are entitled to issue a notice to the owner requiring the owner/occupier:

- to shore up any building adjacent to the building to which the notice relates;
- to weatherproof any surfaces of an adjacent building that are exposed by the demolition;
- to repair and make good any damage to an adjacent building caused by the demolition or by the negligent act or omission of any person engaged in it;
- to remove material or rubbish resulting from the demolition and clear the site;
- to disconnect, seal and remove any sewer or drain in or under the building;
- to make good the surface of the ground that has been disturbed in connection with this removal of drains etc.;
- in accordance with the Water Act 1945 (interference with valves and other apparatus) and the Gas Act 1972 (public safety), arranging with the relevant statutory undertakers (e.g. water board, British Gas or electricity supplier) for the disconnection of gas, electricity and water supplies to the building;
- to leave the site in a satisfactory condition following completion of all demolition work.

Before complying with this notice, the owner must give the local authority 48 hours' notice of commencement.

In certain circumstances, the owner of an adjacent building may be liable to assist in the cost of shoring up their part of the building and waterproofing the surfaces. It could be worthwhile checking this point with the local authority!

Under Section 80 of the Building Act 1984 anyone carrying out demolition work is required to notify the local authority. The local authority then has 6 weeks to respond with appropriate notices and consultation under Sections 81 and 82 of the Act (this does not apply to inner London).

Replacing a demolished building

If you decide to demolish a building, even one that has suffered fire or storm damage, it does not automatically follow that you will get planning permission to build a replacement.

6

Meeting the requirements of the Building Regulations

Background

The Building Regulations 2000 as amended by the Building Amendment Regulations 2001 (SI 2001/3335) replaced the Building Regulations 1991 (SI 1985 No. 1065). Since then, a series of Approved Documents have been endorsed by the Secretary of State that are intended to provide guidance to some of the more common building situations. They also provide a practical guide to meeting the requirements of Schedule 1 and Regulation 7 of the Building Regulations.

Approved Documents

The 2003 list of Approved Documents is given in Table 6.1 below.

Table 6.1 Approved Documents 2004

Section	Title	Edition	Latest amendment
A	Structure	2004	
B	Fire safety	2006	
C	Site preparation and resistance to moisture	2004	
D	Toxic substances	1992	2000
E	Resistance to the passage of sound	2003	2004
F	Ventilation	2006	
G	Hygiene	1992	2000
H	Drainage and waste disposal	2002	
J	Combustion and waste disposal	2002	
K	Protection from falling, collision and impact	1998	2000
L1	Conservation of fuel and power	2006	
M	Access and facilities for disabled people	2004	
N	Glazing – safety in relation to impact, opening and cleaning	1998	2000
P	Electrical safety	2006	
	Approved Document to support Regulation 7 – Materials and workmanship	1999	2000

Note: All of these documents are published by the Stationery Office. For availability and further details, see www.thestationeryoffice.com

Compliance

There is no obligation to adopt any particular solution that is contained in any of these guidance documents especially if you prefer to meet the relevant requirement in some other way. However, should a contravention of a requirement be alleged, if you have followed the guidance in the relevant Approved Documents, that will be evidence tending to show that you have complied with the Regulations. If you have **not** followed the guidance, then that will be seen as evidence tending to show that you have not complied with the requirements and it will then be up to you, the builder, architect and/or client to demonstrate that you have satisfied the requirements of the Building Regulations.

This compliance may be shown in a number of ways such as using:

- a product bearing CE marking (in accordance with the Construction Products Directive (89/106/EEC) as amended by the CE Marking Directive (93/68/EEC) as implemented by the Construction Products Directive 1994 (SI 1994/3051);
- an appropriate technical specification (as defined in the Construction Products Directive – 89/1 06/EEC);
- a recognized British Standard;
- a British Board of Agrément Certificate;
- an alternative, equivalent national technical specification from any member state of the European economic area or Turkey;
- a product covered by a national or European certificate issued by a European Technical Approval issuing body.

Limitation on requirements

Parts A to D, F to K (except for paragraphs H2 and J6), N and P of Schedule 1 do not require anything to be done except for the purpose of securing reasonable standards of health and safety for persons in or about buildings (and any others who may be affected by buildings, or matters connected with buildings).

You may show that you have complied with Regulation 7 in a number of ways, for example, by the appropriate use of a product bearing a CE marking in accordance with the Construction Products Directive (89/106/EEC) as amended by the CE Marking Directive (93/68/EEC), or by following an appropriate technical specification (as defined in that Directive), a British Standard, a British Board of Agrément Certificate, or an alternative national technical specification of any member state of the European Community or Turkey which, in use, is equivalent and/or provides an equivalent level of safety and protection. You will find further guidance in the Approved Document supporting Regulation 7 on materials and workmanship.

Materials and workmanship

As stated in the Building Regulations, '*Any building work which is subject to requirements imposed by Schedule 1 of the Building Regulations should, in accordance with Regulation 7, be carried out with proper materials and in a workmanlike manner*'.

What materials can I use?

Other than the two exceptions below, provided that the materials and components you have chosen to use are from an approved source and are of approved quality (CE marking in accordance with the Construction Products Directive (89/106/EEC), the Low Voltage Directive (73/23/EEC and amendment 93/68 EEC) and the EMC Directive (89/336/EEC) as amended by the CE Marking Directive (93/68/EEC)) then the choice is fairly unlimited.

Short lived materials

Even if a plan for building work complies with the Building Regulations, if this work has been completed using short lived materials (i.e. materials that are, in the absence of special care, liable to rapid deterioration) the local authority can:

- reject the plans;
- pass the plans subject to a limited use clause (on expiration of which they will have to be removed);
- restrict the use of the building.

(Building Act 1984 Section 19)

Unsuitable materials

If, once building work has begun, it is discovered that it has been made using materials or components that have been identified by the Secretary of State (*or his nominated deputy*) as being unsuitable materials, the local authority have the power to:

- reject the plans;
- fix a period in which the offending work must be removed;
- restrict the use of the building.

(Building Act 1984 Section 20)

If the person completing the building work fails to remove the unsuitable material or component(s), then that person is liable to be prosecuted and, on summary conviction, faces a heavy fine.

Technical specifications

Building Regulations may be made for specific purposes such as:

- health and safety;
- welfare and convenience of disabled people;
- conservation of fuel and power;
- prevention of waste or contamination of water.

These are aimed at furthering the protection of the environment, facilitating sustainable development or the prevention and detection of crime.

Although the main requirements for health and safety are now covered by the Building Regulations, there are still some requirements contained in the Workplace (Health, Safety and Welfare) Regulations 1992 that may need to be considered as they could contain requirements which affect building design. For further information see *Workplace (Health, Safety and Welfare) Regulations 1992. Approved Code of Practice L24*, published by HSE Books 1992 (ISBN 0 7176 0413 6).

Standards and technical approvals, as well as providing guidance, also address other aspects of performance such as serviceability and/or other aspects related to health and safety not covered by the Regulations.

When an Approved Document makes reference to a named standard, the relevant version of the standard is the one listed at the end of that particular Approved Document. However, if this version of the standard has been revised or updated by the issuing standards body, the new version may be used as a source of guidance provided it continues to address the relevant requirements of the Regulations.

The Secretary of State has agreed with the British Board of Agrément on the aspects of performance that it needs to assess in preparing its certificates in order that the board may demonstrate the compliance of a product or system that has an Agrément Certificate with the requirements of the Regulations. An Agrément Certificate issued by the board under these arrangements will give assurance that the product or system to which the certificate relates (if properly used in accordance with the terms of the certificate) will meet the relevant requirements.

Independent certification schemes

Within the UK there are many product certification schemes that certify compliance with the requirements of a recognized standard or document that is suitable for the purpose and material being used. Certification Bodies which approve such schemes will normally be accredited by UCAS.

Standards and technical approvals

Standards and technical approvals provide guidance related to the Building Regulations and address other aspects of performance such as serviceability or aspects which, although they relate to health and safety, are not covered by the Regulations.

European pre-standards (ENV)

The British Standards Institution (BSI) will be issuing Pre-standard (ENV) Structural Eurocodes as they become available from the European Standards Organisation, Comité Europeen de Normalisation Electrotechnique (CEN).

DD ENV 1992-1-1: 1992 Eurocode 2: Part 1 and DD ENV 1993-1-1: 1992 Eurocode 3: Part 1-1 *General Rules* and *Rules for Buildings in concrete and steel* have been thoroughly examined over a period of several years and are considered to provide appropriate guidance when used in conjunction with their national application documents for the design of concrete and steel buildings respectively.

When other ENV Eurocodes have been subjected to a similar level of examination they may also offer an alternative approach to Building Regulation compliance and, when they are eventually converted into fully approved EN standards, they will be included as referenced standards in the guidance documents.

Note: If a national standard is going to be replaced by a European harmonized standard, then there will be a coexistence period during which either standard may be referred to. At the end of the coexistence period the national standard will be withdrawn.

House – construction

There are two main types of buildings in common use today: those made of brick and those made of timber. There are many different styles of brick-built houses and, equally there are various methods of construction.

Brickwork, as well as giving a building character, provides the main load bearing element of a brick-built house. Timber-framed houses, on the other hand, are usually built on a concrete foundation with a 'strip' or 'raft' construction to spread the weight and differ from their brick-built counterparts in that the main structural elements are timber frames.

6.1 Foundations

To support the weight of the structure, most brick-built buildings are supported on a solid base called foundations. Timber framed houses are usually built on a concrete foundation with a 'strip' or 'raft' construction to spread the weight.

6.1.1 Requirements

The building shall be constructed so that:

- *the combined dead, imposed and wind loads are sustained and transmitted by it to the ground, safely and without causing any building deflection/deformation or ground movement that will affect the stability of any part of the building;*
- *ground movement caused by swelling, shrinkage or freezing of the subsoil; land-slip or subsidence will not affect the stability of any part of the building.*

(Approved Document A)

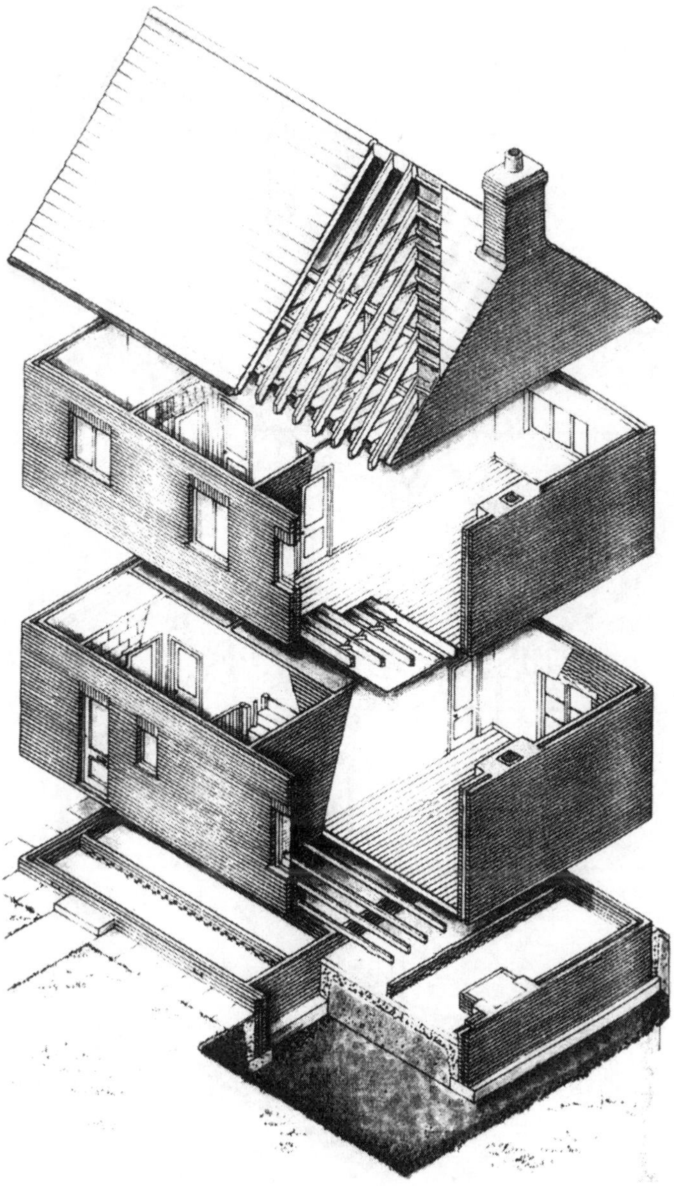

Figure 6.1 Brick built house – typical components

(1) *The ground to be covered by the building shall be reasonably free from any material that might damage the building or affect its stability, including vegetable matter, topsoil and pre-existing foundations.*

(2) *Reasonable precautions shall be taken to avoid danger to health and safety caused by contaminants on or in the ground covered, or to be covered by the building and any land associated with the building.*

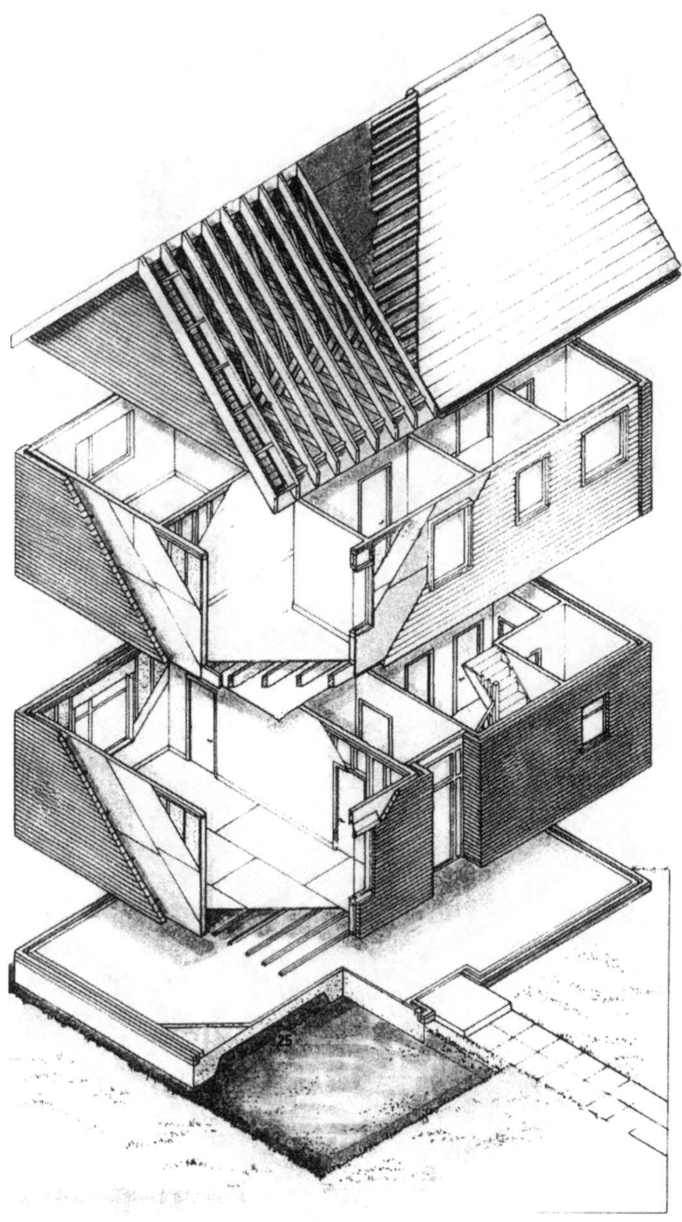

Figure 6.2 Timber framed house – typical components

(3) *Adequate subsoil drainage shall be provided if it is needed to avoid:*
   (a) *the passage of ground moisture to the interior of the building;*
   (b) *damage to the building, including damage through the transport of water-borne contaminants to the foundations of the building.*

(Approved Document C1)

Note: For the purpose of this requirement, 'contaminant' means any substance which is or may become harmful to persons or buildings including substances, which are corrosive, explosive, flammable, radioactive or toxic.

Potential problems

There may be known and/or recorded conditions of ground instability, such as geological faults, landslides or disused mines, or unstable strata of similar nature which affect or may potentially affect a building site or its environs.

There may also be:

- unsuitable material including vegetable matter, topsoil and pre-existing foundations;
- contaminants on or in the ground covered, or to be covered, by the building and any land associated with the building; and
- groundwater.

These conditions should be taken into account before proceeding with the design of a building or its foundations.

What about hazards?

Hazards associated with the ground may include:

- chemical and biological contaminants;
- gas generation from biodegradation of organic matter;
- naturally occurring radioactive radon gas and gases produced by some soils and minerals;
- physical, chemical or biological;
- underground storage tanks or foundations;
- unstable fill or unsuitable hardcore containing sulphate;
- the effects of vegetable matter including tree roots.

In the most hazardous conditions, only the total removal of contaminants from the ground to be covered by the building can provide a complete remedy. In other cases remedial measures can reduce the risks to acceptable levels. These measures should only be undertaken with the benefit of expert advice and where the removal would involve handling large quantities of contaminated materials, then you are advised to seek expert advice.

Even when these actions have been successfully completed, the ground to be covered by the building will **still** need to have at least 100 mm of concrete laid over it!

What about contaminated ground?

Potential building sites which are likely to contain contaminants can be identified at an early stage from planning records or from local knowledge (e.g. previous uses). In addition to solid and liquid contaminants, problems can also arise from natural contamination such as methane and the radioactive radon gas (and its decay product).

The following list are examples of sites that are most likely to contain contaminants:

- asbestos works;
- ceramics, cement and asphalt manufacturing works;
- chemical works;
- dockyards and dockland;
- engineering works (including aircraft manufacturing, railway engineering works, shipyards, electrical and electronic equipment manufacturing works);
- gas works, coal carbonization plants and ancillary by-product works;
- industries making or using wood preservatives;
- landfill and other waste disposal sites;
- metal mines, smelters, foundries, steelworks and metal finishing works;
- munitions production and testing sites;
- oil storage and distribution sites;
- paper and printing works;
- power stations;
- railway land, especially larger sidings and depots;
- road vehicle fuelling, service and repair: garages and filling stations;
- scrap yards;
- sewage works, sewage farms and sludge disposal sites;
- tanneries;
- textile works and dye works.

If any signs of possible contaminants are present, then the local authority's Environmental Health Officer should be told at once. If he confirms the presence of any of these contaminants (see Table 6.2) then he will require their removal or action to be completed before any planning permission for building work can be sought.

What about gaseous contaminants?

Radon is a naturally occurring radioactive colourless and odourless gas which is formed in small quantities by radioactive decay wherever uranium and radium are found. It can move through the subsoil and then into buildings and exposure to high levels over long periods increases the risk of developing lung cancer. Some parts of the country (in particular the West Country) have higher natural levels than elsewhere and precautions against radon may be necessary.

Note: Guidance on the construction of dwellings in areas susceptible to radon has been published by the Building Research Establishment as a Report ('*Radon: guidance on protective measures for new dwellings*').

Landfill gas is generated by the action of anaerobic micro-organisms on biodegradable material in landfill sites and generally consists of methane and carbon dioxide together with small quantities of VOCs (Volatile Organic Compounds) which give the gas its characteristic odour. It can migrate under pressure through the subsoil and through cracks and fissures into buildings.

Table 6.2 Examples of possible contaminants

Signs of possible contaminants	Possible contaminant
Vegetation (absence, poor or unnatural growth)	Metals Metal compounds Organic compounds Gases (landfill or natural source)
Surface materials (unusual colours and contours may indicate wastes and residues)	Metals Metal compounds Oily and tarry wastes Asbestos Other mineral fibres Organic compounds including phenols Combustible material including coal and coke dust Refuse and waste
Fumes and odours (may indicate organic chemicals)	Volatile organic and/or sulphurous compounds from landfill or petrol/solvent spillage Corrosive liquids Faecal animal and vegetable matter (biologically active)
Damage to exposed foundations of existing buildings	Sulphates
Drums and containers (empty or full)	Various

Methane and carbon dioxide can also be produced by organically rich soils and sediments such as peat and river silts and a wide range of VOCs can be present as a result of petrol, oil and solvent spillages.

Site preparation

Site investigation is now the recommended method for determining how much unsuitable material should be removed before commencing building work and this will normally consist of a number of well-defined stages, for example:

Planning stage	scope and requirements
Desktop study	historical, geological and environmental information about the site
Site reconnaissance or walkover survey	identification of actual and potential physical hazards and the design of the main investigation
Main investigation and reporting	intrusive and non-intrusive sampling and testing to provide soil parameters

Risk assessment

The site investigation may identify certain risks which will require a risk assessment, of which there are three types:

- Preliminary (once the need for a risk assessment has been identified, and depending on the situation and the outcome);

- Generic Quantitative Risk Assessment (GQRA);
- Detailed Quantitative Risk Assessment (DQRA).

Each risk assessment should include a:

Hazard identification	developing the conceptual model by establishing contaminant sources, pathways and receptors (this is the preliminary site assessment which consists of a desk study and a site walkover in order to gather sufficient information to obtain an initial understanding of the potential risks. An initial conceptual model for the site can then be based on this information.)
Hazard assessment	identifying what pollutant linkages may be present and analysing the potential for unacceptable risks.
Risk estimation	establishing the scale of the possible consequences by considering the degree of harm that may result and to which receptors.
Risk evaluation	deciding whether the risks are acceptable or unacceptable – review all site data to decide whether estimated risks are unacceptable.

6.1.2 Meeting the requirement

General

Where the site is potentially affected by contaminants, a combined geotechnical and geo-environmental investigation should be considered.	C1.3

Hazard identification and assessment

A preliminary site assessment is required to provide information on the past and present uses of the site and surrounding area that may give rise to contamination (see Table 6.2).	C2.10
The site assessment and risk evaluation should pay particular attention to the area of the site subject to building operations.	C2.11
The planning authority should be informed prior to any intrusive investigations or if any substance is found which was not identified in a preliminary statement about the nature of the site.	C2.12

Risks to buildings, building materials and services

The following hazards shall be considered:

- aggressive substances – including inorganic and organic acids, alkalis, organic solvents and inorganic chemicals such as sulphates and chlorides; C2.23a
- combustible fill – including domestic waste, colliery spoil, coal, plastics, petrol-soaked ground, etc.; C2.23b
- expansive slags – e.g. blast furnace and steel making slag; C2.23c
- floodwater affected by contaminants – substances in the ground, waste matter or sewage. C2.23d

Contaminated ground

The underlying geology of a potential site has to be considered as natural contaminants may be present, for example: C2.3 and 2.4

- naturally occurring heavy metals (e.g. cadmium and arsenic) originating in mining areas;
- gases (e.g. methane and carbon dioxide) originating in coal mining areas;
- organic rich soils and sediments such as peat and river silts;
- radioactive radon gas – which can also be a problem in certain parts of the country.

Possible sulphate attack from some strata on concrete floor slabs and oversite concrete needs to be considered. C2.5

Gaseous contaminants

Radon

All new buildings, extensions and conversions (whether residential or non-domestic), which are built in areas where there may be high radon emissions, may need to incorporate precautions against radon. C2.39

Landfill gas

Methane is an asphyxiant, will burn, and can explode in air. Carbon dioxide is non-flammable and toxic. Many of the other components of landfill gas are flammable and some are toxic. All will require careful analysis. C2

Risk assessment

A risk assessment should be completed for methane and other gases particularly:

- on a landfill site or within 250 m of the boundary of a landfill site; C2.28a
- on a site subject to the wide scale deposition of biodegradable substances (including made ground or fill); C2.28b
- on a site that has been subject to a use that could give rise to petrol, oil or solvent spillages; C2.28c
- in an area subject to naturally occurring methane, carbon dioxide and other hazardous gases (e.g. hydrogen sulphide). C2.28d

During a site investigation for methane and other gases:

- measurements should be taken over a sufficiently long period of time in order to characterize gas emissions fully; C2.30
- should include periods when gas emissions are likely to be higher, e.g. during periods of falling atmospheric pressure. C2.30

Gas risks (i.e. to human receptors) should be considered for:

- gas entering the dwelling through the substructure (and building up to hazardous levels); C2.32
- subsequent householder exposure in garden areas including outbuildings (e.g. garden sheds and greenhouses), extensions and garden features (e.g. ponds). C2.32

When land that is affected by contaminants is being developed, 'receptors' (i.e. buildings, building materials and building services, as well as people) are introduced onto the site and it is necessary to break the pollutant linkages. This can be achieved by: C2.7

- treating the contaminant (e.g. use of physical, chemical or biological processes to eliminate or reduce the contaminant's toxicity or harmful properties);

- blocking or removing the pathway (e.g. isolating the contaminant beneath protective layers or installing barriers to prevent migration);
- protecting or removing the receptor (e.g. changing the form or layout of the development, using appropriately designed building materials, etc.).

A risk assessment based on the concept of a 'source–pathway–receptor' relationship, or pollutant linkage of a potential site (see Figure 6.3) should be carried out to ensure the safe development of land that is affected by contaminants. C2.6

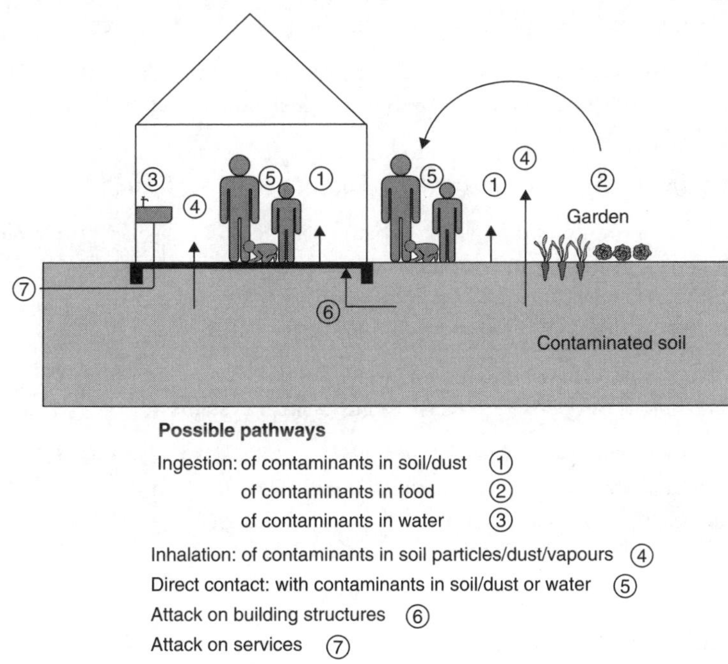

Figure 6.3 Conceptual model of a site showing a source–pathway–receptor

Risk estimation and evaluation

The detailed ground investigation: C2.13

- must provide sufficient information for the confirmation of a conceptual model for the site, the risk assessment and the design and specification of any remedial works;

- is likely to involve collection and analysis of soil, soil gas, surface and groundwater samples by the use of invasive and/or non-invasive techniques.

During the development of land affected by contaminants the health and safety of both the public and workers should be considered. C2.14

Remedial measures

If the risks posed by the gas are unacceptable then these need to be managed through appropriate building remedial measures. C2.36

Site-wide gas control measures may be required if the risks on any land associated with the building are deemed unacceptable. C2.36

Consideration should be given to the design and layout of buildings to maximize the driving forces of natural ventilation. C2.37

For non-domestic buildings, expert advice concerning gas control measures should be sought as the floor area of such buildings can be large and it is important to ensure that gas is adequately dispersed from beneath the floor. C2.38

There is a need for continued maintenance and calibration of mechanical (as opposed to passive) gas control systems. C2.38

Sub-floor ventilation systems should be carefully designed to ensure adequate performance and should not be modified unless subjected to a specialist review of the design. C2.38

Corrective measures

When building work is undertaken on sites affected by contaminants where control measures are already in place, care must be taken not to compromise these measures. C2.15

Depending on the contaminant, three generic types of corrective measures can be considered: treatment, containment and removal.

Note: The containment or treatment of waste may require a waste management licence from the Environmental Agency.

Treatment

The choice of the most appropriate treatment process for a particular site is a highly site-specific decision for which specialist advice should be sought. C2.16

Containment

In-ground vertical barriers may also be required to control lateral migration of contaminants. C2.17

Cover systems involve the placement of one or more layers of materials placed over the site and may be used to: C2.18

- break the pollutant linkage between receptors and contaminants;
- sustain vegetation;
- improve geotechnical properties; and
- reduce exposure to an acceptable level.

Imported fill and soil for cover systems should be assessed at source to ensure that it is not contaminated. C2.20

The size and design of cover systems (particularly soil-based ones used for gardens) should take account of their long-term performance. C2.20

Gradual intermixing due to natural effects and activities such as burrowing animals, gardening, etc., needs to be considered. C2.20

Removal

Imported fill should be assessed at source to ensure that there are no materials that will pose unacceptable risks to potential receptors. C2.21

Site preparation

Vegetable matter such as turf and roots should be removed from the ground that is going to be covered by the building at least to a depth to prevent later growth. C1.4

The effects of roots close to the building need to be assessed. C1.4

Where mature trees are present (particularly on sites with shrinkable clays (see Table 6.3)) the potential damage arising from ground heave to services and floor slabs and oversite concrete should be assessed. C1.5

Table 6.3 Volume change potential for some common clays

Clay type	Volume change potential
Glacial till	Low
London	High to very high
Oxford and Kimmeridge	High
Lower lias	Medium
Gault	High to very high
Weald	High
Mercian mudstone	Low to medium

Building services such as below-ground drainage should be sufficiently robust or flexible to accommodate the presence of any tree roots. C1.6

Joints should be made so that roots will not penetrate them. C1.6

Where roots could pose a hazard to building services, consideration should be given to their removal. C1.6

On sites previously used for buildings, consideration should be given to the presence of other infrastructure (such as existing foundations, services and buried tanks, etc.) that could endanger persons in and about the building and any land associated with the building. C1.7

If the site contains fill or made ground, consideration should be given to its compressibility and its potential to collapse when wet. C1.8

Foundations

Table 6.4 provides guidance on determining the type of soil on which it is intended to lay a foundation.

Subsoil drainage

Where the water table can rise to within 0.25 m of the lowest floor of the building, or where surface water could enter or C3.2

adversely affect the building, either the ground to be covered by the building should be drained by gravity, or other effective means of safeguarding the building should be taken.

If an active subsoil drain is cut during excavation and if it passes under the building it should be either: C3.3

- re-laid in pipes with sealed joints and have access points outside the building; or
- re-routed around the building; or
- re-run to another outfall (see Figure 6.4).

Where contaminants are present in the ground, consideration should be given to subsoil drainage to prevent the transportation of water-borne contaminants to the foundations or into the building or its services. C3.7

Table 6.4 Types of subsoil

Type	Applicable field test
Rock (being stronger/ denser than sandstone, limestone or firm chalk)	Requires at least a pneumatic or other mechanically operated pick for excavation.
Compact gravel and/or sand	Requires a pick for excavation. Wooden peg 50 mm square in cross section hard to drive beyond 150 mm.
Stiff clay or sandy clay	Cannot be moulded with the fingers and requires a pick or pneumatic or other mechanically operated spade for its removal.
Firm clay or sandy clay	Can be moulded by substantial pressure with the fingers and can be excavated with a spade.
Loose sand, silty sand or clayey sand	Can be excavated with a spade. Wooden peg 50 mm square in cross section can be easily driven.
Soft silt, clay, sandy clay or silty clay	Fairly easily moulded in the fingers and readily excavated.
Very soft silt, clay, sandy clay or silty clay	Natural sample in winter conditions exudes between the fingers when squeezed in fist.

Ground movement

Known or recorded conditions of ground instability, such as that arising from landslides, disused mines or unstable strata should be taken into account in the design of the building and its foundations. A1/2 1.9

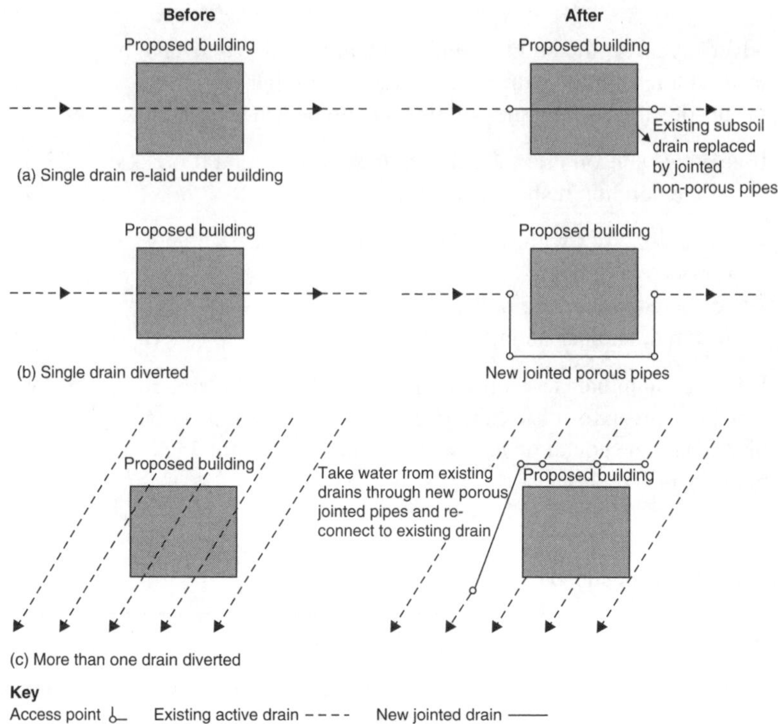

Figure 6.4 Subsoil drain cut during excavation

Foundations – plain concrete

There should **not** be:

- non-engineered fill (see BRE Digest 427) or a wide variation in ground conditions within the loaded area; A1/2 2E1a
- weaker or more compressible ground at such a depth below the foundation as could impair the stability of the structure. A1/2 2E1b

The foundations should be situated centrally under the wall. A1/2 2E2a

In non-aggressive soils, concrete should be composed of Portland cement to BS EN 197 1 & 2 and fine and coarse aggregate conforming to BS EN 12620 and the mix should either be: A1/2 2E2b

- 50 kg of Portland cement to not more than 200 kg (0.1 m^3) of fine aggregate and 400 kg (0.2 m^3) of coarse aggregate, or
- Grade ST2 or Grade GEN I concrete to BS 8500-2.

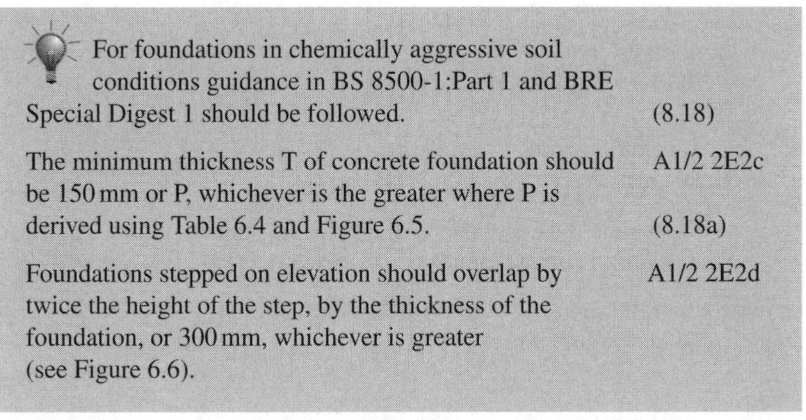

For foundations in chemically aggressive soil conditions guidance in BS 8500-1:Part 1 and BRE Special Digest 1 should be followed. (8.18)

The minimum thickness T of concrete foundation should be 150 mm or P, whichever is the greater where P is derived using Table 6.4 and Figure 6.5. A1/2 2E2c (8.18a)

Foundations stepped on elevation should overlap by twice the height of the step, by the thickness of the foundation, or 300 mm, whichever is greater (see Figure 6.6). A1/2 2E2d

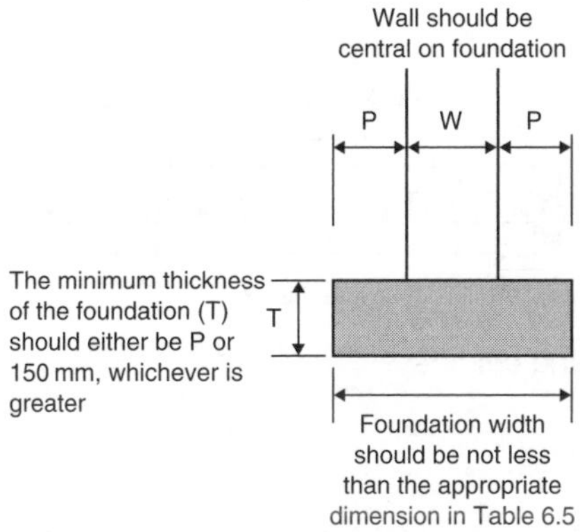

Figure 6.5 Foundation dimensions

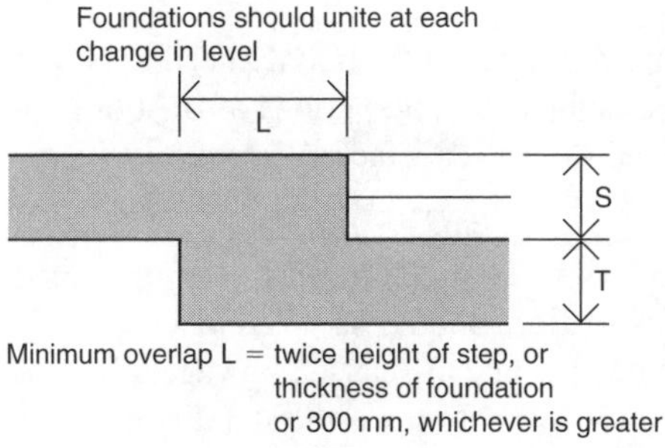

Figure 6.6 Elevation of stepped foundation

The overlap for trench fill foundations should be twice height of step or 1 metre, whichever is greater. A1/2 2E2d

Trench fill foundations may be used as an acceptable alternative to strip foundations. A1/2 2E2c

Steps in foundations should not be of greater height than the thickness of the foundation (see Figure 6.6).

Foundations for piers, buttresses and chimneys should project as shown in Figure 6.7. A1/2 2E2f

The projection X should never be less than the value of P where there is no local thickening of the wall.

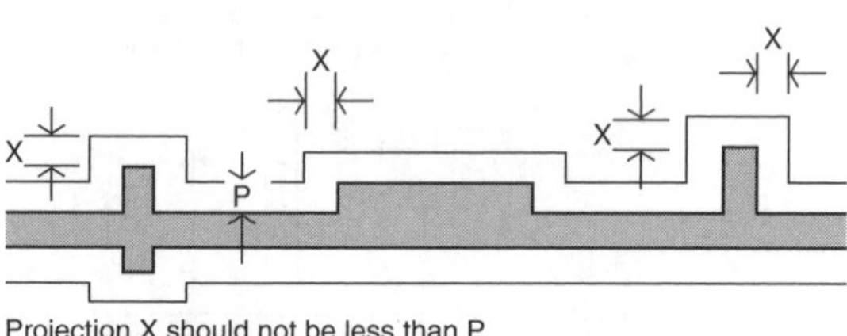

Figure 6.7 Piers and chimneys

Strip foundations

The recommended minimum widths of strip foundations shall be as indicated in Table 6.5. A1/2 2E3

Where strip foundations are founded on rock, the strip foundations should have a minimum depth of 0.45 m to their underside to avoid the action of frost. A1/2 2E4

This depth, however, will commonly need to be increased in areas subject to long periods of frost or in order to transfer the loading onto satisfactory ground.

In clay soils subject to volume change on drying (i.e. 'shrinkable clays' with a Plasticity Index greater than or equal to 10%), strip foundations should be taken to a depth where anticipated ground movements (caused by vegetation and trees on the ground) will not impair the stability of any part of the building. A1/2 2E4

The depth to the underside of foundations on clay soils should not be less than 0.75 m. A1/2 2E4

Although this depth will commonly need to be increased in order to transfer the loading onto satisfactory ground.

Table 6.5 Minimum width of strip footings

Type of ground (including engineered fill)	Condition of ground	Field test applicable	Total load of load-bearing walling not more than (kN/linear metre)					
			20	30	40	50	60	70
			Minimum width of strip foundation (mm)					
I Rock	Not inferior to sandstone, limestone or firm chalk	Requires at least a pneumatic or other mechanically operated pick for excavation	In each case equal to the width of wall					
II Gravel or sand	Medium dense	Requires pick for excavation. Wooden peg 50 mm square in cross-section hard to drive beyond 150 mm	250	300	400	500	600	650
III Clay Sandy clay	Stiff Stiff	Can be indented slightly by thumb	250	300	400	500	600	650
IV Clay Sandy clay	Firm Firm	Thumb makes impression easily	300	350	450	600	750	850
V Sand Silty sand Clayey sand	Loose Loose Loose	Can be excavated with a spade. Wooden peg 50 mm square in cross-section can be easily driven	400	600	**Note:** Foundations on soil types V and VI do not fall within the provisions of this section if the total load exceeds 30 kN/m.			
VI Silt Clay Sandy clay Clay or silt	Soft Soft Soft Soft	Finger pushed in up to 10 mm	450	650				
VII Silt Clay Sandy clay Clay or silt	Very soft Very soft Very soft Very soft	Finger easily pushed in up to 25 mm	Refer to specialist advice					

Disproportionate collapse

All buildings should be built so that their sensitivity to disproportionate collapse in the event of an accident is reduced.

Buildings shall remain sufficiently robust to sustain a limited extent of damage or failure, depending on the class of the building, without collapse (see below). A1/2 5.1

 **Notes:**

(1) Buildings intended for more than one type of use should adopt the most onerous class.
(2) In determining the number of storeys in a building, basement storeys may be excluded provided that they meet the robustness requirements of Class 2B buildings.

Class 1

Building type and occupancy	Requirements
• Houses not exceeding 4 storeys. • Agricultural buildings. • Buildings into which people rarely go, provided no part of the building is closer to another building (or area where people go) than 1.5 times the building height.	Provided the building has been designed and constructed in accordance with Building Regulations and is in normal use, no additional measures are likely to be necessary.

Class 2A

Building type and occupancy	Requirements
• 5 storey single occupancy houses. • Hotels not exceeding 4 storeys. • Flats, apartments and other residential buildings not exceeding 4 storeys. • Offices not exceeding 4 storeys. • Industrial buildings not exceeding 3 storeys. • Retailing premises not exceeding 3 storeys of less than 2000 m^2 floor area in each storey.	Effective horizontal ties (or effective anchorage of suspended floors to walls) is required.

- Single storey educational buildings.
- All buildings not exceeding 2 storeys to which members of the public are admitted and which contain floor areas not exceeding 2000 m^2 at each storey.

Class 2B

Building type and occupancy	Requirements
• Hotels, flats, apartments and other residential buildings greater than 4 storeys but not exceeding 15 storeys. • Educational buildings greater than 1 storey but not exceeding 15 storeys. • Retailing premises greater than 3 storeys but not exceeding 15 storeys. • Hospitals not exceeding 3 storeys. • Offices greater than 4 storeys but not exceeding 15 storeys. • All buildings to which members of the public are admitted which contain floor areas exceeding 2000 m^2 but less than 5000 m^2 at each storey. • Car parking not exceeding 6 storeys.	Effective horizontal ties need to be provided. Effective vertical ties need to be provided in all supporting columns and walls.

Or alternatively check that upon the notional removal of each supporting column and each beam supporting one or more columns, or any nominal length of load-bearing wall (one at a time in each storey of the building) that the building remains stable and that the area of floor at any storey at risk of collapse does not exceed 15% of the floor area of that storey or 70 m^2, whichever is smaller, and does not extend further than the immediate adjacent storeys (see Figure 6.8).

Where the notional removal of such columns and lengths of walls would result in damage in excess of the above limit, then such elements should be designed as a 'key element' (i.e. it should be capable of sustaining an accidental

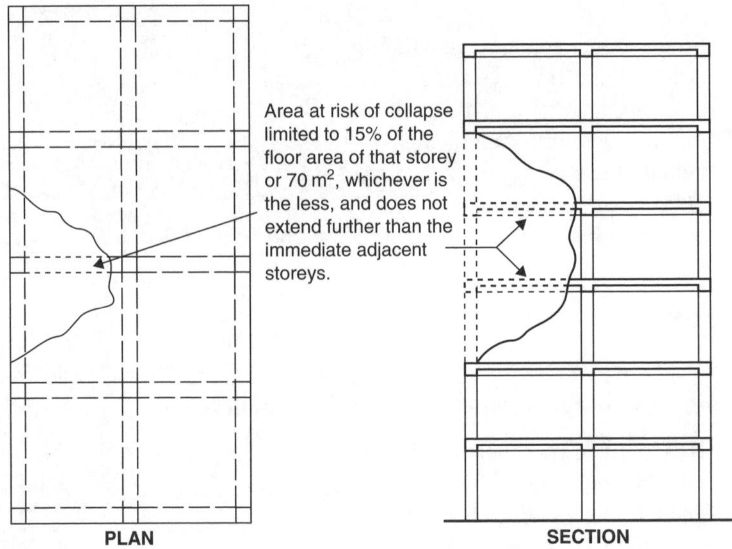
Area at risk of collapse
limited to 15% of the
floor area of that storey
or 70 m², whichever is
the less, and does not
extend further than the
immediate adjacent
storeys.
PLAN
SECTION

Figure 6.8 Area at risk of collapse in the event of an accident

design loading of 34 kN/m²) applied in the horizontal and vertical directions (in one direction at a time) to the member and any attached components (e.g. cladding, etc.).

Class 3

Building type and occupancy	Requirements
• All buildings defined above as Class 2A and 2B that exceed the limits on area and/or number of storeys. • Grandstands accommodating more than 5000 spectators. • Buildings containing hazardous substances and/or processes.	A systematic risk assessment of the building should be undertaken taking into account all the normal hazards that may reasonably be foreseen, together with any abnormal hazards. Critical situations for design should be selected that reflect the conditions that can reasonably be foreseen as possible during the life of the building. Protective measures should be chosen and the detailed design of the structure and its elements

undertaken in accordance with the following recommendations:

- BS 5628: Part 1 – Structural use of unreinforced masonry
- BS 5950: Part 1 – Structural use of steelwork in building
- BS 8110: Parts 1 and 2 – Structural use of plain, reinforced and prestressed concrete.

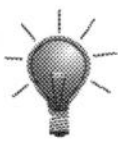

For any building which does not fall into one of the classes listed above, or where the consequences of collapse may warrant particular examination of the risks involved, see one of the following Reports:

'Guidance on Robustness and Provision against Accidental Actions' dated July 1999, together with the accompanying BRE Report No. 200682.
'Calibration of Proposed Revised Guidance on Meeting Compliance with the Requirements of Building Regulation Part A3'.
Both of the above documents are available on the following DCLG website http://www.odpm.gov.uk.

Maximum floor area

No floor enclosed by structural walls on all sides shall exceed 70 m^2 (see Figure 6.9). A1/2 (2C14)

No floor with a structural wall on one side shall exceed 36 m^2 (see Figure 6.9). A1/2 (2C14)

Maximum height of buildings

The maximum height of a building shall not exceed the heights given in Table 6.6 with regard to the relevant wind speed. A1/2 (1C17)

Heights of walls and storeys

The measured height of a wall or a storey should be in accordance with Figure 6.10. A1/2 2C18

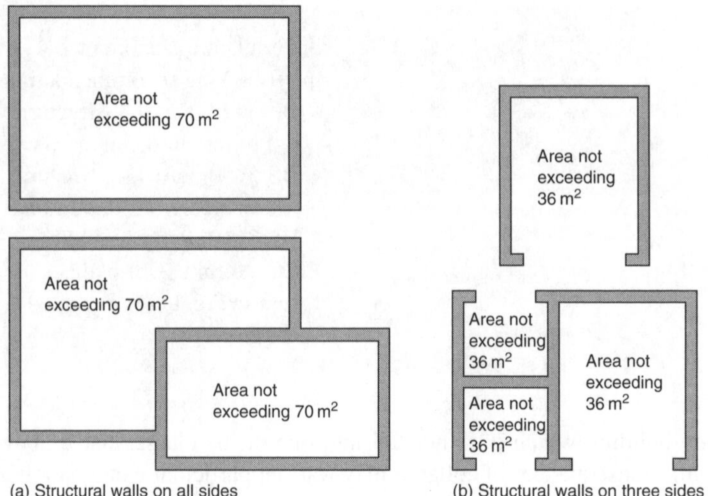

Figure 6.9 Maximum floor area that is enclosed by structural walls

Table 6.6 Maximum allowable building height

Factor S	Country sites			Town sites		
	Distance to the coast			Distance to the coast		
	<10 km	15–50 km	>50 km	<10 km	15–50 km	>50 km
24	15	15	15	15	15	15
25	11.5	14.5	15	15	15	15
26	8	10.5	13	15	15	15
27	6	8.5	10	15	15	15
28	4.5	6.5	8	13.5	15	15
29	3.5	5	6	11	13	14.5
30	3	4	5	9	11	12.5
31		3.5	4	8	9.5	10.5
32		3	3.5	7	8.5	9.5
33			3	6	7.5	8.5
34				5	7	8
35				4	6	7
36				3	5.5	6
37					4.5	5.5
38					4	5
39					3	4
40						3

Imposed loads on roofs, floors and ceilings

The imposed loads on roofs, floors and ceilings shall not exceed those shown in Table 6.7. A1/2 (2C15)

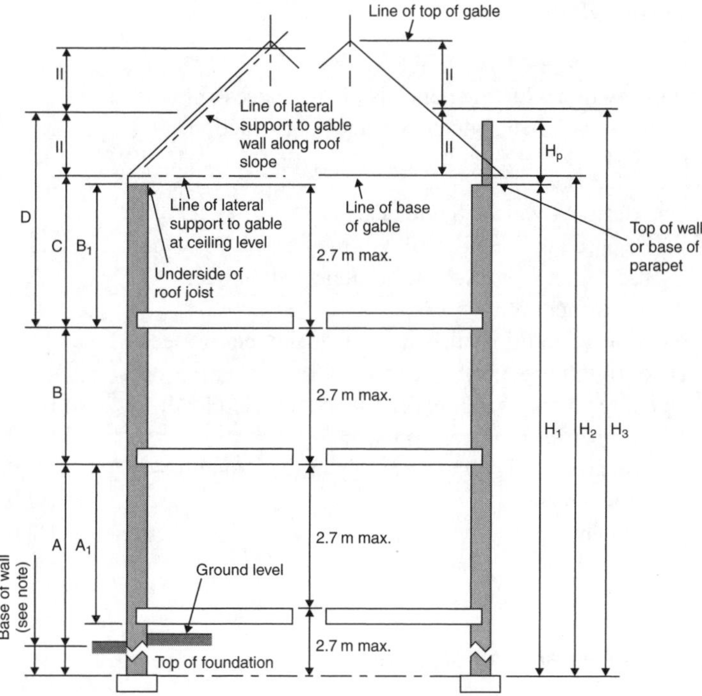

Key

(a) **Measuring Storey Heights**

A_1 is the ground storey height if the ground floor provides effective lateral support to the wall i.e. is adequately tied to the wall or is a suspended floor bearing on the wall.

A is the ground storey height if the ground floor does not provide effective lateral support to the wall.

Note: If the wall is supported adequately and permanently on both sides by suitable compact material, the base of the wall for the purposes of the storey height may be taken as the lower level of this support. (Not greater than 3.7 m ground storey height.)

B is the intermediate storey height.

B_1 is the top storey height for walls which do not include a gable.

C is the top storey height where lateral support is given to the gable at both ceiling level and along the roof slope.

D is the top storey height for the external walls which include a gable where lateral support is given to the gable only along the roof slope.

(b) **Measuring Wall Heights**

H_1 is the height of an external wall that does no include a gable.

H_2 is the height of an internal or separating wall which is built up to the underside of the roof.

H_3 is the height of an external wall which includes a gable.

H_p is the height of a parapet. If Hp is more than 1.2 m add to H_p to H_1.

Figure 6.10 Method for measuring the heights of storeys and walls

Table 6.7 Imposed loads

Element	Distributed loads	Concentrated load
Roofs	1.00 kN/m^2 for spans not exceeding 12 m 1.50 kN/m^2 for spans not exceeding 6 m	
Floors	2.00 kN/m^2	
Ceilings	0.25 kN/m^2	0.9 kN/m^2

Structural safety

The safety of a structure depends on the successful combination of design and completed construction, particularly:

- the design – which should also:
  - be based on identification of the hazards (to which the structure is likely to be subjected) and an assessment of the risks; A1/2 0.2a
  - reflect conditions that can reasonably be foreseen during future use;
- loading – dead load, imposed load and wind load; A1/2 0.2b
- the properties of materials used; A1/2 0.2c
- the detailed design and assembly of the structure; A1/2 0.2d
- safety factors; A1/2 0.2e
- workmanship. A1/2 0.2f

Basic requirements for stability

Adequate provision shall be made to ensure that the building is stable under the likely imposed and wind loading conditions. A1/2 1A2

The overall size and proportioning of the building shall be limited according to the specific guidance for each form of construction. A1/2 1A2a

The layout of walls (both internal and external) forming a robust three-dimensional box structure in plan shall be constructed according to the specific guidance for each form of construction. A1/2 1A2b

The internal and external walls shall be adequately connected by either masonry bonding or by using mechanical connections. A1/2 1A2c

The intermediate floors and roof shall be constructed so that they:

- provide local support to the walls; A1/2 1A2d
- act as horizontal diaphragms capable of transferring the wind forces to buttressing elements of the building.

Note: A traditional cut timber roof (i.e. using rafters, purlins and ceiling joists) generally has sufficient built-in resistance to instability and wind forces (e.g. from either hipped ends, tiling battens, rigid sarking, or the like). However, the need for diagonal rafter bracing equivalent to that recommended in BS 5268: Part 3: 1998 or Annex H of BS 8103: Part 3: 1996 for trussed rafter roofs, should be considered especially for single-hipped and non-hipped roofs of greater than 40° pitch to detached houses.

6.2 Buildings – size

6.2.1 Classification of purpose groups

Many of the provisions in Approved Documents are related to the use of the building. The classifications 'use' are termed purpose groups and represent different levels of hazard. They can apply to a whole building, or (where a building is compartmented) to a compartment in the building and the relevant purpose group should be taken from the main use of the building or compartment. Table 6.8 sets out the purpose group classification.

Table 6.8 Classification of purpose groups

Title	Group	Purpose for which the building or compartment of a building is intended to be used
Residential[(1)] (dwellings)	1(a)	Flat or maisonette.
	1(b)	Dwelling house which contains a habitable storey with a floor level which is more than 4.5 m above ground level.
	1(c)	Dwelling house which does not contain a habitable storey with a floor level which is more than 4.5 m above ground level.
Residential (institutional)	2(a)	Hospital, home, school or other similar establishment used as living accommodation for, or for the treatment, care or maintenance of persons suffering from disabilities due to illness or old age or other physical or mental incapacity, or under the age of five years, or place of lawful detention, where such persons sleep on the premises.
Other	2(b)	Hotel, boarding house, residential college, hall of residence, hostel, and any other residential purpose not described above.
Office	3	Offices or premises used for the purpose of administration, clerical work (including writing, book keeping, sorting papers, filing, typing, duplicating, machine calculating, drawing and the editorial preparation of matter for publication, police and fire service work), handling

Title	Group	Purpose for which the building or compartment of a building is intended to be used
		money (including banking and building society work), and communications (including postal, telegraph and radio communications) or radio, television, film, audio or video recording, or performance (not open to the public) and their control.
Shop and commercial	4	Shops or premises used for a retail trade or business (including the sale to members of the public of food or drink for immediate consumption and retail by auction, self-selection and over-the-counter wholesale trading, the business of lending books or periodicals for gain and the business of a barber or hairdresser) and premises to which the public is invited to deliver or collect goods in connection with their hire, repair or other treatment, or (except in the case of repair of motor vehicles) where they themselves may carry out such repairs or other treatments.
Assembly and recreation	5	Place of assembly, entertainment or recreation; including bingo halls, broadcasting, recording and film studios open to the public, casinos, dance halls; entertainment, conference, exhibition and leisure centres; funfairs and amusement arcades; museums and art galleries; non-residential clubs, theatres, cinemas and concert halls; educational establishments, dancing schools, gymnasia, swimming pool buildings, riding schools, skating rinks, sports pavilions, sports stadia; law courts; churches and other buildings of worship, crematoria; libraries open to the public, non-residential day centres, clinics, health centres and surgeries; passenger stations and termini for air, rail, road or sea travel; public toilets; zoos and menageries.
Industrial	6	Factories and other premises used for manufacturing, altering, repairing, cleaning, washing, breaking-up, adapting or processing any article; generating power or slaughtering livestock.
Storage and other non-industrial[(2)]	7(a)	Place for the storage or deposit of goods or materials (other than described under 7(b)) and any non-residential building not within any of the purpose groups 1 to 6.
	7(b)	Car parks designed to admit and accommodate only cars, motorcycles and passenger or light goods vehicles weighing no more than 2500 kg gross.

Notes:

(1) Includes any surgeries, consulting rooms, offices or other accommodation, not exceeding 50 m^2 in total, forming part of a dwelling and used by an occupant of the dwelling in a professional or business capacity.

(2) A detached garage not more than 40 m^2 in area is included in purpose group 1(c); as is a detached open carport of not more than 40 m^2, or a detached building which consists of a garage and open carport where neither the garage nor open carport exceeds 40 m^2 in area.

(3) 'Room for residential purposes' means a room, or suite of rooms, which is not a dwelling-house or flat and which is used by one or more persons to live and sleep in, including rooms in hotels, hostels, boarding houses, halls of residence and residential homes but not including rooms in hospitals, or other similar establishments, used for patient accommodation.

(4) Modular (i.e. buildings that are made out of sub-assemblies), portable and/or temporary buildings are no different from any other new building and must comply with all requirements of the Building Regulations.

6.2.2 Requirements – size of residential buildings

The building shall be constructed so that the combined dead, imposed and wind loads are sustained and transmitted by it to the ground

- *safely;*
- *without causing such deflection or deformation of any part of the building (or such movement of the ground) as will impair the stability of any part of another building.*

(Approved Document A1)

The building shall be constructed so that ground movement caused by:

- *swelling, shrinkage or freezing of the subsoil; or*
- *landslip or subsidence (other than subsidence arising from shrinkage)*

will not impair the stability of any part of the building.

(Approved Document A2)

The maximum height of the building measured from the lowest finished ground level to the highest point of any wall or roof should be less than 15 m (see Figure 6.11).	A1/2 (2C4i)
The height of the building should not exceed twice the least width of the building (see Figure 6.11).	A1/2 (2C4ii)
The height of the wing H_2 should not be greater than twice the least width of the wing W_2 where the projection P exceeds twice the width W_2.	A1/2 (2C4iii)

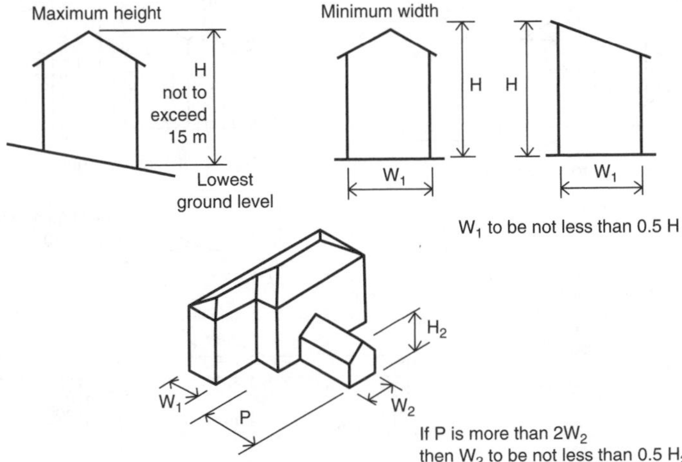

Figure 6.11 Residential buildings not more than three storeys

Small single storey non-residential buildings

The height (H) should not exceed 3 m and the width (or greater length) should not exceed 9 m (see Figure 6.12).	A1/2 (2C4b)

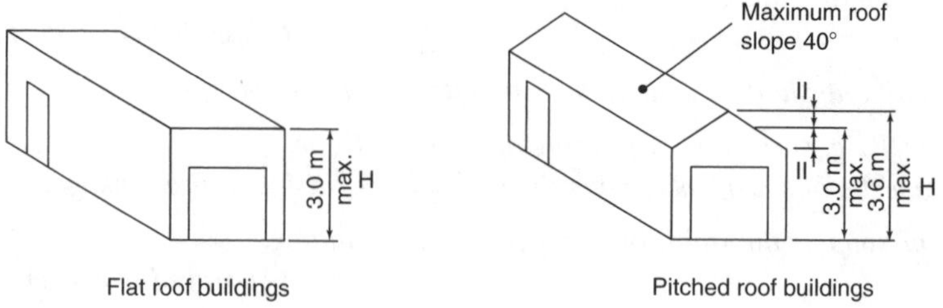

Figure 6.12 Size and proportion of non-residential buildings

Size of annexes

The height H (as variously shown in Figure 6.13) should not exceed 3 m.	A1/2 (2C4b)

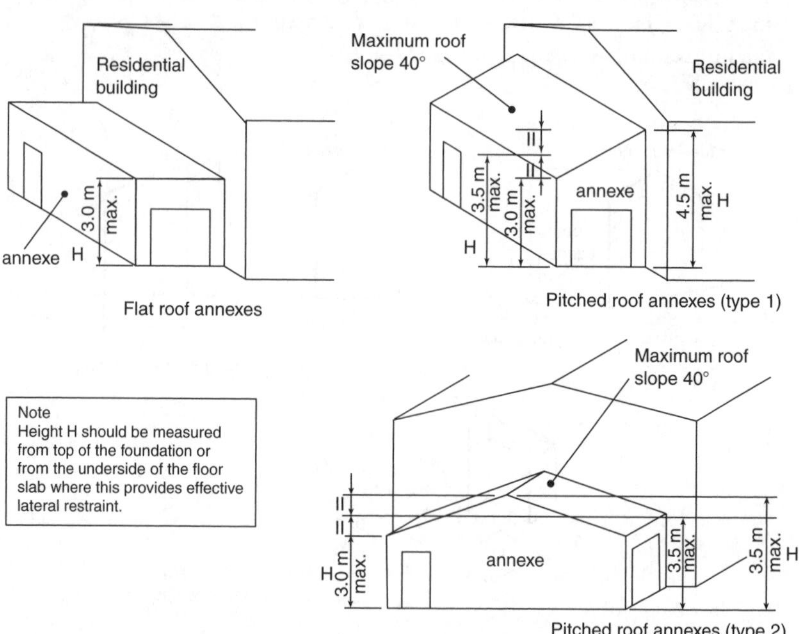

Figure 6.13 Size and proportion of non-residential annexes

6.3 Ventilation

There shall be adequate means of ventilation provided for people in the building.
(Approved Document F)

Ventilation is defined in the Building Regulations as '*the supply and removal of air (by natural and/or mechanical means) to and from a space or spaces in a building*'.

In addition to replacing 'stale' indoor air with 'fresh' outside air, the aim of ventilation is also to:

- limit the accumulation of moisture and pollutants from a building which could, otherwise, become a health hazard to people living and/or working within that building;
- dilute and remove airborne pollutants (especially odours);
- control excess humidity;
- provide air for fuel burning appliances.

Note: The requirements of the 2006 edition of Approved Document F have also been designed to deal with the products of tobacco smoking.

In general terms, all of these aims can be met if the ventilation system:

- disperses residual pollutants and water vapour;
- extracts water vapour from wet areas where it is produced in significant quantities (e.g. kitchens, utility rooms and bathrooms);
- rapidly dilutes pollutants and water vapour produced in habitable rooms, occupiable rooms and sanitary accommodation;
- extracts pollutants from areas where they are produced in significant quantities (e.g. rooms containing processes or activities which generate harmful contaminants);
- is designed, installed and commissioned so that it:
  - is not detrimental to the health of the people living and/or working in the building;
  - helps maintenance and repair;
  - is reasonably secure;
- makes available, over long periods, a minimum supply of outdoor air for the occupants;
- minimizes draughts;
- provides protection against rain penetration.

Ventilation is also a means of controlling thermal comfort (see Annex B, Performance based ventilation).

The aim of Approved Document F is to suggest to the designer the level of ventilation that should be sufficient for a particular situation as opposed to how it should be achieved. The designer is, therefore, free to use whatever ventilation system he considers most suitable for a particular building **provided** that it can be demonstrated that it meets the recommended performance criteria and levels concerning moisture, pollutants and air flow rates standards as shown in Table 6.9.

Table 6.9 Standards for performance-based ventilation

Type	Standard	Part
Intermittent extract fan	BS EN13141-4	Clause 4
Range hood	BS EN13141-3	Clause 4
Background ventilator (non-RH controlled)	BS EN 13141-1	Clauses 4.1 and 4.2
Background ventilator (RH controlled)	PrEN13141-9	Clauses 4.1 and 4.2
Passive stack ventilator	See Appendix D of Approved Document F	
Continuous mechanical extract ventilation (MEV system)	BS EN13141-6	Clause 4
Continuous mechanical supply and extract with heat recovery (MVHR)	PrEN13141-7	Clauses 6.1, 6.2 and 6.2.2
Single room heat recovery ventilator	PrEN13141-8	Clauses 6.1 and 6.2

Note: For further details and example, etc., see Appendix A to Approved Document F.

6.3.1 Background

External pollution

In urban areas, buildings are exposed to a large number of pollution sources from varying heights and upwind distances (i.e. long, intermediate and short range). Internal contamination from these pollution sources can have a detrimental effect on the buildings' occupants and so it is very important to ensure that ventilation system provided is sufficient and, above all, that the air intake cannot be contaminated.

Typical urban pollutants include:

- benzene (C_6H_6)
- butadiene (C_4H_6)
- carbon monoxide (CO)
- lead (Pb)
- nitrogen dioxide (NO_2)
- nitrogen oxide (NO)
- ozone (O_3)
- particles (PM_{10})
- sulphur dioxide (SO_2).

Typical emission sources include:

- building ventilation system exhaust discharges;
- combustion plant (such as heating appliances) running on conventional fuels;
- construction and demolition sites;

- discharges from industrial processes and other sources;
- other combustion type processes (e.g. waste incineration, thermal oxidation abatement schemes);
- road traffic, including traffic junctions and underground car parks;
- uncontrolled ('fugitive') discharges from industrial processes and other sources.

Indoor air pollutants

The maximum permissible level of indoor air pollutants is:

Nitrogen dioxide (NO_2)	not exceeding: • 288 μg/m^3 (150 ppb) – 1 hour average • 40 μg/m^3 (20 ppb) – long-term average
Carbon monoxide (CO)	not exceeding: • 100 mg/m^3 (90 ppm) – 15 minute averaging time • 60 mg/m^3 (50 ppm) – 30 minute averaging time (DOH, 2004) • 30 mg/m^3 (25 ppm) – 1 hour averaging time (DOH, 2004) • 10 mg/m^3 (10 ppm) – 8 hours averaging time (DOH, 2004)
Control of bio-effluents (body odours)	3.5 l/s per person
Total volatile organic compound (TVOC)*	not exceeding: • 300 μg/m^3 averaged over eight hours

Note: TVOC is defined as *any chemical compound based on carbon chains or rings (which also contain hydrogen) with a vapour pressure greater than 2 mm of mercury (0.27 kPa) at 25°C, excluding methane.*

Ventilation extraction rates

Extract ventilation concerns the removal of air directly from a space or spaces to outside. Extract ventilation may be by natural means such as passive stack ventilation (PSV) or by mechanical means (e.g. by an extract fan or central system).

Requirements

All kitchens, utility rooms, bathrooms and sanitary accommodation shall be provided with extract ventilation to the outside, which is capable of operating either intermittently or continuously. The minimum extract airflow rates should be greater than that shown in Table 6.10 below.	F 1.5

Table 6.10 Extract ventilation rates

Room	Minimum intermittent extract rate	Continuous extract	
		Min high rate	Min low rate
Kitchen	30 l/s (adjacent to hob) or 60 l/s elsewhere	13 l/s	Total extract rate must be at least the whole building ventilation rate shown in Table 6.11
Utility room	30 l/s	8 l/s	
Bathroom	15 l/s	8 l/s	
Sanitary accommodation	6 l/s	6 l/s	

The whole building ventilation rate for habitable rooms in a dwelling should be greater than that shown in Table 6.11. F 1.6

Table 6.11 Whole building ventilation rates

	Number of bedrooms				
	1	2	3	4	5
Whole building ventilation rate (l/s)	13	17	21	25	29

The minimum ventilation rate (based on two occupants in the main bedroom and a single occupant in all other bedrooms) should be not less than 0.3 l/s per m^2. F Table 1.1b

 **Note:** For greater occupancy, add 4 l/s per occupant.

Ventilation effectiveness

Ventilation effectiveness is, as the term suggests, a measure of how well a ventilation system supplies air to the building's occupants. From an energy-saving perspective, the higher the level of ventilation effectiveness the more efficient the system will be in reducing pollutant levels at the occupant's breathing zone. As this can result in quite significant energy savings, it has to be considered when designing and installing ventilation systems.

As the designer cannot be absolutely certain of the future occupancy and/or use of the building in terms of seating plan, location of computers and printers etc., a ventilation effectiveness level of 1 (i.e. where the supply air is fully mixed with the room air before it is breathed by the occupants) should, similar to the designs and recommendations of Approved Document F, be assumed in their calculations.

Note: For more details about ventilation effectiveness, see CIBSE Guide A.

Equivalent ventilator area for dwellings

Note: Equivalent area is defined as *the area of a sharp-edged orifice which air would pass at the same volume flow rate, under an identical applied pressure difference.*

Equivalent area is now considered a better measure of the aerodynamic performance of a ventilator instead of the previous free-area sizing of background ventilators. Primarily this is because 'free area' only refers to the physical size of the aperture of the ventilator and does not, therefore, accurately reflect the airflow performance of the ventilator. A new European Standard (BS EN13141-1:2004) has now been published which includes a method of measuring the equivalent area of background ventilator openings.

Designers should use the equivalent ventilator areas shown in Table 6.12 when designing systems using intermittent extract fans and background ventilators for multi-storey dwellings that are more than four storeys above ground level and which have more than one exposed façade.

Table 6.12 Equivalent ventilator area for dwellings

Total floor area (m^2)	Number of bedrooms				
	1	2	3	4	5
≤50	25 000	35 000	45 000		
51–60	25 000	30 000	40 000		
61–70	30 000	30 000	35 000	45 000	55 000
71–80	35 000	35 000	35 000		
81–90	40 000	40 000	40 000		
91–100	45 000	45 000	45 000		
>100	Add 5000 mm^2 for every additional 10 m^2 floor area				

Note:

- For single storey dwellings up to four storeys above ground level, add 5000 mm^2.
- For an occupancy level greater than two persons in the main bedroom and one person in all other bedrooms, assume an extra bedroom for each additional person.
- For more than five bedrooms, add an additional 10 000 mm^2 per bedroom.

Ventilation air intakes

One method of achieving good indoor air quality is to reduce the amount of water vapour and/or air pollutants that are released into the indoor air, particularly those caused from construction and consumer products.

Air intakes that are located on a less polluted side of the building may be used for fresh air.

Note: Further information about control of emissions from construction products is available in BRE Digest 464.

Noise from ventilation systems

As the noise from ventilation systems can disturb the occupants of the building and in doing so affect their work effectiveness, the designer must consider methods of minimizing noise through careful design and use of quieter products. The effect of externally emitted noise on people outside of the building should also be considered.

The installation and use of ventilation systems in buildings will also result in energy being used (e.g. to heat fresh air taken in from outside, to move air into, out of and/or around the building) and so consideration should always be given to using heat recovery devices, efficient types of fan motor and/or energy saving control devices in ventilation systems.

Types of ventilation

Buildings are normally ventilated by a combination of infiltration (from uncontrolled air leakage paths within the building structure) and some form of natural and/or manually controlled air exchange between the inside and the outside of a building.

Approved Document F (2006) recommends a series of controllable ventilation methods that allow for a reasonably high level of air tightness (i.e. air permeability) down to around 3–4 m^3/h per square metre of envelope area at 50 Pa pressure difference.

The three main controllable ventilation methods are listed in Table 6.13.

6.3.2 Purge ventilation

Purge ventilation is a manually controlled type of ventilation that is used in rooms and spaces to rapidly dilute pollutants and/or water vapour. It can be achieved by natural means (e.g. an openable window or an external door) or by mechanical means (e.g. a fan).

Note: For further guidance on purge ventilations, see BS 5925: 1991 *Code of practice for ventilation principles and designing for natural ventilation.*

Requirements

Purge ventilators shall be manually operated.	F Table 1.5
The location of ventilation devices in rooms is not critical.	F Table 1.4

Table 6.13 Ventilation methods

Method	Type	Why used	Remarks
Extract ventilation	Intermittent extract fans	In rooms where most water vapour and/or pollutants are released (e.g. cooking, bathing or photocopying)	This extract may be either intermittent or continuous and is aimed at minimizing the spread of vapour and pollutants to the rest of the building
Whole building ventilation	Trickle ventilators	To provide fresh air to the building, dilute and disperse residual water vapour/pollutants not dealt with by extract ventilation and to remove water vapour and pollutants released by building materials, furnishings, activities and the presence of occupants	This type of ventilation provides continuous air exchange with a ventilation rate that can be reduced or ceased when the building is not occupied. In some cases (e.g. when the building is reoccupied) it may be necessary to purge the air (see below)
Purge ventilation **Note:** Previously referred to as '*rapid*' ventilation in the 1995 edition of Approved Document F	Windows	To assist in the removal of high concentrations of pollutants and water vapour released from occasional activities (such as painting and decorating) or accidental releases (such as smoke from burnt food or water spillage)	Purge ventilation is intermittent and may be used to improve thermal comfort and/or over-heating in summer (see Approved Documents LIA (New dwellings) and L2A (New buildings other than dwellings)

6.3.3 Passive stack ventilation

Passive Stack Ventilation (PSV) is a ventilation device which uses ducts from terminals mounted in the ceiling of rooms to terminals on the roof, to extract air to the outside by a combination of the natural stack effect and the pressure effects of wind passing over the roof of the building.

The so-called 'stack effect' relies on the pressure differential between the inside and the outside of a building caused by differences in the density of the air due to an indoor/outdoor temperature difference.

Table 6.14 Passive stack ventilation

Room	Internal duct diameter (mm)	Internal cross-sectional area (mm)
Kitchen	125	12 000
Utility room	100	8 000
Bathroom	100	8 000
Sanitary accommodation	80	5 000

For sanitary accommodation **only**, purge ventilation may be used **provided** that security is not an issue.

Note: Open-flued appliances may provide sufficient extract ventilation when in operation and can be arranged to provide sufficient ventilation when not firing.

Design

The design and installation of PSV systems is crucial to their operation and Figure 6.14 shows the preferred option for kitchen and bathroom ducts with ridge terminals.

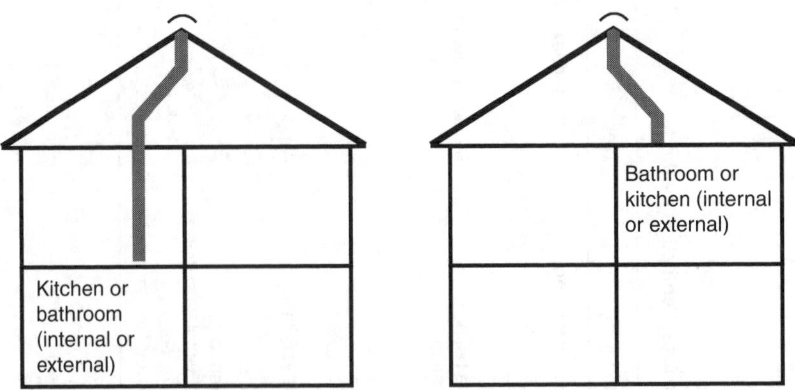

Figure 6.14 Preferred PSV system layouts

Another option (see Figure 6.15) is to have the kitchen and bathroom ducts penetrating the roof and extend its terminals to ridge height.

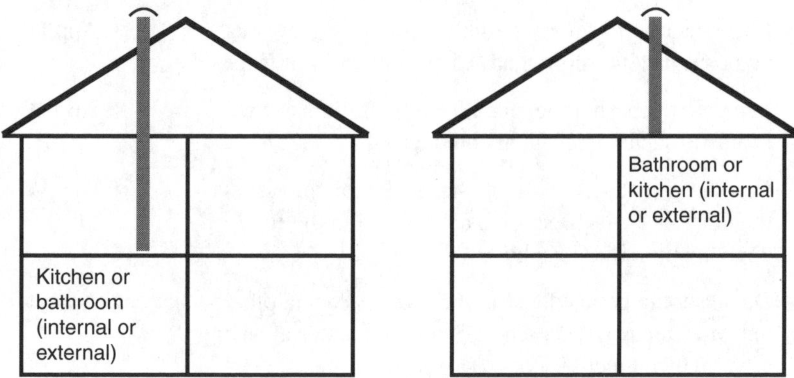

Figure 6.15 Alternative PSV system layouts

Requirements

In designing PSV systems, the following requirements shall be met:

Common outlet terminals and/or branched ducts shall **not** be used for wet rooms (e.g. the kitchen, bathroom, utility room and/or WCs).	F App D
Ducts should have no more than one offset (i.e. bend) and ideally these should be 'swept' at an angle of no more than 40° to the vertical.	F App D

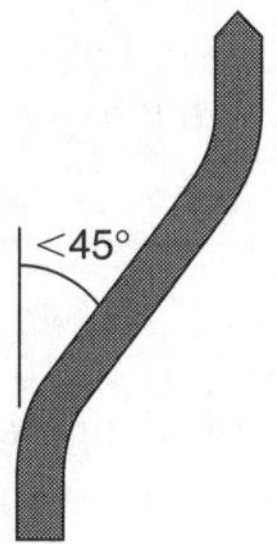

Figure 6.16 PSV offset requirements

If a duct penetrates the roof more than 0.5 m from the roof ridge, then it must extend above the roof slope to at least the height of the roof ridge.	F App D
If tile ventilators are used on the roof slope they must be positioned no more than 0.5 m from the roof ridge.	F App D
Separate ducts shall be taken from the ceilings of wet rooms to separate terminals on the roof.	F App D
Ceiling extract grilles should have a free area, not less than the duct cross-sectional area (when in the fully open position if adjustable).	F App D
Ducts should be insulated in the roof space and other unheated areas with at least 25 mm of a material having a thermal conductivity of 0.04 W/mK.	F App D
If a duct extends above the roof level, then that section of the duct should be insulated or be fitted with a condensation trap just below roof level.	F App D
If a conversion fitting is required to connect the duct to the terminal then the duct cross section area must be maintained (or exceeded) throughout the conversion fitting.	F App D
PSVs for dwellings that are situated near a significantly taller building (i.e. more than 50% taller), should be at least five times the difference in height **away** from the taller building (i.e. if the difference in height is 10 m, then the PSV should not be installed in a dwelling within 50 m of the taller building).	F App D
Outlet terminals should have a free area that is not less than the duct cross-section area.	F App D
Terminals should not allow ingress of large insects or birds and should be designed so that rain is not likely to enter the duct and run down into the dwelling.	F App D
Terminals should be designed so that any condensation forming inside it cannot run down into the dwelling but will run off onto the roof.	F App D

Note: A draft European Standard (i.e. prEN 13141-5) for testing cowls and roof outlets is currently under discussion which (it is anticipated) will suggest that terminals, with any necessary conversion fitting, should have an overall static pressure loss (upstream duct static minus test room static) equivalent to no more than four times the mean duct velocity pressure when measured at a static pressure difference of 10 Pa.

Installation

Location of passive stack ventilators

PSV extract terminals should be located in the ceiling or on a wall less than 400 mm below the ceiling.	F Table 1.4
Background ventilators should not be within the same room as a PSV terminal.	F Table 1.4
If a PSV is located in a protected stairway of a dwelling, it shall not allow smoke or fire to spread into the stairway.	F Table 1.4
The duct length should be just sufficient to fit between the ceiling grille and the outlet terminal.	F App D
Flexible ducting should be fully extended but not taut.	F App D
Allow approximately 300 mm extra to make smooth bends in an offset system.	
Ducting should: • be properly supported along its entire length (remembering that flexible ducting generally requires more support than rigid ducting); • be run straight without any distortion or sagging; • not have any kinks at bends or connections with ceiling grilles and outlet terminals; • be securely fixed to the roof outlet terminal so that it cannot sag or become detached.	F App D
In roof spaces: • ducts should, ideally, be secured to a wooden strut that is securely fixed at both ends; • flexible ducts should be allowed to curve gently at each end of the strut.	F App D
For stability, rigid ducts should be used for any outside part of the PSV system that is above the roof slope. To provide stability, they should also project down into the roof space far enough to allow firm support.	F App D

Fire precautions

In dwellings with three or more storeys and blocks of flats, PSV ducts should not impede fire escape routes.

Common stairs

If a stair serves a place that is a special fire hazard, the lobby or corridor should have not less than 0.4 m^2 permanent ventilation or be protected by a mechanical smoke control system.	B1 2.47 (V2)

External escape stairs

Protected lobbies (with less than 0.4 m^2 permanent ventilation) should be provided between an escape stairway and a place of special fire hazard.	B1 4.35 (V2)

Flues

If a flue (or a duct containing a flue and/or ventilation duct) passes through a compartment wall or compartment floor (or is built into a compartment wall) then each wall of the flue or duct should have a fire resistance of at least half that of the wall or floor.	B3 10.16 (V2)

Mechanical ventilation

Smoke control of common escape routes by mechanical ventilation is permitted provided that it meets the requirements of BS EN 12101-6:2005.	B1 2.27 (V2)
Mechanical ventilation systems should be designed to ensure that: • ductwork does not assist in transferring fire and smoke through the building;	B1 5.46 (V2)

• exhaust points are sited away from final exits, combustible building cladding or roofing materials and openings into the building;	
• recirculated air serving a stairway and/or entrance hall (as well as other areas) should be designed to shut down on the detection of smoke within the system.	B1 2.18 (V2) B3 10.2 (V2)

Protected escape routes

Protected lobbies (with not more than 0.4 m^2 permanent ventilation) should be provided between an escape stairway and a place of special fire hazard.	B1 4.35 (V2)
Ventilation ducts supplying or extracting air directly to or from a protected escape route, should **not** also serve other areas.	B1 5.47 (V2)
Separate ventilation systems should be provided for each protected stairway.	B1 5.47 (V2)

Ventilation ducts

Ventilation ducts supplying or extracting air directly to or from a protected stairway, should not also serve other areas.	B1 2.17 (V1)
Ventilation ducts supplying or extracting air directly to or from a protected stairway or entrance hall, should not also serve other areas.	B1 2.18 (V2) B1 5.47 (V2)
Separate ventilation systems should be provided for each protected stairway.	B1 5.47 (V2)

Noise

If the dwelling is near a busy road or by an airport etc. where the amount of external noise is likely to be intrusive, a sound attenuator duct section may be fitted in the roof space just above the ceiling.

Operation in hot weather

Although PSV units should be capable of extracting sufficient air from wet rooms during the winter, in the summer months (i.e. when the temperature difference between the internal and external air is considerably reduced) they may not. To guard against this happening, purge ventilation should also be provided in these wet rooms.

6.3.4 Installation of fans in dwellings

The three fan types most commonly used in domestic applications are:

- axial fans
- centrifugal fans
- in-line fans.

Axial fans

The axial fan is the most common form of fan which can be mounted on the wall, window (i.e. through a suitable glazing hole) or in the ceiling (e.g. in a bathroom).

For wall and window mounting applications up to 350 mm thick, use a short length of rigid round duct or a flexible duct pulled taut.

For bathrooms, 100 mm diameter fans can be used as an axial fan in the ceiling with a short (1.5 m maximum) length of flexible duct with (a maximum) of two 90 bends.

Note: The duct must be pulled taut and the discharge terminal should have at least 85% free area of the duct diameter.

Centrifugal fans

Centrifugal fans (because they develop greater pressure) permit longer lengths of ducting to be used and so can be used for most wall and/or window applications in high-rise (i.e. above three storeys) buildings or in exposed locations to overcome wind pressure.

Most centrifugal fans are designed with 100 mm diameter outlets which enables them to be connected to a wide variety of duct types.

Requirements

Wall/ceiling-mounted centrifugal fans that are designed to achieve 60 l/s for kitchens and which are fitted with a 100 mm diameter flexible duct or rectangular duct, should not be ducted further than 3 metres and should have no more than one 90 bend.	F App E

Wall/ceiling-mounted centrifugal fans that are designed to achieve 15 l/s for bathrooms which are fitted with 100 mm diameter flexible duct or rectangular duct should not be ducted further than 6 metres and should have no more than two 90 bends. F App E

In-line fans

There are two types of in-line fans available:

- in-line axial fans which have to be installed with the shortest possible duct length to the discharge terminal; and
- in-line mixed flow fans which have the characteristics of both axial and centrifugal fans and can, therefore, be used with longer lengths of ducting.

Both types can be used for bathrooms (100 mm diameter), utility rooms (125 mm diameter) and kitchens (150 mm diameter).

Intermittent extract fans

Minimum extract airflow rates for intermittent extract fans should be greater that that shown in Table 6.15 below. F Table 1.2a

Note: For sanitary accommodation, a purge ventilation system may be used and in wet rooms, a heat recovery ventilator may be used instead of a conventional fan provided that it has the same extract rate.

Table 6.15 Extract ventilation rates

Room	Minimum intermittent extract rate	Continuous extract	
		Min high rate	Min low rate
Kitchen	30 l/s (adjacent to hob) or 60 l/s elsewhere	13 l/s	Total extract rate must be at least the whole building ventilation rate shown in Table 6.10
Utility room	30 l/s	8 l/s	
Bathroom	15 l/s	8 l/s	
Sanitary accommodation	6 l/s	6 l/s	

Fan terminals

When installing fans:

Ensure that the free area of the grill opening of a room terminal extract grille and/or discharge terminal has a minimum of 85% of the free area of the ducting being used.	F App E

Note: In these cases (only), the equivalent area may be assumed to be equal to the free area.

6.3.5 Ventilation systems – dwellings without basements

The following systems may be used in dwellings without basements:

- background ventilators;
- continuous mechanical extract;
- continuous mechanical supply and extract with heat recovery;
- intermittent extract fans;
- passive stack ventilation.

Background ventilators

The need for background ventilators will depend on the air permeability or air tightness of a building.

Note: Air permeability is defined as *the average volume of air (in cubic metres per hour) that passes through unit area of the building envelope (in square metres) when subject to an internal to external pressure difference of 50 Pa.*

Location of background ventilators in rooms

Approved Document F describes four different scenarios concerning background ventilators and their equivalent areas.

Table 6.16 Equivalent areas for background ventilator systems

Location	System 1 All habitable rooms & wet rooms	System 2 All habitable rooms without a passive stack ventilator	System 3 All habitable rooms (other than wet rooms from which air is extracted)	System 4
All habitable rooms	5000 mm^2 (min)			No background ventilators required

(Continued)

Table 6.16 (*Continued*)

Location	System 1 All habitable rooms & wet rooms	System 2 All habitable rooms without a passive stack ventilator	System 3 All habitable rooms (other than wet rooms from which air is extracted)	System 4
All habitable rooms without a passive stack ventilator		5000 mm^2 (min)		
Habitable rooms with no external walls	8000 mm^2 (min)	8000 mm^2 (min)		
Habitable rooms (other than wet rooms from which air is extracted)			2500 mm^2	
Wet rooms with an external wall	2500 mm^2 (min)			

Note: For Systems 1 and 2, additional ventilation may be required during warmer months as stack driving pressures are reduced. The provisions for purge ventilation (e.g. windows) could also be used.

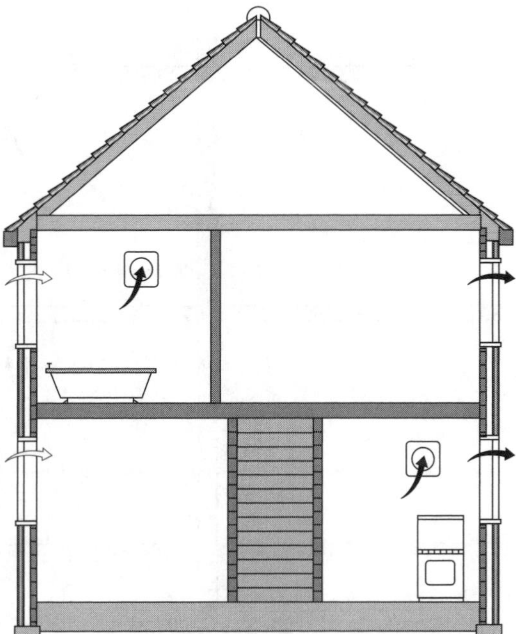

Figure 6.17 Background ventilators and intermittent extract fans

Requirements

Controllable background ventilators with a minimum equivalent area of $2500\,mm^2$ shall be fitted in each room (except wet rooms from which air is extracted).	F Table 1.2c
Background ventilators may be manually adjustable or automatically controlled.	F Table 1.5
Windows with night latches should **not** be used as they are more liable to draughts as well as being a potential security risk.	F 0.17
Background ventilators for dwellings with a single exposed façade should be located at both high (typically 1.7 m above floor level and low positions (i.e. at least 1.0 m below the high ventilators) in the façade (see Figure 6.18 below).	F Table 1.2a

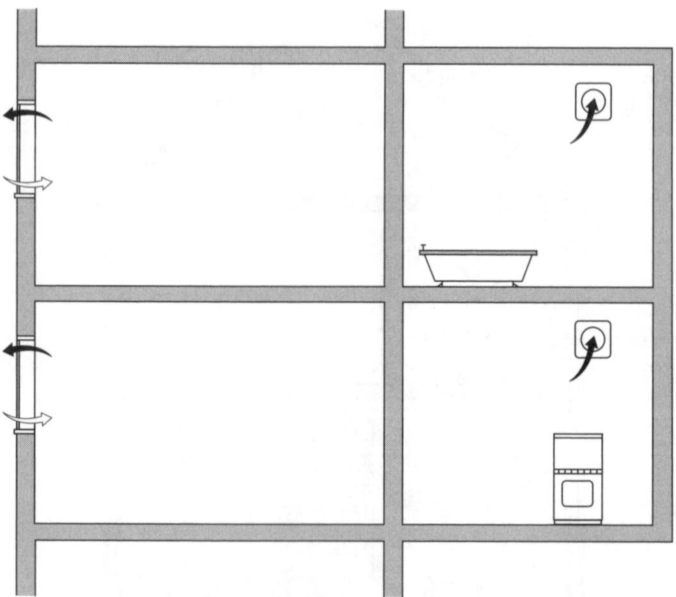

Figure 6.18 Single-sided ventilation

Dwellings with only a single exposed façade should be designed so that the habitable rooms are on the exposed façade in order to achieve cross ventilation.	F Table 1.2b
Background ventilators should be at least 0.5 m from an extract fan.	F Table 1.4
Background ventilators should be located so as to avoid draughts (e.g. typically 1.7 m above floor level).	F Table 1.4

Trickle ventilators

Manually controlled trickle ventilators are widely used for background ventilation and these can be located as shown in Figure 6.19.

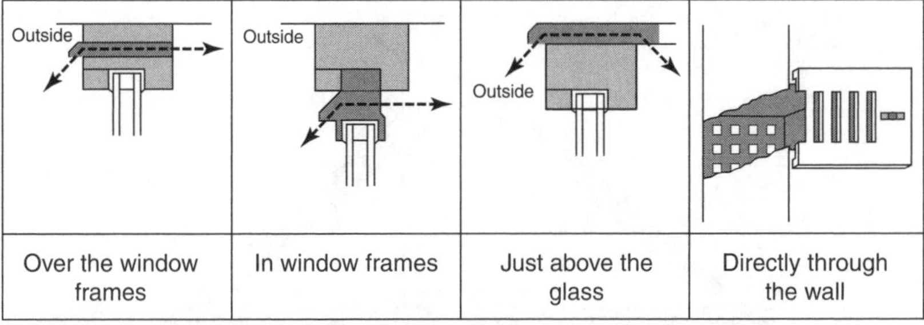

Figure 6.19 Background ventilation systems

To avoid cold draughts, trickle ventilators are normally positioned 1.7 m above floor level and usually include a simple control (such as a flap) to allow users to shut off the ventilation according to personal choice or external weather conditions. Nowadays, pressure-controlled trickle ventilators that reduce the air flow according to the pressure difference across the ventilator are available to reduce draught risks during windy weather.

Trickle ventilators are normally left open in occupied rooms in dwellings.

Requirements

Trickle ventilators that include an automatic control should be capable of being manually overridden so that they can be opened by the occupant when required.	F Table 1.2c F 0.19

Pressure-controlled trickle ventilators that, under normal conditions, are left open (e.g. 1 Pa pressure difference) should only be capable of being manually closed.	F 0.19
Trickle ventilators etc. should be clearly marked with their equivalent area (measured according to BS EN13141–1:2004) either by means of a stamp or an indelibly printed self-adhesive label.	F 0.25
All fans should operate quietly at their minimum (i.e. normal) rate so as not to disturb the occupants of the building.	F Table 1.2c

Continuous mechanical extract

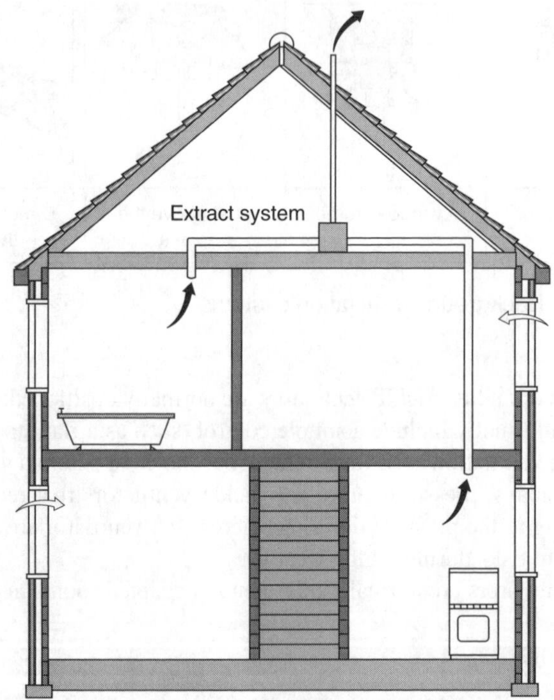

Figure 6.20 Continuous mechanical extract

This system may consist of either a central extract system or individual room fans, or a combination of both.

To calculate the required extract rate first determine the **whole building ventilation rate** from Table 6.17.

Table 6.17 Whole building ventilation rates

	Number of bedrooms				
Whole building ventilation rate (l/s)	1 13	2 17	3 21	4 25	5 29

Then work out the **whole dwelling air extract rate** (at maximum operation) by summing the individual room rates from Table 6.18.

Table 6.18 Whole dwelling air extract rate

Room	Minimum intermittent extract rate	Minimum high continuous extract rate
Kitchen	30 l/s (adjacent to hob) or 60 l/s elsewhere	13 l/s
Utility room	30 l/s	8 l/s
Bathroom	15 l/s	8 l/s
Sanitary accommodation	6 l/s	6 l/s

For sanitary accommodation **only**, purge ventilation may be used provided that security is not an issue.

Then proceed as below:

The maximum ('boost') rate should be the greater of whole building ventilation rate or the whole dwelling air extract rate.	F Table 1.2c
The maximum individual room extract rates should be at least those given in Table 6.17.	F Table 1.2c
The minimum air supply rate should be at least the whole building ventilation rate.	F Table 1.2c

Note: Extract terminals located on the prevailing windward façade should be protected against the effects of wind by using ducting to another façade, using a constant volume flow rate unit or a central extract system.

Requirements

All fans should operate quietly at their minimum (i.e. normal) rate so as not to disturb the occupants of the building.	F Table 1.2c
Ventilation devices designed to work continuously: • shall be set-up to operate without occupant intervention; • may have a manual control to select maximum 'boost'; • may have automatic controls such as humidity control (but not if used for sanitary accommodation), occupancy/usage sensor, moisture/pollutant release detector etc.	F Table 1.5
Automatic controls for ventilators that are designed to work continuously in kitchens must be capable of providing sufficient flow during cooking with fossil fuels (e.g. gas) so as to avoid the build-up of combustion products.	F Table 1.5

Continuous mechanical supply and extract with heat recovery

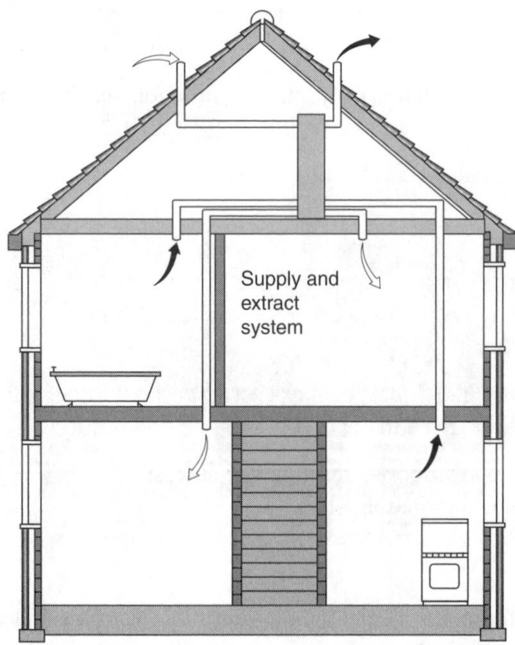

Figure 6.21 Continuous mechanical supply – with heat recovery

To calculate the air flow rate of a building using continuous Mechanical Supply and extract with Heat Recovery (MVHR), first determine the **whole building ventilation rate** from Table 6.17, then, depending on whether it is a multi-storey, or single-storey, subtract the gross internal volume of dwelling heated space (m^2) as follows:

Multi-storey dwelling	whole building ventilation rate − 0.04 × gross internal volume
Single-storey dwelling	whole building ventilation rate − 0.06 × gross internal volume

Next, work out the whole dwelling air extract rate at maximum operation by summing the individual room rates from Table 6.18, and then proceed as below:

The maximum ('boost') rate should be the greater of the whole building ventilation rate or the whole dwelling air extract rate.	F Table 1.2d
The maximum individual room extract rates should be at least those given in Table 6.17.	F Table 1.2d
The minimum air supply rate should be at least the whole building ventilation rate.	F Table 1.2d

Single room heat recovery ventilator

If a Single Room Heat Recovery Ventilator (SRHRV) is used to ventilate a habitable room, to calculate the air flow rate, first determine the **whole building ventilation rate** from Table 6.17, then work out the room supply rate using the following formula:

$$\frac{\text{whole building ventilation rate} \times \text{room volume}}{\text{total volume of all habitable rooms}}$$

When working out the continuous mechanical extract for a whole building which also includes a room ventilated by an SRHRV, the following formula should be used:

$$\frac{\text{whole building ventilation rate} \times \text{room volume} - \text{SRHRV supply rate}}{\text{total volume of all habitable rooms}}$$

Mechanical intermittent extract

As odour is the main pollutant, humidity controls should **not** be used for intermittent extract in sanitary accommodation.	F Table 1.5
Ventilators equipped with intermittent extract shall be capable of being operated manually and/or automatically by a sensor (e.g. humidity sensor, occupancy/usage sensor, moisture/pollutant release detector etc).	F Table 1.5
All ventilator automatic controls **must** be provided with a manual over-ride to allow the occupant to turn the extract on.	F Table 1.5
Automatic controls for ventilators used in kitchens **must** be capable of providing sufficient flow during cooking with fossil fuels (e.g. gas) so as to avoid the build-up of combustion products.	F Table 1.5
If a fan is installed in an internal room without an openable window, then the fan should have a 15 minute over-run.	F Table 1.5
In rooms with no natural light, fans could be controlled by the operation of the main room light switch.	F Table 1.5

Note: In dwellings, humidistat controls should be available to regulate the humidity of the indoor air and to minimize the risk of condensation and mould growth. Humidistats are normally installed as part of an extract ventilator especially in moisture-generating rooms such as a kitchen or a bathroom. They should **not** be used for sanitary accommodation where the dominant pollutant is usually odour.

6.3.6 Ventilation systems – basements

If a basement is connected to the rest of the dwelling by a large permanent opening such as an open stairway then the whole dwelling including the basement should be treated as a multi-storey dwelling and ventilated in a similar manner to dwellings without basements.	F 1.9

If the basement has a single exposed façade, whilst the rest of the dwelling above ground has more than one exposed façade, then passive stack ventilation or continuous mechanical extract should be used. F 1.9

For basements that are not connected to the rest of the dwelling by a large permanent opening, then: F 1.10

- the part of the dwelling above ground should be considered separately;
- the basement should be treated as if it were a single-storey dwelling above ground.

If the part of the dwelling above ground has no bedrooms, then for the purpose of ventilation requirements:

- assume that the dwelling has one bedroom; and
- treat the basement as a single-storey dwelling (with one bedroom) as if it were above ground.

If a dwelling only compromises a basement, then it should be treated as if it were a single-storey dwelling (with one bedroom) above ground. F 1.11

Table 6.19 Ventilation systems for basements

Type of basement	Background ventilators and intermittent extract fans	Passive stack ventilation	Continuous mechanical extract	Continuous mechanical supply and extract with heat
Basement connected to the rest of the dwelling by an open stairway	Yes	Yes	Yes	Yes
Basement with a single exposed façade and dwelling above ground with more than one exposed façade		Yes	Yes	
Basements not connected to the rest of the dwelling by an open stairway	Yes	Yes	Yes	Yes
Dwelling above ground has no bedrooms	Yes	Yes	Yes	Yes
Dwelling comprises just a basement	Yes	Yes	Yes	Yes

6.3.7 Ventilation of habitable rooms through another room or a conservatory

Habitable rooms without an openable window shall be either ventilated through another habitable room or through a conservatory. F 1.12

6.3.8 Ventilation through another room

Habitable rooms without an openable window may be ventilated through another habitable room provided that the other room has: F 1.13

- purge ventilation; and
- an 8000 mm^2 background ventilator; and
- there is a permanent opening between the two rooms.

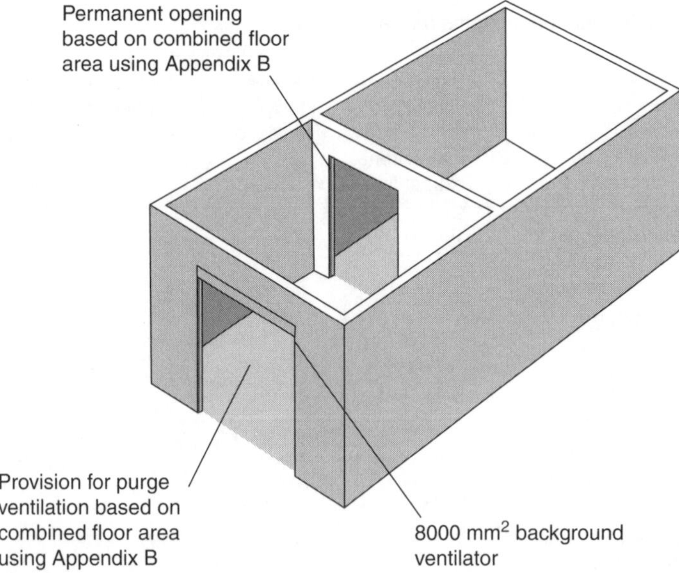

Figure 6.22 Two habitable rooms treated as a single room for ventilation purposes

6.3.9 Ventilation through a conservatory

Habitable rooms without an openable window may be ventilated through a conservatory (see Figure 6.23) provided that that conservatory has: F 1.14

- purge ventilation; and
- an 8000 mm^2 background ventilator; and
- there is a closable opening between the room and the conservatory that is equipped with:
  - purge ventilation and
  - an 8000 mm^2 background ventilator.

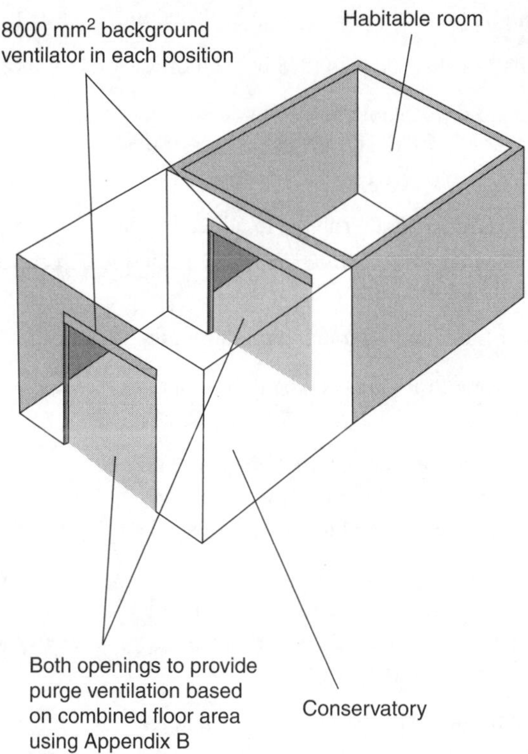

Figure 6.23 A habitable room ventilated through a conservatory

6.3.10 Ventilation systems – buildings other than dwellings

Fresh air supplies should be protected from contaminants that would be injurious to health. F 2.3

Offices

All office sanitary accommodation, washrooms and food and beverage preparation areas shall be provided with intermittent air extract ventilation capable of meeting the requirements of Table 6.20.	F 2.11
Extract fans that are located in an internal room which does not have an openable window, should have a 15 minute over-run.	F Table 2.2c
Extract ventilators should be located as high as practicable and preferably not less than 400 mm below the ceiling level.	F Table 2.2b
PSVs should be located in the ceiling of the room.	F Table 2.2b
Purge ventilation shall be provided in each office.	F 2.13
Purged air should be taken directly to outside and should not be recirculated to any other part of the building.	F 2.13
PSVs can be used as an alternative to a mechanical extract fan for office sanitary and washrooms and food preparation areas.	F Table 2.2a
PSV controls can be either manual or automatic.	F Table 2.2c
The controls for extract fans can be either manual or automatic.	F Table 2.2c
Printers and photocopiers that are being used in large numbers and which are in almost constant use (i.e. greater than 30 minutes per hour) shall: • be located in a separate room; • have extract facilities capable of providing an extract rate greater than 20 l/s per machine, during use (see Table 6.20).	F 2.11
The whole building ventilation rate for the supply of air to the offices should be greater than 10 l/s per person (see Table 6.21).	F 2.12

The following air flow rates can mainly be provided by natural ventilation.

Table 6.20 Extract ventilation rate

Room	Air extract rate
Rooms containing printers and photocopiers in substantial use (greater than 30 minutes per hour)	20 l/s per machine during use
Office sanitary accommodation and washrooms	15 l/s per shower/bath 6 l/s per WC/urinal
Food and beverage preparation areas (not commercial kitchens)	15 l/s with microwave and beverages only 30 l/s adjacent to the hob with cooker(s) 60 l/s elsewhere with cooker(s)
Specialist buildings and spaces (e.g. commercial kitchens, fitness rooms)	See Table 2.3

Table 6.21 Whole building ventilation rate for air supply to offices

	Air supply rate
Total outdoor air supply rate for offices (no-smoking and no significant pollutant sources)	10 l/s per person

Note: The outdoor air supply rates shown above for offices are based on controlling body odours with low levels of other pollutants.

Modular and portable buildings

Other types of buildings

The ventilation requirements for other buildings (e.g. such as assembly halls, broadcasting studios, computer rooms, factories, hospitals, hotels, museums, schools, sports centres and warehouses etc.) are listed in Table 2.3 of Approved Document F which also provides a link to the relevant controlling Acts of Parliament, Statutory Instruments, BS, CIBSE and HSE standards, practices and recommendations.

Sensors

Ventilation in buildings **other than** dwellings is dependent upon occupancy levels and currently there are some very sophisticated automatic control systems such as local passive infra-red detectors and electronic carbon dioxide detectors available.

6.3.11 Ventilation systems – car parks

Ventilation is the important factor and, as heat and smoke cannot be dissipated so readily from a car park that is not open-sided, fewer concessions are made. For more guidance see Section 11 of Part B3 (Volume 2).

Underground car parks, enclosed car parks and multi-storey car parks should be designed to limit the concentration of carbon monoxide to not more than 30 parts per million averaged over an eight hour period and peak concentrations.	F 2.19
Ramps and exits shall not go above 90 parts per million for periods not exceeding 15 minutes.	F 2.19
Naturally ventilated car parks shall have openings at each car parking level: • at least 1/20th of the floor area at that level; • with a minimum of 25% on each of two opposing walls.	F 2.21a
Mechanically ventilated car parks can have either natural ventilation openings that are not less than 1/40th of the floor area or a mechanical ventilation system capable of at least three air changes per hour (ach).	F 2.21b
Mechanically ventilated basement car parks shall be capable of at least six air changes per hour (ach).	F 2.21b
Mechanically ventilated exits and ramps (i.e. where cars queue inside the building with engines running) shall be capable of at least ten air changes per hour (ach).	F 2.21b

6.3.12 General requirements

To ensure good transfer of air throughout the dwelling, there shall be an undercut of 7600 mm^2 (minimum) in all internal doors above the floor finish (equivalent to an undercut of 10 mm for a standard 760 mm width door).	F Table 1.4
Adequate replacement air must also be available (e.g. a 10 mm gap under the door or equivalent).	F App E
All ducting that passes through a fire stopping wall or fire compartment shall meet the requirements of Approved Document B of the Building Regulations.	F App E
Duct runs should be straight, with as few bends and kinks as possible to minimize system resistance.	F App E
Horizontal ducting (including ducting in walls) should be arranged to slope slightly downwards away from the fan to prevent backflow of any moisture.	F App E

Fans and/or ducting placed in, or passing through, an unheated void or loft space should be insulated to reduce the possibility of condensation forming.	F App E
The inner radius of any bend should be greater or equal to the diameter of the ducting being used (see Figure 6.24).	F App E
Vertical duct rises may need to be fitted with a condensation trap in order to prevent the backflow of any moisture.	F App E
The circular profile of a flexible duct should be maintained throughout the full length of the duct run (see Figure 6.24).	F App E

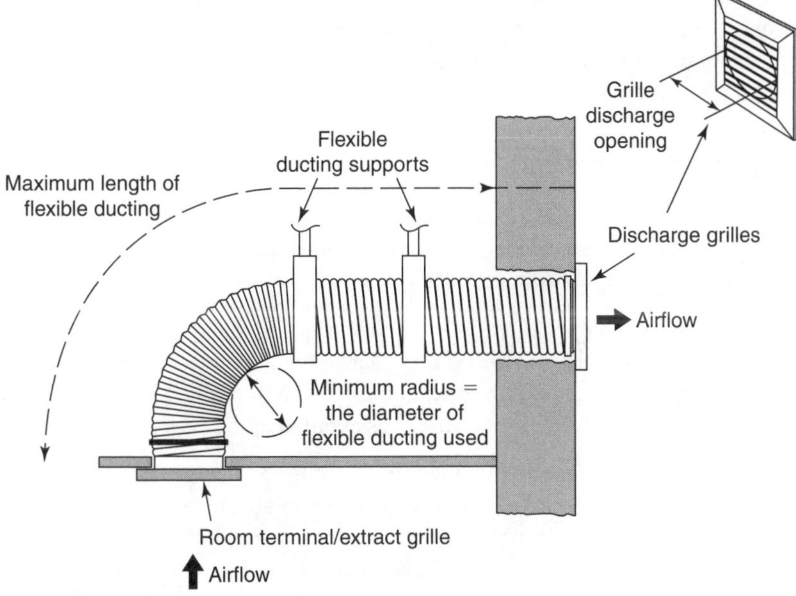

Figure 6.24 Correct installation of ducting

If a back-draught device is used it may be incorporated into the fan itself.	F App E
Flexible ducting should be installed without any peaks or troughs (see Figure 6.25).	F App E

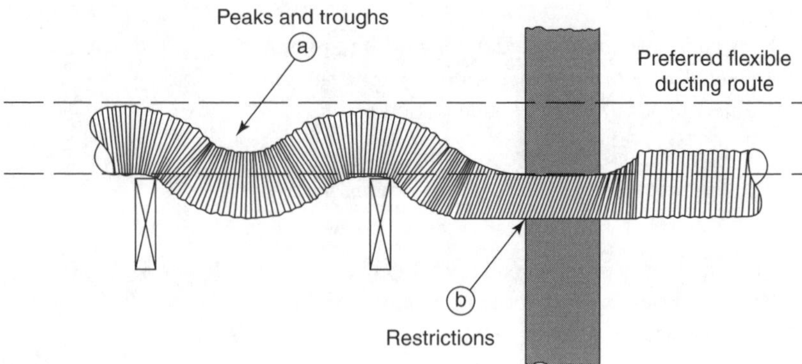

Figure 6.25 Incorrect installation of ducting

Access

There should be reasonable access to ventilation systems to enable changing filters, replacing defective components, cleaning duct work and other maintenance activities.	F 1.2

Accessibility of controls

Ventilators that are provided with manual controls (e.g. pull cords, operating rods etc.) should be: • within reasonable reach of occupants • located in accordance with the guidance for Requirements of Approved Document N3 (*Safe opening and closing of windows*) as detailed below.	F
Where controls can be reached without leaning over an obstruction, they should not be more than 1.9 m above the floor. Where there is an obstruction the control should be lower (e.g. not more than 1.7 m where there is a 600 mm deep obstruction).	N3 3.2
Where controls cannot be positioned within safe reach from a permanent stable surface, a safe means of remote operation, such as a manual or electrical system, should be provided.	N3 3.2
Where there is a danger of the operator or other person falling through a window above ground floor level, suitable opening limiters should be fitted or guarding should be provided.	N3 3.3

Note: Although Requirement N3 only applies to work places, for ventilation purposes this requirement also applies to dwellings.

Room thermostats

Room thermostats located in an individual flat with an internal protected stairway or entrance hall should be mounted in the living room at a height between 1370 mm and 1830 mm and its maximum setting should not exceed 27°C.	B1 2.18 (V2)

Access for maintenance

Buildings other than dwellings should include: • reasonable access for the purpose of replacing filters, fans and coils; and • availability of access points for cleaning duct work.	F 2.6
Central plant rooms should include adequate space for the maintenance of the plant (see Figure 6.26).	F 2.7

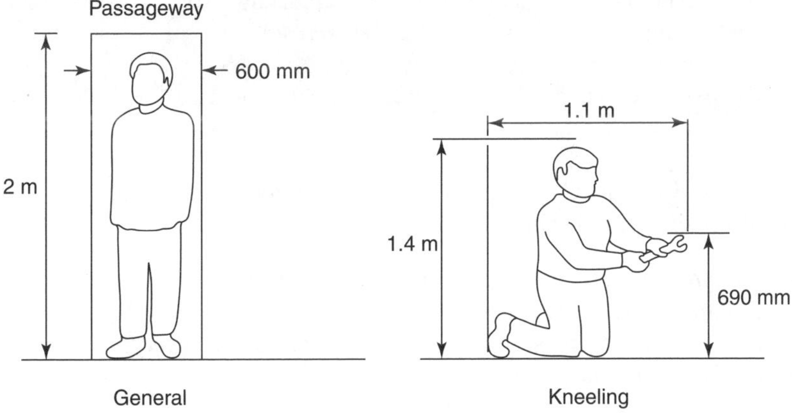

Figure 6.26 Access space in central plant rooms

Combustion appliances

If open-flued combustion appliances and extract fans are going to be installed, then the combustion appliance should be capable of operating safely – whether or not the fans are running.	F 1.3

Exhaust outlets

Exhaust outlets should be located so that re-entry, or ingestion, in to the building and/or other nearby buildings, is minimized. This can be achieved by ensuring that:

exhausts: F App F

- are located downstream of air intakes which are located in a prevailing wind direction;
- do not discharge into courtyards, enclosures or architectural screens.

stacks discharge vertically upwards with sufficient height to clear surrounding buildings and avoid a downwash occurring. F App F

Note: Where possible, pollutants from stacks should be grouped together and discharged vertically upwards.

External doors

The height times width of an external door (including patio doors) should be at least 1/20 of the floor area of the room. F App B

If a room contains more than one external door (or a combination of at least one external door and at least one openable window) then the areas of **all** the opening parts may be added together to achieve the required floor area.

Note: See Appendix C of Approved Document F for example calculations for ventilator sizing for dwellings using:

- background ventilators and intermittent extract fans;
- continuous mechanical extract;
- continuous mechanical supply and extract with heat recovery;
- passive stack ventilation.

Location of ventilation devices in rooms

Openings in compartment walls and/or compartment floors should be limited to those for the passage of pipes, ventilation ducts, service cables, chimneys, appliance ventilation ducts or ducts encasing one or more flue pipes. B3 8.34 (V2)

Cooker hoods should be 650 to 750 mm above the hob surface.	F Table 1.4
Ducts etc. that are located in a protected stairway of a dwelling shall not allow smoke or fire to spread into the stairway.	F Table 1.4
Mechanical extract terminals and fans should be located as high as possible.	F Table 1.4
Mechanical supply terminals should be located and directed to avoid draughts.	F Table 1.4
Recirculation (i.e. by the system) of moist air from wet rooms to habitable rooms should be avoided.	F Table 1.4

Windows

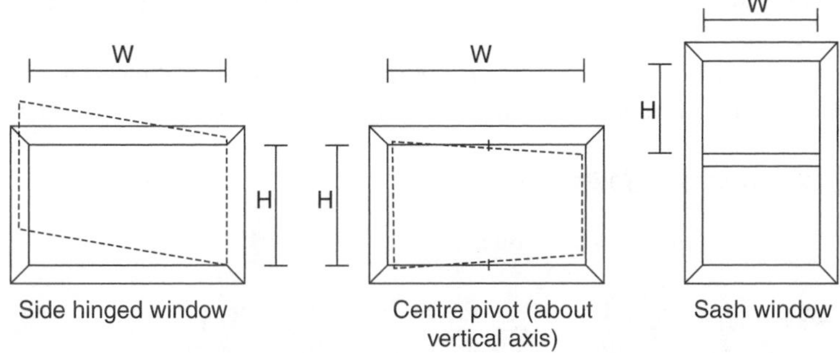

Figure 6.27 Window dimensions

Note: The window opening area is the dimensions of the open area (i.e. height (H) × width (W)).

The height times width of the opening part of hinged or pivot windows that are designed to open **more** than 30° and/or sliding sash windows, should be at least 1/20 of the floor area of the room.	F App B
The height times width of the opening part of hinged or pivot windows designed to open **less** than 30° should be at least 1/10 of the floor area of the room.	F App B

If a room contains more than one openable window, then the areas of **all** the opening Approved Documents may be added together to achieve the required floor area.

Protected shaft

A protected shaft conveying piped flammable gas should have ventilation openings to the outside air at high and low level in the shaft.	B2 8.41 (V2)

Note: Generally speaking, an external wall of a protected shaft does not need to have fire resistance (but see BS 5588-5:2004 for fire resistance of external walls of firefighting shafts).

6.3.13 Work on existing buildings

Under Regulation 3(1) and 3(1A) of the Building Regulations 2000 (as amended), **windows are a controlled fitting**. These clauses, therefore, make it mandatory that when windows in an existing building are replaced the replacement work:

- shall comply with the requirements of Approved Documents L and Approved Document N;
- shall not have a worse level of compliance with other applicable Approved Documents of Schedule 1 (in particular Approved Documents B, F and J).

Replacement windows

All replacement windows should include trickle ventilators or have an equivalent background ventilation opening in the same room.	F 3.4
Ventilation openings should not be smaller than the original opening and it should be controllable.	F 3.6
Where there was no previous ventilation opening, or where the size of the original ventilation opening is not known, the replacement window(s) shall be greater than the minimum requirements shown in Tables 6.22 and 6.23.	F 3.6

Table 6.22 Equivalent areas for replacement windows – dwellings

Type of room	Equivalent area
Habitable rooms	5000 mm^2
Kitchen	2500 mm^2
Utility room	2500 mm^2
Bathroom (without a WC)	2500 mm^2

Table 6.23 Equivalent areas for replacement windows – buildings other than dwellings

Type of room	Equivalent area
Occupiable rooms with floor areas $>10\,m^2$	$2500\,mm^2$
Occupiable rooms with floor areas $<10\,m^2$	$250\,mm^2$ per m^2 of floor area
Kitchens (domestic type)	$2500\,mm^2$
Bathrooms and shower rooms	$2500\,mm^2$ per bath or shower
Sanitary accommodation (and/or washing facilities)	$2500\,mm^2$ per WC

The addition of a habitable room

The general ventilation rates for an additional habitable room (not including a conservatory) to an existing building may be achieved by using background ventilators, heat recovery ventilators and/or purge ventilation.

A single room heat recovery ventilator may be used to ventilate an additional habitable room (F 3.8b).

Additional requirements for background ventilators

If the additional room is connected to an existing habitable room which now has no windows opening to outside, then the ventilation opening (or openings) shall be greater than $8000\,mm^2$ equivalent area. F 3.8a(i)

If the additional room is connected to an existing habitable room which still has windows opening to outside, but with a total background ventilator equivalent area less than $5000\,mm^2$ equivalent area, then the ventilation opening (or openings) shall be greater than $8000\,mm^2$ equivalent area. F 3.8a(ii)

If the additional room is connected to an existing habitable room which still has windows opening to outside, but with a total background ventilator equivalent area of at least $5000\,mm^2$ equivalent area, then there should be: F 3.8a(iii)

- background ventilators of at least $8000\,mm^2$ equivalent area between the two rooms and
- background ventilators of at least $8000\,mm^2$ equivalent area between the additional room and outside.

The addition of a wet room to an existing building

Internal doors between the wet room and the existing building should have an undercut of at least minimum area 7600 mm^2 (equivalent to an undercut of 10 mm above the floor finish for a standard 760 mm width door).	F 3.13
Whole building and extract ventilation can be provided by: • intermittent extract and a background ventilator of at least 2500 mm^2 equivalent area or • single room heat recovery ventilator or • passive stack ventilator or • continuous extract fan.	F 3.12

The addition of a conservatory to an existing building

The general ventilation rate for conservatories with a floor area greater than 30 m^2 conservatory (and adjoining rooms) can be achieved by the use of background ventilators.	F 3.18

Historic buildings

Ventilation systems should **not** introduce new or increased technical risk, or in any other way prejudice the use or character of the building – particularly historic buildings that are:

- listed;
- situated in a conservation area;
- have a local architectural and historical interest;
- are within a national park, an area of outstanding natural beauty or a world heritage site.

Many books have been written about the problems related to restoring historic buildings and before considering any work of this nature, you would be advised to seek the advice of the local planning authority's conservation officer, particularly if you are contemplating:

- the restoration of a historic building that had been subject to previous inappropriate alteration (such as replacement windows, doors and rooflights);

- rebuilding a former historic building following a fire or major demolition;
- making the building's fabric to 'breathe', in order to control moisture and potential long-term decay.

In all cases:

The overall aim should be to improve ventilation of a historic building without: • having a detrimental influence on the character of the building; • increasing the risk of long-term deterioration of the building's fabric or fittings.	F 3.21

6.4 Drainage

6.4.1 The requirement (Building Act 1984 Sections 21 and 22)

All plans for building work need to show that drainage of refuse water (e.g. from sinks) and rainwater (from roofs) have been adequately catered for. Failure to do so will mean that these plans will be rejected by the local authority.

All plans for buildings must include at least one (or more) water or earth closets **unless** the local authority are satisfied that one is not required (for example in a large garage separated from the house).

If you propose using an earth closet, the local authority cannot reject the plans unless they consider that there is insufficient water supply to that earth closet.

What are the rules about drainage? (Building Act 1984 Section 59)

The Building Act requires that all drains are connected either with a sewer (unless the sewer is more than 120 ft away or the person carrying out the building work is not entitled to have access to the intervening land) or is able to discharge into a cesspool, settlement tank or other tank designed for the reception and/or disposal of foul matter from buildings.

The local authorities view this requirement very seriously and will need to be satisfied that:

- satisfactory provision has been made for drainage;
- all cesspools, private sewers, septic tanks, drains, soil pipes, rain water pipes, spouts, sinks or other appliances are adequate for the building in question;
- all private sewers that connect directly or indirectly to the public sewer are not capable of admitting subsoil water;
- the condition of a cesspool is not detrimental to health, or does not present a nuisance;
- cesspools, private sewers and drains previously used, but now no longer in service, do not prejudice health or become a nuisance.

This requirement can become quite a problem if it is not recognized in the early planning stages and so it is always best to seek the advice of the local authority. In certain circumstances, the local authority might even help to pay for the cost of connecting you up to the nearest sewer!

The local authority has the authority to make the owner renew, repair or cleanse existing cesspools, sewers and drains etc.

Can two buildings share the same drainage?

Usually the local authority will require every building to be drained separately into an existing sewer but in some circumstances they may decide that it would be more cost effective if the buildings were drained in combination. On occasions, they might even recommend that a private sewer is constructed.

What about ventilation of soil pipes? (Building Act 1984 Section 60)

A major requirement of the Building Regulations is that all soil pipes from water closets shall be properly ventilated and that no use shall be made of:

- an existing or proposed pipe designed to carry rain water from a roof to convey soil and drainage from a sanitary convenience;
- an existing pipe designed to carry surface water from a premises to act as a ventilating shaft to a drain or a sewer conveying foul water.

What happens if I need to disconnect an existing drain? (Building Act 1984 Section 62)

If, in the course of your building work, you need to:

- reconstruct, renew or repair an existing drain that is joined up with a sewer or another drain;
- alter the position of an existing drain that is joined up with a sewer or another drain;
- seal off an existing drain that is joined up with a sewer or another drain,

then, provided that you give 48 hours' notice to the local authority, the person undertaking the reconstruction may break open any street for this purpose.

You do not need to comply with this requirement if you are demolishing an existing building.

Can I repair an existing water closet or drain? (Building Act 1984 Section 63)

Repairs can be carried out to water closets, drains and soil pipes, but if that repair or construction work is prejudicial to health and/or a public nuisance, then the person who completed the installation or repair is liable, on conviction, to a heavy fine.

In the Greater London area, a 'water closet' can **also** be taken to mean a urinal.

Can I repair an existing drain? (Building Act 1984 Section 61)

Only in extreme emergencies are you allowed to repair, reconstruct or alter the course of an underground drain that joins up with a sewer, cesspool or other drainage method (e.g. septic tank).

If you have to carry out repairs etc. in an emergency, then make sure that you do **not** cover over the drain or sewer without notifying the local authority of your intentions!

Drains – Fire protection

Drains should also provide a degree of fire protection as shown by the following requirement:

- *all openings in fire-separating elements shall be suitably protected in order to maintain the integrity of the continuity of the fire separation,*
- *any hidden voids in the construction shall be sealed and subdivided to inhibit the unseen spread of fire and products of combustion, in order to reduce the risk of structural failure, and the spread of fire.*

(Approved Document B3)

Foul water drainage

The foul water drainage system shall:

- *convey the flow-off foul water to a foul water outfall (i.e. sewer, cesspool, septic tank or settlement (i.e. holding) tank),*
- *minimise the risk of blockage or leakage,*
- *prevent foul air from the drainage system from entering the building under working conditions,*
- *be ventilated,*
- *be accessible for clearing blockages,*
- *not increase the vulnerability of the building to flooding.*

(Approved Document H1)

Wastewater treatment systems and cesspools

Wastewater treatment systems shall:

- *have sufficient capacity to enable breakdown and settlement of solid matter in the wastewater from the buildings;*
- *be sited and constructed so as to prevent overloading of the receiving water.*

Cesspools shall have sufficient capacity to store the foul water from the building until they are emptied.

Wastewater treatment systems and cesspools shall be sited and constructed so as not to:

- *be prejudicial to health or a nuisance;*
- *adversely affect water sources or resources;*

- *pollute controlled waters;*
- *be in an area where there is a risk of flooding.*

Septic tanks and wastewater treatment systems and cesspools are constructed and sited so as to:

- *have adequate ventilation;*
- *prevent leakage of the contents and ingress of subsoil water;*
- *have regard to water table levels at any time of the year and rising groundwater levels.*

Drainage fields are sited and constructed so as to:

- *avoid overloading of the soakage capacity, and*
- *provide adequately for the availability of an aerated layer in the soil at all times.*

(Approved Document H2)

Rainwater drainage

Rainwater drainage systems shall:

- *minimize the risk of blockage or leakage;*
- *be accessible for clearing blockages;*
- *ensure that rainwater soaking into the ground is distributed sufficiently so that it does not damage foundations of the proposed building or any adjacent structure;*
- *ensure that rainwater from roofs and paved areas is carried away from the surface either by a drainage system or by other means;*
- *ensure that the rainwater drainage system carries the flow of rainwater from the roof to an outfall (e.g. a soakaway, a watercourse, a surface water or a combined sewer).*

(Approved Document H3)

Building over existing sewers

Building or extension or work involving underpinning shall:

- *be constructed or carried out in a manner which will not overload or otherwise cause damage to the drain, sewer or disposal main either during or after the construction;*
- *not obstruct reasonable access to any manhole or inspection chamber on the drain, sewer or disposal main;*
- *in the event of the drain, sewer or disposal main requiring replacement, not unduly obstruct work to replace the drain, sewer or disposal main, on its present alignment;*
- *reduce the risk of damage to the building as a result of failure of the drain, sewer or disposal main.*

(Approved Document H4)

Separate systems for drainage

Separate systems of drains and sewers shall be provided for foul water and rainwater where:

(*a*) *the rainwater is not contaminated; and*
(*b*) *the drainage is to be connected either directly or indirectly to the public sewer system and either –*
   (*i*) *the public sewer system in the area comprises separate systems for foul water and surface water; or*
   (*ii*) *a system of sewers which provides for the separate conveyance of surface water is under construction either by the sewerage undertaker or by some other person (where the sewer is the subject of an agreement to make a declaration of vesting pursuant to section 104 of the Water Industry Act 1991).*

(Approved Document H5)

Solid waste storage shall be:

- *designed and sited so as not to be prejudicial to health,*
- *of sufficient capacity having regard to the quantity of solid waste to be removed and the frequency of removal,*
- *sited so as to be accessible for use by people in the building and of ready access from a street for emptying and removal.*

(Approved Document H6)

(Building Act 1984 Section 84)

You are required by the Building Act 1984 to ensure that all courts, yards and passageways giving access to a house, industrial or commercial building (not maintained at public expense) are capable of allowing satisfactory drainage of its surface or subsoil to a proper outfall.

The local authority can require the owner of any of the buildings to complete such works as may be necessary to remedy the defect.

6.4.2 Meeting the requirement

Enclosures for drainage and/or water supply pipes

The enclosure should:	B3 7.6–7.9 (V1)
• be bounded by a compartment wall or floor, an outside wall, an intermediate floor, or a casing;	B3 10.7 (V2)
• have internal surfaces (except framing members) of Class 0;	
• not have an access panel which opens into a circulation space or bedroom;	
• be used only for drainage, or water supply, or vent pipes for a drainage system.	

The casing should:	B3 7.6–7.9 (V1)
• be imperforate except for an opening for a pipe or an access pane	B3 10.7 (V2)
• not be of sheet metal;	
• have (including any access panel) not less than 30 minutes' fire resistance.	
The opening for a pipe, either in the structure or the casing, should be as small as possible and fire-stopped around the pipe.	B3 7.6–7.9 (V1) B3 10.7 (V2)

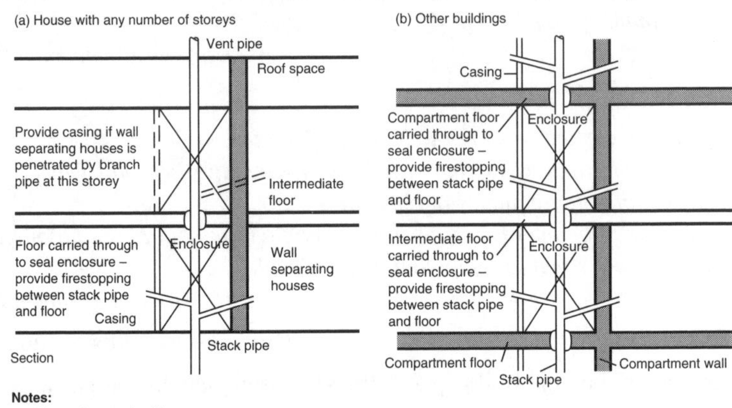

Notes:

1 The enclosure should:
   a be bounded by a compartment wall or floor, an outside wall, an intermediate floor, or a casing (see specification at 2 below), and
   b have internal surfaces (except framing members) of Class 0, and
   c not have an access panel which opens into a circulation space or bedroom, and
   d be used only for drainage, or water supply, or vent pipes for a drainage system

2 The casing should:
   a be imperforate except for an opening for a pipe or an access panel, and
   b not be of sheet metal, and
   c have (including any access panel) not less than 30 minutes fire resistance

3 The opening for a pipe, either in the structure or the casing, should be as small as possible and fire-stopped around the pipe.

Figure 6.28 Enclosure for drainage or water supply pipes

Protection of openings for pipes

Pipes which pass through a compartment wall or compartment floor (unless the pipe is in a protected shaft), or through a cavity barrier, should conform to one of the following alternatives:

Proprietary seals (any pipe diameter) that maintain the fire resistance of the wall, floor or cavity barrier.	B3 7.6–7.7 (V1) B3 10.5–10.6 (V1)
Pipes with a restricted diameter should be used where fire-stopping is used around the pipe, keeping the opening as small as possible	B3 7.6–7.8 (V1) B3 10.5–10.7 (V2)

Sleeving – a pipe of lead, aluminium, aluminium alloy, fibre-cement or UPVC, with a maximum nominal internal diameter of 160 mm, may be used with a sleeving of non-combustible pipe as shown in Figure 6.29.	B3 7.6–7.9 (V1) B3 10.5–10.8(V2)

Foul water drainage

The capacity of the system should be large enough to carry the expected flow at any point (*BS 5572, BS 8301*).	H1 (0.1)
All pipes, fittings and joints should be capable of withstanding an air test of positive pressure of at least 38 mm water gauge for at least 3 minutes.	H1 (1.38)
Every trap should maintain a water seal of at least 25 mm.	H1 (1.38)

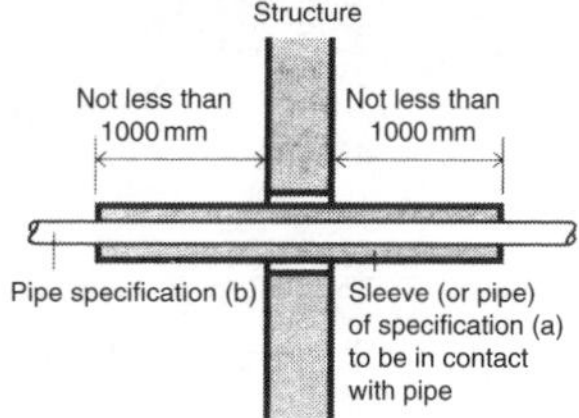

Figure 6.29 Pipes penetrating a structure
Make the opening in the structure as small as possible and provide fire-stopping between pipe and structure.

Traps

All points of discharge into the system should be fitted with a trap (e.g. a water seal) to prevent foul air from the system entering the building.	H1 (1.3–1.4)
All traps should be fitted directly over an appliance and should be removable or be fitted with a cleaning eye.	H1 (1.6)

Branch discharge pipes

Branch pipes should either discharge into another branch pipe or a discharge stack (unless the appliances discharge into a gully on the ground floor or at basement level).	H1 (1.5)
If the appliances are on the ground floor, the pipe(s) may discharge to a stub stack, discharge stack,	H1 (1.5–1.17) H1 (1.11)

directly to a drain, or (if the pipe carries only waste water) to a gully.	H1 (1.30)
A branch pipe from a ground floor closet should only discharge directly to a drain if the depth from the floor to the drain is 1.3 m or less (see Figure 6.30).	H1 (1.9)
A branch pipe serving any ground floor appliance may discharge direct to a drain or into its own stack.	H1 (A5)
A branch pipe should not discharge into a stack in a way which could cause cross flow into any other branch pipe (see Figure 6.31).	H1 (1.10)

Table 6.24 Minimum trap sizes and seal depths

Appliance	Diameter of trap (mm)	Depth of seal (mm of water or equivalent)
Washbasin Bidet	32	75
Bath Shower	40	50
Food waste disposal unit Urinal bowl	40	75
Sink Washing machine Dishwashing machine		
WC pan (outlet < 80 mm) WC pan (outlet > 80 mm)	75 100	50 50

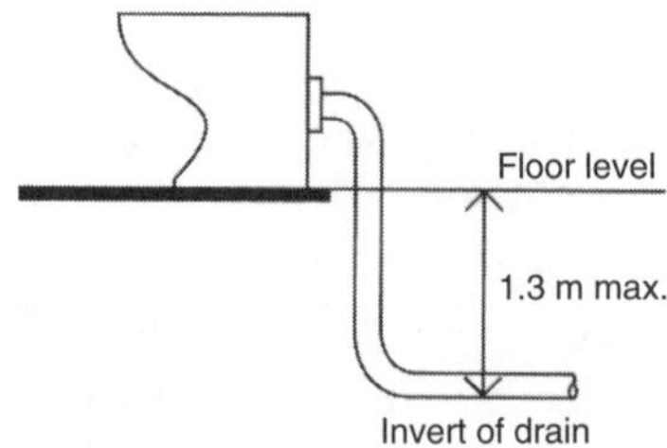

Figure 6.30 Direct connection of ground floor WC to drain

Branch discharge pipes

A branch discharge pipe should not discharge into a stack lower than 450 mm above the invert of the tail of the bend at the foot of the stack in single dwellings up to 3 storeys (see Figure 6.32).	H1 (1.8) H1 (A3, A4) H1 (1.21)

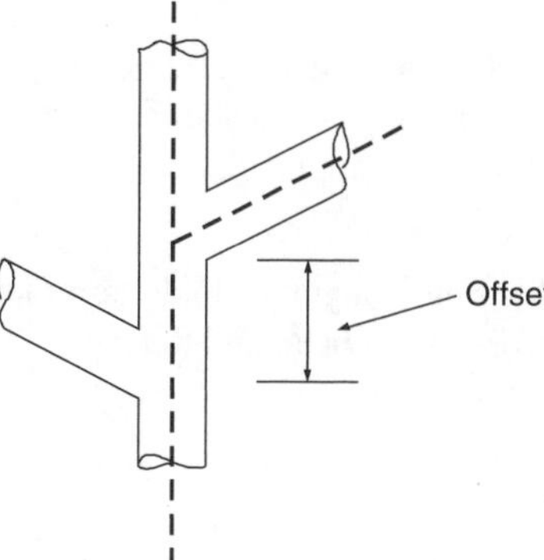

Figure 6.31 Branch connections

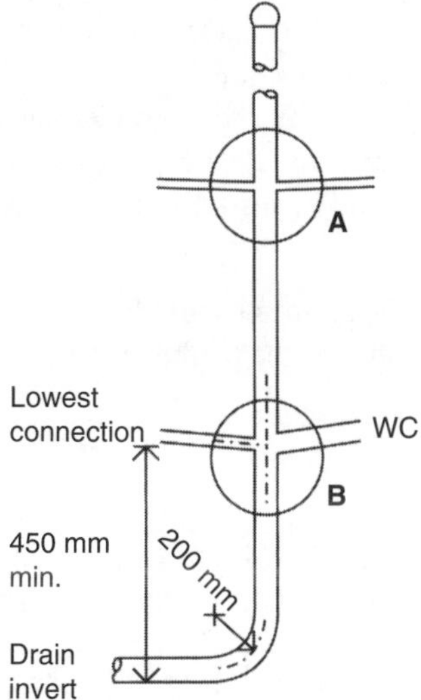

Figure 6.32 Branch discharge stack

Branch discharge pipes

Branch pipes may discharge into a stub stack.	H1 (1.12) H1 (1.30)
A branch pipe discharging to a gully should terminate between the grating or sealing plate and the top of the water seal.	H1 (1.13)
Bends in branch pipes should be avoided if possible.	H1 (1.16)
Junctions on branch pipes should be made with a sweep of 25 mm radius or at 45°.	H1 (1.17)

Rodding points should be provided to give access to any lengths of discharge pipes which cannot be reached by removing traps or appliances with integral traps. H1 (1.25) H1 (1.6)

A branch pipe discharging to a gully should terminate between the grating or sealing plate and the top of the water seal. H1 (1.13)

Condensate drainage from boilers may be connected to sanitary pipework provided: H1 (1.14)

(a) The connection should preferably be made to an internal stack with a 75 mm condensate trap.
(b) If the connection is made to a branch pipe, the connection should be made downstream of any sink waste connection.
(c) All sanitary pipework receiving condensate should be made from materials resistant to a pH value of 6.5 and lower and be installed in accordance with BS 6798.

Pipes serving a single appliance should have at least the same diameter as the appliance trap (see Table 6.24).

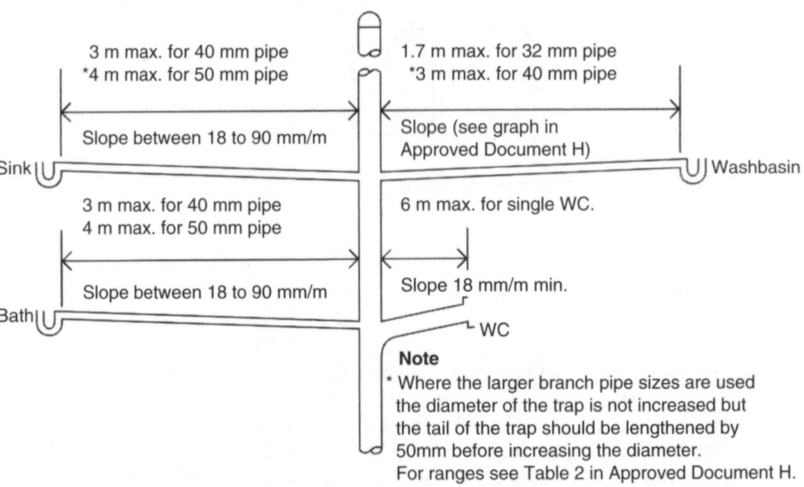

Figure 6.33 Branched connections

A separate ventilating stack is only likely to be preferred where the numbers of sanitary appliances and their distance to a discharge stack are large.

Branch ventilation stacks

Should be connected to the discharge pipe within 750 mm of the trap and should connect to the ventilating stack or the stack vent, above the highest 'spillover' level of the appliances served.	H1 (1.22)
The ventilating pipe should have a continuous incline from the discharge pipe to the point of connection to the ventilating stack or stack vent.	H1 (1.22)
Branch ventilating pipes which run direct to outside air should finish at least 900 mm above any opening into the building nearer than 3 m (see Figure 6.35).	H1 (1.23)
A dry stack may provide ventilation for branch ventilation pipes as an alternative to carrying them to outside air or to a ventilated discharge stack (ventilated system).	H1 (A7 and 1.21)
Ventilation stacks serving buildings with not more than 10 storeys and containing only dwellings should be at least 32 mm diameter (for all other buildings see paragraph H1 (1.29)).	H1 (A8) H1 (1.21 and 1.29)
A separate ventilating stack is only likely to be preferred where the numbers of ventilating pipes and their distance to a discharge stack are large.	H1 (1.19) H1 (Table 2)

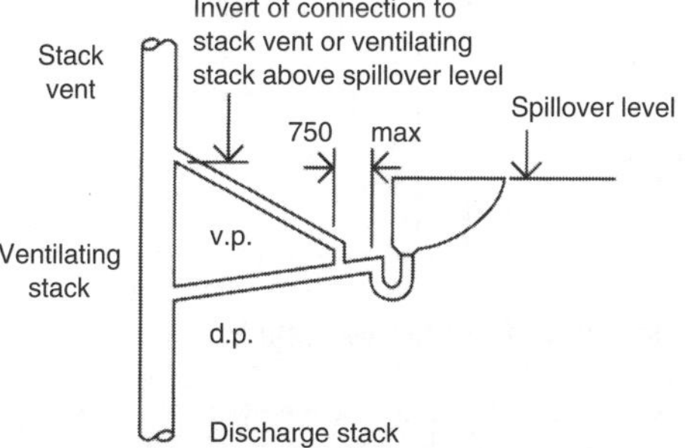

Figure 6.34 Branch ventilation pipes

Discharge stacks

All stacks should discharge to a drain.	H1 (1.26)
The bend at the foot of the stack should have as large a radius (i.e. at least 200 mm) as possible.	H1 (1.26)

Discharge stacks should be ventilated.	H1 (1.29)
Offsets in the 'wet' portion of a discharge stack should be avoided.	H1 (1.27)
Stacks serving urinals should be not less than 50 mm.	H1 (1.28)
Stacks serving closets with outlets less than 80 mm should be not less than 75 mm.	H1 (1.28)
Stacks serving closets with outlets greater than 80 mm should be not less than 100 mm.	H1 (1.28)
The internal diameter of the stack should be not less than that of the largest trap or branch discharge pipe.	H1 (1.28)

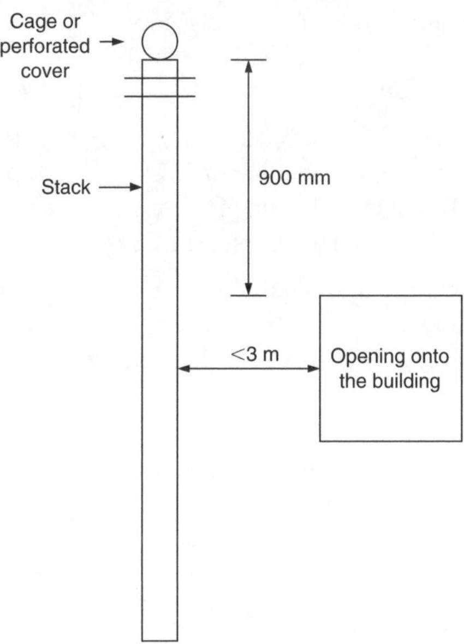

Figure 6.35 Termination of ventilation stacks

Ventilating pipes open to outside air should finish at least 900 mm above any opening into the building within 3 m and should be fitted with a perforated cover or cage (see Figure 6.35) which should be metal if rodent control is a problem.	H1 (1.31)
Ventilating pipes open to outside air should finish at least 900 mm above any opening into the building within 3 m.	H1 (1.31)

Ventilating pipes should be finished with a wire cage (metallic in areas with a rodent problem) or other perforated cover, fixed to the end of the ventilating pipe.	H1 (1.31)
Stack ventilation pipes should be not less than 75 mm.	H1 (1.32)
Ventilated discharge stacks may be terminated inside a building when fitted with air admittance valves complying with prEN 12380.	H1 (1.33)
Discharge stacks may terminate inside a building when fitted with air admittance valves.	H1 (1.29)
Rodding points should be provided to give access to any lengths of pipe that cannot be reached from any other part of the system.	H1 (1.30–1.31)
Pipes should be firmly supported without restricting thermal movement.	H1 (1.31)
Pipes, fittings and joints should be airtight.	H1 (1.32)
A stub stack may be used if it connects into a ventilated discharge stack or into a ventilated drain not subject to surcharging.	H1 (1.30)
Air admittance valves should be located in areas that have adequate ventilation.	H1 (1.33)
Air admittance valves should not be used outside buildings or in dust laden atmospheres.	H1 (1.33)
Rodding points should be provided in discharge stacks.	H1 (1.34)
Pipes should be firmly supported without restricting thermal movement.	H1 (1.35)
Sanitary pipework connected to WCs should not allow light to be visible through the pipe wall, as this is believed to encourage damage by rodents.	H1 (1.36)

Drainage serving kitchens in commercial hot food premises should be fitted with a grease separator complying with prEN 1825-1.

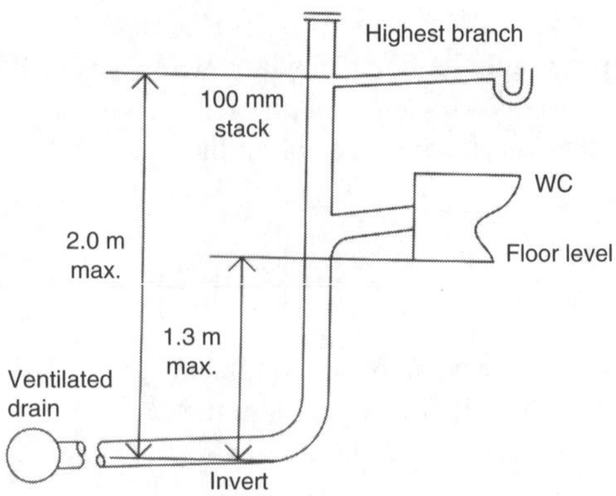

Figure 6.36 Stub stack

Foul drainage

Some public sewers may carry foul water and rainwater in the same pipe. If the drainage system is also to carry rainwater to such a sewer these combined systems should not be capable of discharging into a cesspool or septic tank. H1 (2.1)

Foul drainage should be connected to either:

- a public foul or combined sewer (wherever this is reasonably practicable) H1 (2.3)
- an existing private sewer that connects with a public sewer, or H1 (2.6)
- a wastewater treatment system or cesspool should be provided. H1 (2.7)

Combined and rainwater sewers shall be designed to surcharge (i.e. the water level in the manhole rises above the top of the pipe) in heavy rainfall. H1 (2.8)

Basements containing sanitary appliances, where the risk of flooding due to sewer surcharge of the sewer is possible should either use an anti-flooding valve (if the risk is low) or be pumped. H1 (2.9) H1 (2.36–2.39) H1 (2.10)

For other low lying sites (i.e. not basements) where the risk is considered low, a gully (at least 75 mm below the floor level) can be dug outside the building.

Anti-flooding valves should preferably be a double valve type that complies with prEN 13564.	H1 (2.11)
The layout of the drainage system should be kept simple.	H1 (2.13)
Pipes should (wherever possible) be laid in straight lines. Changes of direction and gradient should be minimized.	
Access points should be provided only if blockages could not be cleared without them.	H1 (2.13)
Connections should be made using prefabricated components.	H1 (2.15)
Connection of drains to other drains or private or public sewers and of private sewers to public sewers should be made obliquely, or in the direction of flow.	H1 (2.14)
The system should be ventilated by a flow of air.	H1 (2.18) H1 (1.27–1.29)
Ventilating pipes should not finish near openings in buildings.	H1 (2.18) H1 (1.31)
Pipes should be laid to even gradients and any change of gradient should be combined with an access point.	H1 (2.19) H1 (2.49)
Pipes should also be laid in straight lines where practicable.	H1 (2.20) H1 (2.49)

Rodent control

If the site has been previously developed, the local authority should be consulted to determine whether any special measures are necessary for control of rodents. Special measures which may be taken include the following:

Sealed drainage – should have access covers to the pipework in the inspection chamber instead of an open channel.	H1 (2.22a)
Intercepting traps – should be of the locking type that can be easily removed from the chamber surface and securely replaced.	H1 (2.22b)
Rodent barriers – including enlarged sections on discharge stacks to prevent rats climbing, flexible	H1 (2.22c)

downward facing fins in the discharge stack, or one-way valves in underground drainage.	
Metal cages on ventilator stack terminals – to discourage rats from leaving the drainage system.	H1 (2.22d) H1 (1.31)
Covers and gratings to gullies – used to discourage rats from leaving the system.	H1 (2.22e)
During construction, drains and sewers that are left open should be covered when work is not in progress to prevent entry by rats.	H1 (2.56)
Disused drains or sewers less than 1.5 m deep that are in open ground should as far as is practicable be removed. Other pipes should be sealed at both ends (and at any point of connection) and grout filled to ensure that rats cannot gain access.	H1 (B18)

Protection from settlement

• A drain may run under a building if at least 100 mm of granular or other flexible filling is provided round the pipe.	H1 (2.23)
• Where pipes are built into a structure (e.g. inspection chamber, manhole, footing, ground beam or wall) suitable measures (such as using rocker joints or a lintel) should be taken to prevent damage or misalignment (see Figures 6.37 and 6.38).	H1 (2.24)
The depth of cover will usually depend on the levels of the connections to the system, the gradients at which the pipes should be laid and the ground levels.	H1 (2.27) H1 (2.41–2.45)
All drain trenches should not be excavated lower than the foundations of any building nearby (see Figure 6.39).	H1 (2.25)

Pipe gradients and sizes

Drains should have enough capacity to carry the anticipated maximum flow (see Table 6.25).	H1 (2.29)

Sewers (i.e. a drain serving more than one property) should have a minimum diameter of 100 mm when serving 10 dwellings or diameter of 150 mm if more than 10.	H1 (2.30)
Drains carrying foul water should have an internal diameter of at least 75 mm.	H1 (2.33)
Drains carrying effluent from a WC or trade effluent should have an internal diameter of at least 100 mm.	H1 (2.33)

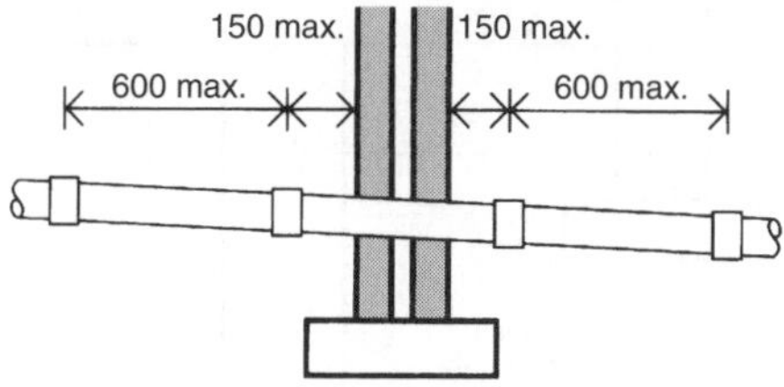

Figure 6.37 Pipe imbedded in the wall. Short length of pipe bedded in a wall with joints 150 mm of either wallface. Additional rocker pipes (max length 600 mm) with flexible joints are then added

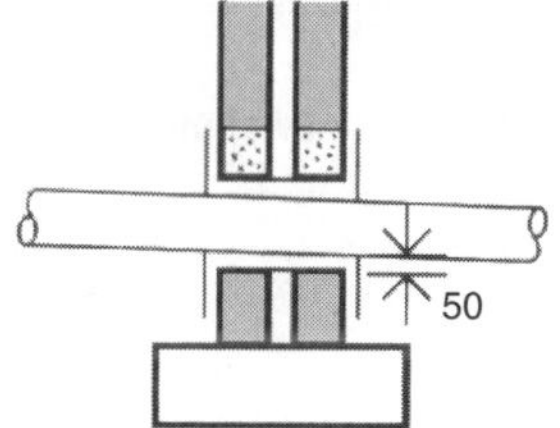

Figure 6.38 Pipe shielded by a lintel. Both sides are masked with rigid sheet material (to prevent entry of fill or vermin) and the void is filled with a compressible sealant to prevent entry of gas

Pumping installations

Where gravity drainage is impracticable, or protection against flooding due to surcharge in downstream sewers is required, a pumping installation will be needed.	H1 (2.36)
Where foul water drainage from a building is to be pumped, the effluent receiving chamber should be sized to contain 24-hour inflow to allow for disruption in service.	H1 (2.39)
The minimum daily discharge of foul drainage should be taken as 150 litres per head per day for domestic use.	H1 (2.39)

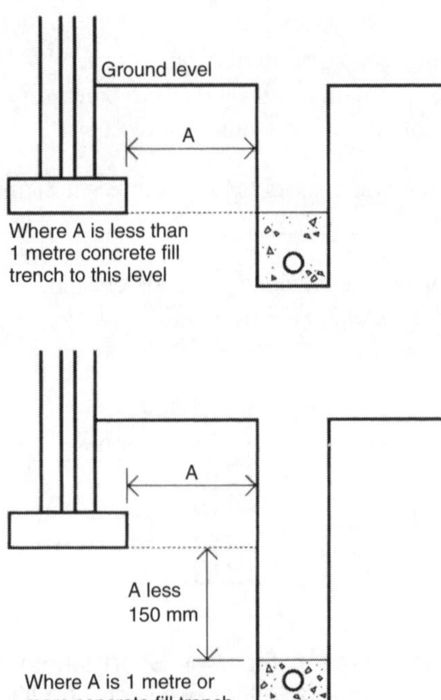

Figure 6.39 Pipe runs near buildings

Table 6.25 Flow rates from dwellings

Number of dwellings	Flow rate (litres/sec)
1	2.5
5	3.5
10	4.1
15	4.6
20	5.1
26	5.4
30	5.8

Table 6.26 Materials for below-ground gravity drainage

Material	British Standard
Rigid pipes	
Vitrified clay	BS 65, BSEN 295
Concrete	BS 5911
Grey iron	BS 437
Ductile iron	BSEN 598
Flexible pipes	
UPVC	BSEN 1401
PP	BSEN 1852
Structured walled plastic pipes	BSEN 13476

Materials for pipes and jointing

To minimize the effects of any differential settlement, pipes should have flexible joints.	H1 (2.40)
All joints should remain watertight under working and test conditions.	H1 (2.40)
Nothing in the pipes, joints or fittings should project into the pipe line or cause an obstruction.	H1 (2.40)
Different metals should be separated by non-metallic materials to prevent electrolytic corrosion.	H1 (2.40)

Bedding and backfill

The choice of bedding and backfill depends on the depth at which the pipes are to be laid and the size and strength of the pipes.	H1 (2.41)
Special precautions should be taken to take account of the effects of settlement where pipes run under or near buildings.	
The depth of the pipe cover will usually depend on the levels of the connections to the system and the gradients at which the pipes should be laid and the ground levels.	
Pipes need to be protected from damage particularly pipes which could be damaged by the weight of backfilling.	
Rigid pipes should be laid in a trench as shown in Figure 6.40.	H1 (2.42)

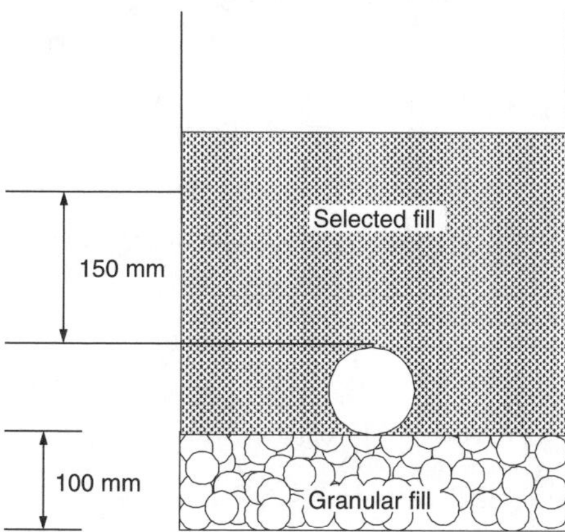

Figure 6.40 Bedding for rigid pipes

Flexible pipes shall be supported to limit deformation under load.	H1 (2.44)
Flexible pipes with very little cover shall be protected from damage by a reinforced cover slab with a flexible filler and at least 75 mm of granular material between the top of the pipe and the underside of the flexible filler below the slabs (see Figure 6.42).	H1 (2.42–2.44)
Trenches may be backfilled with concrete to protect nearby foundations. In these cases a movement joint (as shown in Figure 6.43) formed with a compressible board should be provided at each socket or sleeve joint.	H1 (2.45)

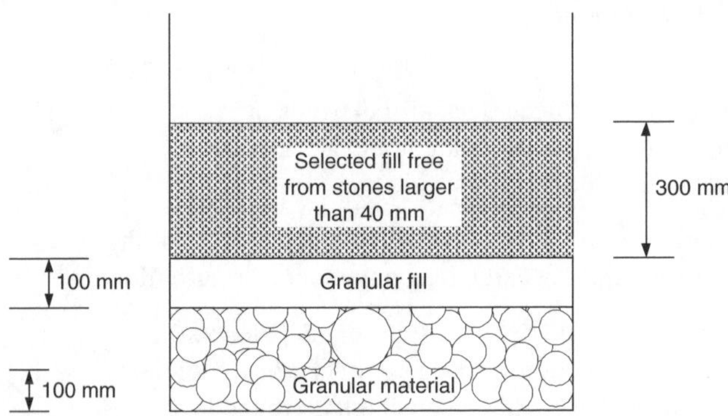

Figure 6.41 Bedding for flexible pipes

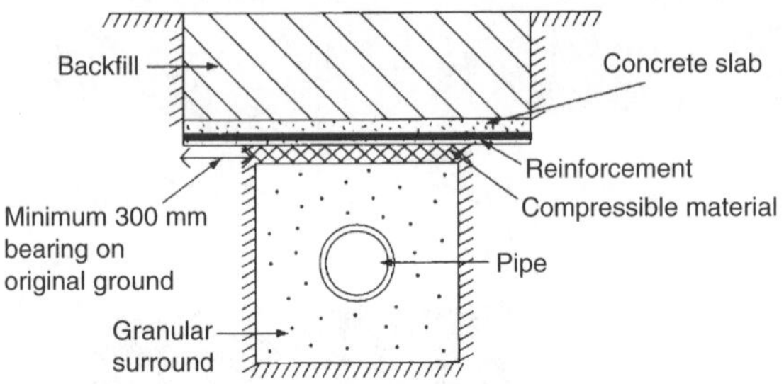

Figure 6.42 Protection of pipes laid in shallow depths

Access points

Access should be provided to long runs.	H1 (2.50)
Sufficient and suitable access points should be provided for clearing blockages from drain runs that cannot be reached by any other means.	H1 (2.46)

Access points should be provided: • on or near the head of each drain run • at a bend • at a change of gradient or pipe size • at a junction.	H1 (2.49)
Access points should be either: • rodding eyes – capped extensions of the pipes; • access fittings – small chambers on (or an extension of) the pipes but not with an open channel; • inspection chambers – chambers with working space at ground level; • manholes – deep chambers with working space at drain level.	H1 (2.48)
Access points should be constructed so as to resist the ingress of ground water or rainwater.	H1 (2.52)
Inspection chambers and manholes should have removable non-ventilating covers of durable material (such as cast iron, cast or pressed steel, precast concrete or UPVC).	H1 (2.54)
Access points to sewers (serving more than one property) should be located in places where they are accessible and apparent for use in an emergency (e.g. highways, public open space, unfenced front gardens, and shared or unfenced driveways).	H1 (2.51)
Inspection chambers and manholes in buildings should have mechanically fixed airtight covers unless the drain itself has watertight access covers.	H1 (2.54)
Manholes deeper than 1 m should have metal step irons or fixed ladders.	H1 (2.54)

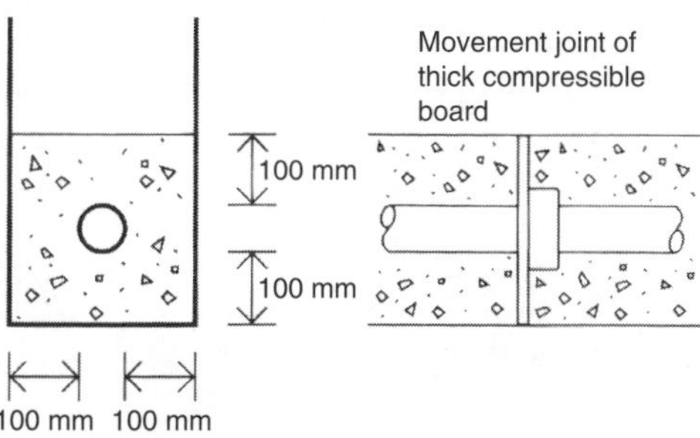

Figure 6.43 Joints for concrete encased pipes

General

Drains and sewers should be protected from damage by construction traffic and heavy machinery.	H1 (2.57)
Heavy materials should not be stored over drains or sewers.	H1 (2.57)
After laying (including any necessary concrete or other haunching or surrounding and backfiring) gravity drains and private sewers should be tested for watertightness.	H1 (2.59)
All pipework carrying greywater for reuse should be clearly marked with the word 'GREYWATER'.	H1 (A11)
Material alterations to existing drains and sewers are subject to (and covered by) the Building Regulations.	H1 (b7)
Repairs, reconstruction and alterations to existing drains and sewers should be carried out to the same standards as new drains and sewers.	H1 (B15)

Wastewater treatment systems and cesspools

A notice giving information as to the nature and frequency of maintenance required for the cesspool or wastewater treatment system to continue to function satisfactorily should be displayed within each of the buildings.

The use of non-mains foul drainage, such as wastewater treatment systems, septic tanks or cesspools, should only be considered where connection to mains drainage is **not** practicable.

Any discharge from a wastewater treatment system is likely to require a consent from the Environment Agency. For the detailed design and installation of small sewage treatment works, specialist knowledge is advisable. Guidance is also given in BS 6297: 1983 *Code of practice for design and installation of small sewage treatment works and cesspools.*

Septic tanks

Septic tanks with some form of secondary treatment (such as from a drainage field/mound or constructed wetland such as a reed bed) will normally be the most economic means of treating wastewater from small developments (e.g. 1 to 3 dwellings). They provide suitable conditions for the settlement, storage and partial decomposition of solids which need to be removed at regular intervals.

Septic tanks should be sited at least 7 m from any habitable parts of buildings, and preferably down a slope.	H2 (1.16)

Septic tanks should only be used in conjunction with a form of secondary treatment (e.g. a drainage field, drainage mound or constructed wetland).	H2 (1.15)
Septic tanks should be sited within 30 m of a vehicle access to enable the tank to be emptied and cleaned without hazard to the building occupants and without the contents being taken through a dwelling or place of work.	H2 (1.17 and 1.64)
Septic tanks and settlement tanks should have a capacity below the level of the inlet of at least 2700 litres (2.7 m^3) for up to 4 users. This size should be increased by 180 litres for each additional user.	H2 (1.18)
Septic tanks may be constructed in brickwork or concrete (roofed with heavy concrete slabs) or factory-manufactured septic tanks (made out of glass reinforced plastics, polyethylene or steel) can be used.	H2 (1.19–20 and 1.65–66)
The brickwork should consist of engineering bricks at least 220 mm thick. The mortar should be a mix of 1:3 cement sand ratio and in-situ concrete should be at least 150 mm thick of C/25/P mix (see BS 5328).	H2 (1.20 and 1.66)
Septic tanks should be ventilated.	H2 (1.21)
Septic tanks should incorporate at least two chambers or compartments operating in series.	H2 (1.22)
Septic tanks should be provided with access for emptying and cleaning.	H2 (1.24)
A notice should be fixed within the building describing the necessary maintenance.	H2 (1.25)
Septic tanks should be inspected monthly to check they are working correctly.	H2 (A.11)
Septic tank should be emptied at least once a year.	H2 (A.13)

Cesspools

A cesspool is a watertight tank, installed underground, for the storage of sewage. No treatment is involved.

Cesspools should be sited at least 7 m from any habitable parts of buildings and preferably downslope.	H2 (1.58)
Cesspools should be provided with access for emptying and cleaning.	H2 (1.60)
Cesspools should be inspected fortnightly for overflow.	H2 (A.20)
Cesspools should be emptied on a monthly basis by a licensed contractor.	H2 (1.60) H2 (A.21)
A filling rate of 150 litres per person per day is assumed and if the cesspool does not fill within the estimated period, the tank should be inspected for leakage.	H2 (A.22)
Cesspools should be ventilated.	H2 (1.63)
The inlet of a cesspool should be provided with access for inspection.	H2 (1.67)
Cesspools and settlement tanks (if they are to be desludged using a tanker) should be sited within 30 m of a vehicle access.	H2 (1.64)
Cesspools and settlement tanks should prevent leakage of the contents and ingress of subsoil water.	H2 (1.63)
Cesspools should have a capacity below the level of the inlet of at least 18 000 litres (18 m^3) for 2 users increased by 6800 litres (6.8 m^3) for each additional user.	H2 (1.61)
Cesspools, septic tanks and settlement tanks may be constructed in brickwork, concrete, or glass reinforced concrete.	H2 (1.65–66)
Factory-made cesspools and septic tanks are available in glass reinforced plastic, polyethylene or steel.	
The brickwork should consist of engineering bricks at least 220 mm thick. The mortar should be a mix of 1:3 cement sand ratio and in-situ concrete should be at least 150 mm thick of C/25/P mix (see BS 5328).	H2 (1.66)
Cesspools should be covered (with heavy concrete slabs) and ventilated.	
Cesspools should have no openings except for the inlet, access for emptying and ventilation.	H2 (1.62)
Cesspools should be inspected fortnightly for overflow and emptied as required.	H2 (A.20)

Packaged treatment works

This term is applied to a range of systems designed to treat a given hydraulic and organic load using prefabricated components which can be installed with minimal site work. They are capable of treating effluent more efficiently than septic tank systems and this normally allows the product to be directly discharged to a watercourse.

The discharge from the wastewater treatment plant should be sited at least 10 m away from watercourses and any other buildings.	H2 (1.54)
Regular maintenance and inspection should be carried out in accordance with the manufacturer's instructions.	H2 (A.17)

Drainage fields and mounds

Drainage fields (or mounds) serving a wastewater treatment plant or septic tank should be located: • at least 10 m from any watercourse or permeable drain • at least 50 m from the point of abstraction of any groundwater supply • at least 15 m from any building • sufficiently far from any other drainage fields, drainage mounds or soakaways so that the overall soakage capacity of the ground is not exceeded.	H2 (1.27)
No water supply pipes or underground services other than those required by the disposal system itself should be located within the disposal area.	H2 (1.29)
No access roads, driveways or paved areas should be located within the disposal area.	H2 (1.30)
The ground water table should not rise to within 1 m of the invert level of the proposed effluent distribution pipes.	H2 (1.33)
An inspection chamber should be installed between the septic tank and the drainage field.	H2 (1.43)
Constructed wetlands should not be located in the shade of trees or buildings.	H2 (1.47)
The drainage field/mound should be checked on a monthly basis to ensure that it is not waterlogged and that the effluent is not backing up towards the septic tank.	H2 (A.15)

Under Section 50 (overflowing and leaking cesspools) of the Public Health Act 1936 action could be taken against a builder who had caused the problem, and **not** just against the owner.

Under Section 59 (drainage of building) of the Building Act 1984, local authorities can require either the owner or the occupier to remove (or otherwise make innocuous) any disused cesspool, septic tank or settlement tank.

Greywater and rainwater tanks

Greywater and rainwater tanks should: H2 (1.70)

- prevent leakage of the contents and ingress of subsoil water;
- be ventilated;
- have an anti-backflow device;
- be provided with access for emptying and cleaning.

Rainwater drainage

The capacity of the drainage system should be large enough to carry the expected flow at any point in the system. H3 (0.3)

Rainwater or surface water should not be discharged to a cesspool or septic tank. H3 (0.6)

Gutters and rainwater pipes

Although this part of the Building Regulations only actually applies to draining the rainfall from areas of 6 m^2 or more (unless they receive a flow from a rainwater pipe or from paved and/or other hard surfaces). Each case should be considered separately and a decision made. This particularly applies to small roofs and balconies. Table 6.27 shows the largest effective area that should be drained into the gutter sizes most often used.

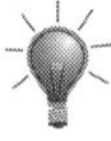

For eaves gutters the design rainfall intensity should be 0.021 litres/second/m^2. In some cases, eaves drop systems may be used (H3 (1.13)).

Gutters should be laid with any fall towards the nearest outlet.

Gutters should be laid so that any overflow in excess of the design capacity (e.g. above normal rainfall) will be discharged clear of the building. H3 (1.7)

Table 6.27 Gutter and outlet sizes

Max effective roof area (m^2)	Gutter size (mm dia)	Outlet size (mm dia)	Flow capacity (litres/sec)
6.0	–	–	–
18.0	75	50	0.38
37.0	100	63	0.78
53.0	115	63	1.11
65.0	125	75	1.37
103.0	150	89	2.16

Rainwater pipes should discharge into a drain or gully (but may discharge to another gutter or onto another surface if it is drained). H3 (1.8)

Any rainwater pipe which discharges into a combined system should do so through a trap. H3 (1.8)

The size of a rainwater pipe should be at least the size of the outlet from the gutter. H3 (1.10)

A down pipe which serves more than one gutter should have an area at least as large as the combined areas of the outlets. H3 (1.10)

On flat roofs, valley gutters and parapet gutters additional outlets may be necessary. H3 (1.7)

Where a rainwater pipe discharges onto a lower roof or paved area, a pipe shoe should be fitted to divert water away from the building. H3 (1.9)

Gutters and rainwater pipes should be firmly supported without restricting thermal movement.

The materials used should be of adequate strength and durability, and H3 (1.16)

- all gutter joints should remain watertight under working conditions
- pipework in siphonic roof drainage systems should be able to resist to negative pressures in accordance with the design
- gutters and rainwater pipes should be firmly supported
- different metals should be separated by non-metallic material to prevent electrolytic corrosion.

Drainage of paved areas

Surface gradients should direct water draining from a paved area away from buildings.	H3 (2.2)
Gradients on impervious surfaces should be designed to permit the water to drain quickly from the surface. A gradient of at least 1 in 60 is recommended.	H3 (2.3)
Paths, driveways and other narrow areas of paving should be free draining to a pervious area such as grassland, provided that: • the water is not discharged adjacent to buildings where it could damage foundations; and • the soakage capacity of the ground is not overloaded.	H3 (2.6)
Where water is to be drained onto the adjacent ground the edge of the paving should be finished above or flush with the surrounding ground to allow the water to run off.	H3 (2.7)
• Where the surrounding ground is not sufficiently permeable to accept the flow, filter drains may be provided.	H3 (2.8 and 3.33)
• Pervious paving should not be used where excessive amounts of sediment are likely to enter the pavement and block the pores.	H3 (2.11)
• Pervious paving should not be used in oil storage areas, or where runoff may be contaminated with pollutants.	H3 (2.12)
• Gullies should be provided at low points where water would otherwise pond.	H3 (2.15)
• Gully gratings should be set approximately 5 mm below the level of the surrounding paved area in order to allow for settlement.	H3 (2.16)
• Provision should be made to prevent silt and grit entering the system, either by provision of gully pots of suitable size, or catchpits.	H3 (2.17)

Surface water drainage

Discharge to a watercourse may require a consent from the Environment Agency, who may limit the rate of discharge. Where other forms of outlet are not practicable, discharge should be made to a sewer (H3 (3.2–3.3)). For design purposes a rainfall interval of 0.014 litres/second/m^2 can be assumed as normal.

Some drainage authorities have sewers that carry both foul water and rainwater (i.e. combined systems) in the same pipe. Where they do, they can allow rainwater to discharge into the system if the sewer has enough capacity to take the added flow. Some private sewers (drains serving more than one property) also carry both foul water and rainwater. If a sewer (or private sewer) operated as a combined system does not have enough capacity, the rainwater should be run in a separate system with its own outfall.

Surface water drainage should discharge to a soakaway or other infiltration system where practicable.	H3 (3.2)
Surface water drainage connected to combined sewers should have traps on all inlets.	H3 (3.7)
Drains should be at least 75 mm diameter.	H3 (3.14)
Where any materials that could cause pollution are stored or used, separate drainage systems should be provided.	H3 (3.21)
On car parks, petrol filling stations or other areas where there is likely to be leakage or spillage of oil, drainage systems should be provided with oil interceptors.	H3 (3.22) H3 (A)
Separators should be leak tight and comply with the requirements of the Environmental Agency and prEN858.	H3 (A.9–10)
Infiltration devices (including soakaways, swales, infiltration basins, and filter drains) should not be built: • within 5 m of a building or road or in areas of unstable land; • in ground where the water table reaches the bottom of the device at any time of the year; • sufficiently far from any drainage fields, drainage mounds or other soakaways; • where the presence of any contamination in the runoff could result in pollution of groundwater source or resource.	H3 (3.23–26)
Soakaways should be designed to a return period of once in ten years.	H3 (3.27)
Soakaways for areas less than 100 m^2 shall consist of square or circular pits, filled with rubble or lined with dry jointed masonry or perforated ring units. Soakaways serving larger areas shall be lined pits or trench type soakaways.	H3 (3.26)

The storage volume should be calculated so that, over the duration of a storm, it is sufficient to contain the difference between the inflow volume and the outflow volume.	H3 (3.29)
Soakaways serving larger areas should be designed in accordance with BS EN 752-4.	H3 (3.30)

Under Section 85 (offences concerning the polluting of controlled waters) of the Water Resources Act 1991 it is an offence to discharge any noxious or polluting material into a watercourse, coastal water, or underground water. Most surface water sewers discharge to watercourses.

Under Section 111 (restrictions on use of public sewers) of the Water Industry Act 1991 it is an offence to discharge petrol into any drain or sewer connected to a public sewer.

Building over existing sewers

Where it is proposed to construct a building over or near a drain or sewer shown on any map of sewers, the developer should consult the owner of the drain or sewer.	H4 (0.3)
A building constructed over or within 3 m of any • rising main • drain or sewer constructed from brick or masonry • drain or sewer in poor condition shall not be constructed in such a position unless special measures are taken.	H4 (1.2)
Buildings or extensions should not be constructed over a manhole or inspection chamber or other access fitting on any sewer (serving more than one property).	H4 (1.3)
A satisfactory diversionary route should be available so that the drain or sewer could be reconstructed without affecting the building.	H4 (1.4)
The length of drain or sewer under a building should not exceed 6 m except with the permission of the owners of the drain or sewer.	H4 (1.5)
Buildings or extensions should not be constructed over or within 3 m of any drain or sewer more than 3 m deep, or greater than 225 mm in diameter except with the permission of the owners of the drain or sewer.	H4 (1.60)

Where a drain or sewer runs under a building at least 100 mm of granular or other suitable flexible filling should be provided round the pipe.	H4 (1.9)
Where a drain or sewer running below a building is less than 2 m deep, the foundation should be extended locally so that the drain or sewer passes through the wall.	H4 (1.10)
Where the drain or sewer is more than 2 m deep to invert and passes beneath the foundations, the foundations should be designed with a lintel spanning over the line of the drain or sewer. The span of the lintel should extend at least 1.5 m either side of the pipe and should be designed so that no load is transmitted onto the drain or sewer.	H4 (1.12)
A drain trench should not be excavated lower than the foundations of any building nearby.	H4 (1.13)

Separate systems for drainage

Separate systems of drains and sewers shall be provided for foul water and rainwater where:

(a) the rainwater is not contaminated; and
(b) the drainage is to be connected either directly or indirectly to the public sewer system, which has separate systems for foul water and surface water.

Solid waste storage

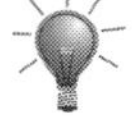

Although the requirements of the Building Regulations do not cover the recycling of household and other waste, H6 sets out general requirements for solid waste storage.

For domestic developments space should be provided for storage of containers for separated waste (i.e. waste that can be recycled is stored separately from waste that cannot) and having a combined capacity of 0.25 m^2 per dwelling.	H6 (1.1)
In low-rise domestic developments (houses, bungalows and flats up to the 4th floor) any dwelling should have, or have access to, a location with at least two movable, individual or communal waste containers.	H6 (1.2)

In multistorey domestic developments, dwellings above the 4th storey should share a container fed by a chute unless siting or operation of a chute is impracticable. In such a case a satisfactory management arrangement for conveying refuse to the storage area should be assured.	H6 (1.6)
In multistorey domestic developments, dwellings up to the 4th floor may each have their own waste container or may share a waste container.	H6 (1.5)
For waste containers up to 250 litres, steps should be avoided between the container store and collection point wherever possible.	H6 (1.10)
Containers and chutes should be sited so that householders are not required to carry refuse further than 30 m.	H6 (1.8)
Containers should be within 25 m of the vehicle access.	H6 (1.8)
Containers should be sited so that they can be collected without being taken through a building, unless it is a garage, carport or other open covered space.	H6 (1.10)
This provision applies only to new buildings.	
The collection point should be reasonably accessible to the size of waste collection vehicles typically used by the waste collection authority.	H6 (1.11)
External storage areas for waste containers should be away from windows and ventilators and preferably be in the shade or under a shelter.	H6 (1.12)
Storage areas should not interfere with pedestrian or vehicle access to buildings.	H6 (1.12)
Where enclosures, compounds or storage rooms are provided they should allow room for filling and emptying and provide a clear space of 150 mm between and around the containers.	H6 (1.13)
• Enclosures, compounds or storage rooms for communal containers should be a minimum of 2 m high.	H6 (1.13)
• Enclosures for individual containers should be sufficiently high to allow the lid to be opened for filling.	H6 (1.13)
• The enclosure should be permanently ventilated at the top and bottom and should have a paved impervious floor.	H6 (1.13)
• Communal storage areas should have provision for washing down and draining the floor into a system suitable for receiving a polluted effluent.	H6 (1.14)

• Gullies should incorporate a trap that maintains a seal even during prolonged periods of disuse.	H6 (1.14)
• Any room (or compound) for the open storage of waste should be secure to prevent access by vermin.	H6 (1.15)
• Where storage rooms are provided, separate rooms should be provided for the storage of waste that cannot be recycled, and waste that can be recycled.	H6 (1.16)
• Where the location for storage is in a publicly accessible area or in an open area around a building (e.g. a front garden) an enclosure or shelter should be considered.	H6 (1.17)
• In high-rise domestic developments, where chutes are provided they should be at least 450 mm in diameter and should have a smooth non-absorbent surface and close-fitting access doors at each storey that has a dwelling and be ventilated at the top and bottom.	H6 (1.18)

6.5 Water supplies

6.5.1 The requirement (Building Act 1984 Sections 25 and 69)

The Building Act stipulates that plans for proposed buildings will ensure that all occupants of the house will be provided with a supply of '*wholesome water, sufficient for their domestic purposes*'. This can be achieved by either:

- connecting the house to water supplies from the local water authority (normally referred to as the '*statutory water undertaker*');
- by otherwise taking water into the house by means of a pipe (e.g. from a local recognized supply);
- by providing a supply of water within a reasonable distance from the house (e.g. such as from a well).

If an occupied house is not within a reasonable distance of a supply of 'wholesome water' or if the local authority is not satisfied that the water supply is capable of supplying 'wholesome water', then they can give notice that the owner of the building must provide water within a specified time. They also have the authority to prohibit the building from being occupied.

What happens if there is more than one property?

Where the local authority are satisfied that two or more houses can most conveniently be met by means of a joint supply, they may give notice accordingly.

Can I ask the local authority to provide me with a supply of water?

If you are unable to provide a suitable supply of water, the local authority can themselves provide, or secure the provision of, a supply of water to the house or houses in question and then recover any expenses reasonably incurred from the owner of the house, or (where two or more houses are concerned), the owners of those houses.

The maximum amount that a local authority can charge for providing a suitable supply of water is £3000 in respect of any one house.

Where a supply of water is provided to a house by statutory water undertakers, water rates will be included in the normal rateable value of the house.

Where two or more houses are supplied with water by a common pipe belonging to the owners or occupiers of those houses, the local authority may, when necessary, repair or renew the pipe and recover any expenses reasonably incurred by them from the owners or occupiers of the houses.

6.5.2 Meeting the requirement

Protection of openings for pipes

Pipes that pass through a compartment wall or compartment floor (unless the pipe is in a protected shaft), or through a cavity barrier, should be one of the following alternatives:

Enclosures for drainage and/or water supply pipes

Proprietary seals (any pipe diameter) that maintain the fire resistance of the wall, floor or cavity barrier.	B3 7.6–7.7 (V1) B3 10.6 (V2)
Pipes with a restricted diameter where fire-stopping is used around the pipe, keeping the opening as small as possible.	B3 7.6 and 7.8 (V1) B3 10.5 and 10.7 (V2)
Sleeving – a pipe of lead, aluminium, aluminium alloy, fibre-cement or UPVC, with a maximum nominal internal diameter of 160 mm, may be used with a sleeving of non-combustible pipe as shown in Figure 6.15.	B3 7.6 and 7.9 (V1) B3 10.5 and 10.8 (V2)

6.6 Cellars and basements

6.6.1 The requirement (Building Act 1984 Section 74)

Unless you have the consent of the local authority, you are not allowed to construct a cellar or room *in (or as part of) a house, an existing cellar, a shop, inn, hotel or office if the floor level of the cellar or room is lower than the ordinary level of the subsoil water on, under or adjacent to the site of the house, shop, inn, hotel or office.*

This does not apply to:

- the construction of a cellar or room carried out in accordance with plans deposited on an application under the Licensing Act 1964;
- the construction of a cellar or room in connection with a shop, inn, hotel or office that forms part of a railway station.

If the owner of the house, shop, inn, hotel or office allows a cellar or room forming part of it to be used in a manner that he knows to be in contravention with the Building Regulations, he is liable, on summary conviction, to a fine.

Fire precautions

The building shall be provided with:

- sufficient internal fire mains and other facilities to assist firefighters in their tasks;
- adequate means for venting heat and smoke from a fire in a basement.

6.6.2 Meeting the requirements

Owing to the risk that a single stairway may be blocked by smoke from a fire in the basement or ground storey:

- basement storeys that contain a habitable room shall be provided with either: B1 2.13 (V1)
  - an external door or window suitable for egress from the basement; or B1 2.6 (V2)
  - a protected stairway leading from the basement to a final exit;
- final exits shall be sited so that they are clear of any risk from fire or smoke in a basement; B1 5.34 (V2)
- in non-residential, purpose group buildings (such as Office, Shop and Commercial, Assembly and Recreation, Industrial, Storage etc.), the following floors shall be constructed as compartment walls and compartment floors: B2 8.18c (V2)
  - the floor of every basement storey (except the lowest floor) greater than 10 m below ground level (see Figure 6.44 and 6.45(a)); B2 8.18d (V2)
  - the floor of the ground storey (see Figure 6.44 and B2 8.18c (V2) 6.45(b)).

If the building has one or more basements and with the exception of small premises (see paragraph 3.1 of V2).

Emergency egress windows and external doors

The window or door should enable the person escaping to reach a place free from danger of fire (e.g. a courtyard or back garden which is at least as deep as the dwelling house is high – see Figure 6.46). B1 2.8 (V1) B1 2.9 (V2)

The window should be at least 450 mm high and 450 mm wide and have an unobstructed openable area of at least 0.33 m^2.

The bottom of the openable area should be not more than 1100 m above the floor.

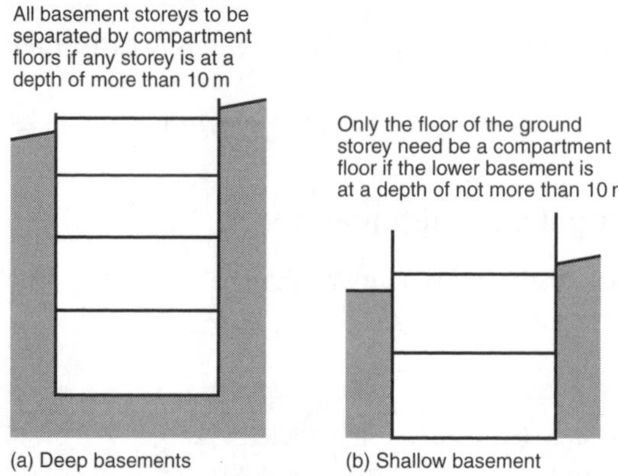
All basement storeys to be separated by compartment floors if any storey is at a depth of more than 10 m
Only the floor of the ground storey need be a compartment floor if the lower basement is at a depth of not more than 10 m
(a) Deep basements
(b) Shallow basement

Figures 6.44 and 6.45 Compartment floors

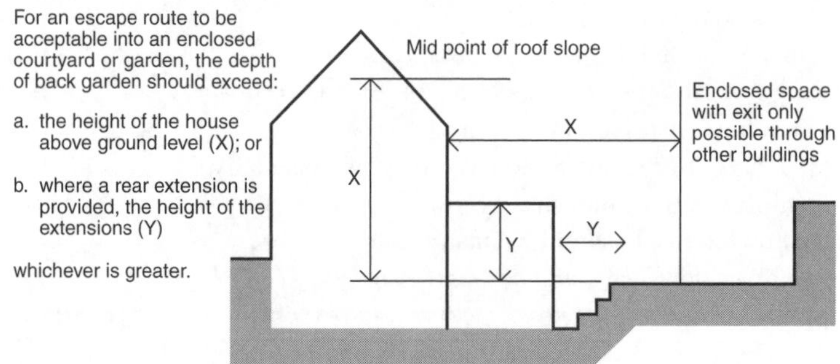
For an escape route to be acceptable into an enclosed courtyard or garden, the depth of back garden should exceed:
a. the height of the house above ground level (X); or
b. where a rear extension is provided, the height of the extensions (Y)
whichever is greater.
Mid point of roof slope
Enclosed space with exit only possible through other buildings
X
X
Y
Y

Figure 6.46 Ground or basement storey exit into an enclosed space

Notes:

(1) Approved Document K (Protection from falling, collision and impact) specifies a minimum guarding height of 800 mm, except in the case of a window in a roof where the bottom of the opening may be 600 mm above the floor.

(2) Locks (with or without removable keys) and stays may be fitted to egress windows, provided that the stay is fitted with a child resistant release catch.
(3) Windows should be designed so that they remain in the open position without needing to be held open by the person making their escape.

Basement stairways

Because basement stairways are more likely to be filled with smoke and heat than stairs in ground and upper storeys:

• the flights and landings of an escape stair shall be constructed using materials of limited combustibility particularly if it is within a basement storey;	B1 5.19 (V2)
• the basement should be served by a separate stair.	B1 2.44 (V2) B1 4.42 (V2)

Note: If an escape stair forms part of the **only** escape route from an upper storey of a large building, it should **not** be continued down to serve any basement storey. Other stairs may connect with the basement storey(s) if there is a protected lobby or a protected corridor between the stair(s) and accommodation at each basement level.

Lifts

In basements:

• the lift should be approached only by a protected lobby or protected corridor (unless it is within the enclosure of a protected stairway);	B1 5.43 (V2)
• lift entrances should be separated from the floor area on every storey by a protected lobby;	B1 5.42 (V2)
• lift shafts should not be continued down to serve any basement storey if it is: – in a building served by only one escape stair – within the enclosure to an escape stair which is terminated at ground level.	B1 5.44 (V2)

Access and facilities for the fire service

Buildings with a basement more than 10 m below the fire and rescue service vehicle access level, should be provided with at least two firefighting shafts containing firefighting lifts (see Figure 6.47).	B5 17.2 (V2) B5 17.8 (V2)

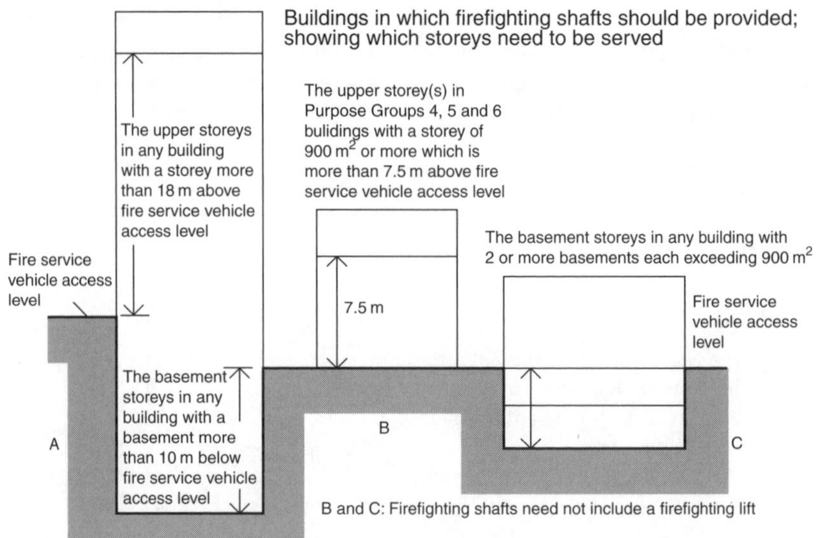

Note: Height excludes any top storey(s) consisting exclusively of plant rooms.

Figure 6.47 Provision of firefighting shafts

Buildings with two or more basement storeys, each exceeding 900 m^2 in area, should be provided with firefighting shaft(s), which need not include firefighting lifts.	B5 17.4 (V2)

Venting of heat and smoke from basements

The building should be provided with adequate means for venting heat and smoke from a fire in a basement.	B5 (V2)
Where practicable each basement space should have one or more smoke outlets (see Figure 6.44).	B5 18.3 (V2)
Outlet ducts or shafts, including any bulkheads over them (see Figure 6.48), should be enclosed in non-combustible construction having a greater fire resistance than the element that they pass through.	B5 18.15 (V2)
Smoke outlets connected to the open air should be provided from every basement storey, except for a basement in a single family dwelling.	B5 18.4 (V2)

Smoke outlets (also referred to as smoke vents) should be:

- available so as to provide a route for heat and smoke to escape to the open air from the basement area; B5 18.2 (V2)
- sited at high level, either in the ceiling or in the wall of the space they serve; B5 18.7 (V2) B5 18.3 (V2)
- evenly distributed around the perimeter to discharge in the open air outside the building. B5 18.3 (V2)

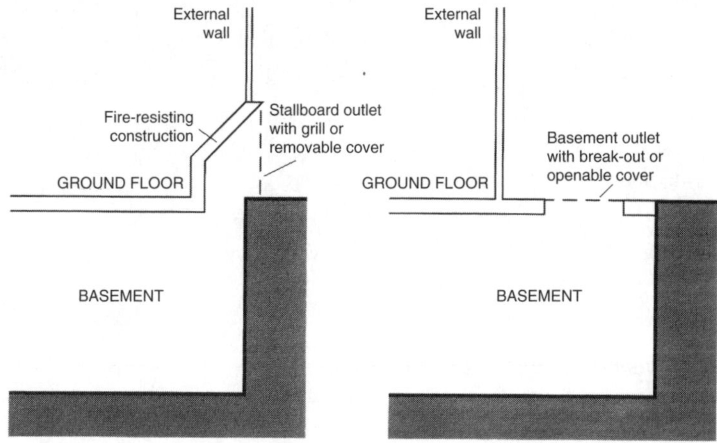

Figure 6.48 Fire-resisting construction for smoke outlet shafts

A system of mechanical extraction may be provided as an alternative to natural venting to remove smoke and heat from basements, **provided** that the basement storey(s) is fitted with a sprinkler system. B5 18.13 (V2)

Where there are natural smoke outlet shafts from different compartments to the same or different basement storeys, they should be separated from each other by a non-combustible construction. B5 18.16 (V2)

Ventilation

If a basement is connected to the rest of the dwelling by a large permanent opening such as an open stairway then the whole dwelling including the basement should be treated as a multi-storey dwelling and ventilated in a similar manner to dwellings without basements. F 1.9

If the basement has a single exposed façade, whilst the rest of the dwelling above ground has more than one exposed façade, then passive stack ventilation (PSV) or continuous mechanical extract should be used.	F 1.9
If the basement is not connected to the rest of the dwelling by a large permanent opening, then: • the part of the dwelling above ground should be considered separately and • the basement should be treated as a single-storey dwelling, as if it were above ground. If the part of the dwelling above ground has no bedrooms, then for the purpose of ventilation requirements: • assume that the dwelling has one bedroom and • treat the basement as a single-storey dwelling (with one bedroom) as if it were above ground.	F 1.10
If a dwelling only compromises a basement, then it should be treated as if it were a single-storey dwelling (with one bedroom) above ground.	F 1.11

Table 6.28 Ventilation systems for basements

Type of basement	Background ventilators and intermittent extract fans	Passive stack ventilation	Continuous mechanical extract	Continuous mechanical supply and extract with heat
Basement connected to the rest of the dwelling by an open stairway	Yes	Yes	Yes	Yes
Basement with a single exposed façade and dwelling above ground with more than one exposed façade		Yes	Yes	
Basements not connected to the rest of the dwelling by an open stairway	Yes	Yes	Yes	Yes
Dwelling above ground has no bedrooms	Yes	Yes	Yes	Yes
Dwelling comprised just a basement	Yes	Yes	Yes	Yes

Mechanically ventilated basement car parks shall be capable of at least six air changes per hour (ach).	F 2.21b

6.7 Floors

The ground floor of a building is either solid concrete or a suspended timber type. With a concrete floor, a Damp Proof Membrane (DPM) is laid between walls. With timber floors, sleeper walls of honeycomb brickwork are built on oversite concrete between the base brickwork; a timber sleeper plate rests on each wall and timber joists are supported on them. Their ends may be similarly supported, let into the brickwork or suspended on metal hangers. Floorboards are laid at right angles to joists. First-floor joists are supported by the masonry or hangers.

Similar to a brick built house, the floors in a timber-framed house are either solid concrete or suspended timber. In some cases, a concrete floor may be screeded or surfaced with timber or chipboard flooring. Suspended timber floor joists are supported on wall plates and surfaced with chipboard.

6.7.1 Requirements

The building shall be constructed so that the combined dead, imposed and wind loads are sustained and transmitted by it to the ground

- *safely;*
- *without causing such deflection or deformation of any part of the building (or such movement of the ground) as will impair the stability of any part of another building.*

(Approved Document A1)

The building shall be constructed so that ground movement caused by:

- *swelling, shrinkage or freezing of the subsoil; or*
- *landslip or subsidence (other than subsidence arising from shrinkage) will not impair the stability of any part of the building.*

(Approved Document A2)

Fire precautions

As a fire precaution, the spread of flame over the internal linings of a building and the amount of heat released from internal linings shall be restricted.

- *all loadbearing elements of structure of the building shall be capable of withstanding the effects of fire for an appropriate period without loss of stability;*
- *ideally the building should be subdivided by elements of fire-resisting construction into compartments;*
- *all openings in fire-separating elements shall be suitably protected in order to maintain the integrity of the continuity of the fire separation;*
- *any hidden voids in the construction shall be sealed and subdivided to inhibit the unseen spread of fire and products of combustion, in order to reduce the risk of structural failure, and the spread of fire.*

(Approved Document B3)

The floors of the building shall adequately protect the building and people who use the building from harmful effects caused by:

- *ground moisture;*
- *precipitation and wind-driven spray;*
- *interstitial and surface condensation; and*
- *spillage of water from or associated with sanitary fittings or fixed appliances.*

(Approved Document C2)

Airborne and impact sound

Dwellings shall be designed so that the noise from domestic activity in an adjoining dwelling (or other parts of the building) is kept to a level that:

- *does not affect the health of the occupants of the dwelling;*
- *will allow them to sleep, rest and engage in their normal activities in satisfactory conditions.*

(Approved Document E1)

Dwellings shall be designed so that any domestic noise that is generated internally does not interfere with the occupants' ability to sleep, rest and engage in their normal activities in satisfactory conditions.

(Approved Document E2)

Domestic buildings shall be designed and constructed so as to restrict the transmission of echoes.

(Approved Docume nt E3)

Schools shall be designed and constructed so as to reduce the level of ambient noise (particularly echoing in corridors).

(Approved Document E4)

Note: The normal way of satisfying Requirement E4 will be to meet the values for sound insulation, reverberation time and internal ambient noise which are given in Section 1 of Building Bulletin 93 '*The Acoustic Design of Schools*', produced by DFES and published by the Stationery Office (ISBN: 0 11 271105 7).

Conservation of fuel and power

Reasonable provision shall be made for the conservation of fuel and power in buildings by:

(a) limiting heat gains and losses:
  - *(i) through thermal elements and other parts of the building fabric; and*
  - *(ii) from pipes, ducts and vessels used for space heating, space cooling and hot water services;*

(b) *providing and commissioning energy-efficient fixed building services with effective controls; and*

(c) *providing to the owner sufficient information about the building, the fixed building services and their maintenance requirements so that the building can be operated in such a manner as to use no more fuel and power than is reasonable in the circumstances.*

(Approved Document L1)

6.7.2 The use of Robust Standards

Background

The 2003 edition of Part E of the Building Regulations (i.e. 'Resistance to the passage of sound') involves Pre-Completion Sound Testing (PCT) for certain types of homes. In an attempt to eliminate the risk of any remedial work being required to completed floor and/or wall constructions (together with the potential for delays in completing the property) the House Builders Federation (HBF) suggested that a series of construction solutions (called Robust Details) should be developed as an alternative to PCT.

This approach was agreed and in May 2003, the (then) Office of the Deputy Prime Minister (ODPM) published the first batch of Robust Detail proposals for public consultation. At the same time, the introduction of PCT in new homes was postponed until July 2004 – on the assumption that the Robust Details scheme would eventually receive ministerial approval.

In January 2004 the Minister responsible for Building Regulations announced that he would allow Robust Details to be used as an alternative to PCT and that it would take effect from 1st July 2004 (i.e. so as to coincide with the introduction of PCT). Under a Memorandum of Understanding with the ODPM, a Limited Company (Robust Details Ltd) was set up to approve, manage, monitor and promote the use of Robust Details as a method of satisfying Building Regulations.

What is a Robust Detail?

Robust Details provide builders with a choice of possible construction solutions that have been proven to outperform the standards of Part E, thus eliminating the need for routine pre-completion sound testing!

A Robust Detail is only used in connection with Part E and is defined as *'a separating wall or floor (of concrete, masonry, timber, steel or steel-concrete composite) construction, which has been assessed and approved by Robust Details Limited'*.

How are Robust Details approved?

In order to be approved, each Robust Detail must:

- be capable of consistently exceeding the performance standards given in Approved Document E to the Building Regulations;

- be practical to construct on site;
- be reasonably tolerant to workmanship.

How can Robust Details be used?

Builders are only permitted to use Robust Details instead of PCT **if** the plots concerned have been registered in advance with Robust Details Ltd.

Once a plot has been registered, Robust Details Ltd will provide the relevant registration documentation (which will be accepted by all building control bodies as evidence that the builder is entitled to use Robust Details instead of PCT). The builder will then need to select the Robust Detail specific to the walls and/or floors they wish to build from the Robust Details Handbook (available from Robust Details Ltd) and produce a sitework checklist to show how they are going to ensure that building work is carried out exactly in accordance with the Robust Detail specifications.

Will there be more Robust Details?

Trade associations, manufacturers or other interested parties may submit applications for new Robust Details which will be evaluated and, if found acceptable, approved and published.

Where can I obtain more information?

Robust Details Limited
PO Box 7289
Milton Keynes
MK14 6ZQ
Telephone/Fax:
Business line: 0870 240 8210
Technical support line: 0870 240 8209
Fax: 0870 240 8203
e-mail Support:
Technical email support (technical@robustdetails.com)
Administrative email support (administration@robustdetails.com)
Other support (customerservice@robustdetails.com)

6.7.3 Meeting the requirements

General

Floors next to the ground should: C4.2

- resist the passage of ground moisture to the upper surface of the floor;
- not be damaged by moisture from the ground;

- not be damaged by groundwater;
- resist the passage of ground gases.

Floors next to the ground and floors exposed from below should be designed and constructed so that their structural and thermal performance are not adversely affected by interstitial condensation. C4.4

All floors should be designed so they do not promote surface condensation or mould growth. C4.5

Ground supported floors exposed to moisture from the ground

Unless subjected to water pressure, the ground of a ground supported floor should be covered with dense concrete laid on a hardcore bed and a damp-proof membrane as shown in Table 6.29.

Note: Suitable insulation may also be incorporated.

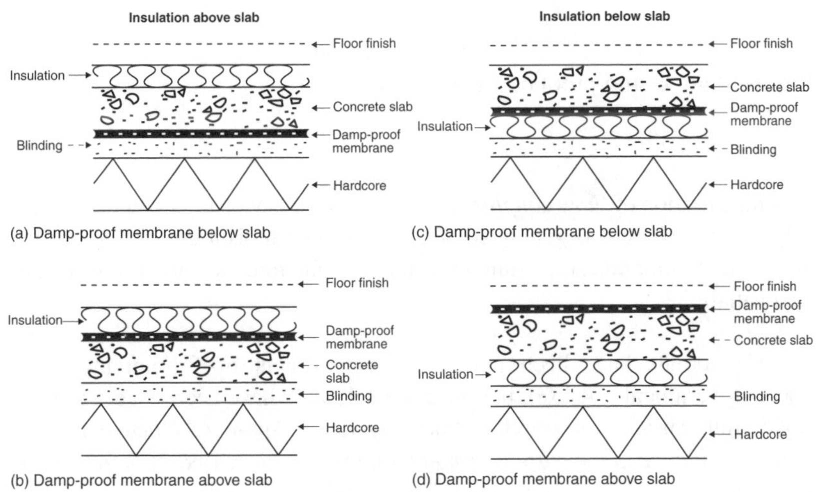

Figure 6.49 Ground supported floor – construction

Note: Some schools of thought believe that there is also a need for an additional DPM on top of the insulation to combat interstitial condensation, but then this

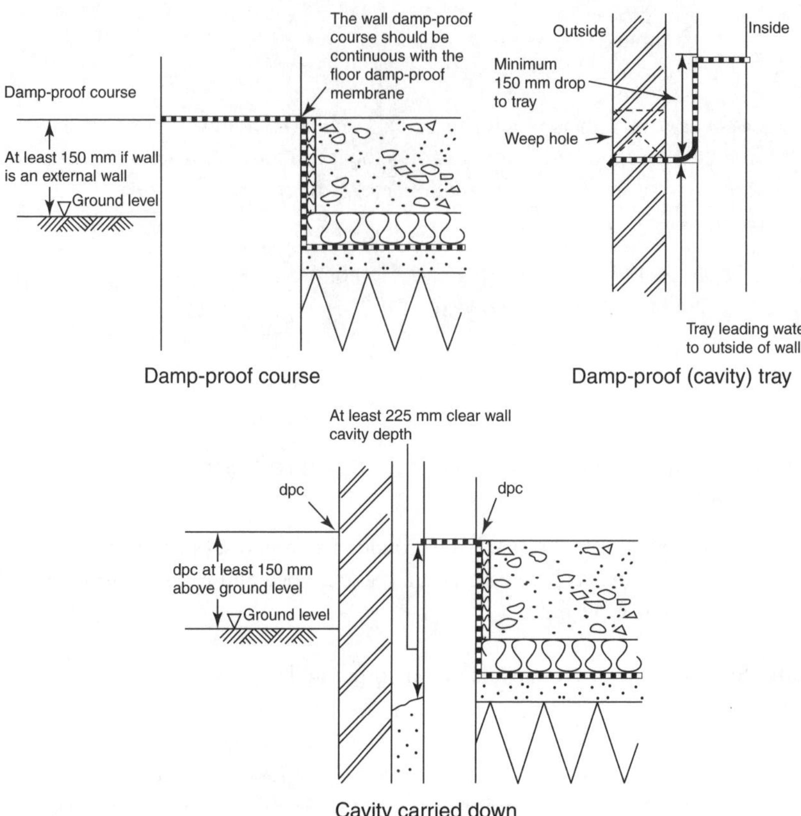

Figure 6.50 Damp-proof courses

begs the question of '*how can this moisture escape?*'! Moisture would, presumably, just sit where it is generated and if interstitial moisture is not controlled by a vapour membrane, then surely it will eventually migrate into the concrete or the insulation?

These points have been put to the Department of Communities and Local Government (DCLG), but unfortunately they have been unable to offer any definite answer – saying that '*the intention of Approved Documents is to provide guidance to the more common building situations and as there may be alternative ways of achieving compliance with the requirements, there is no obligation to adopt any particular solution contained in an Approved Document – if the builder prefers to meet the relevant requirement in some other way.*' One of my readers has said that he prefers to employ the insulation below the slab and place a DPM between the insulation and the blinding wherever possible (if only for ease of construction), which sounds like a very logical solution.

Table 6.29 Ground supported floor – construction

Hardcore	Well compacted, no greater than 600 mm deep, of clean, broken brick or similar inert material, free from materials including water-soluble sulphates in quantities which could damage the concrete.	C4.7a
Concrete	At least 100 mm thick to mix ST2 in BS 8500 or (if there is embedded reinforcement) to mix ST4 in BS 8500.	C4.7b
Damp-proof membrane	Above or below the concrete which is continuous with the damp-proof courses in walls and piers etc.	C4.7c
	If below the concrete, the membrane could be formed with a sheet of polyethylene, at least 300 mm thick with sealed joints and laid on a bed of material that will not damage the sheet.	C4.8
	If laid above the concrete, the membrane may be either polyethylene (but without the bedding material) or three coats of cold applied bitumen solution or similar moisture and water vapour resisting material.	C4.9
	In each case it should be protected either by a screed or a floor finish, unless the membrane is pitchmastic or similar material which will also serve as a floor finish.	C4.9
Insulation	If placed beneath floor slabs should have sufficient strength to resist the weight of the slab, the anticipated floor loading as well as any possible overloading during construction.	C4.10
	If placed below the damp-proof membrane, then it should have low water absorption and (if considered necessary) should be resistant to contaminants in the ground.	C4.10
Timber floor finish	If laid directly on concrete, it may be bedded in a material which can also serve as a damp-proof membrane.	C4.11
	Timber fillets that are laid in the concrete as a fixing for a floor finish should be treated with an effective preservative unless they are above the damp-proof membrane.	C4.11

Suspended timber ground floors exposed to moisture from the ground

Any suspended timber floor next to the ground should:

- ensure that the ground is covered so as to resist moisture and prevent plant growth; C4.13a
- have a ventilated air space between the ground covering and the timber; C4.13b
- have damp-proof courses between the timber and any material which can carry moisture from the ground. C4.13c

Unless covered with a highly vapour resistant floor finish, a suspended timber floor next to the ground may be built as shown in Figure 6.51 and as follows:

Hardcore	A bed of clean, broken brick or any other inert material free from materials including water-soluble sulphates in quantities which could damage the concrete.	
Concrete	A ground covering of unreinforced concrete at least 100 mm thick to mix ST 1 in BS 8500 or	C4.14a(i)
	laid on at least 300 μm polyethylene sheet with sealed joints (and itself laid on a bed of material which will not damage the sheet).	C4.14a(ii)
	Note: To prevent water collecting on the ground covering, either the top should be entirely above the highest level of the adjoining ground or, on sloping sites, consideration should be given to installing drainage on the outside of the upslope side of the building.	
Ventilation	There should be a ventilated air space at least 75 mm from the ground (and covering the underside of any wall plates) and at least 150 mm from the underside of the suspended timber floor.	C4.14b
	Two opposing external walls should have ventilation openings placed so that the ventilating air will have a free path between opposite sides and to all parts.	C4.14b
	Ventilating openings should be not less than either 1500 mm^2/m run of external wall or 500 mm^2/m^2 of floor area – whichever gives the greater opening area.	C4.14b
	Any pipes needed to carry ventilating air should have a diameter of at least 100 mm.	C4.14b
	Ventilation openings should incorporate suitable grilles to prevent the entry of vermin to the subfloor.	C4.14b
	If floor levels need to be nearer to the ground to provide level access, subfloor ventilation can be provided through offset (periscope) ventilators.	C4.14b

Damp-proof membrane	DPMs should be of impervious sheet material, engineering brick or slates in cement mortar or other material which will prevent the passage of moisture.	C4.14c
Insulation		
Timber floor finish	In areas such as kitchens, utility rooms and bathrooms where water may be spilled, any board used as a flooring, irrespective of the storey, should be moisture resistant.	C4.15
	In the case of chipboard it should be of one of the grades with improved moisture resistance specified in BS 7331: 1990 or BS EN 312 Part 5: 1997.	C4.15
	Identification marks should be facing upwards.	C4.15
	Any softwood boarding should be at least 20 mm thick and from a durable species or treated with a suitable preservative.	C4.15

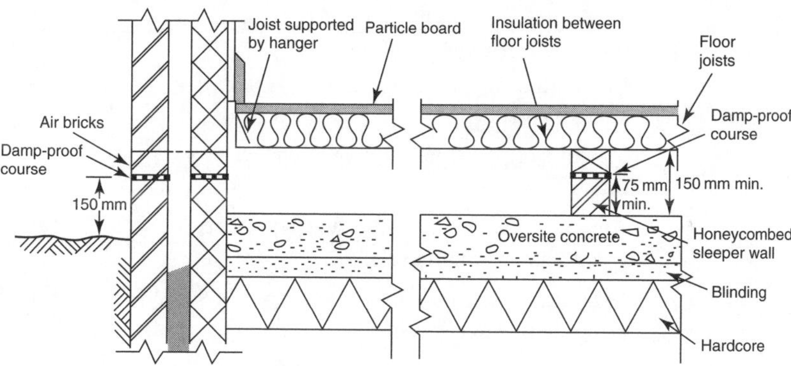

Figure 6.51 Suspended timber floor – construction

Suspended concrete ground floors exposed to moisture from the ground

Concrete suspended floors (including beam and block floors) that are next to the ground should: • adequately prevent the passage of moisture to the upper surface; • be reinforced to protect against moisture.	C4.17

There should be a facility for inspecting and clearing out the subfloor voids beneath suspended floors – particularly in localities where flooding is likely.	C4.20

Hardcore		
Concrete	In-situ concrete at least 100 mm thick containing at least 300 kg of cement for each m^3 of concrete; or precast concrete construction (with or without infilling slabs).	C4.1
	Reinforcing steel should be protected by a concrete cover of at least 40 mm (if the concrete is in situ) and at least the thickness required for a moderate exposure, if the concrete is precast.	C4.1
Ventilation	There should be a ventilated air space at least 150 mm clear from the ground to the underside of the floor (or insulation if provided).	C4.19b
	Two opposing external walls should have ventilation openings placed so that the ventilating air will have a free path between opposite sides and to all parts.	C4.19b
	Ventilating openings should be not less than either 1500 mm^2/m run of external wall or 500 mm^2/m^2 of floor area – whichever gives the greater opening area.	C4.19b
	Any pipes needed to carry ventilating air should have a diameter of at least 100 mm.	C4.19b
	Ventilation openings should incorporate suitable grilles to prevent the entry of vermin to the subfloor.	C4.19b
Damp-proof membrane	A suspended concrete floor should contain a damp-proof membrane (if the ground below the floor has been excavated below the lowest level of the surrounding ground and will not be effectively drained).	C4.19a

Ground floors and floors exposed from below (resistance to damage from interstitial condensation)

A ground floor (or floor exposed from below such as above an open parking space or passageway – see Figure 6.52) shall be designed in accordance with Clause 8.5 and Appendix D of BS 5250: 2002.

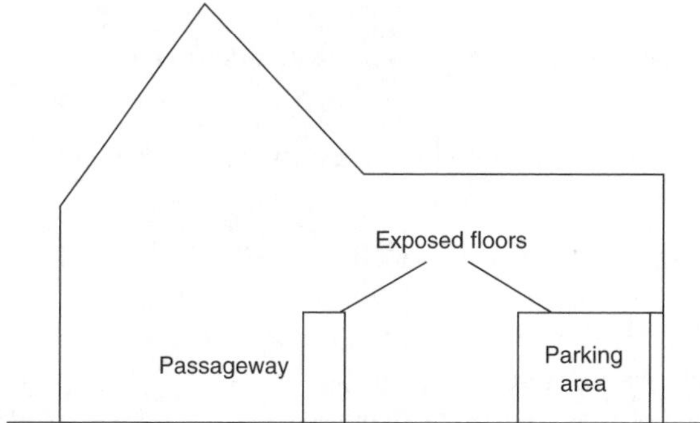

Figure 6.52 Typical floors exposed from below

Floors (resistance to surface condensation and mould growth)

Ground floors should be designed and constructed so that the thermal transmittance (U-value) does not exceed 0.7 $W/m^2 \cdot K$ at any point.	C4.22a
Junctions between elements should be designed in accordance with robust construction recommendations.	C4.22b
Junctions between elements should be designed in accordance with robust construction recommendations.	C4.22b

Small single-storey non-residential buildings and annexes

The floor area of the building or annexe shall not exceed 36 m^2.	A1/2 2C38(1)a
Where the floor area of the building or annexe exceeds 10 m^2, the walls shall have a mass of not less than 130 kg/m^2.	A1/2 2C38(1)c

Tension straps

Tension straps (conforming to BS EN 845-1) should be used to strap walls to floors above ground level, at intervals not exceeding 2 m.	A1/2 2C35
For corrosion resistance purposes, the tension straps should be material reference 14 or 16.1 or 16.2 (galvanized steel) or other more resistant specifications	A1/2 2C35

including material references 1 or 3 (austenitic stainless steel).

The declared tensile strength of tension straps should not be less than 8 kN.

Tension straps need **not** be provided:

- in the longitudinal direction of joists in houses of not more than 2 storeys if the joists:
  - are at not more than 1.2 m centres; A1/2 2C35a
  - have at least 90 mm bearing on the supported walls or 75 mm bearing on a timber wallplate at each end; or A1/2 2C35a
  - are carried on the supported wall by joist hangers (in accordance with BS EN 845-1 and BS 5628 – see Figure 6.53(c)); A1/2 2C35b
  - and are incorporated at not more than 2 m centres; A1/2 2C35b
- when a concrete floor has at least 90 mm bearing on the supported wall (see Figure 6.53(d)); and A1/2 2C35c
- where floors are at or about the same level on each side of a supported wall, and contact between the floors and wall is either continuous or at intervals not exceeding 2 m. Where contact is intermittent, the points of contact should be in line or nearly in line on plan (see Figure 6.53(e)). A1/2 2C35d

Interruption of lateral support

Where an opening in a floor or roof for a stairway or the like adjoins a supported wall and interrupts the continuity of lateral support:

- the maximum permitted length of the opening is to be 3 m, measured parallel to the supported wall; A1/2 2C37a
- connections (if provided by means other than by anchor) should be throughout the length of each portion of the wall situated on each side of the opening; A1/2 2C37b
- connections via mild steel anchors should be spaced closer than 2 m on each side of the opening to provide the same number of anchors as if there were no opening; A1/2 2C37c
- there should be no other interruption of lateral support. A1/2 2C37d

Lateral support by floors

Floors should:

- act to transfer lateral forces from walls to buttressing walls, piers or chimneys; A1/2 2C33a
- be secured to the supported wall by connections (see Figure 6.53). A1/2 2C33b

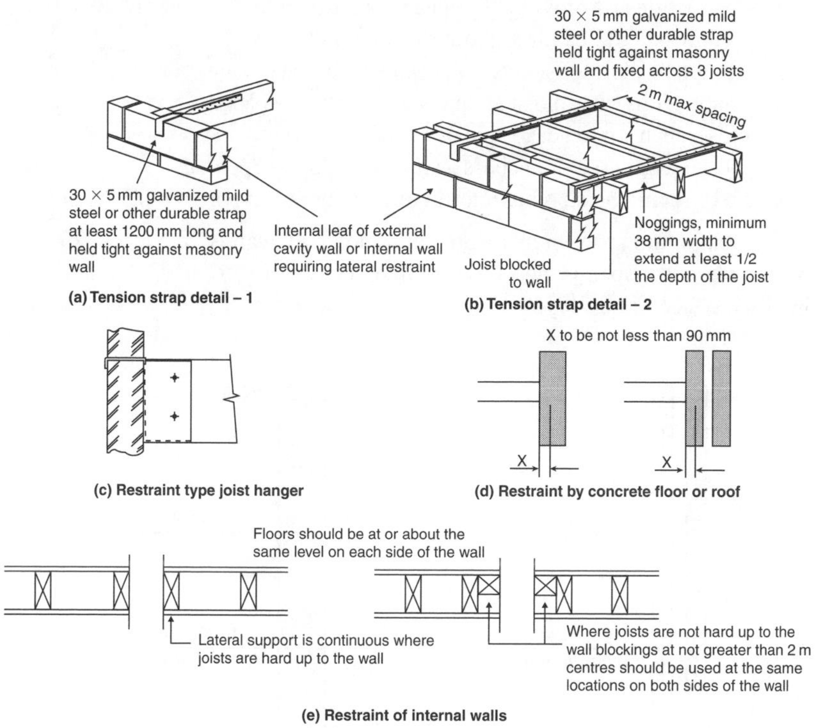

Figure 6.53 Lateral support by floors

Intermediate floors and roof shall be constructed so that they provide local support to the walls and act as horizontal diaphragms capable of transferring the wind forces to buttressing elements of the building. A1/2 1A2d

A wall in each storey of a building should:

- extend to the full height of that storey; A1/2 2C32
- have horizontal lateral supports to restrict movement of the wall at right angles to its plane.

Walls should be strapped to floors above ground level, at intervals not exceeding 2 m and as shown in Figure 6.53 by tension straps conforming to BS EN 845-1.	A1/2 2C35
Where an opening in a floor for a stairway or the like adjoins a supported wall and interrupts the continuity of lateral support:	
• the maximum permitted length of the opening is to be 3 m, measured parallel to the supported wall;	A1/2 2C37a
• connections (if provided by means other than by anchor) should be throughout the length of each portion of the wall situated on each side of the opening;	A1/2 2C37b
• connections via mild steel anchors should be spaced closer than 2 m on each side of the opening to provide the same number of anchors as if there were no opening;	A1/2 2C37c
• there should be no other interruption of lateral support.	A1/2 2C37d
The maximum span for any floor supported by a wall is 6 m where the span is measured centre to centre of bearing (see Figure 6.54).	A1/2 2C23

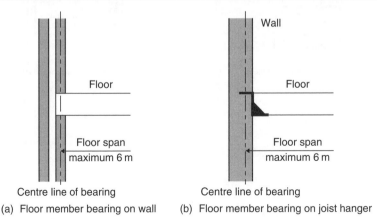

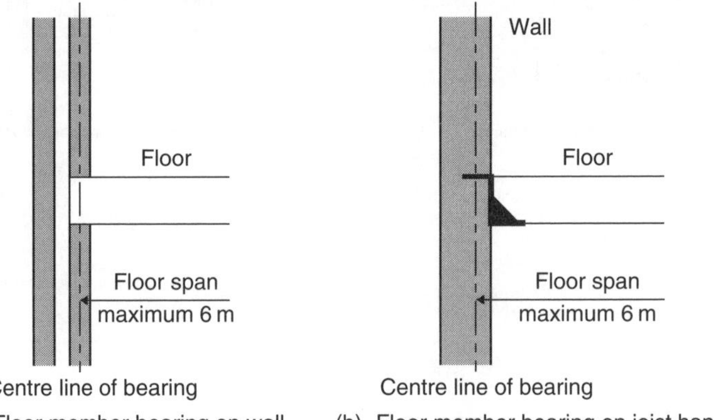

Figure 6.54 Maximum span of floors

Internal fire spread (structure)

Loadbearing elements of structure

All loadbearing elements of a structure shall have a minimum standard of fire resistance.	B3 4.1 (V1) B3 7.1 (V2)
Structural frames, beams, floor structures and gallery structures should have at least the fire	B3 4.2 (V1) B3 7.2 (V2)

resistance given in Appendix A of Approved Document B.	
When altering an existing two-storey, single-family dwelling house to provide additional storeys, the floor(s), both old and new, shall have the full 30 minute standard of fire resistance.	B3 4.7 (V1)

Fire resistance – compartmentation

To prevent the spread of fire within a building, whenever possible, the building should be sub-divided into compartments separated from one another by walls and/or floors of fire-resisting construction.	B3 5.1 (V1) B3 8.1 (V2)
Parts of a building that are occupied mainly for different purposes, should be separated from one another by compartment walls and/or compartment floors.	B3 5.3 (V1) B3 8.11 (V2)
The wall and any floor between the garage and the house shall have a 30 minute fire resistance. Any opening in the wall to be at least 100 mm above the garage floor level with an FD30 door.	B3
In buildings containing flats or maisonettes compartment walls or compartment floors shall be constructed between: • every floor (unless it is within a maisonette) • one storey and another within one dwelling • lvery wall separating a flat or maisonette from any other part of the building • every wall enclosing a refuse storage chamber.	B3 8.13 (V2)
Every compartment floor should: • form a complete barrier to fire between the compartments they separate; and • have the appropriate fire resistance as indicated in Appendix A of Approved Document B, Tables A1 and A2.	B3 5.6 (V1) B3 8.20 (V2)
Where a compartment wall or compartment floor meets another compartment wall, or an external wall, the junction should maintain the fire resistance of the compartmentation.	B3 5.9 (V1) B3 8.25 (V2)

Junctions between a compartment floor and an external wall that has no fire resistance (such as a curtain wall) should be restrained at floor level to reduce the movement of the wall away from the floor when exposed to fire.	B2 8.26 (V2)
Compartment walls should be able to accommodate the predicted deflection of the floor above by either: • having a suitable head detail between the wall and the floor, that can deform but maintain integrity when exposed to a fire; or • the wall may be designed to resist the additional vertical load from the floor above as it sags under fire conditions and thus maintain integrity.	B2 8.27 (V2)

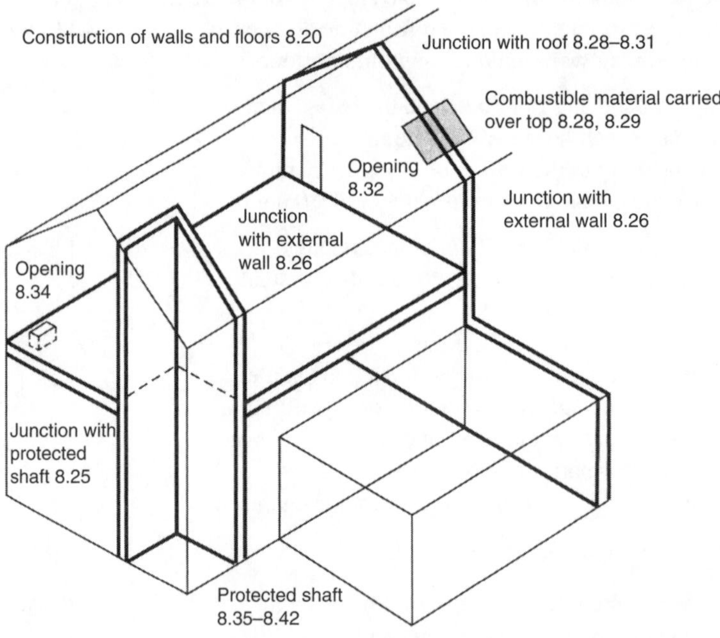

Figure 6.55 Compartment walls and compartment floors with reference to relevant paragraphs in Part B

Under floor voids

Extensive cavities in floor voids should be subdivided with cavity barriers.	B3

Concrete

With a concrete intermediate floor:

The ground floor may be a solid slab, laid on the ground, or a suspended concrete floor.	E2.51 E2.88 E2.126
A concrete slab floor on the ground may be continuous under a solid separating wall but may not be continuous under the cavity masonry core of the separating wall.	E2.51 E2.88 E2.127 E2.130
An internal concrete floor slab may only be carried through a separating wall if the floor base has a mass of at least 365 kg/m^2.	E2.46 E2.121
Note: Internal concrete floors should generally be built into a separating wall and carried through to the cavity face of the leaf.	
The cavity should not be bridged (E2.85, E2.122). Internal hollow-core concrete plank floors (and concrete beams with infilling block floors) should **not** be continuous through or under a separating wall (E2.47, E2.53, E2.129).	
A suspended concrete floor may only pass under a separating wall if the floor has a mass of at least 365 kg/m^2.	E2.52 E2.89
Note: A suspended concrete floor should **not** be carried through to the cavity face of the leaf and the cavity should not be bridged (E2.89, E2.132).	E2.128 E2.131

Floors – general

Floors that separate a dwelling from another dwelling (or part of the same building) shall resist the transmission of airborne sounds.	E

Floors above a dwelling that separate it from another dwelling (or another part of the same building) shall resist: • the transmission of impact sound (such as speech, musical instruments and loudspeakers and impact sources such as footsteps and furniture moving);	E

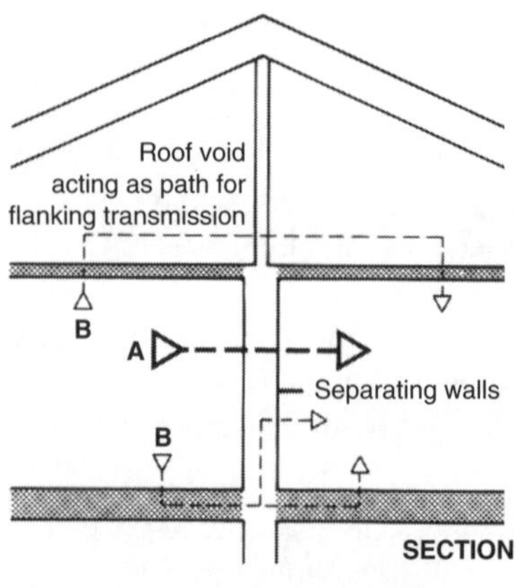

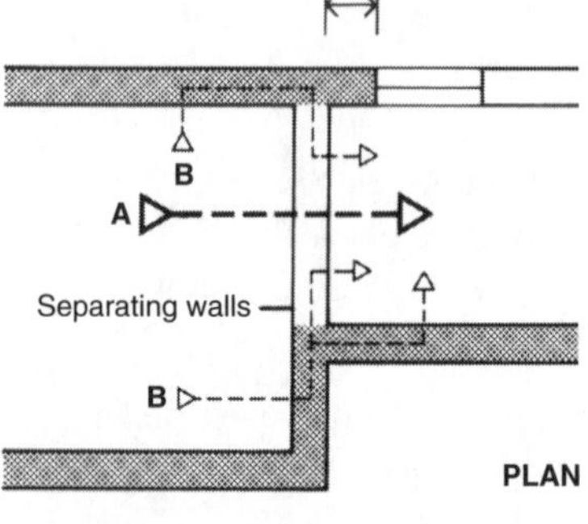

Figure 6.56 Direct and flanking transmission
For clarity not all flanking paths have been shown.

- the flow of sound energy through walls and floors;
- the level of airborne sound.

Air paths, including those due to shrinkage, must be avoided – porous materials and gaps at joints in the structure must be sealed. E

The possibility of resonance in parts of the structure (such as a dry lining) should be avoided. E

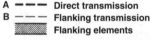

Flanking transmission (i.e. the indirect transmission of sound from one side of a wall or floor to the other side) should be minimized. E

Requirement E1

Figure 6.57 illustrates the relevant parts of the building that should be protected from airborne and impact sound in order to satisfy Requirement E1.

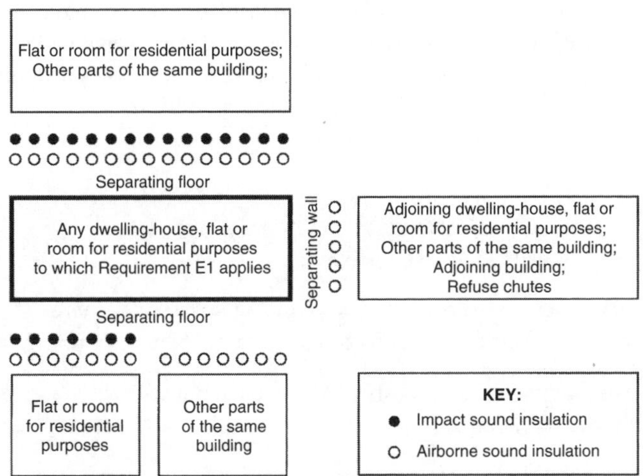

Figure 6.57 Requirement E1 – resistance to sound

All new floors constructed within a dwelling-house (flat or room used for residential purposes) – whether purpose built or formed by a material change of use – shall meet the laboratory sound insulation values set out in Table 6.33. E0.9

Floors that have a separating function should achieve the sound insulation values: E0.1

- for rooms for residential purposes as set out in Table 6.30;
- for dwelling-houses and flats as set out in Table 6.30.

Table 6.30 Dwelling-houses and flats – performance standards for separating floors and stairs that have a separating function

	Airborne sound insulation $D_{nT,w} + C_{tr}$ dB (minimum values)	Impact sound insulation $L_{nT,w}$ dB (maximum values)
Purpose built rooms for residential purposes	45	62
Purpose built dwelling houses and flats	45	62
Rooms for residential purposes formed by material change of use	43	64
Dwelling-houses and flats formed by material change of use	43	64

Notes:

(1) The sound insulation values in this table include a built-in allowance for 'measurement uncertainty' and so if any these test values are not met, then that particular test will be considered as failed.

(2) Occasionally a higher standard of sound insulation may be required between spaces used for normal domestic purposes and noise generated in and to an adjoining communal or non-domestic space. In these cases it would be best to seek specialist advice before committing yourself.

Flanking transmission from walls and floors connected to the separating wall shall be controlled.	E2
Tests should be carried out between rooms or spaces that share a common area formed by a separating wall or separating floor.	E1
Impact sound insulation tests should be carried out without a soft covering (e.g. carpet, foam backed vinyl etc.) on the floor.	E1
When a separating floor is used, a minimum mass per unit area of 120 kg/m^2 (excluding finish) shall always apply, irrespective of the presence or absence of openings.	E2
Care should be taken to correctly detail the junctions between the separating floor and other elements such as external walls, separating walls and floor penetrations.	E3
Spaces between floor joists should be sealed with full depth timber blocking.	E2
If the floor joists are to be supported on the separating wall then they should be supported on hangers and should not be built in.	E2
If the joists are at right angles to the wall, spaces between the floor joists should be sealed with full depth timber blocking.	E3
The floor base (excluding any screed) should be built into a cavity masonry external wall and carried through to the cavity face of the inner leaf.	E
The floor base should be continuous or above an internal masonry wall.	E3
All pipes and ducts that penetrate a floor separating habitable rooms in different flats should: • be enclosed for their full height in each flat; • have fire protection to satisfy Building Regulation Part B – Fire safety.	E3

Notes:

(1) Where any building element functions as a separating element (e.g. a ground floor that is also a separating floor for a basement flat) then the separating element requirements should take precedence.

(2) In some circumstances (for example, when a historic building is undergoing a material change of use) it may not be practical to improve the sound insulation to the standards set out in Approved Document E1 particularly if the special characteristics of such a building need to be recognized. In these circumstances the aim should be to improve sound insulation to the 'extent that it is practically possible'.

(3) BS 7913:1998 *The principles of the conservation of historic buildings* provides guidance on the principles that should be applied when proposing work on historic buildings.

Requirement E2

Constructions for new floors within a dwelling-house (flat or room for residential purposes) – whether purpose built or formed by a material change of use – shall meet the laboratory sound insulation values set out in Table 6.31.	E0.9

Table 6.31 Laboratory values for new internal walls within dwelling-houses, flats and rooms for residential purposes – whether purpose built or formed by a material change of use

	Airborne sound insulation R_W dB (minimum values)
Purpose built dwelling houses and flats	
Floors	40

Figure 6.58 illustrates the relevant parts of the building that should be protected from airborne and impact sound in order to satisfy Requirement E2.

Requirement E3

Sound absorption measures described in Section 7 of Approved Document N shall be applied.	E0.11

Requirement E4

The values for sound insulation, reverberation time and indoor ambient noise as described in Section 1 of Building	E0.12

Bulletin 93 *'The Acoustic Design of Schools'* (produced by DFES and published by the Stationery Office (ISBN 0 11 271105 7)) shall be satisfied.

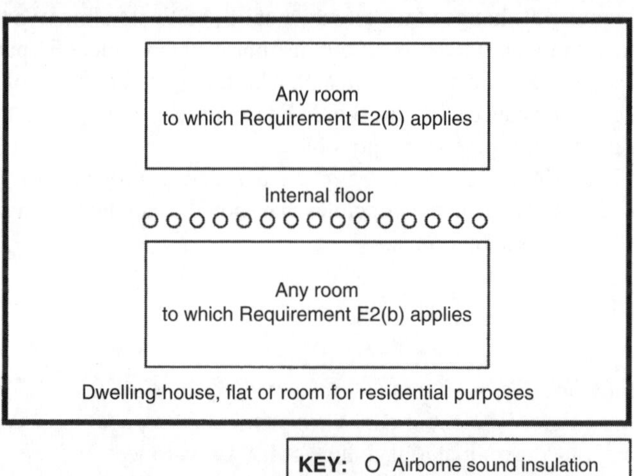

Figure 6.58 Requirement E2(b) – internal floors

Separating floors and associated flanking constructions for new buildings

There are three groups of floors as shown below:

Floor type 1
Concrete base with ceiling and soft floor covering

Soft covering
Ceiling
Concrete base

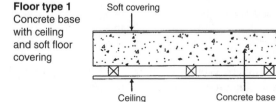

The resistance to airborne sound depends mainly on the mass per unit area of the concrete base and partly on the mass per unit area of the ceiling. The soft floor covering reduces impact sound at source.

Floor type 2
Concrete base with ceiling and floating floor (three types of floating floor are available see p. 271)

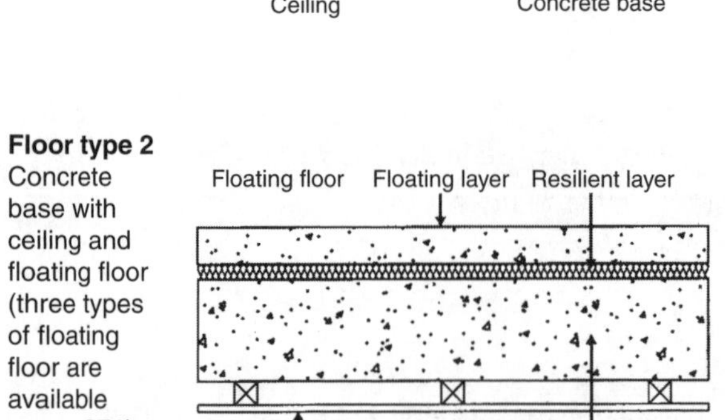

The resistance to airborne and impact sound depends on the mass per unit area of the concrete as well as the mass per unit area and isolation of the floating layer and ceiling. The floating floor reduces impact sound at source.

Floor type 3 Timber frame base with ceiling and platform floor

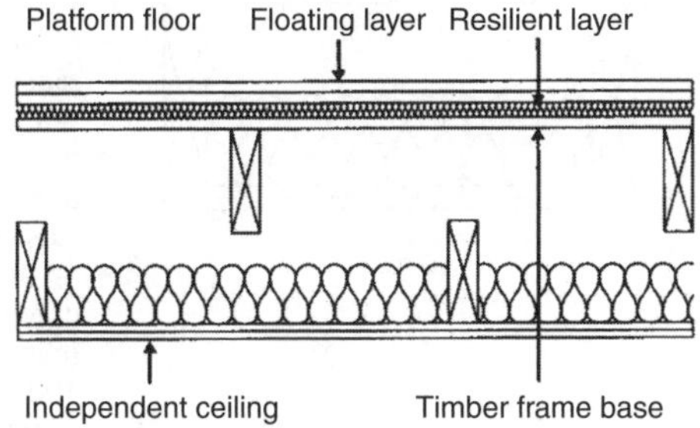

The resistance to airborne and impact sound depends on the structural floor base and the isolation of the platform floor and the ceiling. The platform floor reduces impact sound at source.

General requirements

Floor types should be capable of achieving the performance standards shown in Table 6.30.	E3.1
Care should be taken to correctly detail the junctions between the separating floor and other elements such as external walls, separating walls and floor penetrations.	E3.10

Note: Where any building element functions as a separating element (e.g. a ground floor that is also a separating floor for a basement flat) then the separating element requirements should take precedence.

Ceiling treatments

Each floor type should use one of the following three ceiling treatments which are ranked in order of sound insulation performance from A to C. E3.17 to E3.18

Note: Use of a better performing ceiling than that described in the guidance should improve the sound insulation of the floor provided there is no significant flanking transmission.

Ceiling treatment A Independent ceiling with absorbent material	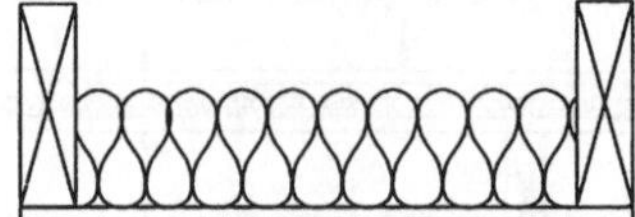	• at least 2 layers of plasterboard with staggered joints; • minimum total mass per unit area of plasterboard 20 kg/m^2; • an absorbent layer of mineral wool (minimum thickness 100 mm, minimum density 10 kg/m^3) laid in the cavity formed above the ceiling.
The type of ceiling support depends on the floor type	**For floor types 1, 2 and 3** Use independent joists fixed only to the surrounding walls **For floor type 3** Use independent joists fixed to the surrounding walls with additional support provided by resilient hangers attached directly to the floor	Always ensure • that you seal the perimeter of the independent ceiling with tape or sealant; • you do not create a rigid or direct connection between the independent ceiling and the floor base.
Ceiling treatment B Plasterboard on proprietary resilient bars with absorbent material	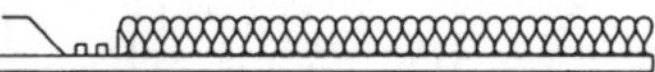	• single layer of plasterboard, minimum mass per unit area of plasterboard 10 kg/m^2; • fixed using proprietary resilient metal bars; • an absorbent layer of mineral wool (minimum density 10 kg/m^3) should fill the ceiling void.
Ceiling treatment C Plasterboard on timber battens or proprietary resilient channels with absorbent material	 If timber battens are used 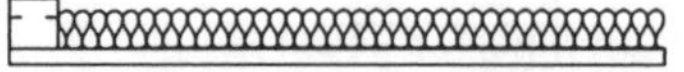 If resilient channels are used	• single layer of plasterboard, minimum mass per unit area 10 kg/m^2; • fixed using timber battens or proprietary resilient channels; • if resilient channels are used, incorporate an absorbent layer of mineral wool (minimum density 10 kg/m^3) that fills the ceiling void.

Notes: (1) Electrical cables give off heat when in use and special precautions may be required when they are covered by thermal insulating materials. See BRE BR 262, Thermal Insulation: avoiding risks, Section 2.3. (2) Installing recessed light fittings in ceiling treatments A to C can reduce their resistance to the passage of airborne and impact sound.

Floor type 1 Concrete base with ceiling and soft floor covering

A floor type 1 consists of a concrete floor base with a soft floor covering and a ceiling. Its resistance to airborne sound mainly depends on:

- the mass per unit area of the concrete base;
- the mass per unit area of the ceiling;
- the soft floor covering (which helps to reduce the source of the impact sound).

General requirements

To allow for future replacements, the soft floor covering should be fixed or glued to the floor.	E3.27a
To avoid air paths all joints between parts of the floor should be filled.	E3.27b
To reduce flanking transmission, air paths should be avoided at all points where a pipe or duct penetrates the floor.	E3.27c
A separating concrete floor should be built into the walls (around its entire perimeter) if the walls are masonry.	E3.27d
All gaps between the head of a masonry wall and the underside of the concrete floor should be filled with masonry.	E3.27e
Flanking transmission from walls connected to the separating floor should be controlled.	E3.27f
The floor base shall not bridge the cavity in a cavity masonry wall.	E3.27a2
Non-resilient floor finishes (such as ceramic floor tiles and wood block floors that are rigidly connected to the floor base) shall not be used.	E3.27b2
Any soft floor covering that is used should be of a resilient material with an overall uncompressed thickness of at least 4.5 mm (also see BS EN ISO 140-8:1998).	E3.28

Two floor types (see below) will meet these requirements.

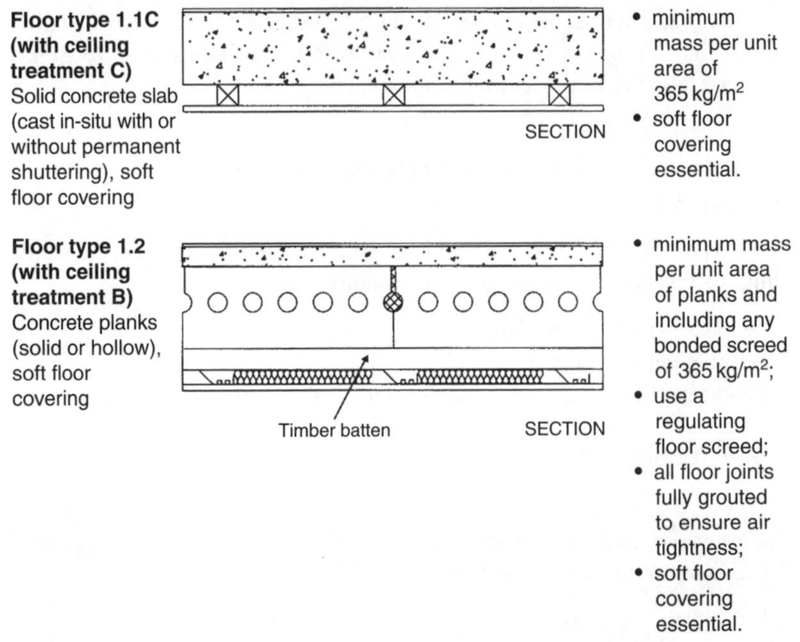

Floor type 1.1C (with ceiling treatment C) Solid concrete slab (cast in-situ with or without permanent shuttering), soft floor covering	SECTION	• minimum mass per unit area of 365 kg/m^2 • soft floor covering essential.
Floor type 1.2 (with ceiling treatment B) Concrete planks (solid or hollow), soft floor covering	Timber batten SECTION	• minimum mass per unit area of planks and including any bonded screed of 365 kg/m^2; • use a regulating floor screed; • all floor joints fully grouted to ensure air tightness; • soft floor covering essential.

Junction requirements for floor type 1

Junctions with an external cavity wall with masonry inner leaf

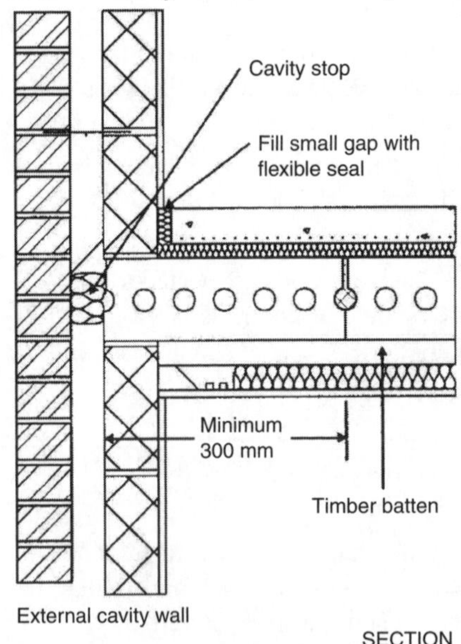

Figure 6.59 Junctions with an external cavity wall with masonry inner leaf

If the external wall is a cavity wall:

- the outer leaf of the wall may be of any construction; E3.31a
- the cavity should be stopped with a flexible closer to ensure adequate drainage. E3.31b

The masonry inner leaf of an external cavity wall should have a mass per unit area of at least 120 kg/m^2 excluding finish. E3.32

The floor base (excluding any screed) should be built into a cavity masonry external and carried through to the cavity face of the inner leaf. E3.33

Junctions with an external cavity wall with timber frame inner leaf

Where the external wall is a cavity wall:

- the outer leaf of the wall may be of any construction; E3.36a
- the cavity should be stopped with a flexible closer; E3.36b
- the wall finish of the inner leaf of the external wall should be two layers of plasterboard, each sheet of plasterboard to be a minimum mass per unit area 10 kg/m^2; E3.36c
- all joints should be sealed with tape or caulked unenclosed. E3.36c

Junctions with an external solid masonry wall

No official guidance currently available about junctions with a solid masonry wall. Best to seek specialist advice. E3.37

Junctions with internal framed walls

There are no restrictions on internal walls meeting a type 1 separating floor. E3.38

Junctions with internal masonry walls

The floor base should be continuous through or above an internal masonry wall. E3.39

The mass per unit area of any load bearing internal wall (or any internal wall rigidly connected to a separating floor) should be at least 120 kg/m^2 excluding finish. E3.40

Junctions with floor penetrations (excluding gas pipes)

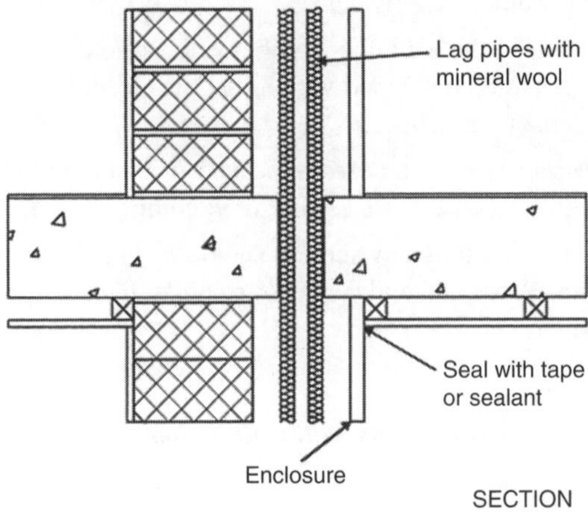

Figure 6.60 Floor type 1 floor penetrations

Pipes and ducts should be in an enclosure (both above and below the floor). In all cases:

The enclosure should be constructed of material having a mass per unit area of at least 15 kg/m^2.	E3.32
The enclosure should either be lined or the duct (or pipe) within the enclosure wrapped with 25 mm unfaced mineral fibre.	E3.42
Penetrations through a separating floor by ducts and pipes should have fire protection to satisfy Building Regulation Part B – Fire safety.	E3.43
Fire stopping should be flexible to prevent a rigid contact between the pipe and the floor.	E3.43
Gas pipes may be contained in a separate (ventilated) duct or can remain unenclosed.	E3.43
If a gas service is installed it shall comply with the Gas Safety (Installation and Use) Regulations 1998, S1 1998 No 2451.	E3.43
If the pipes and ducts penetrate a floor separating habitable rooms in different flats, then they should be enclosed for their full height in each flat.	E3.41

Junctions with separating wall type 1 – solid masonry

For floor types 1.1C and 1.2C, two possibilities exist:

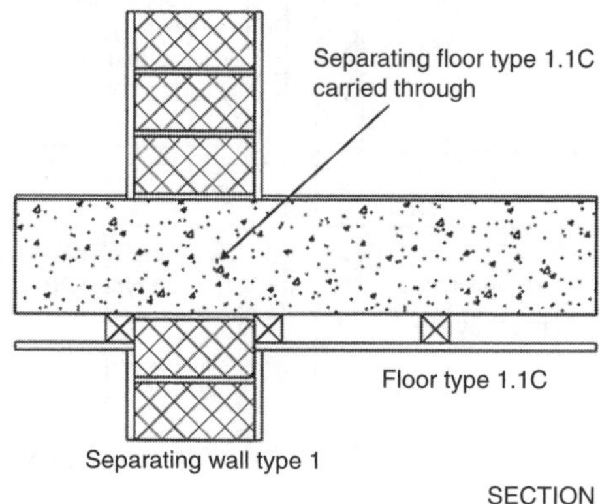

Figure 6.61 Floor type 1.1C – wall type 1

> A separating floor type 1.1 C base (excluding any screed) **should** pass through a separating wall type 1 (for flats where there are separating walls, guidance on p. 318 may also apply). E3.44

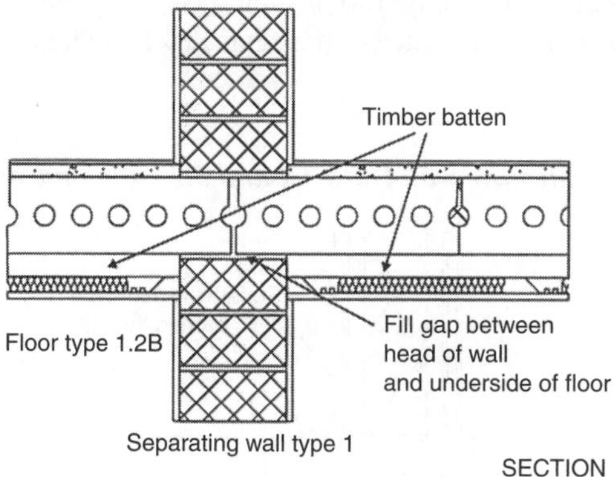

Figure 6.62 Floor type 1.1C – wall type 1

> A separating floor type 1.2B base (excluding any screed) should **not** pass through a separating wall type 1 (for flats where there are separating walls, guidance on p. 318 may also apply). E3.44

Junctions with separating wall type 2 cavity masonry

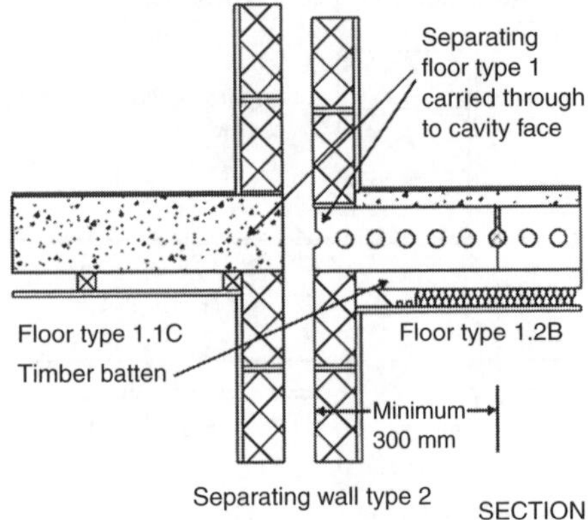

Figure 6.63 Floor types 1.1C and 1.2B – wall type 2

The mass per unit area of any leaf that is supporting or adjoining the floor should be at least 15 kg/m^2 excluding finish.	E3.46
The floor base (excluding any screed) should be carried through to the cavity face of the leaf.	E3.47
The wall cavity should not be bridged.	E3.47
Where floor type 1.2B is used (and the planks are parallel to the separating wall) the first joint should be a minimum of 300 mm from the inner face of the adjacent cavity leaf.	E3.48

Junctions with separating wall types 3.1 and 3.2 (solid masonry core)

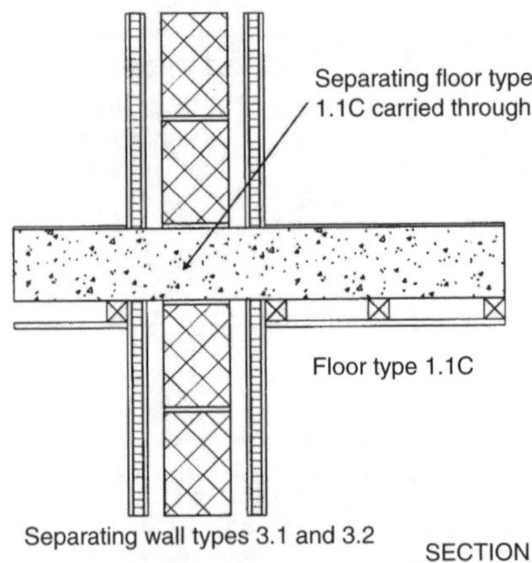

Figure 6.64 Floor type 1.1C – wall types 3.1 and 3.2

A separating floor type 1.1C base (excluding any screed) should pass through separating wall types 3.1 and 3.2.	E3.49
A separating floor type 1.2B base (excluding any screed) should not be continuous through a separating wall type 3.	E3.50
Where a separating wall type 3.2 is used with floor type 1.2B (and the planks are parallel to the separating wall) the first joint should be a minimum of 300 mm from the centreline of the masonry core.	E3.51

Junctions with separating wall type 3.3 (cavity masonry core)

The mass per unit area of any leaf that is supporting or adjoining the floor should be at least 120 kg/m^2 excluding finish.	E3.52
The floor base (excluding any screed) should be carried through to the cavity face of the leaf of the core.	E3.53
The cavity should not be bridged.	E3.53
Where floor type 1.2B is used (and the planks are parallel to the separating wall) the first joint should be a minimum of 300 mm from the inner face of the adjacent cavity leaf of the masonry core.	E3.54

Junctions with separating wall type 4 timber frames with absorbent material

No official guidance currently available. Best to seek specialist advice.	E3.55

Floor type 2: Concrete base with ceiling and floating floor

A floor type 2 consists of a concrete floor base with a floating floor (which in turn consists of a floating layer and a resilient layer) and a ceiling. Its resistance to airborne and impact sound depends on:

- the mass per unit area of the concrete base;
- the mass per unit area and isolation of the floating layer and the ceiling;
- the floating floor (which reduces impact sound at source).

General requirements

All joints between parts of the floor should be filled to avoid air paths.	E3.61a
To reduce flanking transmission, air paths should be avoided at all points where a pipe or duct penetrates the floor.	E3.61b
A separating concrete floor should be built into the walls (around its entire perimeter) if the walls are masonry.	E3.61c
All gaps between the head of a masonry wall and the underside of the concrete floor should be filled with masonry.	E3.61d
Flanking transmission from walls connected to the separating floor should be controlled.	E3.61e
The floor base shall not bridge a cavity in a cavity masonry wall.	E3.61

Two floor types (consisting of a floating layer and resilient layer – see below) will meet these requirements. A performance-based approach (type C) is also available.

Floating floor (a)
Timber raft floating layer with resilient layer

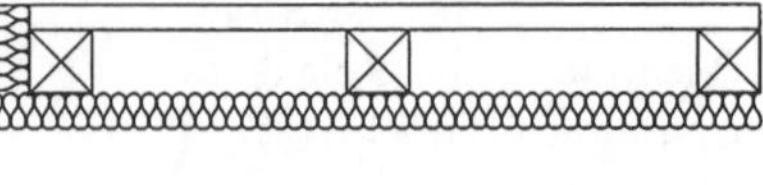

- timber raft of board material (with bonded edges, e.g. tongued and grooved);
- minimum mass per unit area 12 kg/m^2;
- fixed to 45 mm × 45 mm battens laid loose on the resilient layer (but **not** along any joints in the resilient layer);
- resilient layer of mineral wool (which may be paper faced on the underside) with density 36 kg/m^{-3} and minimum thickness 25 mm.

Floating floor (b)
Sand cement screed floating layer with resilient layer

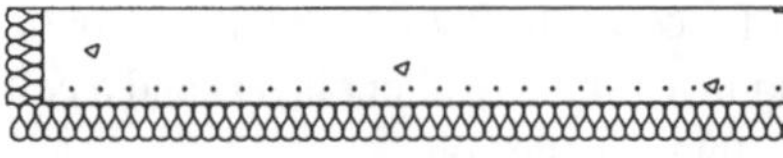

Floating layer

- of 65 mm sand cement screed with a mass per unit area of at least 80 kg/m^2.

Resilient layer

- protected while the screed is being

Floating floor (b) – *continued*		laid (e.g. by a 20–50 mm wire mesh) and consisting of either: • a layer of mineral wool of minimum thickness 25 mm with density 36 kg/m^{-3} (paper faced on the upper side); • an alternative type of resilient layer with maximum dynamic stiffness of 15 kg/m^3; • an alternative type of resilient layer with minimum thickness of 5 mm (see BS EN ISO 29052-1:1992).
Floating floor (c) Performance-based approach	—	• rigid boarding above a resilient and/or damping layer; • weighted reduction in impact sound pressure level of not less than 29 dB (see BS EN ISO 717-2:1997 and BS EN ISO 140-8:1998).

A small gap filled with a flexible sealant should be left between the floating layer and wall at all room edges.	E3.63a
A small gap (approx. 5 mm and filled with a flexible sealant) should be left between skirting and floating layer.	E3.63b
Resilient materials should be laid in rolls or sheets either with lapped joints or with joints tightly butted and taped.	E3.63c
Paper facing should be used on the upper side of fibrous materials to prevent screed entering the resilient layer.	E3.63d
The floating layer and the base or surrounding walls shall not be bridged (e.g. with services or fixings that penetrate the resilient layer).	E3.63a2
The floating screed shall create a bridge (for example, through a gap in the resilient layer) to the concrete floor base or surrounding walls.	E3.63b2

Depending on the type of ceiling two options can be used:

Floor type 2.1C(a) (with ceiling treatment C and floating floor (a))	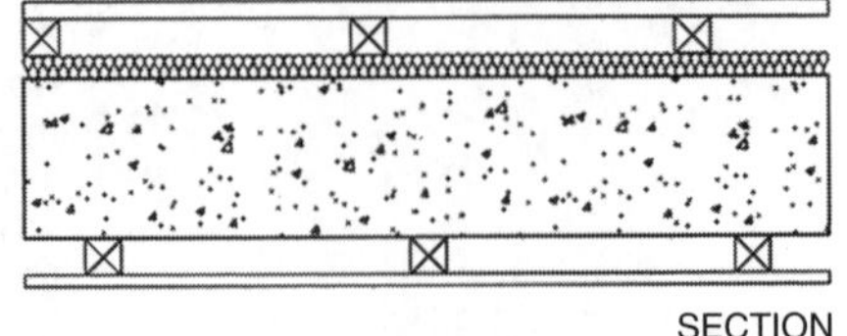 **Floor type a**	• minimum mass per unit area of 300 kg/m^2;
Solid concrete slab (cast in-situ with or without permanent shuttering), floating floor, ceiling treatment	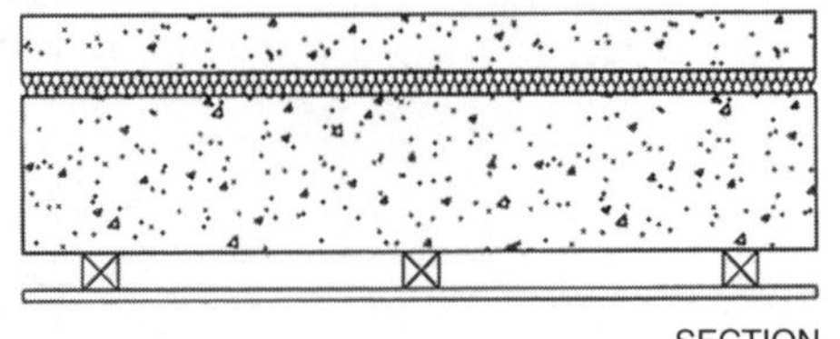 **Floor type b**	• regulating floor screed optional; • floating floor (a), (b) or (c) essential; • ceiling treatment C (or better) essential.
Floor type 2.1C(b) (with ceiling treatment C and floating floor (b)) Concrete planks (solid or hollow), floating floor, ceiling treatment B	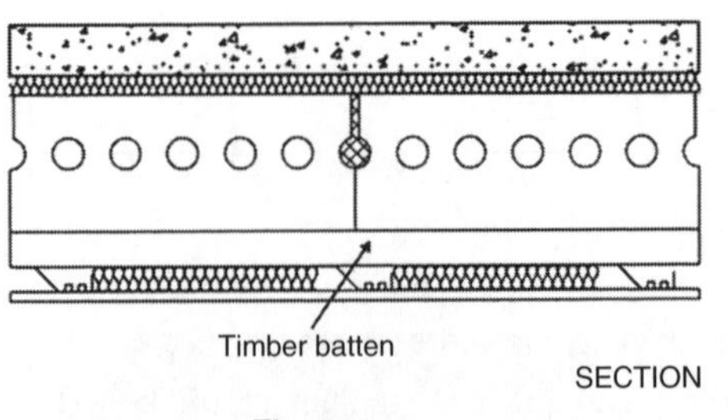 **Floor type a** 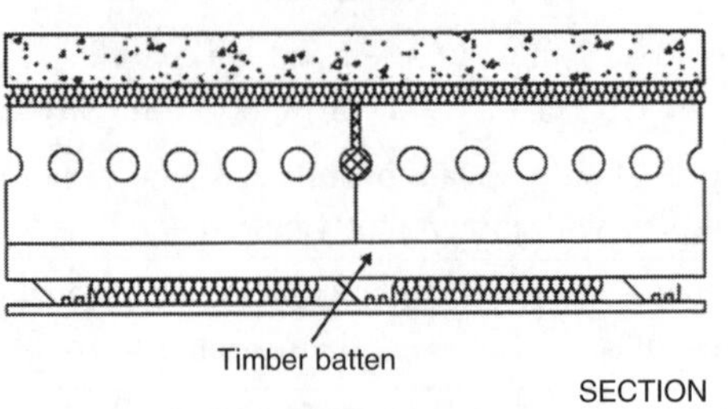 **Floor type b**	• minimum mass per unit area of planks and any bonded screed of 300 kg/m^2; • use a regulating floor screed; • all floor joints fully grouted to ensure air tightness; • floating floor (a), (b) or (c) essential; • ceiling treatment B (or better) essential.

Junction requirements for floor type 2

Junctions with an external cavity wall with type 2 timber frame inner leaf

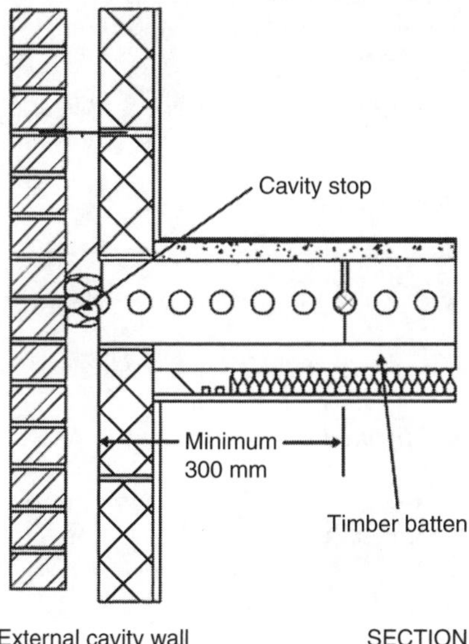

Figure 6.65 Floor type 2 – external cavity wall with masonry internal leaf

Where the external wall is a cavity wall:

- the outer leaf of the wall may be of any construction; E3.69a
- the cavity should be stopped with a flexible closer. E3.69b

The masonry inner leaf of an external cavity wall should have a mass per unit area of at least 120 kg/m^2. E3.70

The floor base (excluding any screed) should be built into a cavity masonry external wall and carried through to the cavity face of the inner leaf. E3.71

The cavity should not be bridged. E3.71

If a floor 2.2B is used (and the planks through, or above, an internal masonry wall are parallel to the external wall) the first joint should be a minimum of 300 mm from the cavity face of the inner leaf. E3.72

Junctions with an external cavity wall with timber frame inner leaf

Where the external wall is a cavity wall:

- the outlet leaf of the wall may be of any construction; E3.74a
- the cavity should be stopped with a flexible closer; E3.74b

- the wall face of the inner leaf of the external wall should be two layers of plasterboard; E3.74c
- each sheet of plasterboard to be of minimum mass per unit area 10 kg/m^2; E3.74c
- all joints should sealed or caulked with sealant. E3.74c

Junctions with an external solid masonry wall

No official guidance currently available. Best to seek specialist advice. E3.75

Junctions with internal framed walls

There are no restrictions on internal walls meeting a type 4 separating wall. E3.76

Junctions with internal masonry walls

The floor base should be continuous or above an internal masonry wall. E3.77

The mass per unit area of any load bearing internal wall or any internal wall rigidly connected to a separating floor should be at least 120 kg/m^2 excluding finish. E3.78

Junctions with floor penetrations (excluding gas pipes)

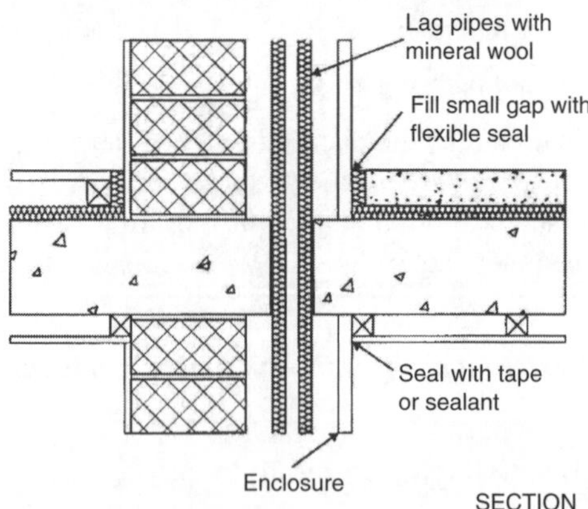

Figure 6.66 Floor type 2 – floor penetrations

Pipes and ducts that penetrate a floor separating habitable rooms in different flats should be enclosed for their full height in each flat.	E3.79
The enclosure should be constructed of material having a mass per unit area of at least 15 kg/m^2.	E3.80
Either line the enclosure, or wrap the duct or pipe within the enclosure, with 25 mm unfaced mineral wool.	E3.80
A small gap (sealed with sealant or neoprene) of about 5 mm should be left between the enclosure and the floating.	E3.81
Where floating floor (a) or (b) is used the enclosure may go down to the floor base (provided that the enclosure is isolated from the floating layer).	E3.81
Penetrations through a separating floor by ducts and pipes should have fire protection to satisfy Building Regulation Part B – Fire safety.	E3.82
Gas pipes may be contained in a separate (ventilated) duct or can remain unenclosed.	E3.82
If a gas service is installed it shall comply with the Gas Safety (Installation and Use) Regulations 1998, S1 1998 No 2451.	E3.82

Junctions with a separating wall type 1 – solid masonry

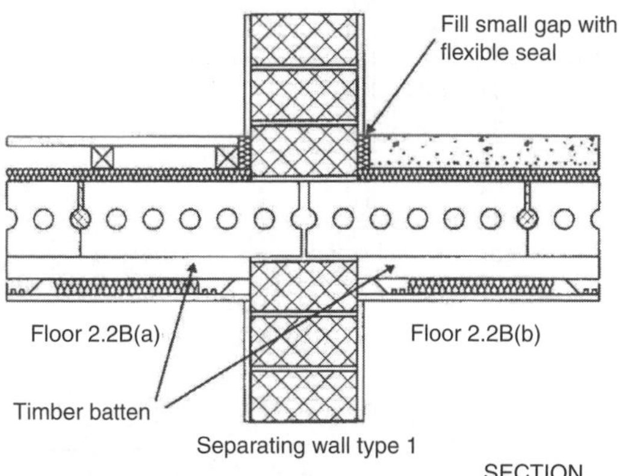

Figure 6.67 Floor type 2.1C – wall types 3.1 and 3.2

A separating floor type 2.1C base (excluding any screed) should pass through a separating wall type 1.	E3.84
A separating floor type 2.2B base (excluding any screed) should **not** be continuous through a separating wall type 1.	E3.84

Junctions with a separating wall type 2 cavity masonry

The floor base (excluding any screed) should be carried through to the cavity face of the leaf.	E3.85
The cavity should not be bridged.	E3.85
If a floor type 2.2B is used (and the planks are parallel to the separating wall) the first joint should be a minimum of 300 mm from the cavity face of the leaf.	E3.86

Junctions with separating wall type 3.1 and 3.2 (solid masonry core)

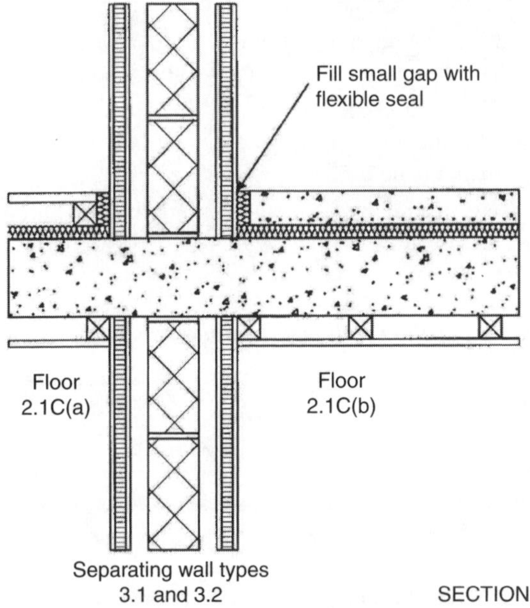

Figure 6.68 Floor type 2.1C – wall types 3.1 and 3.2

A separating floor type 2.1C base (excluding any screed) should pass through separating wall types 3.1 and 3.2.	E3.87
A separating floor type 2.2B base (excluding any screed) should not be continuous through a separating wall type 3.	E3.88
If a separating wall type 3.2 is used with floor type 2.2B (and the planks are parallel to the separating wall) the first joint should be a minimum of 300 mm from the centreline of the masonry core.	E3.89

Junctions with separating wall type 3.3 (cavity masonry core)

The mass per unit area of any leaf that is supporting or adjoining the floor should be at least 120 kg/m^2 excluding finish.	E3.90
The floor base (excluding any screed) should be carried through to the cavity face of the leaf of the core.	E3.91
The cavity should not be bridged.	E3.91
If a floor type 2.2B is used (and the planks are parallel to the separating wall) the first joint should be a minimum of 300 mm from the inner face of the adjacent cavity leaf of the masonry core.	E3.92

Junctions with separating wall type 4 timber frames with absorbent material

No official guidance currently available. Best to seek specialist advice.	E3.93

Floor type 3: Timber frame base with ceiling and platform floor

A floor type 3 consists of a timber frame structural floor base with a deck, platform floor (consisting of a floating layer and a resilient layer) and ceiling treatment. Its resistance to airborne and impact sound depends on:

- the structural floor base;
- the isolation of the platform floor and the ceiling;
- the platform floor (which reduces impact sound at source).

General requirements

To reduce flanking transmission, air paths should be avoided at all points where the floor is penetrated.	E3.99a
Flanking transmission from walls connected to the separating floor should be as described in the following junction requirements for floor type 3	E3.99b
There should be no bridge (e.g. formed by services or fixings that penetrate the resilient layer) between the floating layer and the base or surrounding walls.	E3.99

For the platform floor, ensure that:

- the correct density of resilient layer is used; E3.99c
- the layer can carry the anticipated load; E3.99c
- during construction a gap is maintained between the wall and the floating layer (filled with a flexible sealant, expanded or extruded polystyrene strip); E3.99d
- resilient materials are laid in sheets with all joints tightly butted and taped. E3.99e

The following floor type (floor type 3.1A) will meet these requirements.

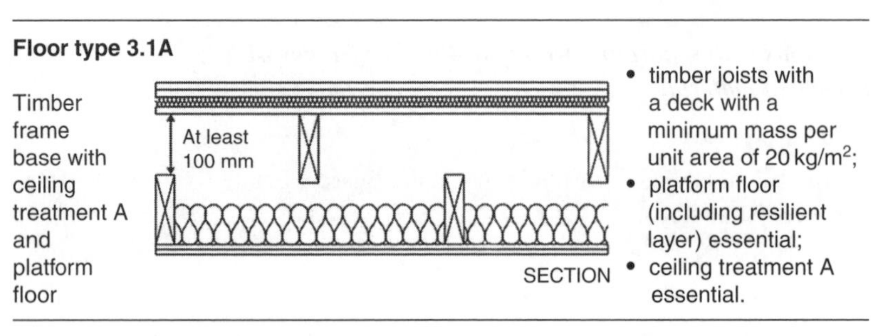

Floor type 3.1A
Timber frame base with ceiling treatment A and platform floor
At least 100 mm
SECTION
• timber joists with a deck with a minimum mass per unit area of 20 kg/m²;
• platform floor (including resilient layer) essential;
• ceiling treatment A essential.

Platform floor

The floating layer should: E3.101

- be a minimum of two layers of board material;
- be minimum total mass per unit area 25 kg/m^2;
- have layers of minimum thickness 8 mm;
- be fixed together with joints staggered;
- be laid loose on a resilient layer.

Resilient layer

The resilient layer should be of mineral wool: E3.102

- minimum thickness 25 mm;
- density 60 to 100 kg/m^3;
- paper faced on the underside.

Junction requirements for floor type 3

Junctions with an external cavity wall with masonry inner leaf

Where the external wall is a cavity wall:	
• the outer leaf of the wall may be of any construction;	E3.103a
• the cavity should be stopped with a flexible closer.	E3.103b
The masonry inner leaf of a cavity wall should be lined with an independent panel.	E3.104
The ceiling should be taken through to the masonry.	E3.105
The junction between the ceiling and the independent panel should be sealed with tape or caulked with sealant.	E3.105
Air paths between floor and wall cavities should be blocked.	E3.106

Note:

(1) Any normal method of connecting floor base to wall may be used.
(2) Independent panels are not required if the mass per unit area of the inner leaf is greater than 375 kg/m^2.

Junctions with an external cavity wall with timber frame inner leaf

Where the external wall is a cavity wall:	
• the outer leaf of the wall may be of any construction;	E3.109a
• the cavity should be stopped with a flexible closer.	E3.109b
The wall finish of the inner leaf of the external wall should:	
• be two layers of plasterboard;	E3.110a
• be each sheet of plasterboard of minimum mass per unit area 10 kg/m^2;	E3.110b
• have all joints sealed with tape or caulked with sealant.	E3.110c
Any normal method of connecting floor base to wall may be used.	E3.111
If the joists are at right angles to the wall, spaces between the floor joists should be sealed with full depth timber blocking.	E3.112
The junction between the ceiling and wall lining should be sealed with tape or caulked with sealant.	E3.113

Junctions with an external solid masonry wall

No official guidance currently available. Best to seek specialist advice.	E3.113

Junctions with internal framed walls

The spaces between joists are at right angles and should be sealed with full depth timber blocking.	E3.114
The junction between the ceiling and the internal framed wall should be sealed with tape or caulked with sealant.	E3.115

Junctions with internal masonry walls

No official guidance currently available. Best to seek specialist advice.	E3.116

Junctions with floor penetrations (excluding gas pipes)

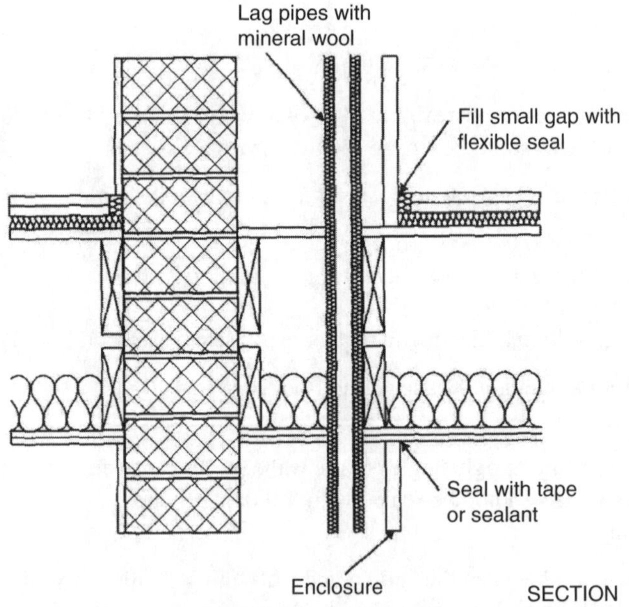

Figure 6.69 Floor type 3 – floor penetrations

Pipes and ducts that penetrate a floor separating habitable rooms in different flats should be enclosed for their full height in each flat.	E3.117
The enclosure should:	
• be constructed of material having a mass per unit area of at least 15 kg/m^2;	E3.118
• have a small, sealed (with sealant or neoprene) 5 mm gap between the enclosure and floating layer;	E3.119
• go down to the floor base;	E3.119
• be isolated from the floating layer.	E3.119
The duct or pipe within the enclosure should be lined or wrapped with 25 mm unfaced mineral wool.	E3.118
Penetrations through a separating floor by ducts and pipes should have fire protection to satisfy Building Regulation Part B – Fire safety.	E3.120
Fire stopping should be flexible and also prevent rigid contact between the pipe and floor.	E3.121
Gas pipes may be contained in a separate (ventilated) duct or can remain unenclosed.	E3.120
If a gas service is installed it shall comply with the Gas Safety (Installation and Use) Regulations 1998, S1 1998 No 2451.	E3.120

Junctions with a separating wall type 1 solid masonry

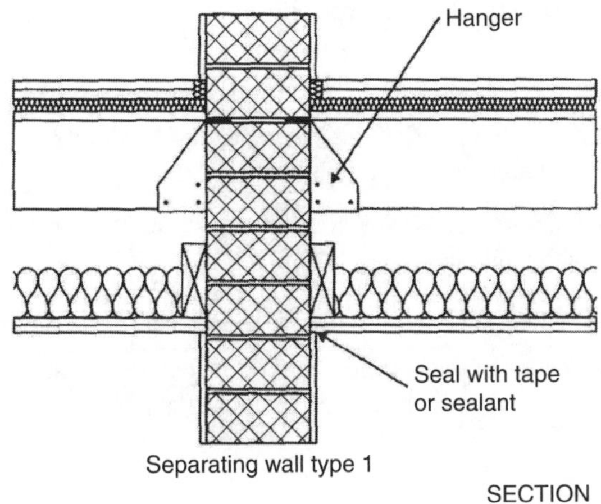

Figure 6.70 Floor type 3 – wall type 1

Floor joists supported on a separating wall should be supported on hangers as opposed to being built in.	E3.121
The junction between the ceiling and wall should be sealed with tape or caulked with sealant.	E3.122

Note: The above is particularly relevant for flats where there are separating walls.

Junctions with a separating wall type 2 – cavity masonry

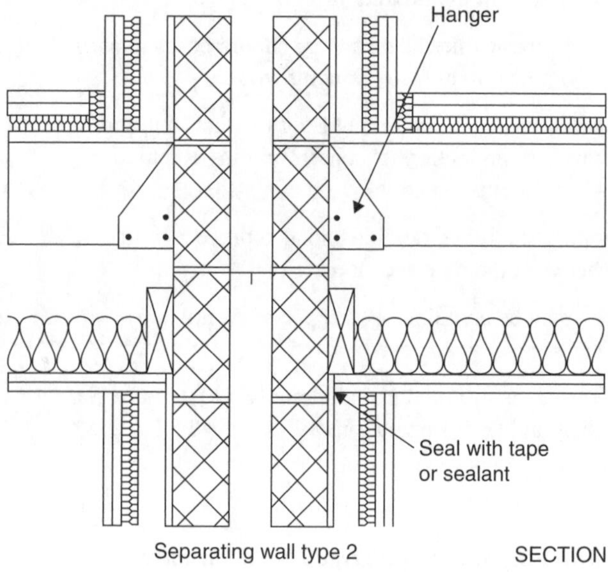

Figure 6.71 Floor type 3 – wall type 2

Floor joists that are supported on a separating wall should be supported on hangers and not built in.	E3.123
The adjacent leaf of a cavity separating wall should be lined with an independent panel.	E3.124
The ceiling should be taken through to the masonry.	E3.125
The junction between the ceiling and the independent panel should be sealed with tape or caulked with sealant.	E3.125

Note: Independent panels are not required if the mass per unit area of the inner leaf is greater than 375 kg/m^2.

Junctions with a separating wall type 3 – masonry between independent panels

Floor joists that are supported on a separating wall should be supported on hangers and not built in.	E3.127
The ceiling should be taken through to the masonry.	E3.128
The junction between the ceiling and the independent panel should be sealed with tape or caulked with sealant.	E3.128

Junctions with a separating wall type 4 – timber frames with absorbent material

Spaces between the floor joists that are at right angles to the wall should be sealed with full depth timber blocking.	E3.129
The junction of the ceiling and wall lining should be sealed with tape or caulked with sealant.	E3.130

6.8 Walls

In a brick built house, the external walls are loadbearing elements that support the roof, floors and internal walls. These walls are normally cavity walls comprising of two leaves braced with metal ties but older houses will have solid walls, at least 225 mm (9″) thick. Bricks are laid with mortar in overlapping bonding patterns to give the wall rigidity and a Damp-Proof Course (DPC) is laid just above ground level to prevent the moisture rising. Window and door openings are spanned above with rigid supporting beams called lintels. The internal walls of a brick built house are either non-loadbearing divisions (made from lightweight blocks, manufactured boards or timber studding) or loadbearing structures made of brick or block.

Modern timber-framed house walls are constructed of vertical timber studs with horizontal top and bottom plates nailed to them. The frames, which are erected on a concrete slab or a suspended timber platform supported by cavity brick walls, are faced on the outside with plywood sheathing to stiffen the structure. Breather paper is fixed over the top to act as a moisture barrier. Insulation quilt is used between studs. Rigid timber lintels at openings carry the weight of the upper floor and roof. Brick cladding is typically used to cover the exterior of the frame. It is attached to the frame with metal ties. Weatherboarding often replaces the brick cladding on upper floors.

When reading this section, you will probably notice that a few of the requirements have already been covered in Section 6.6 Floors and ceilings. This has been done in order to save the reader having to constantly turn back and re-read a previous page.

6.8.1 Requirements

Fire precautions

Materials and/or products used for the internal linings of walls shall restrict:

- *the spread of flame:*
- *the amount of heat released.*

(Approved Document B2)

B3 Internal Fire spread (structure)

A wall common to two or more buildings shall be designed and constructed so that it adequately resists the spread of fire between those buildings.

(Approved Document B3)

External walls shall be constructed so as to have a low rate of heat release and thereby be capable of reducing the risk of ignition from an external source and the spread of fire over their surfaces.

The amount of unprotected area in the sides of the building shall be restricted so as to limit the amount of thermal radiation that can pass through the wall.

(Approved Document B4)

The walls of the building shall adequately protect the building and people who use the building from harmful effects caused by:

- *ground moisture;*
- *precipitation and wind-driven spray;*
- *interstitial and surface condensation; and*
- *spillage of water from or associated with sanitary fittings or fixed appliances.*

(Approved Document C2)

Cavity insulation

Fumes given off by insulating materials such as by Urea Formaldehyde (UF) foams should not be allowed to penetrate occupied parts of buildings to an extent where it could become a health risk to persons in the building by becoming an irritant concentration.

(Approved Document D)

Airborne and impact sound

Dwellings shall be designed so that the noise from domestic activity in an adjoining dwelling (or other parts of the building) is kept to a level that:

- *does not affect the health of the occupants of the dwelling;*
- *will allow them to sleep, rest and engage in their normal activities in satisfactory conditions.*

(Approved Document E1)

Dwellings shall be designed so that any domestic noise that is generated internally does not interfere with the occupants' ability to sleep, rest and engage in their normal activities in satisfactory conditions.

(Approved Document E2)

Domestic buildings shall be designed and constructed so as to restrict the transmission of echoes.

(Approved Document E3)

Schools shall be designed and constructed so as to reduce the level of ambient noise (particularly echoing in corridors).

(Approved Document E4)

Conservation of fuel and power

Reasonable provision shall be made for the conservation of fuel and power in buildings by:

(a) limiting heat gains and losses:
- *(i) through thermal elements and other parts of the building fabric; and*
- *(ii) from pipes, ducts and vessels used for space heating, space cooling and hot water services;*

(b) providing and commissioning energy-efficient fixed building services with effective controls; and

(c) providing to the owner sufficient information about the building, the fixed building services and their maintenance requirements so that the building can be operated in such a manner as to use no more fuel and power than is reasonable in the circumstances.

(Approved Document L1)

6.8.2 Meeting the requirements

General

Walls should comply with the relevant requirements of BS 5628: Part 3:2001.	A1/2 2C2c

Basic requirements for stability

The layout of walls (both internal and external) shall: A1/2 1A2b

- form a robust three-dimensional box structure in plan;
- be constructed according to the specific guidance for each form of construction.

Internal and external walls shall be adequately connected by either masonry bonding or by using mechanical connections. A1/2 1A2c

Building height

For residential buildings, the maximum height of the building measured from the lowest finished ground level adjoining the building to the highest point of any wall or roof should not be greater than 15 m. A1/2 2C4i

Types of wall shown in Table 6.32 must extend to the full storey height. A1/2 2C2

Table 6.32 Wall types considered in this section

Residential buildings of up to three storeys	Small single storey non-residential buildings and annexes
External walls Internal load-bearing walls Compartment walls Separating walls	External walls Internal load-bearing walls

Thickness of walls

The thickness of the wall depends on the general conditions relating to the building of which the wall forms a part (e.g. floor area, roof loading, wind speed, etc.) and the design conditions relating to the wall (e.g. type of materials, loading, end restraints, openings, recesses, overhangs and lateral floor support requirements, etc.).

Note: Where walls are constructed of bricks or blocks, they shall be in accordance with BS 6649: 1985.

Masonry units

Walls should be properly bonded and solidly put together with mortar and constructed of masonry units conforming to:

- clay bricks or blocks conforming to BS 3921: 1985 or BS 6649: 1985 or BS EN 771-1; A1/2 2C20a
- calcium silicate bricks conforming to BS 187: 1978 or BS 6649: 1985 or BS EN 771-2; A1/2 2C20b
- concrete bricks or blocks conforming to BS 6073: Part 1: 1981 or BS EN 771-3 or 4; A1/2 2C20c
- square dressed natural stone conforming to the appropriate requirements described in BS EN 771-6 or BS 5628: Part 3: 2001; A1/2 2C20d
- Manufactured stone complying with BS 6457: 1984 and BS EN 771-5. A1/2 2C20e

Note: See BS 3921, BS 6073-1, BS 187, BS 5390 and BS 6649 for further details about the minimum compressive strength requirements for masonry units.

Mortar

Mortar should be equivalent to (or of greater strength and durability) than:

- mortar designation (iii) according to BS 5628: Part 3: 2001; A1/2 2C22a
- strength class M4 according to BS EN 998-2; A1/2 2C22b
- 1:1:5 or 6 CEM 1, lime and fine aggregate measured by volume of dry materials. A1/2 2C22c

Tension straps

Tension straps (conforming to BS EN 845-1) should be used to strap walls to floors above ground level, at intervals not exceeding 2 m and as shown in Figure 6.72. A1/2 2C35

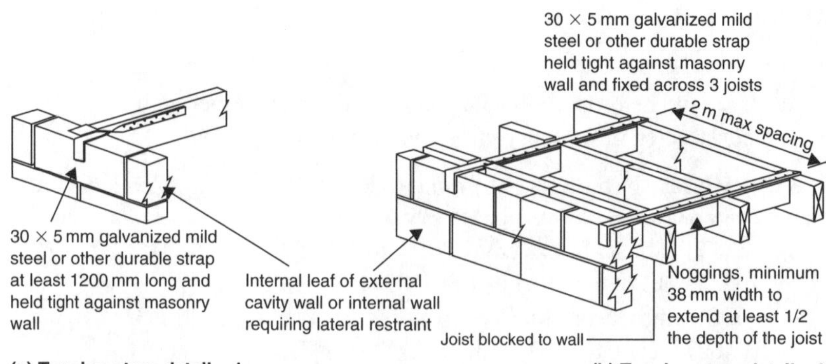

Figure 6.72 Lateral support by floors

Gable walls should be strapped to roofs as shown in Figure 6.73(a) and (b) by tension straps. A1/2 2C36

For corrosion resistance purposes, tension straps should be material reference 14 or 16.1 or 16.2 (galvanized steel) or other more resistant specifications including material references 1 or 3 (austenitic stainless steel). A1/2 2C35

The declared tensile strength of tension straps should not be less than 8 kN.

Tension straps need **not** be provided:

- in the longitudinal direction of joists in houses of not more than two storeys if the joists:
  - are at not more than 1.2 m centres; A1/2 2C35a
  - have at least 90 mm bearing on the supported walls or 75 mm bearing on a timber wall-plate at each end; or A1/2 2C35a
  - are carried on the supported wall by joist hangers (in accordance with BS EN 845-1 and BS 5628 – see Figure 6.74); A1/2 2C35b
  - and are incorporated at not more than 2 m centres; A1/2 2C35b
- when a concrete floor has at least 90 mm bearing on the supported wall (see Figure 6.75); and A1/2 2C35c
- where floors are at or about the same level on each side of a supported wall, and contact between the floors and wall is either continuous or at intervals not exceeding 2 m. Where contact is intermittent, the points of contact should be in line or nearly in line on plan (see Figure 6.76). A1/2 2C35d

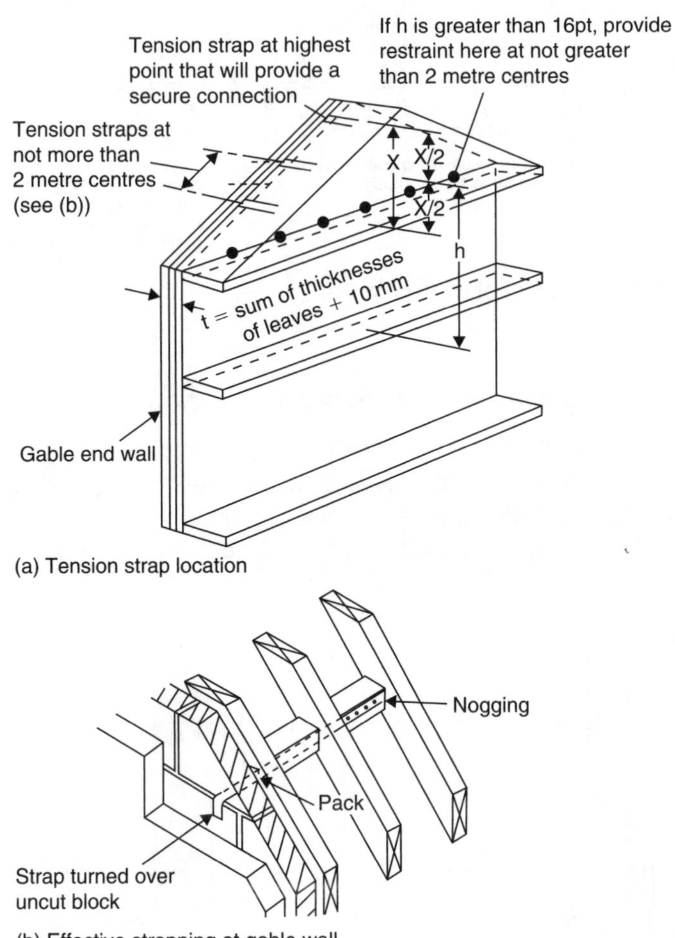

(a) Tension strap location

(b) Effective strapping at gable wall

Figure 6.73 Lateral support at roof level

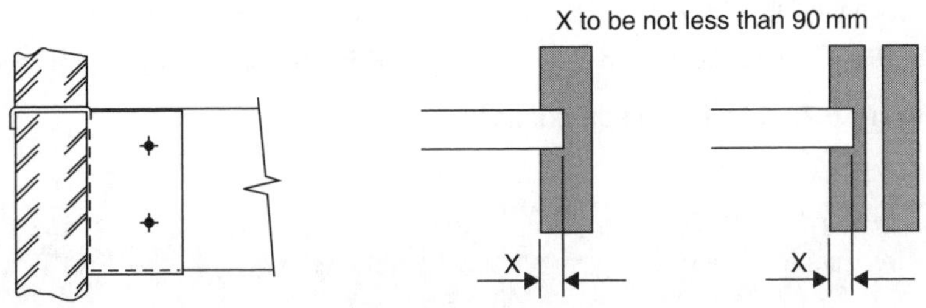

Figure 6.74 Restraint type joist hanger

Figure 6.75 Restraint by concrete floor or roof

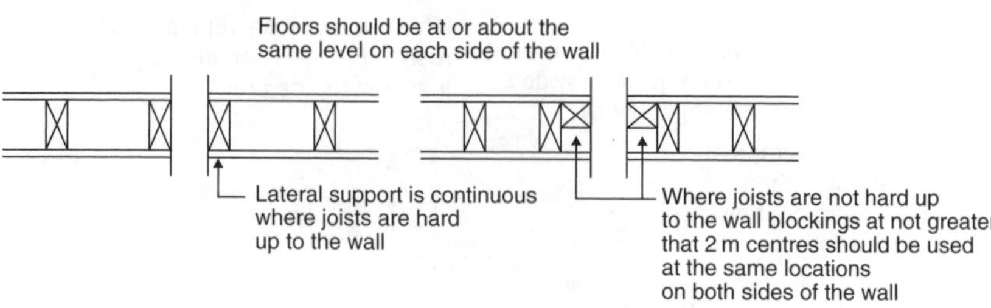
Floors should be at or about the same level on each side of the wall
Lateral support is continuous where joists are hard up to the wall
Where joists are not hard up to the wall blockings at not greater that 2 m centres should be used at the same locations on both sides of the wall

Figure 6.76 Restraint of internal walls

Internal load-bearing walls in brickwork or blockwork

The maximum span for any floor supported by a wall is 6 m where the span is measured centre to centre of bearing (see Figure 6.77). A1/2 2C23

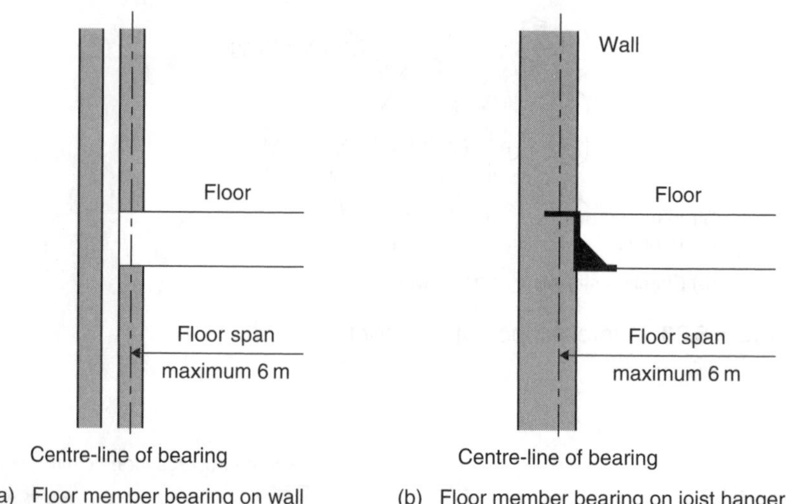
Floor
Floor span
maximum 6 m
Centre-line of bearing
(a) Floor member bearing on wall
Wall
Floor
Floor span
maximum 6 m
Centre-line of bearing
(b) Floor member bearing on joist hanger

Figure 6.77 Maximum span of floors

Vertical loading on walls should be distributed. A1/2 2C23a

Differences in level of ground or other solid construction between one side of the wall and the other should be less than four times the thickness of the wall as shown in Figure 6.78. A1/2 2C23b

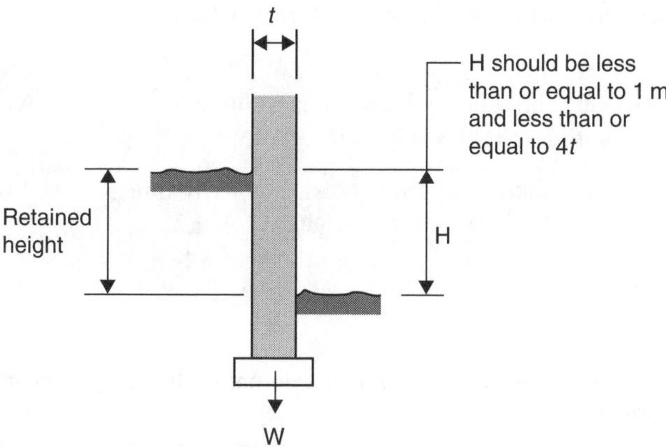

Figure 6.78 Maximum difference in permitted level

Dead load, imposed load and wind load should be in accordance with current codes of practice.	A1/2 0.2b
Loads used in calculations should allow for possible dynamic, concentrated and peak load effects that may occur.	A1/2 0.2
All walls (except compartment and/or separating walls) should have a thickness not less than:	A1/2 2C10

$$\frac{\text{Specified thickness from Table 6.33}}{2} - 5\,\text{mm}$$

Note: Except for a wall in the lowest storey of a three storey building, carrying load from both upper storeys, which should have a thickness as determined by the equation or 140 mm whichever is the greatest.

Solid external walls, compartment walls and separating walls in coursed brickwork or blockwork

Solid walls constructed of coursed brickwork or blockwork should be at least as thick as 1/16 of the storey height.	A1/2 2C6

Solid external walls, compartment walls and separating walls in uncoursed stone, flints, etc.

The thickness of walls constructed in uncoursed stone, flints, clunches, of bricks or other burnt or vitrified material should not be less than 1.33 times the thickness of the storey height.	A1/2 2C7

Cavity walls in coursed brickwork or blockwork

All cavity walls should have leaves at least 90 mm thick and cavities at least 50 mm wide. A1/2 2C8

The combined thickness of the two leaves plus 10 mm should be at least 1/16 of the storey height (see as per Table 6.33). A1/2 2C8

Table 6.33 Minimum thickness of certain external walls, compartment walls and separating walls

Height of wall	Length of wall	Minimum thickness of wall
not exceeding 3.5 m	not exceeding 12 m	190 mm for whole of its height
exceeding 3.5 m but not exceeding 9 m	not exceeding 9 m	190 mm for whole of its height
	exceeding 9 m	290 mm from the base for the height of one storey and 190 mm for the rest of its height
exceeding 9 m but not exceeding 12 m	not exceeding 9 m	290 mm from the base for the height of one storey and 190 mm for the rest of its height
	exceeding 9 m but not exceeding 12 m	290 mm from the base for the height of two storeys and 190 mm for the rest of its height

Wall ties should either comply with BS 1243, DD 140 or BS EN 845-1. A1/2 2C19

Wall ties should have a horizontal spacing of 900 mm and a vertical spacing of 450 mm. A1/2 2C8

Equivalent to 2.5 ties per square metre.

Wall ties should also be provided, spaced not more than 300 mm apart vertically, within a distance of 225 mm from the vertical edges of all openings, movement joints and roof verges. A1/2 2C8

For external walls, compartment walls and separating walls in cavity construction, the combined thickness of the two leaves plus 10 mm should be at least as thick as 1/16 of the storey height. A1/2 2C8

Walls providing vertical support to other walls

Irrespective of the material used in the construction, a wall should not be less than the thickness of any part of the wall to which it gives vertical support.	A1/2 2C9

Parapet walls

The minimum thickness and maximum height of parapet walls should be as shown in Figure 6.79.	A1/2 2C11

Single leaves of certain external walls

The single leaf of external walls of small single storey non-residential buildings and of annexes need be only 90 mm thick.	A1/2 2C12

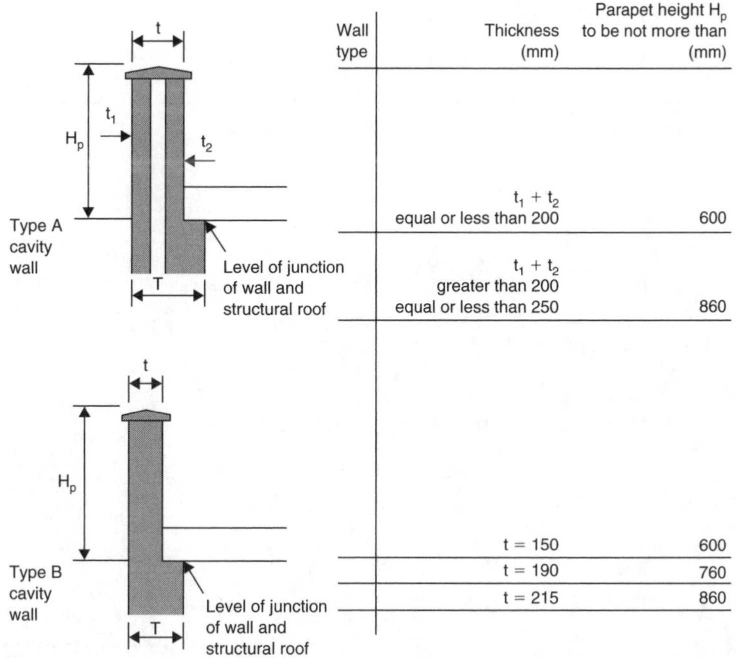

Wall type	Thickness (mm)	Parapet height H_p to be not more than (mm)
Type A cavity wall	$t_1 + t_2$ equal or less than 200	600
	$t_1 + t_2$ greater than 200 equal or less than 250	860
Type B cavity wall	t = 150	600
	t = 190	760
	t = 215	860

Figure 6.79 Height of parapet walls

The combined dead and imposed load should not exceed 70 kN/m at base of wall (see Figure 6.80).	A1/2 2C23c
Walls should not be subjected to lateral load other than from wind.	A1/2 2C23c

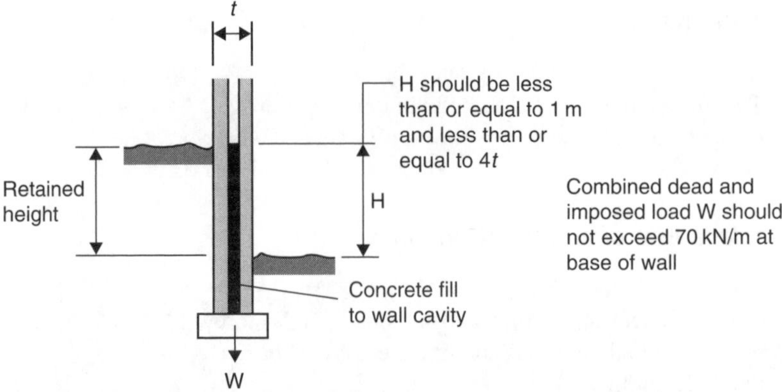

Figure 6.80 Combined and imposed dead load

Vertical lateral restraint to walls

The ends of every wall should be bonded or otherwise securely tied throughout their full height to a buttressing wall, pier or chimney.	A1/2 2C25

Long walls may be provided with intermediate buttressing walls, piers or chimneys dividing the wall into distinct lengths within each storey.	A1/2 2C25
Note: Each distinct length is considered to be a supported wall for the purposes of the Building Regulations.	
Intermediate buttressing walls, piers or chimneys should provide lateral restraint to the full height of the supported wall.	A1/2 2C25
They may be staggered at each storey.	
A wall in each storey of a building should:	A1/2 2C32
• extend to the full height of that storey;	

- have horizontal lateral supports to restrict movement of the wall at right angles to its plane.

The requirements for lateral restraint are shown in Table 6.34. A1/2 2C34

Table 6.34 Lateral support for walls

Wall type	Wall length	Lateral support required
Solid or cavity: external compartment separating	Any length	Roof lateral support by every roof forming a junction with the supported wall
	Greater than 3 m	Floor lateral support by every floor forming a junction with the supported wall
Internal load-bearing wall (not being a compartment or separating wall)	Any length	Roof or floor lateral support at the top of each storey

Walls should be strapped to floors above ground level, at intervals not exceeding 2 m and as shown in Figure 6.81 by tension straps conforming to BS EN 845-1. A1/2 2C35

Buttressing walls

If the buttressing wall is not itself a supported wall its thickness T_2 should not be less than:

- half the thickness required for an external or separating wall of similar height and length less 5 mm; or A1/2 2C26a
- 75 mm if the wall forms part of a dwelling-house and does not exceed 6 m in total height and 10 m in length; and A1/2 2C26b
- 90 mm in other cases. A1/2 2C26c
- The length of the buttressing wall should be:
  - at least 1/6 of the overall height of the supported wall
  - be bonded or securely tied to the supporting wall and at the other end to a buttressing wall, pier or chimney.
- The size of any opening in the buttressing wall should be restricted as shown in Figure 6.81. A1/2 2C26c

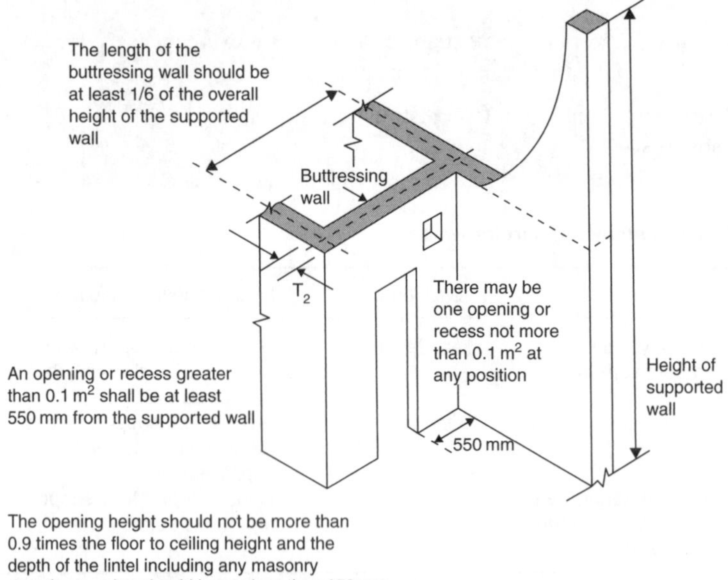
The length of the buttressing wall should be at least 1/6 of the overall height of the supported wall
Buttressing wall
T_2
There may be one opening or recess not more than 0.1 m² at any position
An opening or recess greater than 0.1 m² shall be at least 550 mm from the supported wall
550 mm
Height of supported wall
The opening height should not be more than 0.9 times the floor to ceiling height and the depth of the lintel including any masonry over the opening should be not less than 150 mm

Figure 6.81 Openings in a buttressing wall

Gable walls

Gable walls should be strapped to roofs as shown in Figure 6.82(a) and (b) by tension straps.	A1/2 2C36
Vertical strapping at least 1 m in length should be provided at eaves level at intervals not exceeding 2 m as shown in Figure 6.82(c) and (d).	A1/2 2C36
Vertical strapping may be omitted if the roof:	A1/2 2C36a–d

- has a pitch of 15° or more; and
- is tiled or slated; and
- is of a type known by local experience to be resistant to wind gusts; and
- has main timber members spanning onto the supported wall at not more than 1.2 m centres.

Piers

Piers should have a minimum width of 190 mm (see Figure 6.83).	A1/2 2C27a
Piers should measure at least three times the thickness of the supported wall.	A1/2 2C27a

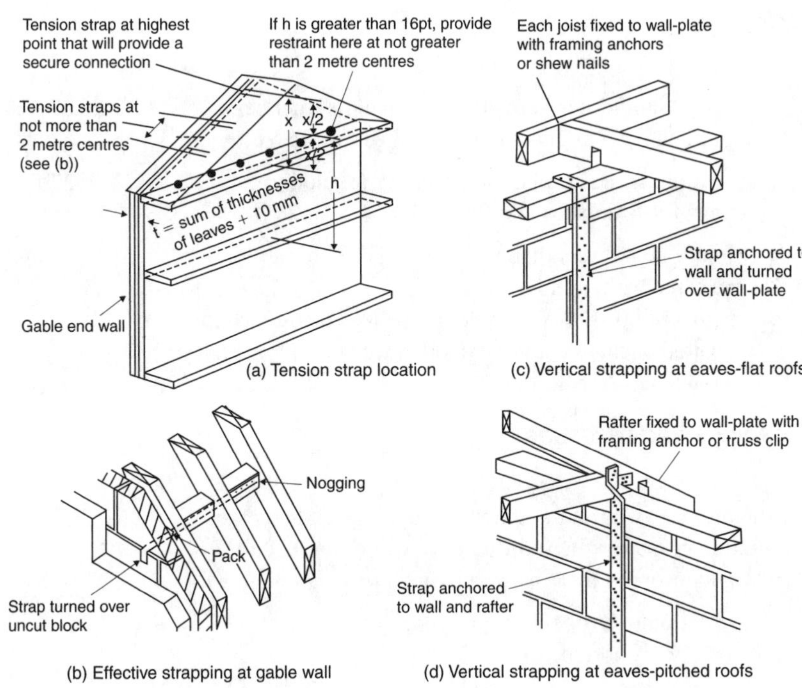

Figure 6.82 Lateral support at roof level

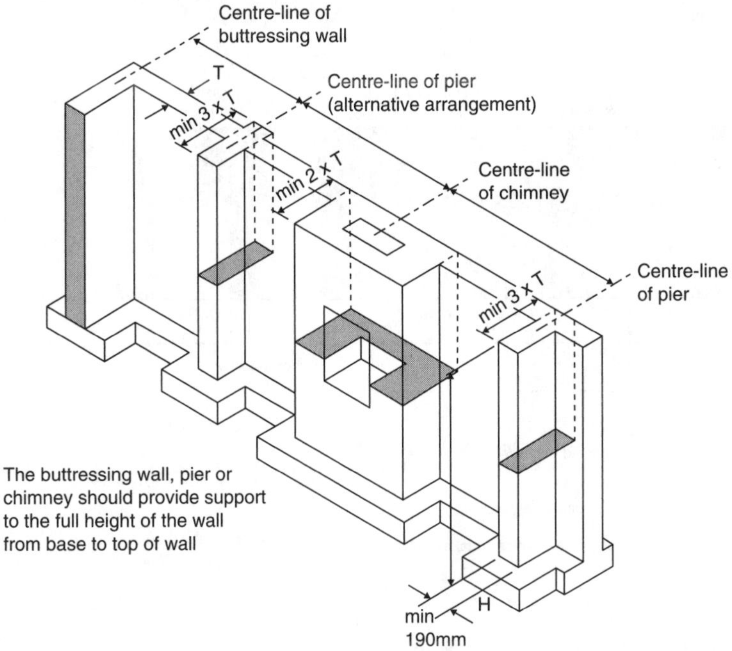

Figure 6.83 Buttressing

Chimneys

Chimneys should measure at least twice the thickness, measured at right angles to the wall (see Figure 6.83). A1/2 2C27a

The sectional area on plan of chimneys (excluding openings for fireplaces and flues): A1/2 2C27b

- should be not less than the area required for a pier in the same wall; and
- the overall thickness should not be less than twice the required thickness of the supported wall (see Figure 6.83).

Openings and recesses

The number, size and position of openings and recesses should not impair the stability of a wall or the lateral restraint afforded by a buttressing wall to a supported wall. A1/2 2C28

Construction over openings and recesses should be adequately supported. A1/2 2C28

No openings should be provided in walls below ground floor except for small holes for services and ventilation, etc. which should be limited to a maximum area of 0.1 m^2 at not less than 2 m centres (see Figure 6.84 and Table 6.35). A1/2 2C29

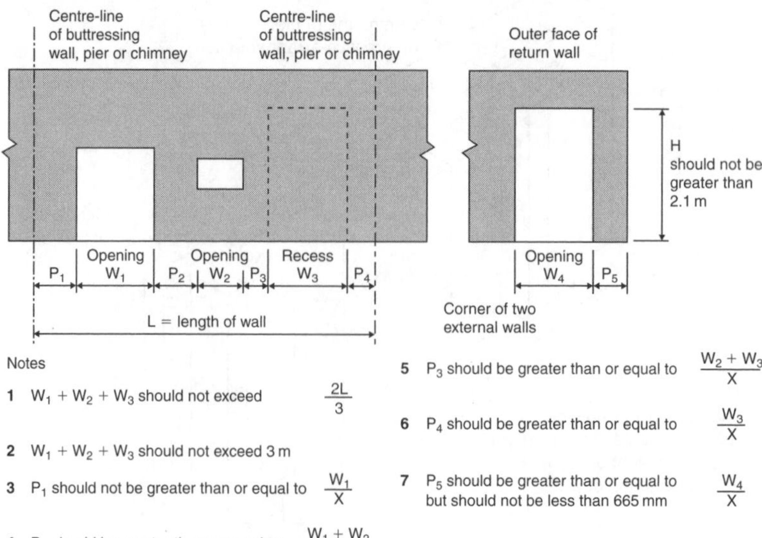

Figure 6.84 Sizes of openings and recesses

Table 6.35 Value of 'X' factor for Figure 6.84

Nature of roof span	Maximum roof span (m)	Minimum thickness of wall inner (mm)	Span of floor is parallel to wall	Span of timber floor into wall		Span of concrete floor into wall	
				Max 4.5 m	Max 6.0 m	Max 4.5 m	Max 6.0 m
			Value of factor 'X'				
Roof spans parallel to wall	Non-applicable	100	6	6	6	6	6
		90	6	6	6	6	5
Timber roof spans into wall	9	100	6	6	5	4	3
		90	6	4	4	3	3

Overhangs

The amount of any projection should not impair the stability of the wall. A1/2 2C31

Chases

Vertical chases should not be deeper than 1/3 of the wall thickness. A1/2 2C30a

Note: Or, in cavity walls, 1/3 of the thickness of the leaf.

Horizontal chases should not be deeper than 1/6 of the thickness of the leaf of the wall. A1/2 2C30b

Chases should not be so positioned as to impair the stability of the wall (particularly where hollow blocks are used). A1/2 2C30c

Small single-storey non-residential buildings and annexes

General

The walls shall be solidly constructed in brickwork or blockwork. A1/2 2C38(i)b

Where the floor area of the building or annexe exceeds 10 m^2, the walls shall have a mass of not less than 130 kg/m^2. A1/2 2C38(i)c

The only lateral loads are wind loads.	A1/2 2C38(i)e
The maximum length or width of the building or annexe shall not exceed 9 m.	A1/2 2C38(i)f
The height of the building or annexe shall not exceed the lower value derived from Figure 6.85.	A1/2 2C38(i)g

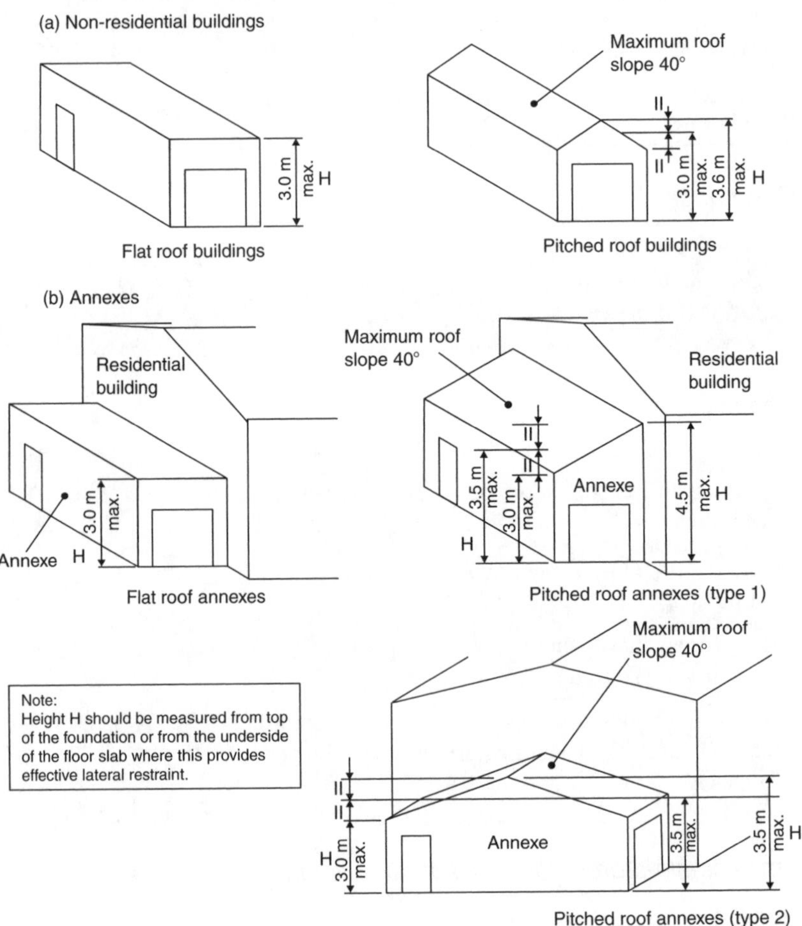

Figure 6.85 Size and proportions of non-residential buildings and annexes

Walls shall be tied to the roof structure vertically and horizontally and have a horizontal lateral restraint at roof level.	A1/2 2C38(i)i

Size and location of openings

One or two major openings not more than 2.1 m in height are permitted in one wall of the building or annexe only.	A1/2 2C38(ii)
The width of a single opening or the combined width of two openings should not exceed 5 m.	A1/2 2C38(ii)
The only other openings permitted in a building or annexe are for windows and a single leaf door.	A1/2 2C38(ii)
The size and location of these openings should be in accordance with Figure 6.86.	A1/2 2C38(ii)
Major openings should be restricted to one wall only. Their aggregate width should not exceed 0.5 m and their height should not be greater than 2.1 m.	A1/2 2C38(ii)
There should be no openings within 2.0 m of a wall containing a major opening.	A1/2 2C38(ii)
The aggregate size of the openings in a wall not containing a major opening should not exceed 2.4 m.	A1/2 2C38(ii)
There should not be more than one opening between piers.	A1/2 2C38(ii)
Unless there is a corner pier the distance from a window or a door to a corner should not be less than 390 mm.	A1/2 2C38(ii)

Wall thicknesses and recommendations for piers

The walls should have a minimum thickness of 90 mm.	A1/2 2C38(iii)
Walls which do not contain a major opening but exceed 2.5 m in length or height should be bonded or tied to piers for their full height at not more than 3 m centres as shown in Figure 6.87.	A1/2 2C38(iii)
Walls which contain one or two major openings should in addition have piers as shown in Figure 6.87(b) and (c).	A1/2 2C38(iii)

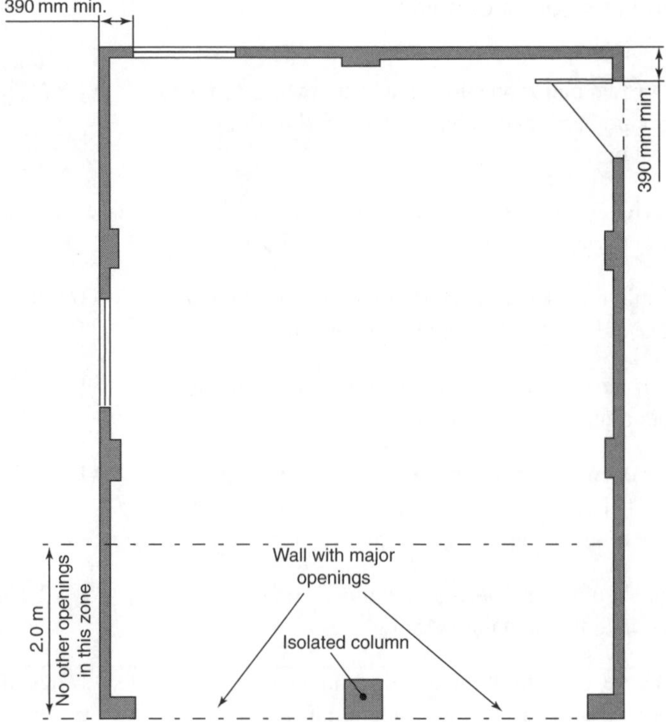

Figure 6.86 Size and location of openings

Where ties are used to connect piers to walls they should be: A1/2 2C38(iii)

- flat;
- 20 mm × 3 mm in cross-section;
- stainless steel;
- placed in pairs;
- spaced at not more than 300 mm centre vertically.

Walls should be tied horizontally at no more than 2 m centres to the roof structure at eaves level, base of gables and along roof slopes (as shown in Figure 6.88) with straps. A1/2 2C38(iv)

Where straps cannot pass through a wall, they should be adequately secured to the masonry using suitable fixings. A1/2 2C38(iv)

Isolated columns should also be tied to the roof structure (see Figure 6.88). A1/2 2C38(iv)

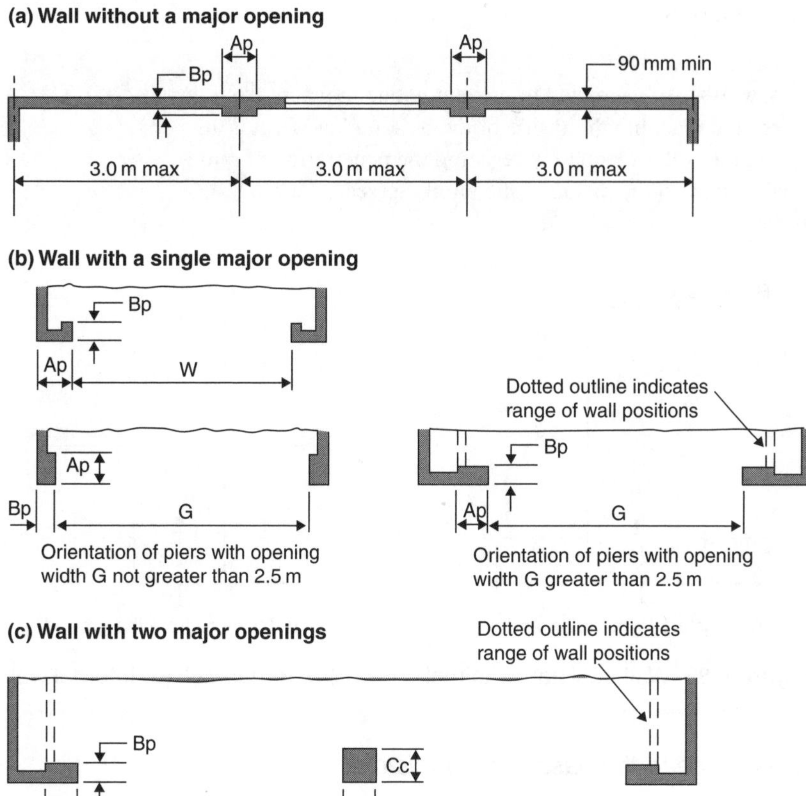

Figure 6.87 Wall thicknesses

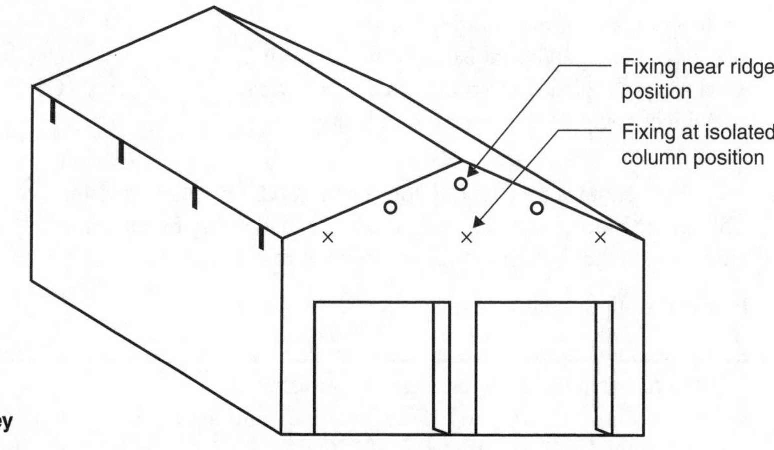

Key

| denotes fixings at eaves level. × denotes fixings at base of gable.

o denotes fixings along roof slope.

Figure 6.88 Lateral restraint at roof level

Foundations

A wall shall be erected to prevent undue moisture from the ground reaching the inside of the building and (if it is an outside wall) adequately resisting the penetration of rain and snow to the inside of the building (see Figure 6.89). C3

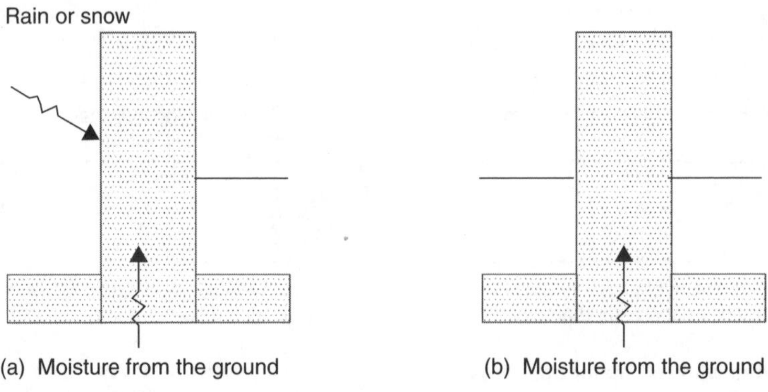

Figure 6.89 Wall, resistance to moisture. (a) External wall. (b) Internal wall

Resistance to the passage of moisture

Walls should:

- resist the passage of moisture from the ground to the inside of the building*; C5.2a
- not be damaged by moisture from the ground; C5.2b
- not carry moisture from the ground to any part which would be damaged by moisture. C5.2b

For buildings used wholly for storing goods or where provisions C5.3 put in place do not increase the health and safety of persons employed in that building, this requirement may not apply.

External walls should:

- resist rain penetrating components of the structure that might be damaged by moisture; C5.2c
- resist rain penetrating to the inside of the building*; C5.2d
- be designed and constructed so that their structural and thermal performance are not adversely affected by interstitial condensation; C5.2e
- not promote surface condensation or mould growth. C5.2f

For buildings used wholly for storing goods or where provisions put in place do not increase the health and safety of persons employed in that building, this requirement may not apply. C5.3

Internal and external walls exposed to moisture from the ground

Internal and external walls (subject to moisture from the ground) shall have a damp-proof course of bituminous material, polyethylene, engineering bricks or slates in cement mortar or any other material that will prevent the passage of moisture. C5.5a

The damp-proof course should be continuous with any damp-proof membrane in the floors. C5.5a

If the wall is an external wall, the damp-proof course should be at least 150 mm above the level of the adjoining ground (see Figure 6.90) unless the design is such that a part of the building will protect the wall. C5.5b

If the wall is an external cavity wall (see Figure 6.91) the cavity should either:

- be taken down at least 225 mm below the level of the lowest damp-proof course; or C5.5c
- a damp-proof tray should be provided so as to prevent precipitation passing into the inner leaf (see Figure 6.92), with weep holes every 900 mm to assist in the transfer of moisture through the external leaf. C5.5c

Where the damp-proof tray does not extend the full length of the exposed wall (i.e. above an opening) stop ends and at least two weep holes should be provided. C5.5c

As well as giving protection against moisture from the ground, an external wall should give protection against precipitation. C5.7

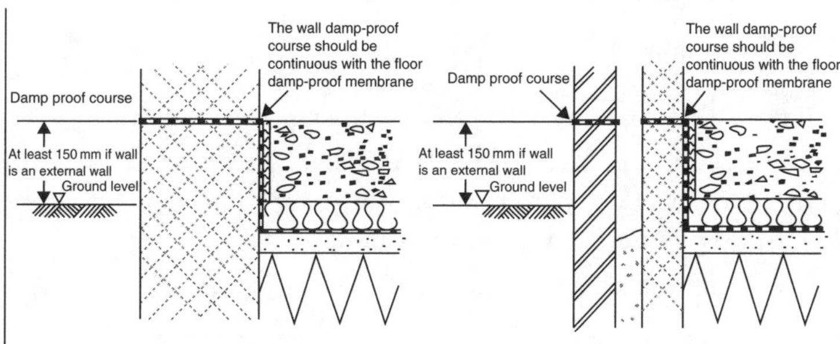

Figure 6.90 Damp-proof courses

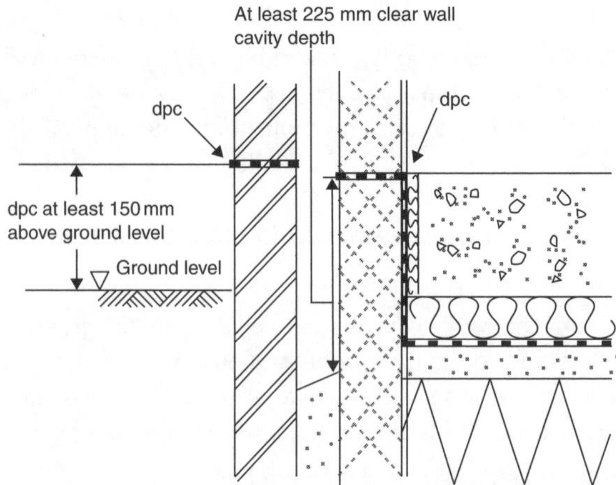

Figure 6.91 Cavity carried down

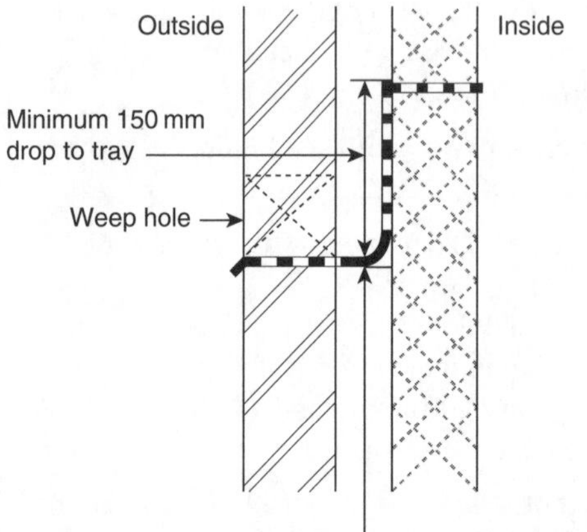

Figure 6.92 Damp-proof (cavity) tray

Solid external walls

Solid walls shall hold moisture arising from rain and snow until it can be released in a dry period without penetrating to the inside of the building, or causing damage to the building. C5.8

Solid external walls exposed to **very severe** conditions should be protected by external impervious cladding. C5.9

Solid external walls exposed to **severe** conditions may be built with: C5.9a

- brickwork (or stonework) at least 328 mm thick;
- dense aggregate concrete blockwork at least 250 mm thick; or

- lightweight aggregate (aerated autoclaved concrete blockwork) at least 215 mm thick.

Solid external walls exposed to **severe** conditions may be built, providing:

- the rendering is in two coats with a total thickness of at least 20 mm and has a scraped or textured finish; C5.9b
- the strength of the mortar is compatible with the strength of the bricks or blocks; C5.9b
- the joints (if the wall is to be rendered) are raked out to a depth of at least 10 mm; C5.9b
- the rendering mix is 1 part of cement, 1 part of lime and 6 parts of well-graded sharp sand (nominal mix 1:1:6) unless the blocks are of dense concrete aggregate, in which case the mix may be 1:½. C5.9b

Adequate protection should be provided at the top of walls, etc. (see Figure 6.93). C5.9c

Unless the protection and joints are a complete barrier to moisture, a damp-proof course should also be provided. C5.9c

Damp-proof courses, cavity trays and closers should be provided and designed to ensure that water drains outwards:

- where the downward flow will be interrupted by an obstruction (e.g. from some types of lintel); C5.9d(i)
- under openings – unless there is a sill and the sill and its joints will form a complete barrier; C5.9d(ii)
- at abutments between walls and roofs. C5.9d(iii)

A solid external wall may be insulated on the inside or on the outside. C5.10

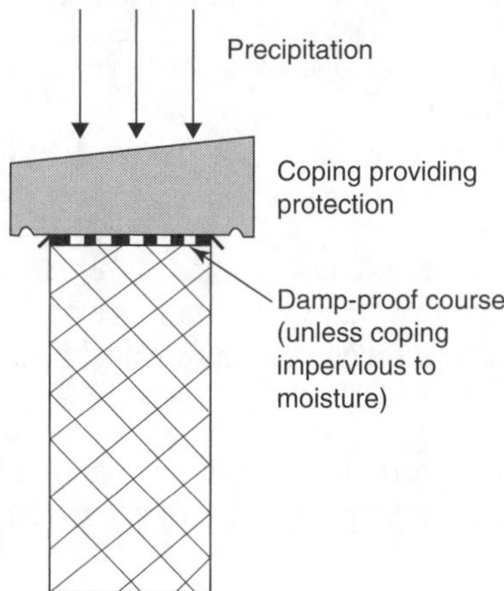

Figure 6.93 Projection of wall head from precipitation

Where the insulation is on the inside, a cavity should be provided to give a break in the path for moisture.	C5.10
Where the insulation is on the outside, it should provide some resistance to the ingress of moisture to ensure the wall remains relatively dry (see Figure 6.94).	C5.10

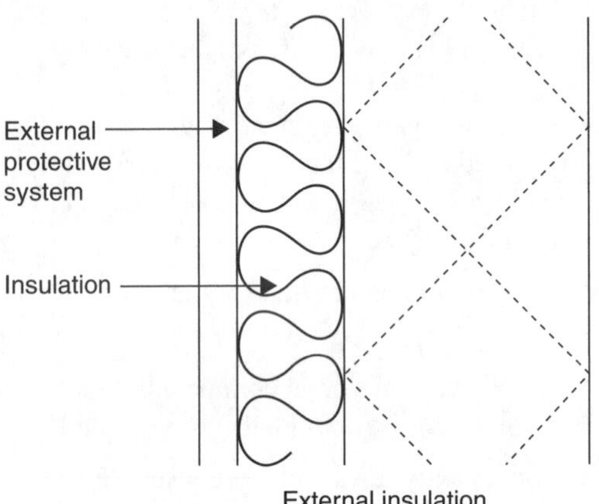

Figure 6.94 Insulated (solid) external wall

Cavity external walls

The outer leaf shall be separated from the inner leaf by a drained air space (or in any other way which will prevent precipitation from being carried to the inner leaf).	C5.12
The construction of a cavity external wall could include:	
• outer leaf masonry (bricks, blocks, stone or manufactured stone);	C5.13a
• a cavity at least 50 mm wide;	C5.13b
• inner leaf masonry or frame with lining.	C5.13b
Masonry units should be laid on a full bed of mortar with the cross joints substantially and continuously filled to ensure structural robustness and weather resistance.	C5.13c
Where a cavity is to be partially filled, the residual cavity should not be less than 50 mm wide (see Figure 6.95).	C5.13c

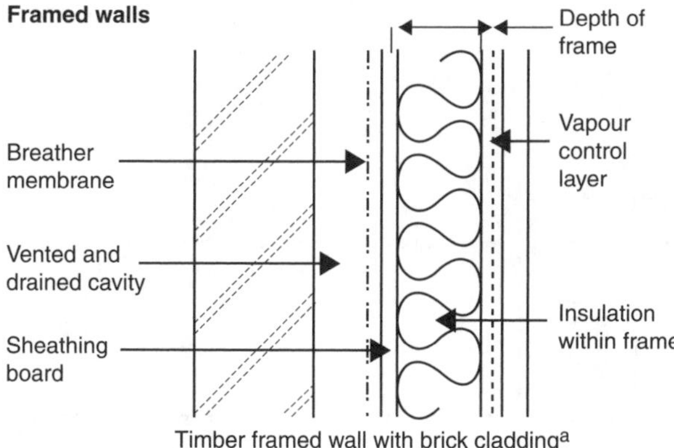

Figure 6.95 Insulated framed wall

Cavity insulation

The suitability of the wall for installing insulation material(s) is to be assessed before the work is carried out.	C5.15a and d
When the cavity of an existing house is being filled, attention should be given to the condition of the external leaf of the wall, e.g. its state of repair and type of pointing.	C5.15e
A full or partial fill insulating material may be placed in the cavity between the outer leaf and an inner leaf of masonry subject to the suitability of a wall for installing insulation into the cavity (see Table 6.36).	C5.15a
The insulating material should be the subject of current certification from an appropriate body or a European Technical Approval.	C5.15c
When partial fill materials are used, the residual cavity should not be less than 50 mm nominal.	C5.15b
Rigid (board or batt) thermal insulating material built into the wall must be certified as being in conformance by an approved installer.	C5.15b
Urea-formaldehyde foam inserted into the cavity should be: • in accordance with BS 5617: 1985; • installed in accordance with BS 5618: 1985.	C5.15d
The person undertaking installation work should operate under an Approved Installer Scheme.	C5.15c

Table 6.36 Maximum recommended exposure zones for insulated masonry walls

Wall construction		Maximum recommended exposure zone for each construction						
Insulation method	Min. width of filled or clear cavity (mm)	Impervious cladding		Rendered finish		Facing masonry		Flush skills and copings
		Full height of wall	Above facing masonry	Full height of wall	Above facing masonry	Tooled flush joints	Recessed mortar joints	
Built-in full fill	50	4	3	3	3	2	1	1
	75	4	3	4	3	3	1	1
	100	4	4	4	3	3	1	2
	125	4	4	4	3	3	1	2
	150	4	4	4	4	4	1	2
Injected fill *not* UF foam	50	4	2	3	2	2	1	1
	75	4	3	4	3	3	1	1
	100	4	3	4	3	3	1	1
	125	4	4	4	3	3	1	2
	150	4	4	4	4	4	1	2
Injected fill UF foam	50	4	2	3	2	1	1	1
	75	4	2	3	2	2	1	1
	100	4	2	3	2	2	1	1
Partial fill								
Residual 50 mm cavity	50	4	4	4	4	3	1	1
Residual 75 mm cavity	75	4	4	4	4	4	1	1
Residual 100 mm cavity	100	4	4	4	4	4	2	1
Internal insulation								
Clear cavity 50 mm	50	4	3	4	3	3	1	1
Clear cavity 100 mm	100	4	4	4	4	4	2	2
Fully filled								
Cavity 50 mm	50	4	3	3	3	2	1	1
Cavity 100 mm	100	4	4	4	3	3	1	2

Framed external walls

The cladding shall be separated from the insulation or sheathing by a vented and drained cavity with a membrane that is vapour open, but resists the passage of liquid water, on the inside of the cavity (see Figure 6.95). C5.17

Cracking of external walls

The possibility of severe rain penetration occurring through cracks in masonry external walls should be taken into account when designing a building. C5.18

Impervious cladding systems for walls

Cladding systems for walls should:

- resist the penetration of precipitation to the inside of the building; C5.19a
- not be damaged by precipitation; C5.19b
- not carry precipitation to any part of the building which would be damaged by it. C5.19b

Cladding that is designed to protect a building from precipitation shall be:

- jointless or have sealed joints; C5.21a
- impervious to moisture. C5.21a

If the cladding has overlapping dry joints it shall be:

- impervious or weather resisting; C5.21b
- backed by a material which will direct precipitation which enters the cladding towards the outer face. C5.21b

Materials that can deteriorate rapidly without special care should only be used as the weather-resisting part of a cladding system. C5.22

Cladding may be:

- impervious – e.g. metal, plastic, glass and bituminous products; C5.23a
- weather-resisting – e.g. natural stone or slate, cement-based products, fired clay and wood; C5.23b
- moisture-resisting – e.g. bituminous and plastic products lapped at the joints; C5.23c

- jointless materials and sealed joints – i.e. to allow for structural and thermal movement. C5.23d

Dry joints between cladding units should be designed so that:

- precipitation will not pass through them; C5.24
- precipitation which enters the joints will be directed towards the exposed face without it penetrating beyond the back of the cladding. C5.24

Note: Whether dry joints are suitable will depend on the design of the joint or the design of the cladding and the severity of the exposure to wind and rain.

Each sheet, tile and section of cladding should be securely fixed (as per guidance contained in BS 8000-6: 1990). C5.25

Particular care should be taken with detailing and workmanship at the junctions between cladding and window and door openings as they are vulnerable to moisture ingress. C5.25

Insulation may be incorporated into the construction provided it is either protected from moisture or is unaffected by it. C5.26

Where cladding is supported by timber components (or is on the façade of a timber framed building) the space between the cladding and the building should be ventilated to ensure rapid drying of any water that penetrates the cladding. C5.27

Joint between walls and doors/window frames

The joint between walls and doors and window frames should:

- resist the penetration of precipitation to the inside of the building; C5.29a
- not be damaged by precipitation; C5.29b
- not permit precipitation to reach any part of the building which would be damaged by it. C5.29c

Damp-proof courses should be provided to direct moisture towards the outside, particularly:

- where the downward flow of moisture would be interrupted at an obstruction, e.g. at a lintel; C5.30a
- where sill elements (including joints) do not form a complete barrier to the transfer of precipitation, e.g. under openings, windows and doors; C5.30b

- where reveal elements, including joints, do not form a complete barrier to the transfer of rain and snow, e.g. at openings, windows and doors. C5.30c

Direct plastering of the internal reveal of any window frame should only be used with a backing of expanded metal lathing or similar. C5.31

In areas of the country that are exposed to very severe driving rain:

- checked rebates should be used in all window and door reveals; C5.32
- the frame should be set back behind the outer leaf of masonry as shown in Figure 6.96; C5.32
- alternatively an insulated finned cavity closer may be used. C5.32

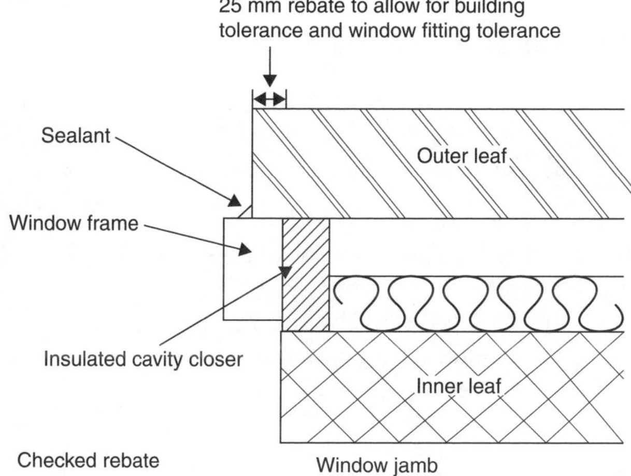

Figure 6.96 Window reveals for use in areas subject to very severe driving rain

Door thresholds

Where an accessible threshold is provided to allow unimpeded access (as specified in Part M):

- the external landing (see Figure 6.97) should be laid to a fall between 1 in 40 and 1 in 60 in a single direction away from the doorway; C5.33a
- the sill leading up to the door threshold has a maximum slope of 15. C5.33b

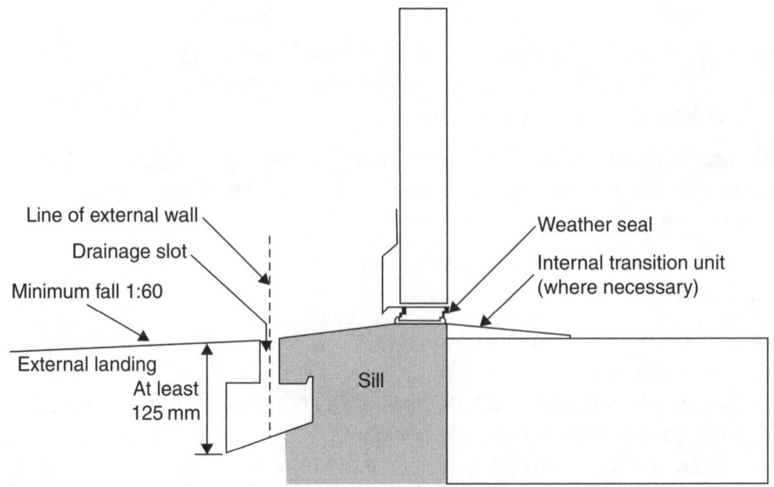

Figure 6.97 Accessible threshold for use in exposed areas

Interstitial condensation (external doors)

External walls shall be designed and constructed in accordance with Clause 8.3 of BS 5250: 2002. C5.34

Specialist advice should be sought when designing swimming pools and other buildings where interstitial condensation in the walls (caused by high internal temperatures and humidities) can cause high levels of moisture being generated. C5.35

Surface condensation and mould growth (external doors)

External walls shall be designed and constructed so that the:

- thermal transmittance (U-value) does not exceed 0.7 W/m^{2}K at any point; C5.36a
- junctions between elements and details of openings (such as doors and windows) meet with the recommendations in the report on robust construction details. C5.36b

Wall cladding

Wall cladding presents a hazard if it becomes detached from the building. An acceptable level of safety can be achieved depending on the type and location of the cladding.

The guidance given below relates to all forms of cladding, including curtain walling and glass façades.

Requirement	Reference
Cladding shall be capable of safely sustaining and transmitting (to the supporting structure of the building) all dead, imposed and wind loads.	A1/2 3.2a
Provision shall be made, where necessary, to accommodate differential movement of the cladding and the supporting structure of the building.	A1/2 3.2c
Wind loading on the cladding should be derived from BS 6399, Part 2: 2001.	A1/2 3.3
Due consideration shall be given to local increases in wind suction arising from funnelling of the wind through gaps between buildings.	A1/2 3.3
Note: Guidance on funnelling effects is given in BRE Digest 436 '*Wind loading on buildings – Brief guidance for using BS 6399-2: 1997*' available from BRE, Bucknalls Lane, Garston, Watford, Herts WD2 7JR.	
The cladding shall be securely fixed to, and supported by, the structure of the building using both vertical support and horizontal restraint.	A1/2 3.2b
The cladding and its fixings (including any support components) shall be of durable materials.	A1/2 3.2d
The design life of the fixings shall not be less than that of the cladding.	A1/2 3.2d
Fixings shall be corrosion resistant and of a material type appropriate for the local environment.	A1/2 3.2d
Fixings for supporting cladding should be determined from a consideration of the proven performance of the fixing and the risks associated with the particular application.	A1/2 3.7
The strength of fixings should be derived from tests using materials representative of the material into which the fixing is to be anchored, taking account of any inherent weaknesses that may affect the strength of the fixing, e.g. cracks in concrete due to shrinkage and flexure, or voids in masonry construction.	A1/2 3.8
Where the cladding is required to support other fixtures (e.g. handrails or fittings such as antennae and signboards)	A1/2 3.4

account should be taken of the loads and forces arising from such fixtures and fittings.

Where the wall cladding is required to function as pedestrian guarding to stairs, ramps, vertical drops of 600 mm or greater or as a vehicle barrier, account should be taken of the additional imposed loading as stipulated in Part K. A1/2 3.5

Where wall cladding is required to safely withstand lateral pressures from crowds, an appropriate design loading is given in BS 6399 Part 1 and the '*Guide to Safety at Sports Grounds*' (4th Edition, 1997).

Applications should be designated as being either non-redundant (where the failure of a single fixing could lead to the detachment of the cladding) or redundant (where failure or excessive movement of one fixing results in load sharing by adjacent fixings) and the required reliability of the fixing determined accordingly. A1/2 3.7

All cladding (used to protect the building from rain or snow) shall be jointless or have sealed joints. C4 (5.1–5.6)

Note: Large glass panels in cladding of walls and roofs (where the cladding is not divided into small areas by load bearing framing) needs special consideration. Guidance is given in the following documents:

The Institution of Structural Engineers' Report on '*Structural use of glass in buildings*' dated 1999, available from 11 Upper Belgrave Street, London SW1X 8BH.
'*Nickel sulfide in toughened glass*' published by the Centre for Window Cladding and Technology, dated 2000.
Further guidance on cladding is given in the following documents:
The Institution of Structural Engineers' Report on '*Aspects of cladding*', dated 1995.
The Institution of Structural Engineers' Report on '*Guide to the structural use of adhesives*', dated 1999.
BS 8297 '*Code of practice for the design and installation of non-load bearing precast concrete cladding*'.
BS 8298 '*Code of practice for the design and installation of natural stone cladding and lining*'.

Internal fire spread (linings)

The choice of materials for walls and ceilings can significantly affect the spread of a fire and its rate of growth, even though they are not likely to be the materials first ignited. Although furniture and fittings can have a major effect on fire spread it is not possible to control them through Building Regulations.

The surface linings of walls should meet the classifications shown in Table 6.37.

Table 6.37 Classification of linings

Location	Class*
Small rooms with an area of not more than 4 m^2 (in residential accommodation) or 30 m^2 (in non-residential accommodation)	3
Domestic garages not more than 40 m^2	3
Other rooms (including garages)	1
Circulation spaces within buildings	1
Other circulation spaces (including the common area of flats and maisonettes)	0

* Classifications are based on tests as per BS 476 and as described in Appendix A of Approved Document B.

Any flexible membrane covering a structure (other than an air supported structure) should comply with the recommendations given in Appendix A of BS 7157.	B2 6.8 (V2)
The wall and any floor between the garage and the house shall have a 30 minute standard of fire resistance.	B2

Loadbearing elements of structure

All loadbearing elements of a structure shall have a minimum standard of fire resistance.	B3 4.1 (V1) B3 7.1 (V2)
Structural frames, beams, columns, loadbearing walls (internal and external), floor structures and gallery structures, should have at least the fire resistance given in Appendix A of Part B.	B3 4.2 (V1) B3 7.2 (V2)

Compartmentation

To prevent rapid fire spread and to reduce the chance of fires becoming large, the spread of fire within a building can be restricted by sub-dividing that building into compartments that are separated from one another by walls and/or floors of fire-resisting construction.

The appropriate degree of sub-division depends on:

- the use of and fire load in the building;
- the height to the floor of the top storey in the building; and
- the availability of a sprinkler system.

General

To prevent the spread of fire within a building, whenever possible, the building should be sub-divided into compartments separated from one another by walls and/or floors of fire-resisting construction.	B3 5.1 (V1) B3 8.1 (V2)
Walls separating semi-detached houses, or houses in terraces, should be constructed as a compartment wall and the houses should be considered as separate buildings.	B3 5.3 (V1)
Compartment walls that are common to two or more buildings should:	B3 8.10 (V2)
• be constructed as a compartment wall;	
• run the full height of the building in a continuous vertical plane.	B3 5.7 (V1)

The lowest floor in a building does not need to be constructed as a compartment floor.

Compartment walls should:	B3 5.6 (V1)
• form a complete barrier to fire between the compartments they separate; and	B2 8.20a (V2)
• have the appropriate fire resistance as indicated in Appendix A, Tables Al and A2.	B3 5.6 (V1) B2 8.20b (V2)

Note: Adjoining buildings should only be separated by walls, not floors.

Junction of compartment wall with other walls

Where a compartment wall meets another compartment wall, the junction should maintain the fire resistance of the compartmentation.	B3 5.9 (V1) B3 8.25 (V2)
At the junction of a compartment floor with an external wall that has no fire resistance (such as a curtain wall) the external wall should be restrained at floor level to reduce the movement of the wall away from the floor when exposed to fire.	B3 5.10 (V1)

Junction of compartment wall with roof

If a fire penetrates a roof near a compartment wall there is a risk that it will spread over the roof to the adjoining compartment. To reduce this risk the wall should be:

- taken up to meet the underside of the roof; B3 5.11 (V1)
- covered with fire-stopping (where necessary) at the wall/roof junction to maintain the continuity of fire resistance; B3 8.28 (V2)
- continued across any eaves;
- either extended up through the roof for a height of at least 375 mm above the top surface of the adjoining roof covering (see Figure 6.98); or a 1500 mm wide zone on either side of the wall should have a suitable covering (see Figure 6.98). B3 5.12 (V1) B3 8.29 (V2) B3 8.30 (V2)

Compartment walls in a top storey beneath a roof should be continued through the roof space. B3 5.8 (V1) B3 8.24 (V2)

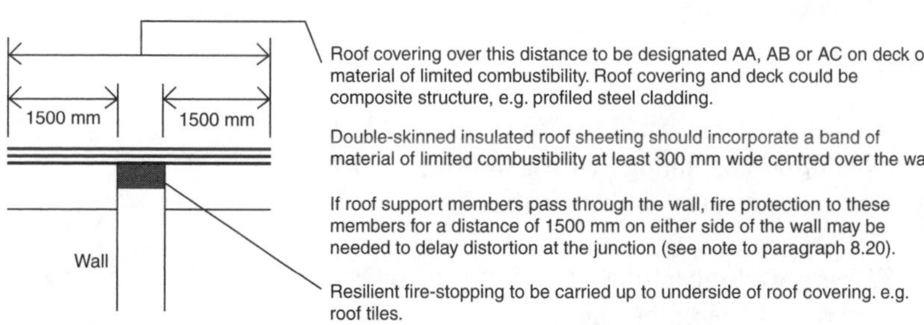

1500 mm
1500 mm
Wall
Roof covering over this distance to be designated AA, AB or AC on deck of material of limited combustibility. Roof covering and deck could be composite structure, e.g. profiled steel cladding.
Double-skinned insulated roof sheeting should incorporate a band of material of limited combustibility at least 300 mm wide centred over the wall
If roof support members pass through the wall, fire protection to these members for a distance of 1500 mm on either side of the wall may be needed to delay distortion at the junction (see note to paragraph 8.20).
Resilient fire-stopping to be carried up to underside of roof covering. e.g. roof tiles.

Figure 6.98 Junction of compartment wall with roof

Openings in compartment walls separating buildings or occupancies

Any openings in a compartment wall which is common to two or more buildings should be limited to those for a door which is providing 'means of access' in case of fire (and which has the same fire resistance as that required for the wall). B3 5.13 (V1) B3 8.32 (V2)

All other openings in compartment walls or compartment floors should be limited to those for: B3 8.34 (V2)

- doors which have the appropriate fire resistance;
- the passage of pipes, ventilation ducts, service cables, chimneys, appliance ventilation ducts or ducts encasing one or more flue pipes;
- refuse chutes of non-combustible construction;
- atria designed in accordance with BS 5588-7: 1997; and
- protected shafts (see B3 8.35 V2 for details of the relevant requirements).

All purpose groups

Parts of a building that are used and/or occupied for different purposes should be separated from one another by compartment walls and/or compartment floors (also see Appendix D to Part B (V2). B2 8.11 (V2)

Walls that are common to two or more buildings should be constructed as a compartment wall. B2 8.10 (V2)

Flats

In buildings containing flats, the following should be constructed as compartment walls: B2 8.13 (V2)

- every wall separating a feat from any other part of the building; and
- every wall enclosing a refuse storage chamber.

Non-residential buildings

In non-residential, purpose group buildings (such as Office, Shop and Commercial, Assembly and Recreation, Industrial, Storage etc.), the following walls should be constructed as compartment walls: B2 8.18a (V2)

- walls that are required to sub-divide buildings in order to meet the size limits on compartments given in Table 12 of Volume 2;

• the walls of a building that form part of a shopping complex;	B2 8.18e (V2)
• walls that divide a building into separate occupancies, (i.e. spaces used by different organizations whether they fall within the same Purpose Group or not).	B2 8.18f (V2)

Construction of compartment walls

Adjoining buildings should only be separated by walls, not floors.	B2 8.21 (V2)
Compartment walls should be able to accommodate the predicted deflection of the floor above by either: • having a suitable head detail between the wall and the floor, that can deform (but still maintain its integrity) when exposed to a fire; or • the wall may be designed to resist the additional vertical load from the floor above as it sags under fire conditions and thus maintain integrity.	B2 8.27 (V2)
Compartment walls that are common to two or more buildings should run the full height of the building in a continuous vertical plane.	B2 8.21 (V2)
Compartment walls used to form a separated part of a building should run the full height of the building in a continuous vertical plane.	B2 8.22 (V2)
If trussed rafters bridge the wall, they should be designed so that failure of any part of the truss due to a fire in one compartment will not cause failure of any part of the truss in another compartment.	B3 5.6 (V1) B3 8.20 (V2)
Junctions between a compartment floor and an external wall that has no fire resistance (such as a curtain wall) should be restrained at floor level to reduce the movement of the wall away from the floor when exposed to fire.	B2 8.26 (V2)

Loadbearing walls (internal and external) should have at least the fire resistance given in Appendix A Table A1 of Part B.	B3 7.2 (V2)
Timber beams, joists, purlins and rafters may be built into or carried through a masonry or concrete compartment wall if the openings for them are kept as small as practicable and then fire-stopped.	B3 5.6 (V1) B3 8.20 (V2)
There should be continuity at the junctions of the fire resisting elements enclosing a compartment.	B3 8.6 (V2)

Note: Generally speaking, an external wall of a protected shaft does **not** need to have fire resistance (but see BS 5588-5:2004 for fire resistance of external walls of firefighting shafts).

Garages

If a domestic garage is attached to (or forms an integral part of) a house: • the wall between the garage and the house shall have a 30 minute standard of fire resistance; • any opening in the wall should be at least 100 mm above the garage floor level with an FD30 door.	B3 5.4 (V1)

Protection of openings for pipes

Openings in compartment walls should be limited to those for: • doors which have the appropriate fire resistance; • the passage of pipes, ventilation ducts, service cables, chimneys, appliance ventilation ducts or ducts encasing one or more flue pipes;	B3 8.34 (V2)

- refuse chutes of non-combustible construction;
- atria designed in accordance with BS 5588-7:1997; and
- protected shafts (see B3 8.35 V2 for details of the relevant requirements).

Ventilation ducts and flues etc.

Air circulation system transfer grilles should not be fitted in any wall enclosing a protected stairway.	B3 7.10 (V1) B1 2.17 (V1)
If a flue (or a duct containing flues and/or ventilation duct(s)), passes through a compartment wall, or is built into a compartment wall, each wall of the flue or duct should have a fire resistance of at least half that of the wall or floor (see Figure 6.99).	B1 7.11 (V1)

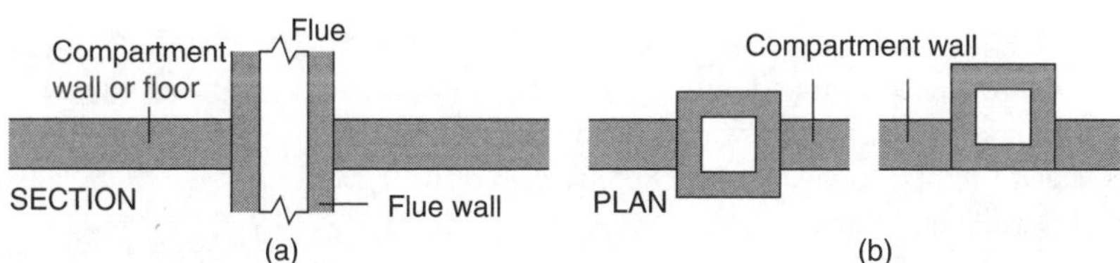

Figure 6.99 Flues penetrating compartment walls or floors

Fire resistance

Proprietary fire-stopping and sealing systems (including those designed for service penetrations) which have been shown by test to maintain the fire resistance of the wall or other element, are available and may be used. Other fire-stopping materials include:

- cement mortar;
- gypsum-based plaster;
- cement or gypsum-based vermiculite/perlite mixes;

- glass fibre, crushed rock, blast furnace slag or ceramic-based products (with or without resin binders), and
- intumescent mastics (B3 11.14).

Joints between fire separating elements should be fire-stopped.	B3 7.12a (V1) B3 10.17a (V2)
All openings for pipes, ducts, conduits or cables to pass through any part of a fire separating element should be: • kept as few in number as possible; • kept as small as practicable; • fire-stopped (which in the case of a pipe or duct, should allow thermal movement).	B3 7.12b (V1) B3 10.17b (V2)
To prevent displacement, materials used for fire-stopping should be reinforced with (or supported by) materials of limited combustibility.	B3 7.13 (V1) B3 10.18 (V2)

Construction of an external wall

Where a portal framed building is near a relevant boundary, the external wall near the boundary may need fire resistance to restrict the spread of fire between buildings.	B4 12.4 (V2)
In cases where the external wall of the building cannot be wholly unprotected, the rafter members of the frame, as well as the column members, may need to be fire protected.	B4 12.4 (V2)
The external surfaces of walls should meet the provisions shown in Figure 6.100.	B4 8.4 (V1) B4 12.4 (V2)

It should be noted that the use of combustible materials for cladding framework, or the use of combustible thermal insulation as an overcladding may be risky in tall buildings, even though the provisions for external surfaces in Figure 6.100 may have been satisfied.

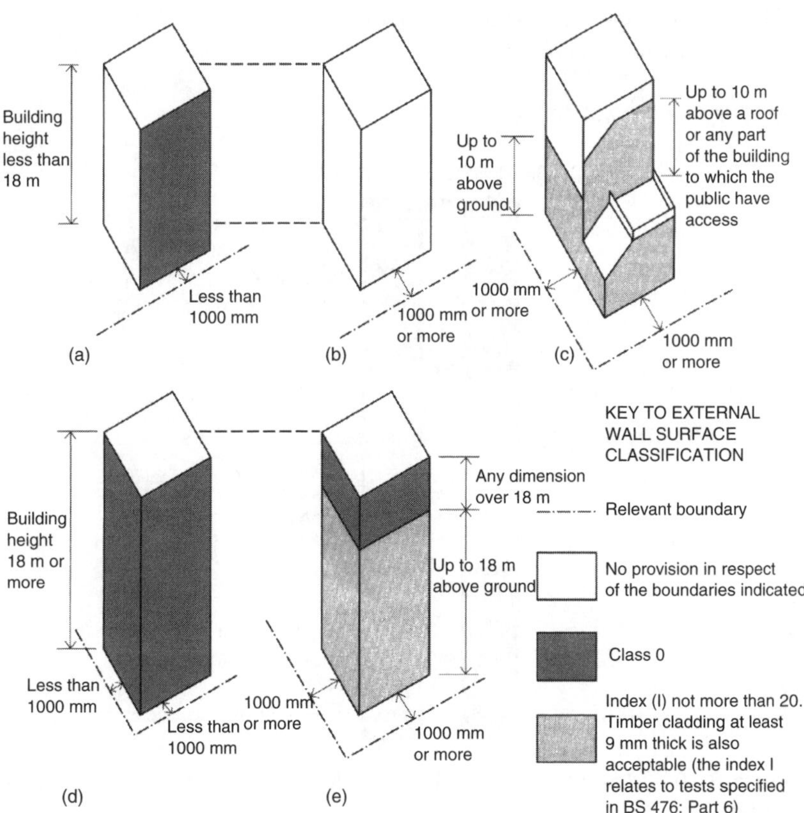

Figure 6.100 Provisions for external surfaces of walls. (a), (d), (e) Any building. (b) Any building other than (c). (c) Assembly or recreation building of more than one storey

The external envelope of a building should not provide a medium for fire spread if it is likely to be a risk to health or safety.	B4 12.5 (V2)
In a building with a storey 18 m or more above ground level, insulation material used in ventilated cavities in the external wall construction should be of limited combustibility (this restriction does not apply to masonry cavity wall construction).	B4 12.5 (V2)

Combustible material should not be placed in or exposed to the cavity, except for:

- timber lintels, window or door frames, or the end stairway of timber joists;
- pipes, conduits or cables;
- DPC, flashing, cavity closer or wall ties;

- fire-resisting thermal insulating material;
- a domestic meter cupboard.

Masonry wall construction

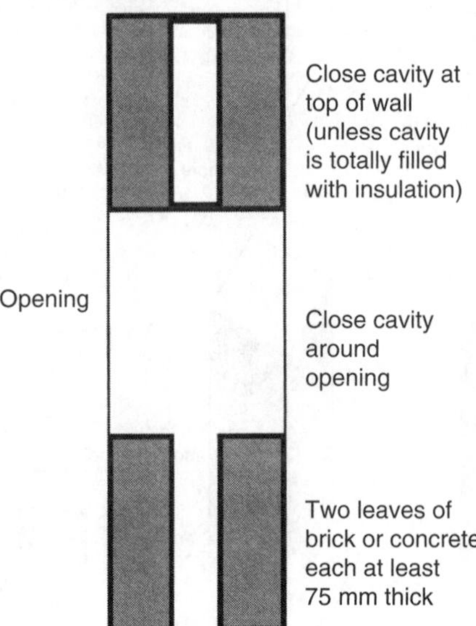

Figure 6.101 Masonry cavity walls excluded from the previous for cavity barriers

Cavity insulation

The outer leaf of the wall should be built of masonry or concrete.	D1 (1.1–1.2)
The inner leaf of the wall should be built of masonry (bricks or blocks).	D1 (1.1–1.2)
The wall being insulated with UF (area formaldehyde) shall be assessed (in accordance with BS 8208) for suitability before any work commences.	D1 (1.1–1.2)
The person carrying out the work needs to hold (or operate under) a current BSI Certificate of Registration of Assessed Capability for the work he is doing.	D1 (1.1–1.2)
The installation shall be in accordance with BS 5618: 1985.	D1 (1.1–1.2)
The material shall be in accordance with the relevant recommendations of BS 5617: 1985.	D1 (1.1–1.2)

Airborne sound

The flow of sound energy through walls should be restricted.	E
Walls should reduce the level of airborne sound.	E
Walls that separate a dwelling from another building (or another dwelling) shall resist the transmission of airborne sound.	E
Habitable rooms (or kitchens) within a dwelling shall resist the transmission of airborne sound.	E
Air paths, including those due to shrinkage, must be avoided.	E
Porous materials and gaps at joints in the structure must be sealed.	E
Flanking transmission (i.e. the indirect transmission of sound from one side of a wall to the other side) should be minimized.	E
The possibility of resonance in parts of the structure (such as a dry lining) should be avoided.	E

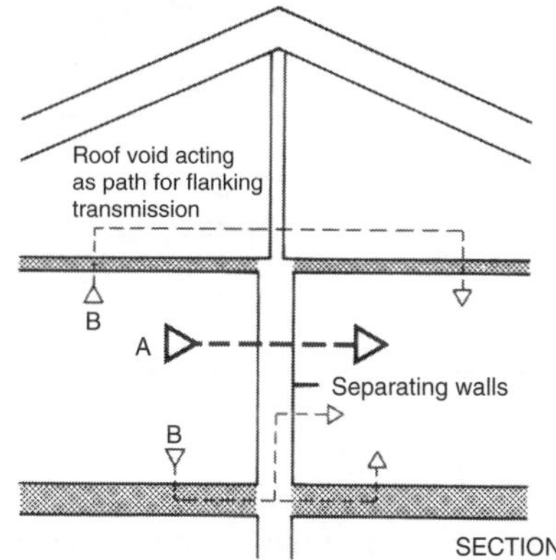

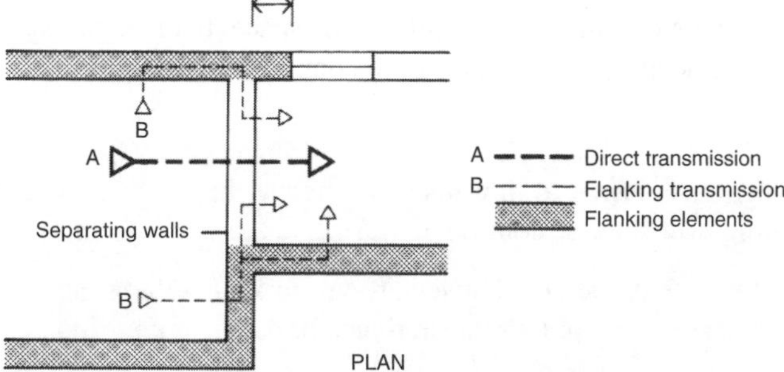

Figure 6.102 Direct and flanking transmission
For clarity not all flanking paths have been shown.

Separating walls (new buildings)

Walls – general

All new walls constructed within a dwelling-house (flat or room used for residential purposes) – whether purpose built or formed by a material change of use – shall meet the laboratory sound insulation values set out in Table 6.38. E0.9

Walls that have a separating function should achieve the sound insulation values: E0.1

- for rooms for residential purposes as set out in Table 6.38;
- dwelling-houses and flats as set out in Table 6.38.

Table 6.38 Dwelling houses and flats – performance standards for separating walls that have a separating function

	Airborne sound insulation $D_{nT,W} + C_{tr}$ dB (minimum values)
Purpose built rooms for residential purposes	43
Purpose built dwelling-houses and flats	45
Rooms for residential purposes formed by material change of use	43
Dwelling-houses and flats formed by material change of use	43

Notes:

(1) The sound insulation values in this table include a built-in allowance for 'measurement uncertainty' and so if any of these test values are not met, then that particular test will be considered as failed.

(2) Occasionally a higher standard of sound insulation may be required between spaces used for normal domestic purposes and noise generated in and to an adjoining communal or non-domestic space. In these cases it would be best to seek specialist advice before committing yourself.

Flanking transmission from walls connected to the separating wall shall be controlled. E2

Tests should be carried out between rooms or spaces that share a common area formed by a separating wall or separating floor. E1

Impact sound insulation tests should be carried out without a soft covering (e.g. carpet, foam backed vinyl etc.) on the floor. E1

If the floor joists are to be supported on the separating wall then they should be supported on hangers and should not be built in. E2

If the joists are at right angles to the wall, spaces between the floor joists should be sealed with full depth timber blocking. E3

The floor base (excluding any screed) should be built into a cavity masonry external wall and carried through to the cavity face of the inner leaf. E

Walls that separate a dwelling from another dwelling (or part of the same building) shall resist:

- the level (and transmission) of airborne sounds; E
- the transmission of impact sound (such as speech, musical instruments and loudspeakers and impact sources such as footsteps and furniture moving);
- the flow of sound energy through walls and floors.

Requirements

Requirement E1

Figure 6.103 illustrates the relevant parts of the building that should be protected from airborne and impact sound in order to satisfy Requirement E1.

In some circumstances (for example, when a historic building is undergoing a material change of use) it may not be practical to improve the sound insulation to the standards set out in Approved Document E1 particularly if the special characteristics of such a building need to be recognized. In these circumstances the aim should be to improve sound insulation to the '*extent that it is practically possible*'.

Note: BS 7913:1998 *The principles of the conservation of historic buildings* provides guidance on the principles that should be applied when proposing work on historic buildings.

Requirement E2

Constructions for new walls within a dwelling-house (flat or room for residential purposes) – whether purpose built or formed by a material change of use – shall meet the laboratory sound insulation values set out in Table 6.39. E0.9

Figure 6.104 illustrates the relevant parts of the building that should be protected from airborne and impact sound in order to satisfy Requirement E2.

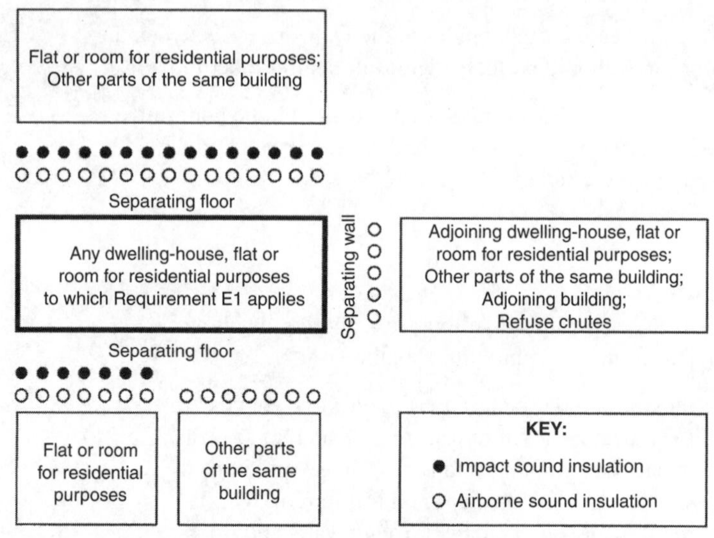

Figure 6.103 Requirement E1 – resistance to sound

Table 6.39 Laboratory values for new internal walls within dwelling-houses, flats and rooms for residential purposes – whether purpose built or formed by a material change of use

	Airborne sound insulation R_W dB (minimum values)
Purpose built dwelling-houses and flats	40

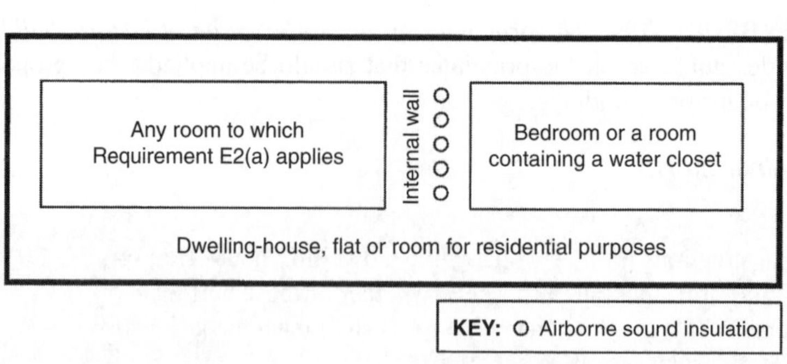

Figure 6.104 Requirement E2a – internal walls

Requirement E3

Sound absorption measures described in Section 7 of Approved Document N shall be applied.	E0.11

Requirement E4

The values for sound insulation, reverberation time and indoor ambient noise as described in Section 1 of Building Bulletin 93 *'The Acoustic Design of Schools'* (produced by DFES and published by the Stationery Office (ISBN 0 11 271105 7)) shall be satisfied.	E0.12

Types of wall

As shown in Figure 6.105 there are four main types of separating walls that can be used in order to achieve the required performance standards shown in Table 6.40.

Solid masonry
(Wall type 1)

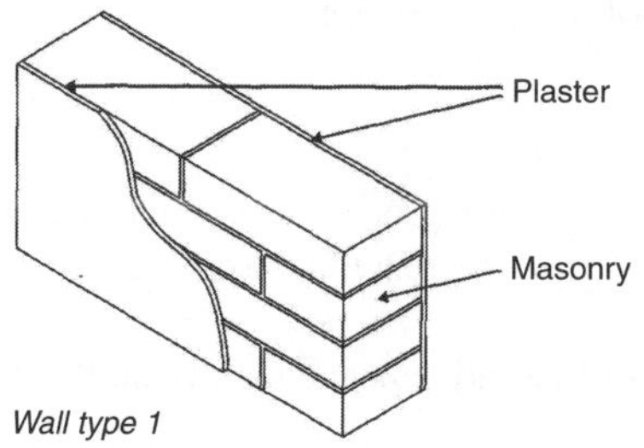

The resistance to airborne sound depends mainly on the mass per unit area of the wall.

Cavity masonry
(Wall type 2)

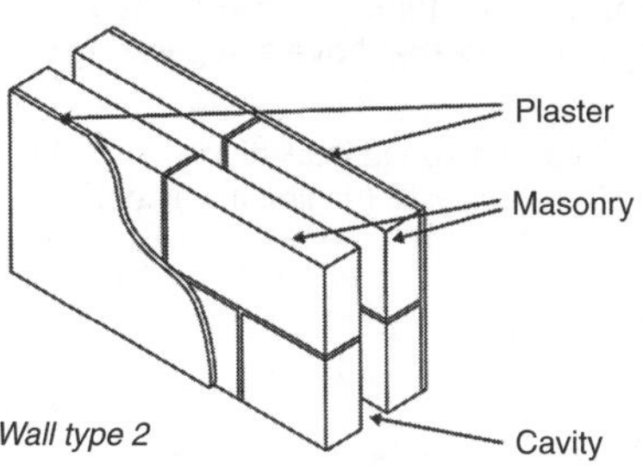

The resistance to airborne sound depends on the mass per unit area of the leaves and on the degree of isolation achieved. The isolation is affected by connections (such as wall ties and foundations) between the wall leaves and by the cavity width.

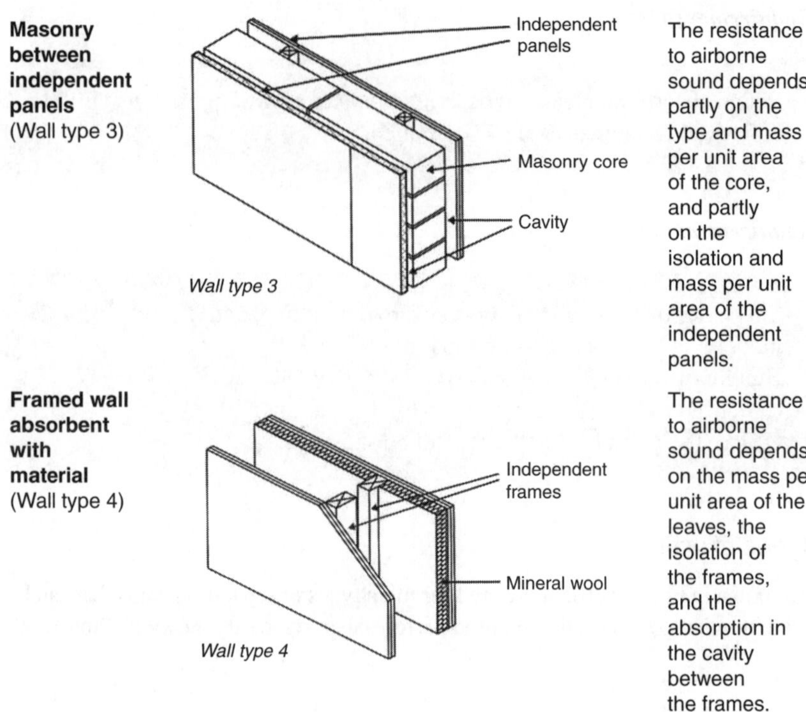
Masonry between independent panels (Wall type 3)
Independent panels
Masonry core
Cavity
Wall type 3
The resistance to airborne sound depends partly on the type and mass per unit area of the core, and partly on the isolation and mass per unit area of the independent panels.
Framed wall absorbent with material (Wall type 4)
Independent frames
Mineral wool
Wall type 4
The resistance to airborne sound depends on the mass per unit area of the leaves, the isolation of the frames, and the absorption in the cavity between the frames.

Figure 6.105 Types of separating walls

Other designs, materials and/or products may also be available and so it is always worthwhile talking to the manufacturers and/or suppliers first.

The resistance to airborne sound depends mainly on the mass of the wall.

Table 6.40 Dwelling-houses and flats – performance standards for separating walls, separating floors and stairs that have a separating function

	Airborne sound insulation $D_{nT,w} + C_{tr}$ dB (minimum values)	Impact sound insulation $L'_{nT,w}$ dB (maximum values)
Purpose built dwelling-houses and flats		
Walls	45	–
Floors and stairs	45	62
Dwelling-houses and flats formed by material change of use		
Walls	43	–
Floors and stairs	43	64

Junctions between separating walls and other building elements

Care should be taken to correctly detail the junctions between the separating wall and other elements, such as floors, roofs, external walls and internal walls.	E2.9

Note: Where any building element functions as a separating element (e.g. a ground floor that is also a separating floor for a basement flat) then the separating element requirements should take precedence.

Mass per unit area of walls

The mass per unit area of a wall is expressed in kilograms per square metre (kg/m^2) and is equivalent to:

$$\text{mass per unit area of a wall} = \frac{\text{mass of co-ordinating area}}{\text{co-ordinating area}} \tag{6.1}$$

Mass per unit area of a wall can be calculated as follows:

$$\text{mass per unit area of a wall} = \frac{M_B + \rho_m\,[Td(l + h - d) + V]}{LH}\,\text{kg/m}^2 \tag{6.2}$$

Where:

M_B = brick/block mass (kg) at appropriate moisture content
ρ_m = density of mortar (kg/m^3) at appropriate mortar content
T = the brick/block finish without surface finish (m)
d = mortar thickness (m)
L = co-ordinating length (m)
H = co-ordinating height (m)
V = volume of any frog/void filled with mortar (m^3)

Note: The method for calculating mass per unit area is provided in Annex A to Part E of the Regulations together with some examples.

Density of the materials

The density of the materials used (and on which the mass per unit area of the wall depends) is expressed in kilograms per cubic metre (kg/m^3).

Plasterboard linings on separating and external masonry walls

Wherever plasterboard is recommended (or the finish is not specified) a drylining laminate of plasterboard with mineral wool may be used.	E2.15
Plasterboard linings should be fixed according to manufacturer's instructions.	E2.16

Note: Recommended cavity widths in separating cavity masonry walls are minimum values.

Wall ties in separating and external cavity masonry walls

There are two types of wall ties that can be used in masonry cavity walls, type A (butterfly ties), which are normal and type B (double triangle ties), which are used only in external masonry cavity walls where tie type A does not satisfy the requirements of Building Regulation Part A – Structure.

Notes:

(1) Recommended cavity widths in separating cavity masonry walls are minimum values.

(2) In external cavity masonry walls, tie type B may decrease the airborne sound insulation due to flanking transmission via the external wall leaf compared to tie type A.

Stainless steel cavity wall ties are specified for all houses regardless of their location.	A1/2
Wall ties should have a horizontal spacing of 900 mm and a vertical spacing of 450 mm.	A1/2 2C8
 Equivalent to 2.5 ties per square metre.	
Wall ties should be spaced not more than 300 mm apart vertically, within a distance of 225 mm from the vertical edges of all openings, movement joints and roof verges.	A1/2 2C8
Wall ties should either comply with BS 1243, DD 140, or BS EN 845-1.	A1/2 2C19
Wall ties should be selected in accordance with Table 6.	A1/2 2C19
The leaves of a cavity masonry wall construction should be connected by either butterfly ties or double-triangle ties spaced as per BS 5628-3:2001 which limits this tie type and spacing to cavity widths of 50 mm to 75 mm with a minimum masonry leaf thickness of 90 mm.	E2.19
Note: Wall ties may be used provided that they have the measured dynamic stiffness for the cavity width (see E2.20 and E2.21 for details of the relevant formula for measuring the dynamic stiffness).	
In conditions of severe exposure, austenitic stainless steel or suitable non-ferrous ties should be used.	A1/2 (1C20)

The number of ties per square metre, n, shall be calculated from the horizontal (S_x) and vertical (S_y) tie spacing distances (in metres) using the formula $n = 1/(S_x \cdot S_y)$. E2.22

All wall ties and spacings specified using the dynamic stiffness parameter should also satisfy the requirements of Building Regulation Part A – Structure. E2.24

Corridor walls and doors

Separating walls should be used between corridors and rooms in flats, in order to control flanking transmission and to provide the required sound insulation. E2.25

Note: It is highly likely that the amount of sound insulation gained by using a separating wall will be reduced by the presence of a door.

Noisy parts of the building should preferably have a lobby, double door or high performance doorset to contain the noise. E2.27

All corridor doors shall have a good perimeter sealing (including the threshold where practical). E2.26

All corridor doors shall have a minimum mass per unit area of 25 kg/m^2. E2.26

All corridor doors shall have a minimum sound reduction index of 29 dB Rw (measured according to BS EN ISO 140-3:1995 and rated according to BS EN ISO 717-1:1997). E2.26

All corridor doors shall meet the requirements for fire safety (see Building Regulations Part B – Fire Safety). E2.26

Refuse chutes

A wall separating a habitable room (or kitchen) from a refuse chute should have a mass per unit area (including any finishes) of at least 1320 kg/m^2. E2.28

A wall separating a non-habitable room from a refuse chute should have a mass per unit area (including any finishes) of at least 220 kg/m^2. E2.28

Wall type 1 (solid masonry)

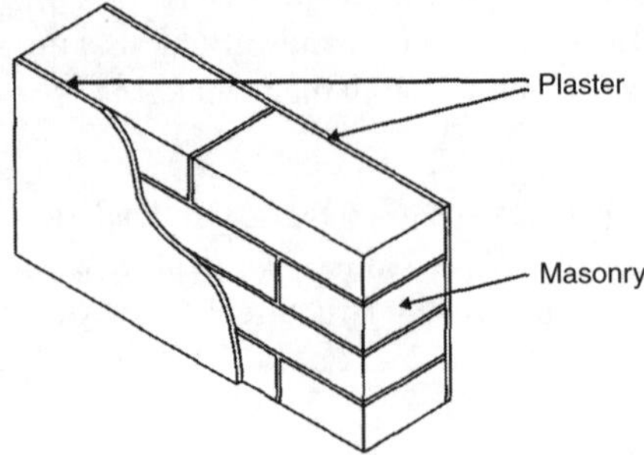

When using a solid masonry wall, the resistance to airborne sound depends mainly on the mass per unit area of the wall. As shown below, there are three different categories of solid masonry walls:

Table 6.41 Wall type 1 – categories

Wall type 1 **Category 1.1 Solid masonry** Dense aggregate concrete block, plaster on both room faces	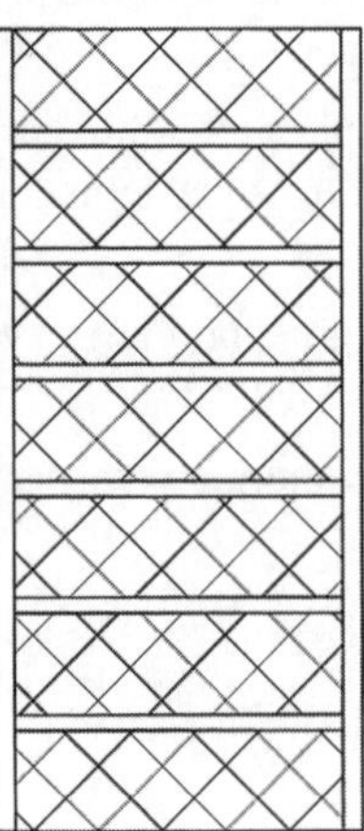	Minimum mass per unit area (including plaster) 415 kg/m^2 Plaster on both room faces Blocks laid flat to the full thickness of the wall For example: Size 215 mm laid flat Density 1840 kg/m^3 Coursing 110 mm Plaster 13 mm lightweight
Wall type 1 **Category 1.2 Dense aggregate concrete** Dense aggregate concrete cast in-situ, plaster on both room faces	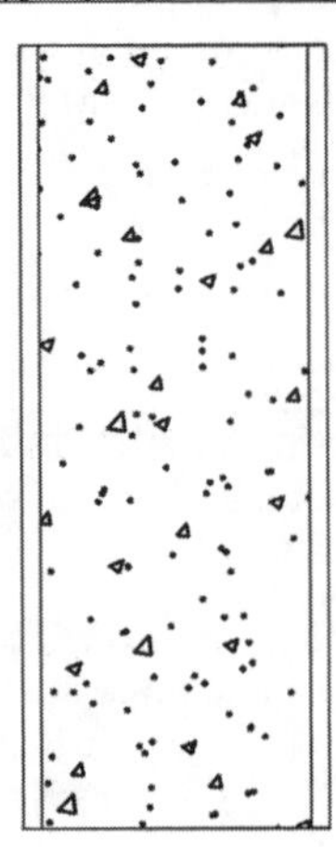	Minimum mass per unit area (including plaster) 415 kg/m^2 Plaster on both room faces For example: Concrete 190 mm Density 2200 kg/m^3 Plaster 13 mm lightweight

Wall type 1

Category 1.3
Brick

Brick, plaster on both room faces

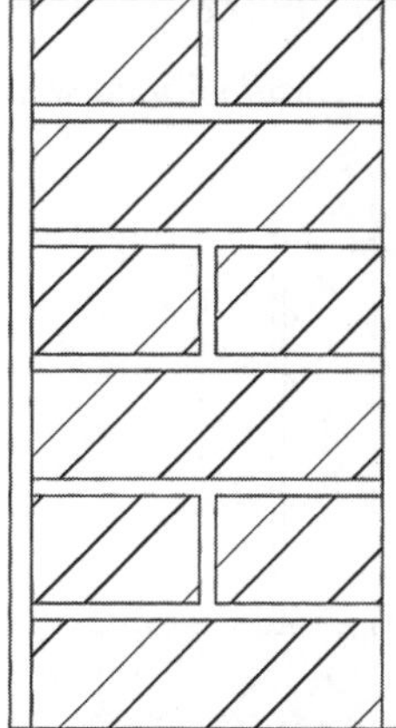

Minimum mass per unit area (including plaster) 375 kg/m^2

Bricks to be laid frog up, coursed with headers

For example:

Size	215 mm laid flat
Density	1610 kg/m^3
Coursing	75 mm
Plaster lightweight	13 mm

General requirements

Fill and seal all masonry joints with mortar.	E2.32a
Lay bricks frog up to achieve the required mass per unit area and avoid air paths.	E2.32b
Use bricks/blocks that extend to the full thickness of the wall.	E2.32c
Ensure that an external cavity wall is stopped with a flexible closer at the junction with a separating wall.	E2.32d
Unless the cavity is fully filled with mineral wool or expanded polystyrene beads.	
Control flanking transmission from walls and floors connected to the separating wall (see guidance on junctions).	E2.32e
Deep sockets and chases should **not** be used in separating walls.	E2.32
Stagger the position of sockets on opposite sides of the separating wall.	E2.32f
Ensure flue blocks: • will not adversely affect the sound insulation; • use a suitable finish.	E2.32g
A cavity separating wall may **not** be changed into a solid masonry (i.e. type 1) wall by filling in the cavity with mortar and/or concrete.	E2.32
When the cavity wall is bridged by the solid wall, ensure that there is no junction between the solid masonry wall and a cavity wall.	E2.32

Wall type 1 – Junction requirements

Junctions with an external cavity wall with masonry inner leaf

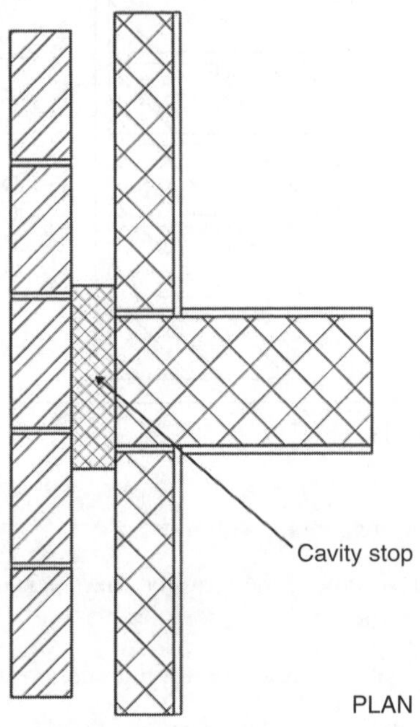

Where the external wall is a cavity wall:

- the outer leaf of the wall may be of any construction; E2.36a
- the cavity should be stopped with a flexible closer. E2.36b

The masonry inner leaf should have a mass per unit area of at least 120 kg/m^2 excluding finish unless there are openings in the external wall (see Figure 6.106) that are: E2.38a

- not less than 1 metre high; E2.38b
- on both sides of the separating wall at every storey; E2.38c
- not more than 700 mm from the face of the separating wall on both sides. E2.39

Note: If there is also a separating floor, then the minimum mass per unit area of 120 kg/m^2 (excluding finish) will always apply, irrespective of the presence or absence of openings.

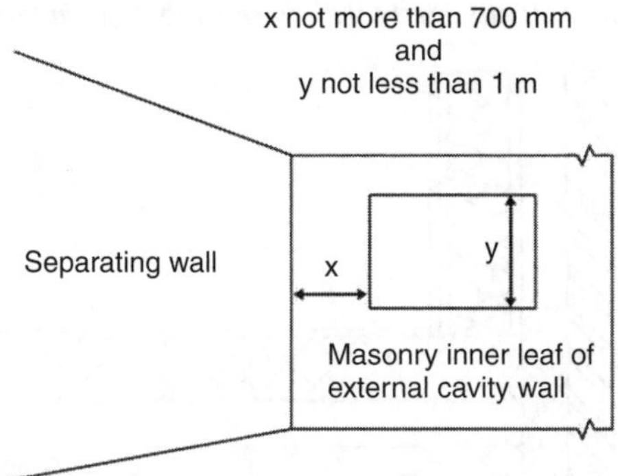

Figure 6.106 Wall type 1 – position of openings in a masonry inner leaf of an external cavity wall

The separating wall should be joined to the inner leaf of the external cavity wall by one of the following methods:

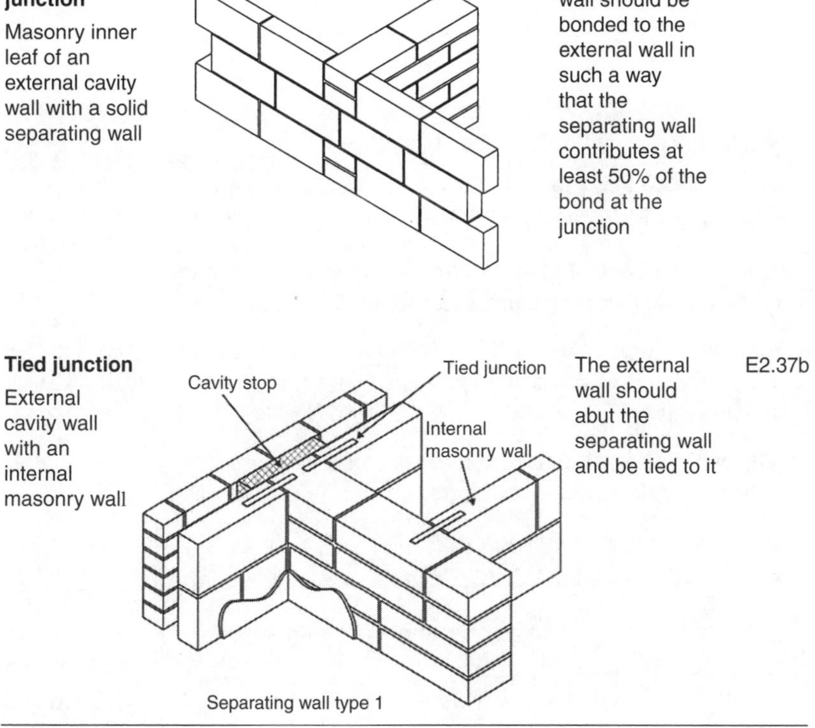

Bonded junction Masonry inner leaf of an external cavity wall with a solid separating wall		The separating wall should be bonded to the external wall in such a way that the separating wall contributes at least 50% of the bond at the junction	E2.37a
Tied junction External cavity wall with an internal masonry wall		The external wall should abut the separating wall and be tied to it	E2.37b

Figure 6.107 Separating wall junctions for a type 1 wall

Junctions with an external cavity wall with timber frame inner leaf

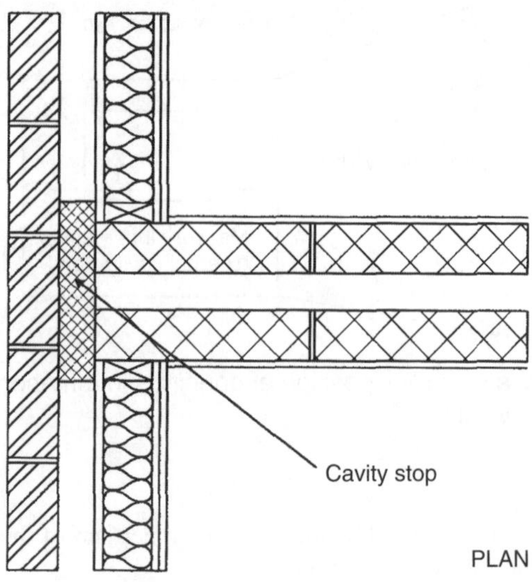

Where the external wall is a cavity wall:

- the outer leaf of the wall may be of any construction; E2.40a
- the cavity should be stopped with a flexible closer. E2.40b

Where the inner leaf of an external cavity wall is of framed construction, the framed inner leaf should:

- abut the separating wall; E2.41a1
- be tied to it with ties at no more than 300 mm centres vertically. E2.41b1

The wall finish of the framed inner leaf of the external wall should be:

- one layer of plasterboard; or E2.41a2
- two layers of plasterboard where there is a separating floor; E2.41b2
- each sheet of plasterboard should be of minimum mass per unit area 10 kg/m^2; E2.41c
- all joints should be sealed with tape or caulked with sealant. E2.41d

Junctions with internal timber floors

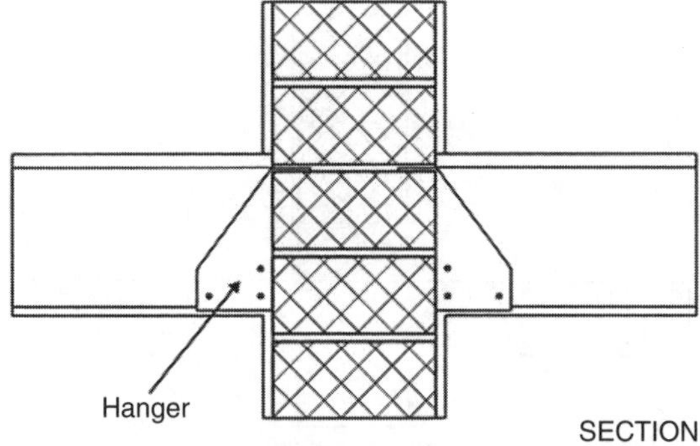
Hanger
SECTION

If the floor joists are to be supported on a type 1 separating wall then they should be supported on hangers as opposed to being built in. E2.45

Junctions with internal concrete floors

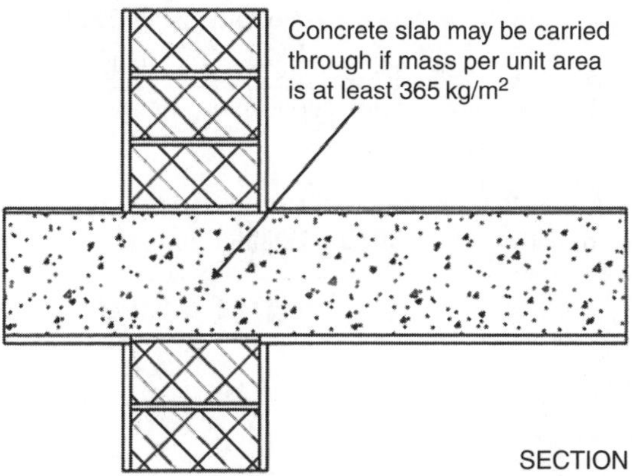
Concrete slab may be carried through if mass per unit area is at least 365 kg/m2
SECTION

An internal concrete floor slab may only be carried through a type 1 separating wall if the floor base has a mass per unit area of at least 365 kg/m^2. E2.46

Internal hollow-core concrete plank floors and concrete beams with infilling block floors should not be continuous through a type 1 separating wall. E2.47

Note: For internal floors of concrete beams with infilling blocks, avoid beams built into the separating wall unless the blocks in the floor fill the space between the beams where they penetrate the wall.

Junctions with concrete ground floors

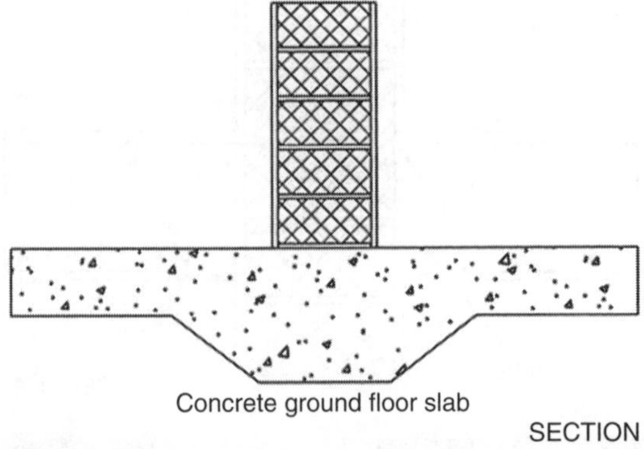

The ground floor may be a solid slab, laid on the ground, or a suspended concrete floor. E2.51

A concrete slab floor on the ground may be continuous under a type 1 separating wall. E2.51

A suspended concrete floor may only pass under a type 1 separating wall if the floor has a mass of at least 365 kg/m^2. E2.52

Hollow core concrete plank and concrete beams with infilling block floors should not be continuous under a type 1 separating wall. E2.53

Note: See also Building Regulation Part C – Site preparation and resistance to moisture, and Building Regulation Part L – Conservation of fuel and power.

Junctions with ceiling and roof

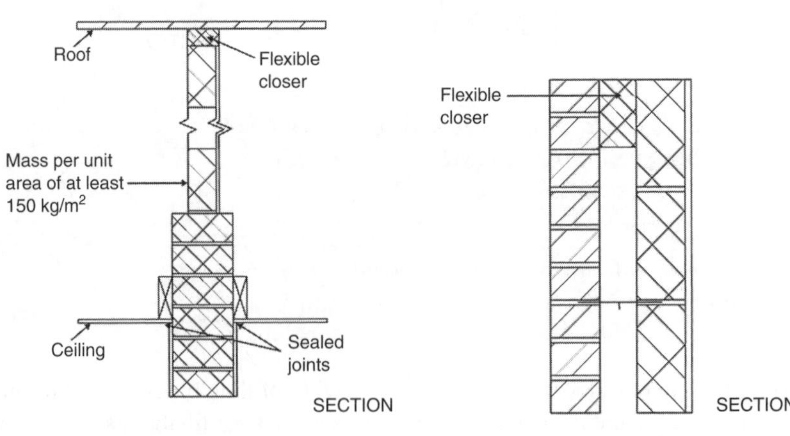

Where a type 1 separating wall is used it should be continuous to the underside of the roof.	E2.55
The junction between the separating wall and the roof should be filled with a flexible closer which is also suitable as a fire stop.	E2.56
Where the roof or loft space is not a habitable room (and there is a ceiling with a minimum mass per unit area of 10 kg/m^2 with sealed joints) then the mass per unit area of the separating wall above the ceiling may be reduced to 150 kg/m^2.	E2.57
If lightweight aggregate blocks of density less than 1200 kg/m^3 are used above ceiling level, then one side should be sealed with cement paint or plaster skim.	E2.58
Where there is an external cavity wall, the cavity should be closed at eaves level with a suitable flexible material (e.g. mineral wool).	E2.59
A rigid connection between the inner and external wall leaves should be avoided.	Ep23
If a rigid material is used, then it should only be rigidly bonded to one leaf.	Ep23

Guidance for other types of wall type 1 junctions

Junctions with an external solid masonry wall	No guidance available (seek specialist advice).	E2.42
Junctions with internal framed walls	There are no restrictions on internal framed walls meeting a type 1 separating wall.	E2.43
Junctions with internal masonry walls	Internal masonry walls that abut a type 1 separating wall should have a mass per unit area of at least 120 kg/m^2 excluding finish.	E2.44
Junctions with timber ground floors	If the floor joists are to be supported on a type 1 separating wall then they should be supported on hangers and should not be built in.	E2.49

Wall type 2 (cavity masonry)

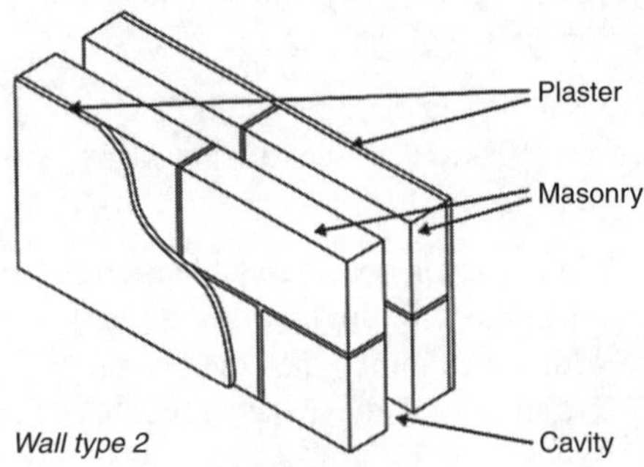

Wall type 2

When using a cavity masonry wall, the resistance to airborne sound depends on the mass per unit area of the leaves and on the degree of isolation achieved. The isolation is affected by connections (such as wall ties and foundations) between the wall leaves and by the cavity width.

As shown below, there are four different categories of cavity masonry walls:

Table 6.42 Wall type 2 – categories

Wall type 2 **Category 2.1** **Two leaves of dense aggregate concrete block with 50 mm cavity incorporated at the separating wall.**	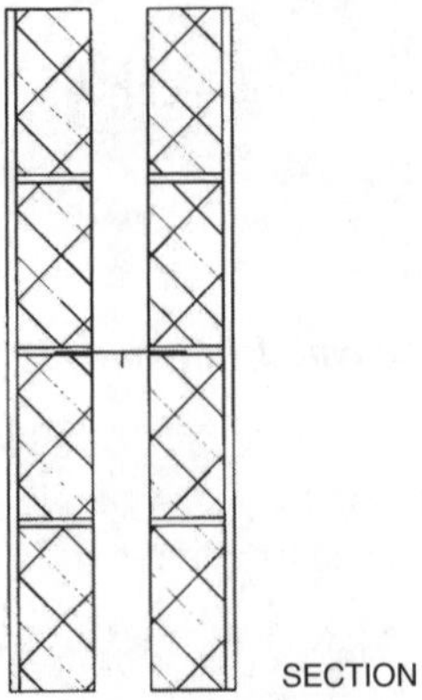 SECTION	Minimum mass per unit area (including plaster) 415 kg/m² Plaster on both room faces Minimum cavity width 50 mm For example: Block leaves 100 mm Density 1990 kg/m³ Coursing 225 mm Plaster 13 mm lightweight
Wall type 2 **Category 2.2** **Two leaves of lightweight aggregate block with 75 mm cavity, plaster on both room faces**	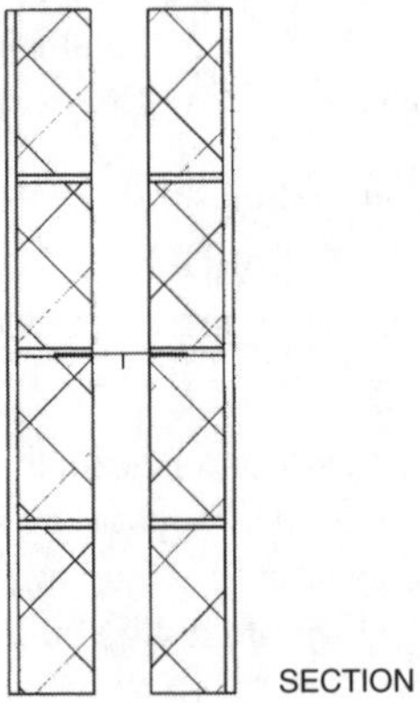 SECTION	Minimum mass per unit area (including plaster) 300 kg/m² Plaster on both room faces Minimum cavity width of 75 mm For example: Block leaves 100 mm Density 1375 kg/m³ Coursing 225 mm Plaster 13 mm lightweight

Wall type 2

Category 2.3

Two leaves of lightweight aggregate block with 75 mm cavity and step/stagger plasterboard on both room faces

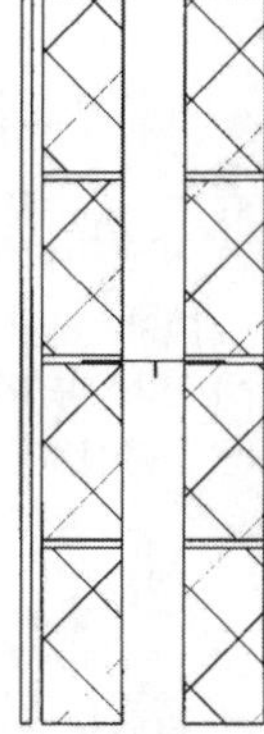

Minimum mass per unit area (including plaster) 290 kg/m^2

Lightweight aggregate blocks should have a density in the range 1350 to 1600 m^3

Minimum cavity width of 75 mm

Wall type 2.3 should **only** be used where there is a step and/or stagger of at least 300 mm

Increasing the size of the step or stagger in the separating wall tends to increase the airborne sound insulation

Plasterboard (lightweight) each sheet of minimum mass per unit area 10 kg/m^2 on both room faces

For example:

Block leaves	100 mm
Density	1375 kg/m^3
Coursing	225 mm

Lightweight plasterboard (minimum mass per unit area 10 kg/m^2) on both room faces

Wall type 2

Category 2.4

Two leaves of Aircrete block with 75 mm cavity and step/stagger plasterboard or plaster on both room faces

SECTION

Minimum mass per unit area (including plaster) 150 kg/m^2

Lightweight aggregate blocks should have a density in the range 1350 to 1600 kg/m^3

Minimum cavity width of 75 mm

Plasterboard (lightweight) minimum mass per unit area 10 kg/m^2 on both room faces or 13 mm plasterboard on both faces

Wall type 2.4 should **only** be used where there is a step and/or stagger of at least 300 mm

Increasing the size of the step or stagger in the separating wall tends to increase the airborne sound insulation

For example:

Aircrete block leaves	100 mm
Density	650 kg/m^3
Coursing	225 mm

Plaster (lightweight) minimum mass per unit area 10 kg/m^2 on both room faces

General requirements

Fill and seal all masonry joints with mortar.	E2.65a
Keep the cavity leaves separate below ground floor level.	E2.65b
Ensure that any external cavity wall is stopped with a flexible closer at the junction with the separating wall.	E2.65c
Control flanking transmission from walls and floors connected to the separating wall.	E2.65d
Stagger the position of sockets on opposite sides of the separating wall.	E2.65e
Ensure that flue blocks will not adversely affect the sound insulation and that a suitable finish is used over the flue blocks.	E2.65f
The cavity separating wall should **not** be converted to a type 1 (solid masonry) separating wall by inserting mortar or concrete into the cavity between the two leaves.	E2.65a2
A solid wall construction in the roof space should **not** be changed.	E2.65b2
Cavity walls should **not** be built off a continuous solid concrete slab floor.	E2.65c2
Deep sockets and chases should **not** be used in a separating wall.	E2.65d2
Deep sockets and chases in a separating wall should **not** be placed back to back.	E2.65d2
Wall ties used to connect the leaves of a cavity masonry wall should be tie type A.	E2.66

Wall type 2 – Junction requirements

Junctions with an external cavity wall with masonry inner leaf

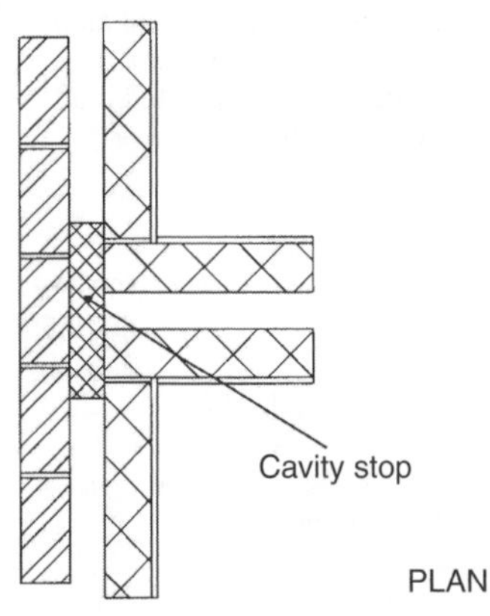

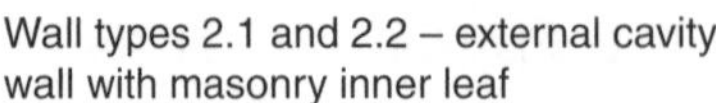
Wall types 2.1 and 2.2 – external cavity wall with masonry inner leaf

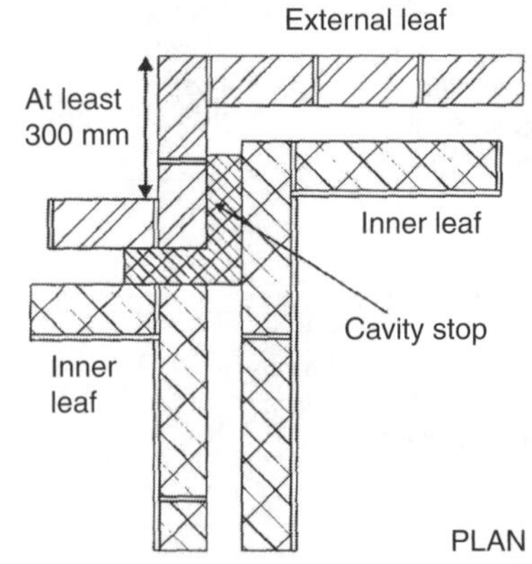

Wall types 2.3 and 2.4 – external cavity wall with masonry inner leaf – stagger

Where the external wall is a cavity wall:

- the outer leaf of the wall may be of any construction; E2.73a
- the cavity should be stopped with a flexible closer. E2.73b

The separating wall should be joined to the inner leaf of the external cavity wall by one of the following methods: E2.74

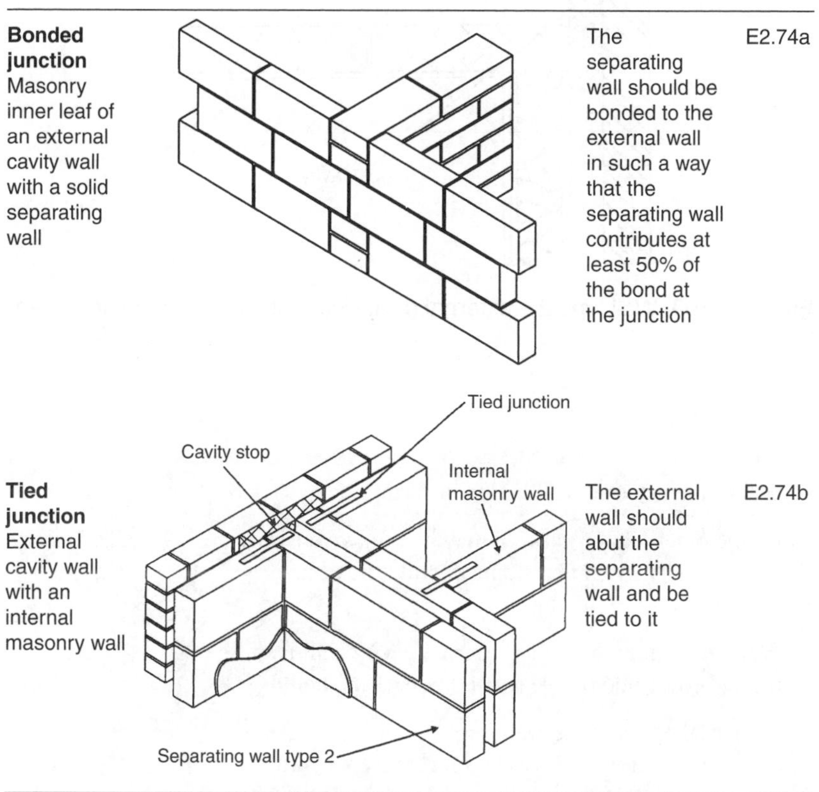

Bonded junction
Masonry inner leaf of an external cavity wall with a solid separating wall
The separating wall should be bonded to the external wall in such a way that the separating wall contributes at least 50% of the bond at the junction
E2.74a
Tied junction
External cavity wall with an internal masonry wall
Tied junction
Cavity stop
Internal masonry wall
Separating wall type 2
The external wall should abut the separating wall and be tied to it
E2.74b

Figure 6.108 Separating wall junctions for a type 2 wall

The masonry inner leaf should have a mass per unit area of at least 120 kg/m^2 excluding finish. E2.75

There is no minimum mass requirement where separating wall type 2.1, 2.3 or 2.4 is used **unless** there is also a separating floor. E2.76

Junctions with an external cavity wall with timber frame inner leaf

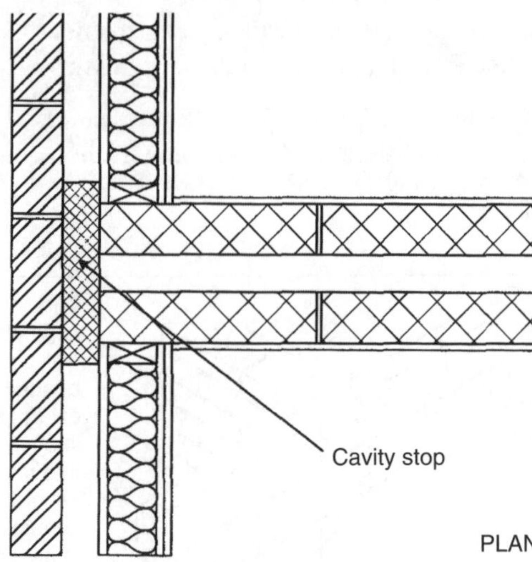

Figure 6.109 Wall type 2 – external cavity wall with timber frame inner leaf

Where the external wall is a cavity wall:

- the outer leaf of the wall may be of any construction; E2.77a
- the cavity should be stopped with a flexible closer (see Figure 6.109). E2.77b

Where the inner leaf of an external cavity wall is of framed construction, the framed inner leaf should:

- abut the separating wall; E2.78a
- be tied to it with ties at no more than 300 mm centres vertically. E2.78b

The wall finish of the inner leaf of the external wall should be:

- one layer of plasterboard; E2.98a2
- two layers of plasterboard where there is a separating floor; E2.78b2
- each sheet of plasterboard to be of minimum mass per unit area 10 kg/m^2; E2.78c2
- all joints should be sealed with tape or caulked with sealant. E2.78d2

Junctions with internal timber floors

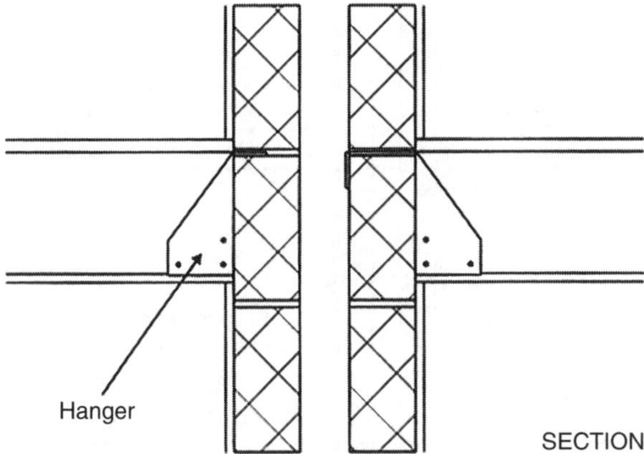

Figure 6.110 Wall type 2 – internal timber floor

If the floor joists are to be supported on the separating wall then they should be supported on hangers as opposed to being built in. E2.84

Junctions with internal concrete floors

Internal concrete floors should generally be built into a type 2 separating wall and carried through to the cavity face of the leaf. E2.85

The cavity should not be bridged.

Junctions with concrete ground floors (see Figure 6.111)

The ground floor may be a solid slab, laid on the ground, or a suspended concrete floor. E2.88

A concrete slab floor on the ground should not be continuous under a type 2 separating wall. E2.88

A suspended concrete floor should not be continuous under a type 2 separating wall. E2.89

A suspended concrete floor should be carried through to the cavity face of the leaf. E2.89

The cavity should not be bridged.

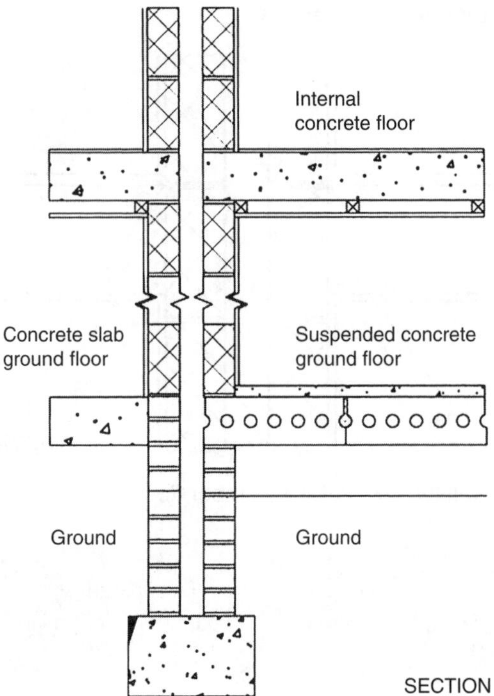

Figure 6.111 Wall type 2 – internal concrete floor and concrete ground floor

Junctions with ceiling and roof space

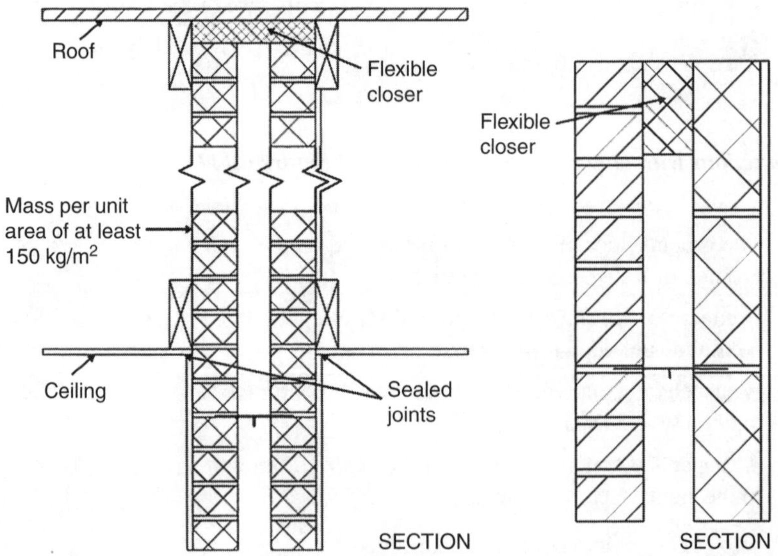

Wall type 2 – ceiling and roof junction

External cavity wall at eaves level

A type 2 separating wall should be continuous to the underside of the roof.	E2.91
The junction between the separating wall and the roof should be filled with a flexible closer which is also suitable as a fire stop.	E2.92
If lightweight aggregate blocks (with a density less than 1200 kg/m^3 are used) above ceiling level, then one side should be sealed with cement paint or plaster skim.	E2.94
The cavity of an external cavity wall should be closed at eaves level with a suitable flexible material (e.g. mineral wool).	E2.95
A rigid connection between the inner and external wall leaves should be avoided.	E2.95
If a rigid material has to be used, then it should **only** be rigidly bonded to one leaf.	E2.95

Note: If the roof or loft space is not a habitable room (and there is a ceiling with a minimum mass per unit area of 10 kg/m^2 with sealed joints) then the mass per unit area of the separating wall above the ceiling may be reduced to 150 kg/m^2 – **but** it should still be a cavity wall.

Guidance for other wall type 2 junctions

Junctions with internal masonry walls	• Internal masonry walls that abut a type 2 separating wall should have a mass per unit area of at least 120 kg/m^2 excluding finish;	E2.81
	• When there is a separating floor, the internal masonry walls should have a mass per unit area of at least 120 kg/m^2 excluding finish.	E2.82
Junctions with internal framed walls	There are no restrictions on internal framed walls meeting a type 2 separating wall.	E2.80
Junctions with an external solid masonry wall	No guidance available (seek specialist advice).	E2.79
Junctions with timber ground floors	If the floor joists are to be supported on a type 1 separating wall then they should be supported on hangers and should not be built in.	E2.49

Wall type 3 (masonry between independent panels)

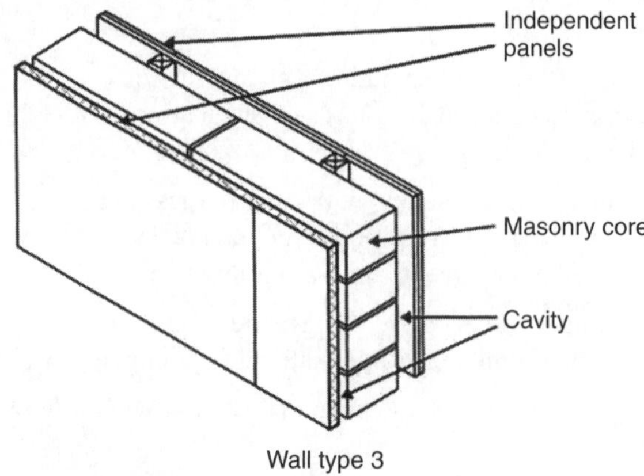

Wall type 3

Wall type 3 provides a high resistance to the transmission of both airborne sound and impact sound on the wall. As shown below there are three different categories of Wall type 3 which comprise either a solid or a cavity masonry core wall with independent panels on both sides. Their resistance to sound depends partly on the type (and mass) of the core and partly on the isolation and mass of the panels.

General requirements

Fill and seal all masonry joints with mortar.	E2.101a
Control flanking transmission from walls and floors connected to the separating wall.	E2.101b
The panels and any frame should not be in contact with the core wall.	E2.99
The panels and/or supporting frames should be fixed to the ceiling and floor only.	E101c
All joints should be taped and sealed.	E2.101d
Flue blocks shall not adversely affect the sound insulation.	E2.101e
A suitable finish is used over the flue blocks (see BS 1289-1:1986).	E2.101e
Free-standing panels and/or the frame should not be fixed, tied or connected to the masonry core.	E2.101
Wall ties in cavity masonry cores used to connect the leaves of a cavity masonry core together should be tie type A.	E2.102

Table 6.43 Wall type 3 – categories

Category	Section	Requirements
Wall type 3 **Category 3.1** Solid masonry core (dense aggregate concrete block), independent panels on both room faces	 SECTION	• Minimum mass per unit area (including plaster) 300 kg/m²; • Independent panels on both room faces; • Minimum core width determined by structural requirements. For example: Size 140 mm core block Density 2200 kg/m³ Coursing 110 mm Independent panels – each panel of mass per unit area 20 kg/m², to be two sheets of plasterboard with joints staggered
Wall type 3 **Category 3.2** Solid masonry core (lightweight concrete block), independent panels on both room faces	SECTION	• Minimum mass per unit area (including plaster) 150 kg/m²; • Independent panels on both room faces; • Minimum core width determined by structural requirements. For example: Size 140 mm core block Density 1400 kg/m³ Coursing 225 mm Independent panels – each panel of mass per unit area 20 kg/m², to be two sheets of plasterboard with joints staggered
Wall type 3 **Category 3.3** Cavity masonry core (brickwork or block work), 50 mm cavity, independent panels on both room faces	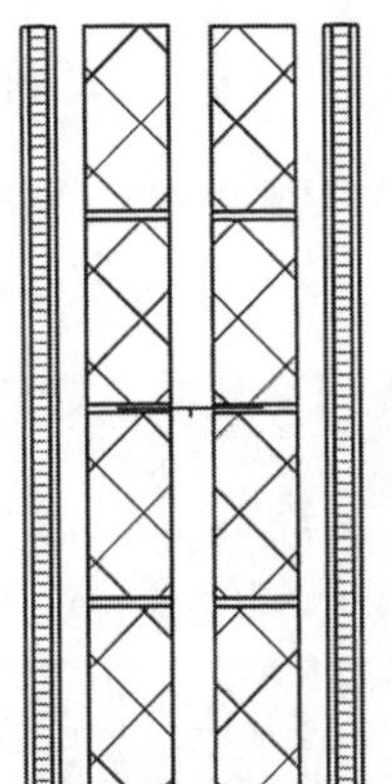 SECTION	• Core mass – unrestricted; • Minimum cavity width of 50 mm; • Independent panels on both room faces; • Minimum core width determined by structural requirements. For example: Concrete block – two leaves (each leaf at least 100 mm thick) Minimum cavity width – 50 mm Independent panels – each panel of mass per unit area 20 kg/m², to be two sheets of plasterboard with joints staggered

The minimum mass per unit area of independent panel (excluding any supporting framework) should be 20 kg/m^2.	E2.104
Panels should be either at least two layers of plasterboard with staggered joints or a composite panel consisting of two sheets of plasterboard separated by a cellular core.	E2.104
Panels that are not supported on a frame should be at least 35 mm from the masonry core.	E2.104
Panels which are supported on a frame should have a gap of at least 10 mm between the frame and the masonry core.	E2.104

Junction requirements for wall type 3

Junctions with an external cavity wall with masonry inner leaf

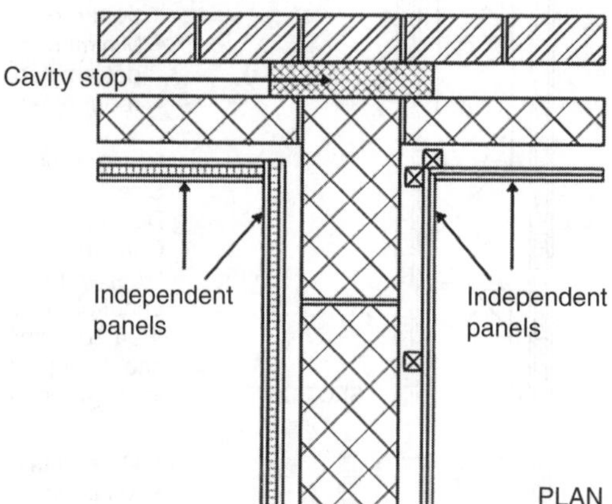

Figure 6.112 Wall type 3 – external cavity wall with masonry inner leaf

If the external wall is a cavity wall: E2.108

- the outer leaf of the wall may be of any construction;
- the cavity should be stopped with a flexible closer.

If the inner leaf of an external cavity wall is masonry: E2.109

- the inner leaf of the external wall should be bonded or tied to the masonry core;
- the inner leaf of the external wall should be lined with independent panels.

If there is a separating floor, the masonry inner leaf (of the external wall) should have a minimum mass per unit area of at least 120 kg/m^2 excluding finish. E2.110

If there is no separating floor:

- the external wall may be finished with plaster or plasterboard of minimum mass per unit area 10 kg/m^2 (provided the masonry inner leaf of the external wall has a mass of at least 120 kg/m^2 excluding finish); E2.111
- there is no minimum mass requirement on the masonry inner leaf (provided that the masonry inner leaf of the external wall is lined with independent panels in the same manner as the separating walls); E2.112

Junctions with internal framed walls

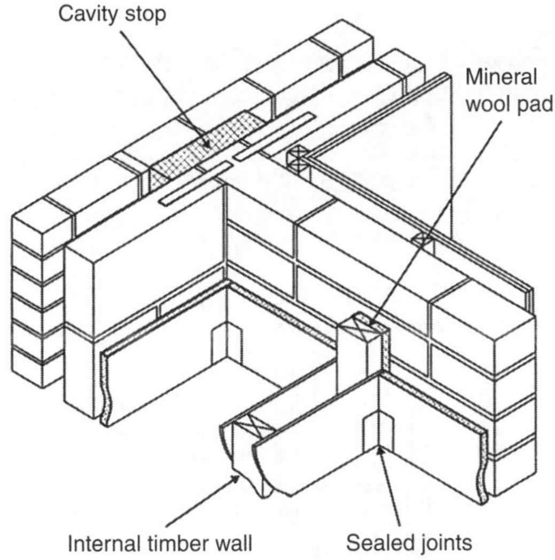

Figure 6.113 Wall type 3 – external cavity wall with internal timber wall

Load bearing (framed) internal walls should be fixed to the masonry core through a continuous pad of mineral wool. E2.115

Non-load bearing internal walls should be butted to the independent panels. E2.116

All joints between internal walls and panels should be sealed with tape or caulked with sealant. E2.117

Junctions with internal timber floors

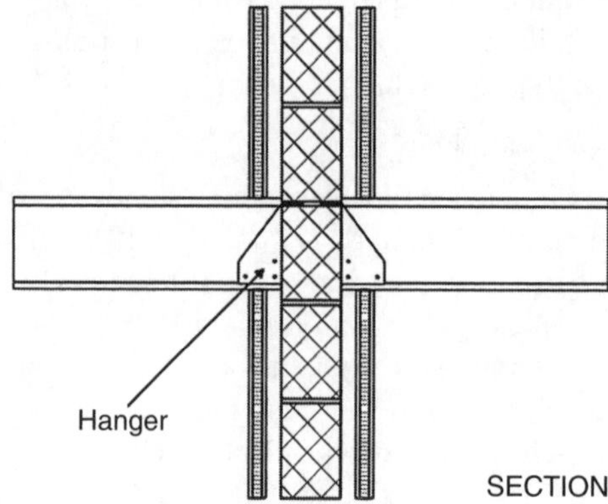
Hanger
SECTION

Figure 6.114 Wall type 3 – internal timber floor

Junctions with internal masonry walls

If the floor joists are to be supported on the separating wall then they should be supported on hangers as opposed to being built in.	E2.119
Spaces between the floor joists should be sealed with full depth timber blocking.	E2.120

Junctions with internal concrete floors

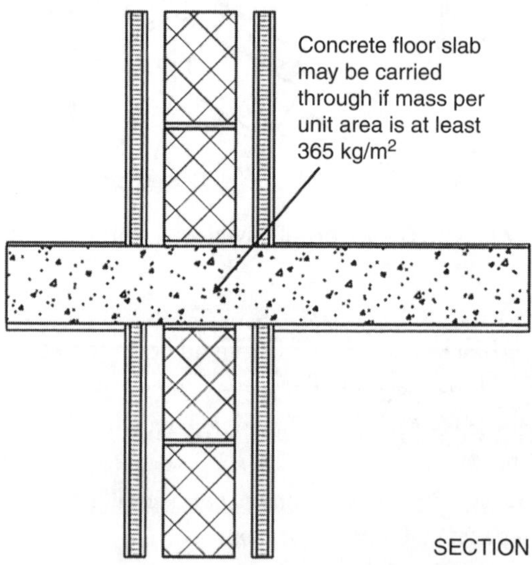
Concrete floor slab may be carried through if mass per unit area is at least 365 kg/m2
SECTION

Figure 6.115 Wall types 3.1 and 3.2 – internal concrete floor

For wall types 3.1 and 3.2 (i.e. those with solid masonry cores) internal concrete floor slabs may only be carried through a solid masonry core if the floor base has a mass per unit area of at least 365 kg/m^2. E2.121

For wall type 3.3 (cavity masonry core): E2.122

- internal concrete floors should generally be built into a cavity masonry core and carried through to the cavity face of the leaf;
- the cavity should not be bridged.

Junctions with ceiling and roof space

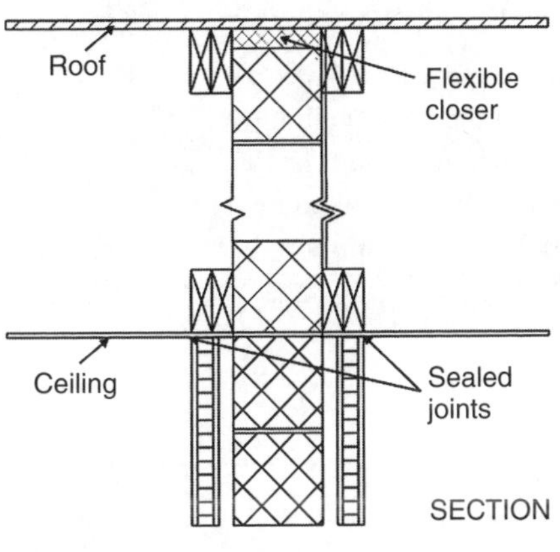

Wall types 3.1 and 3.2 – ceiling and roof junction

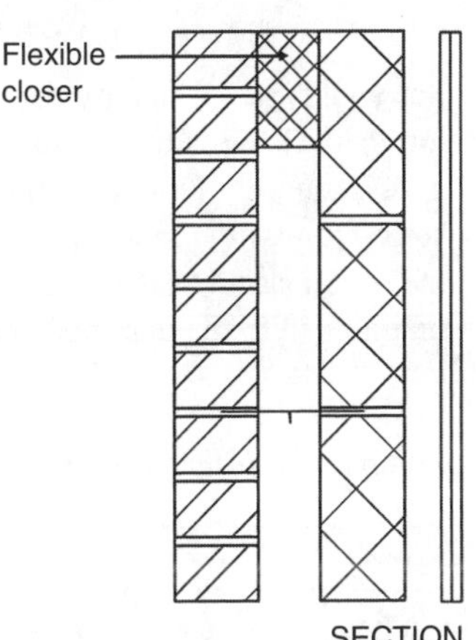

External cavity wall at eaves level

The masonry core should be continuous to the underside of the roof. E2.133

The junction between the separating wall and the roof should be filled with a flexible closer which is also suitable as a fire stop. E2.134

The junction between the ceiling and independent panels should be sealed with tape or caulked with sealant. E2.135

If there is an external cavity wall, the cavity should be closed at eaves level with a suitable flexible material (e.g. mineral wool). E2.136

Rigid connections between the inner and external wall leaves should be avoided where possible. E2.136

If a rigid material is used, then it should only be rigidly bonded to one leaf. E2.136

For wall types 3.1 and 3.2 (solid masonry core):

- if the roof or loft space is not a habitable room (and there is a ceiling with a minimum mass per unit area 10 kg/m^2 and it has sealed joints) the independent panels may be omitted in the roof space and the mass per unit area of the separating wall above the ceiling may be a minimum of 150 kg/m^2; E2.137
- if lightweight aggregate blocks with a density less than 1200 kg/m^3 are used above ceiling level, then one side should be sealed with cement paint or plaster skim. E2.138

For wall type 3.3 (cavity masonry core) if the roof or loft space is not a habitable room (and there is a ceiling with a minimum mass per unit area 10 kg/m^2 and it has sealed joints) the independent panels may be omitted in the roof space, but the cavity masonry core should be maintained to the underside of the roof. E2.139

Junctions with internal masonry floors

Internal walls that abut a type 2 separating wall should not be of masonry construction. E2.118

Junctions with timber ground floors

Floor joists supported on a separating wall should be supported on hangers as opposed to being built in. E2.123

The spaces between floor joists should be sealed with full depth timber blocking. E2.124

Junctions with an external cavity wall with timber frame inner leaf

No official guidance currently available. Best to seek specialist advice. E2.113

Junctions with an external solid masonry wall

No official guidance currently available. Best to seek specialist advice.	E2.114

Junctions with concrete ground floors

The ground floor may be a solid slab, laid on the ground, or a suspended concrete floor.	E2.126
For wall types 3.1 and 3.2 (solid masonry core):	
• a concrete slab floor on the ground may be continuous under the solid masonry core of the separating wall;	E2.127
• a suspended concrete floor may only pass under the solid masonry core if the floor has a mass per unit area of at least 365 kg/m^2;	E2.128
• hollow core concrete plank (and concrete beams with infilling block floors) should **not** be continuous under the solid masonry core of the separating wall.	E2.129
For wall type 3.3 (cavity masonry core):	
• a concrete slab floor on the ground should **not** be continuous under the cavity masonry core of the separating wall;	E2.130
• a suspended concrete floor should **not** be continuous under the cavity masonry core of a Type 3.3 separating wall;	E2.131
• a suspended concrete floor **should** be carried through to the cavity face of the lea but the cavity should not be bridged.	E2.132

Junctions with internal masonry walls

Internal walls that abut a type 3 separating wall should not be of masonry construction.	E2.118

Wall type 4 (framed walls with absorbent material)

A wall type 4 consists of a timber frame with a plasterboard lining on the room surface with an absorbent material between the frames. Its resistance to airborne sound depends on:

- the mass per unit area of the leaves;
- the isolation of the frames;
- the absorption in the cavity between the frames.

General requirements

If a fire stop is required in the cavity between frames, then it should either be flexible or only be fixed to one frame.	E2.146a
Layers of plasterboard should:	
• be independently fixed to the stud frame;	E2.146c
• not be chased.	E2.146b2
If two leaves have to be connected together for structural reasons, then:	
• the cross-section of the ties shall be less than 40 mm × 3 mm;	E2.146a2
• ties should be fixed to the studwork at or just below ceiling level;	E2.146a2
• ties should not be set closer than 1.2 m centres.	E2.146a2
Sockets should:	
• be positioned on opposite sides of a separating wall;	E2.146b
• not be connected back to back;	E2.146b2
• be staggered a minimum of 150 mm edge to edge.	E2.146b2
The flanking transmission from walls and floors connected to a separating wall should be controlled (see guidance on junctions).	E2.146d

Wall type 4.1 (double leaf frames with absorbent material)

General requirements

The lining shall be two or more layers of plasterboard with a minimum sheet mass per unit area 10 kg/m^2 and with staggered joints.	E2.147
If a masonry core is used for structural purposes, then the core should only be connected to one frame.	E2.147
The minimum distance between inside lining faces shall be 200 mm.	E2.147
Plywood sheathing may be used in the cavity if required for structural reasons.	E2.147
Absorbent material:	
• shall have a minimum density of 10 kg/m^3;	E2.147
• shall be unlaced mineral wool batts (or quilt);	

- may be wire reinforced;
- shall have a minimum thickness of between 25 and 50 mm as shown in Figure 6.116.

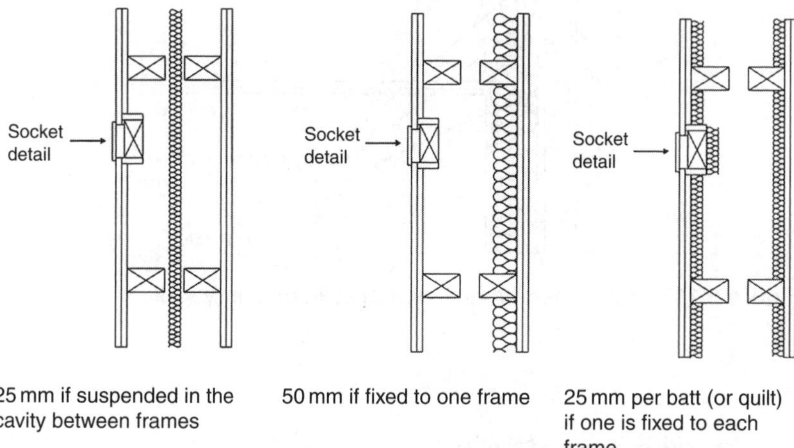

Figure 6.116 Wall type 4.1 – minimum thickness of absorbent material

Junction requirements for wall type 4

Junctions with an external cavity wall with timber frame inner leaf

If the external wall is a cavity wall:

- the outer leaf of the wall may be of any construction; E2.149
- the cavity should be stopped between the ends of the separating wall and the outer leaf with a flexible closer. E2.149

The wall finish of the inner leaf of the external wall should be one layer of plasterboard (or two layers of plasterboard if there is a separating floor). E2.150a and b

Each sheet of plasterboard to be of minimum mass per unit area 10 kg/m^2. E2.150c

All joints should be sealed with tape or caulked with sealant. E2.150

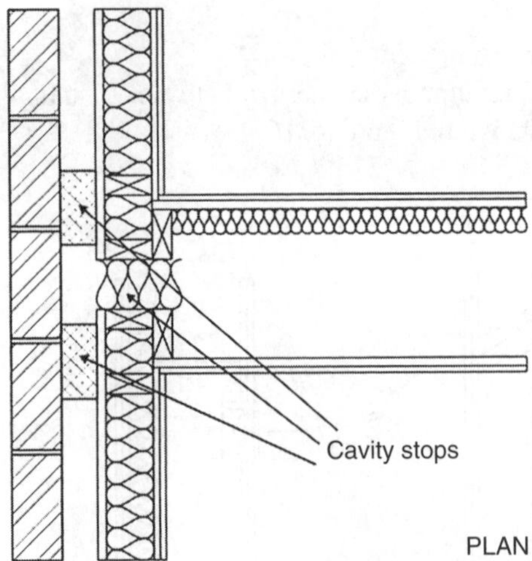

Figure 6.117 Junctions with an external solid masonry wall

Junction with ceiling and roof space

The wall should preferably be continuous to the underside of the roof.	E2.160
The junction between the separating wall and the roof should be filled with a flexible closer.	E2.161
The junction between the ceiling and the wall linings should be sealed with tape or caulked with sealant.	E2.162
If the roof or loft space is not a habitable room (and there is a ceiling with a minimum mass per unit area of 10 kg/m^2 with sealed joints), then:	
• either the linings on each frame may be reduced to two layers of plasterboard, each sheet with a minimum mass per unit area of 10 kg/m^2; or	E2.162a
• the cavity may be closed at ceiling level without connecting the two frames rigidly together.	E2.162b
Note: In which case there need only be one frame in the roof space provided there is a lining of two layers of plasterboard, each sheet of minimum mass per unit area of 10 kg/m^2, on both sides of the frame.	
External wall cavities should be closed at eaves level with a suitable material.	E2.163

Junctions with timber ground floors

Air paths through the wall into the cavity shall be blocked using solid timber blockings, continuous ring beam or joists.	E2.156

See also Building Regulation Part C – Site preparation and resistance to moisture, and Building Regulation Part L – Conservation of fuel and power.

Junctions with concrete ground floors

If the ground floor is a concrete slab laid on the ground, it may be continuous under a type 4 separating wall. If the ground floor is a suspended concrete floor, it may only pass under a wall type 4 if the floor has a mass per unit area of at least 365 kg/m^2.	E2.158

See also Building Regulation Part C – Site preparation and resistance to moisture, and Building Regulation Part L – Conservation of fuel and power.

Junctions with internal timber floors

Air paths through the wall into the cavity shall be blocked using solid timber blockings, continuous ring beam or joists.	E2.154

Junctions with internal concrete floors

No official guidance currently available. Best to seek specialist advice.	E2.155

Junctions with internal framed walls

There are no restrictions on internal framed walls meeting a type 4 separating wall.	E2.152

Junctions with internal masonry walls

There are no restrictions on internal masonry walls meeting a type 4 separating wall.	E2.153

Junctions with an external solid masonry wall

No official guidance currently available. Best to seek specialist advice. E2.151

Junctions with an external cavity wall with masonry inner leaf

No official guidance currently available. Best to seek specialist advice. E2.148

Walls adjacent to hearths

Walls that are not part of a fireplace recess or a prefabricated appliance chamber but are adjacent to hearths or appliances also need to protect the building from catching fire. A way of achieving the requirement is shown in Figure 6.118. J (2.31)

See also p. 327, Appendix A (The Use of Robust Standards)

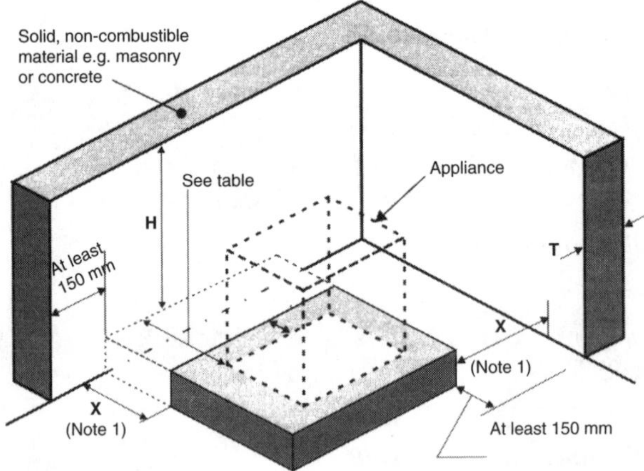

Location of hearth or appliance	Solid, non-combustible material	
	Thickness (T)	Height (H)
Where the hearth abuts a wall and the appliance is not more than 50 mm from the wall	200 mm	At least 300 mm above the appliance and 1.2 m above the hearth
Where the hearth abuts a wall and the appliance is more than 50 mm but not more than 300 mm from the wall	75 mm	At least 300 mm above the appliance and 1.2 m above the hearth
Where the hearth does not abut a wall and is no more than 150 mm from the wall (see Note 1)	75 mm	At least 1.2 m above the hearth
Note 1: There is no requirement for protection of the wall where X is more than 150 mm		

Figure 6.118 Walls adjacent to hearths

6.8.3 Conservation of energy and power

The walls (doors and windows) between the building and the extension should be insulated and weather-stripped to at least the same extent as in the existing building.	L1B 22a L2B 32a
The area-weighted U-value for each element shall be less than 0.35 W/m^{2}K and no more than 0.70 for the worst individual sub element.	L1B 18
Newly constructed thermal elements that are part of an extension should be less than 0.30 W/m^{2}K.	L1B 50 L2B 70a
Any retained thermal element with a U-value worse than the threshold value of 0.70 W/m^{2}K shall be upgraded to achieve 0.55 W/m^{2}K.	L1B 57
Thermal elements constructed as replacements for existing elements (or elements that are being renovated) should be less than 0.35 W/m^{2}K.	L1B 51 and 54 L2B 86 and 88
When fixed building services are provided or extended in an extension:	L2B 29b
• the area weighted U-value for each element type shall be less than 0.35 W/m^{2}K and	
• the U-value of any individual element should be no worse than 0.70 W/m^{2}K.	L2B 29c

6.9 Ceilings

6.9.1 The requirement

As a fire precaution, all materials used for internal linings of a building should have a low rate of surface flame spread and (in some cases) a low rate of heat release.

(Approved Document B2)

Dwellings shall be designed so that the noise from domestic activity in an adjoining dwelling (or other parts of the building) is kept to a level that:

- *does not affect the health of the occupants of the dwelling;*
- *will allow them to sleep, rest and engage in their normal activities in satisfactory conditions.*

(Approved Document E1)

Dwellings shall be designed so that any domestic noise that is generated internally does not interfere with the occupants' ability to sleep, rest and engage in their normal activities in satisfactory conditions.

(Approved Document E2)

Domestic buildings shall be designed and constructed so as to restrict the transmission of echoes.

(Approved Document E3)

Schools shall be designed and constructed so as to reduce the level of ambient noise (particularly echoing in corridors).

(Approved Document E4)

6.9.2 Meeting the requirements

Suspended ceilings

Table 6.44 sets out criteria appropriate to the suspended ceilings that can be accepted as contributing to the fire resistance of a floor.

Table 6.44 Limitations on fire protected suspended ceilings

Height of building or separated part	Type of floor	Provision for fire resistance of floor	Description of suspended ceiling
<18 m	Not compartment	60 mins or less	Type A, B, C or D
	Compartment	<60 mins	Type A, B, C or D
		60 mins	Type B, C or D
18 m or more	Any	60 mins or less	Type C or D
No limit	Any	60 mins	Type D

Ceilings – General

The resistance to airborne and impact sound depends on the independence and isolation of the ceiling and the type of material used. E

Three ceiling treatments (which are ranked in order of sound insulation) may be used: E3

- Ceiling treatment A – independent ceiling with absorbent material;
- Ceiling treatment B – plasterboard on proprietary resilient bars with absorbent material;
- Ceiling treatment C – plasterboard on timber battens (or proprietary resilient channels) with absorbent material.

If the roof or loft space is not a habitable room (and provided that there is a ceiling with a minimum mass per unit area of 10 kg/m^2 with sealed joints and the cavity masonry core is maintained to the underside of the roof) then: E

- the mass per unit area of the separating wall above the ceiling may be reduced to 150 kg/m^2;
- the independent panels may be omitted in the roof space;

- the linings on each frame may be reduced to two layers of plasterboard or the cavity may be closed at ceiling level without connecting the two frames rigidly together.

All junctions between ceilings and independent panels (and joints between casings and ceiling) should be sealed with tape or caulked with sealant. E

At junctions with external cavity walls (with masonry inner leaf) the ceiling should be taken through to the masonry. E3

The ceiling void and roof space detail can only be used where the requirements of Building Regulation Part B – Fire safety can also be satisfied. E3

If there is an existing lath and plaster ceiling it should be retained as long as it satisfies Building Regulation Part B – Fire safety. E3

If the existing ceiling is not lath and plaster, it should be upgraded to provide: E4

- at least two layers of plasterboard with staggered joints;
- a minimum total mass per unit area 20 kg/m^2;
- an absorbent layer of mineral wool laid on the ceiling (minimum thickness 100 mm, minimum density 10 kg/m^3);
- plasterboard with joints staggered, total mass per unit area 20 kg/m^2.

Care should be taken at the design stage to ensure that adequate ceiling height is available in all rooms to be treated.

The ceiling should be supported by either: E2

- independent joists fixed only to the surrounding walls; or
- independent joists fixed to the surrounding walls with additional support provided by resilient hangers attached directly to the existing floor base.

Note: A clearance of at least 25 mm should be left between the top of the independent ceiling joists and the underside of the existing floor construction.

Where a window head is near to the existing ceiling, the new independent ceiling may be raised to form a pelmet recess. E4

A rigid or direct connection should not be created between an independent ceiling and the floor base. E4

Where the roof or loft space is not a habitable room (and there is a ceiling with a minimum mass per unit area of 10 kg/m^2 with sealed joints) then the mass per unit area of the separating wall above the ceiling may be reduced to 150 kg/m^2. E2.57

If lightweight aggregate blocks of density less than 1200 kg/m^3 are used above ceiling level, then one side should be sealed with cement paint or plaster skim.	E2.58 E2.94 E2.138
Where the external wall is a cavity wall with a masonry inner leaf (or a simple cavity masonry wall or masonry between independent panels), the ceiling should be taken through to the masonry.	E3.105 E3.125 E3.125
Where a window head is near to the existing ceiling, the new independent ceiling may be raised to form a pelmet recess.	E4.29
A rigid or direct connection should not be created between the independent ceiling and the floor base.	E4.30

Ceiling joists

Softwood timber used for roof construction or fixed in the roof space (including ceiling joists within the void spaces of the roof) should be adequately treated to prevent infestation by the house longhorn beetle (*Hylotrupes bajulus* L.), particularly in the following areas: A1/2 2B2

- the Borough of Bracknell Forest, in the parishes of Sandhurst and Crowthorne;
- the Borough of Elmbridge
- the District of Hart, in the parishes of Hawley and Yateley;
- the District of Runnymede;
- the Borough of Spelthorne;
- the Borough of Surrey Heath;
- the Borough of Rushmoor, in the area of the former district of Farnborough;
- the Borough of Woking.

Note: Guidance on suitable preservative treatments is given within the *British Wood Preserving and Damp-Proofing Association's Manual* (2000 revision), available from 1 Gleneagles House, Vernongate, South Street, Derby DE1 1UP.

Note: Guidance on the sizing of certain members in floors and roofs is given in BS 5268. Part 2: 2002 and Part 3: 1998 as 'Span tables for solid timber members in floors, ceilings and roofs (excluding trussed rafter roofs) for dwellings', published by TRADA, available from Chiltern House, Stocking Lane, Hughenden Valley, High Wycombe, HP14 4ND, Bucks.

Suspended ceilings

Table 6.45 shows criteria appropriate to the suspended ceilings that can be accepted as contributing to the first resistance of a floor.

Table 6.45 Limitations on fire protected suspended ceilings

Height of building or separated part	Type of floor	Provision for fire resistance of floor	Description of suspended ceiling
<18 m	Not compartment	60 mins or less	Type A, B, C or D
	Compartment	<60 mins	Types A, B, C, or D
		60 mins	Type B, C, or D
18 m or more	Any	60 mins or less	Type C, or D
No limit	Any	60 mins	Type D

Air circulation systems

Transfer grilles of air circulation systems should **not** be fitted in any ceiling enclosing a protected stairway.	B1 2.17 (V1) B1 2.18 (V2) B3 7.10 (V1) B3 10.2 (V2)

Ceiling Linings

To inhibit the spread of fire within the building, ceiling internal linings shall:

- adequately resist the spread of flame over their surfaces; and
- have, if ignited, a rate of heat release or a rate of fire growth, which is reasonable in the circumstances.

Note: Flame spread over wall or ceiling surfaces is controlled by ensuring that the lining materials or products meet given performance levels which are measured in terms of performance with reference to Tables A1 and A3 of Part B.

For the purpose of this requirement, ceilings:

Include	Do not include
• the surface of glazing • any part of a wall which slopes at an angle of 70° or less to the horizontal • the underside of a gallery • the underside of a roof exposed to the room below	• trap doors and their frames • the frames of windows or rooflights and frames in which glazing is fitted • architraves, cover moulds, picture rails • exposed beams and similar narrow members

Classification of linings

In general terms, (but see paragraphs 3.2 to 3.14 of Part B V1 and 6.2 to 6.14 of Part B V2 for more details) the surface linings for ceilings should meet the following classifications:

Table 6.46 Classification of linings

Location	National Class	European Class
Small rooms less than 4 m^2	3	D-s3, d2
Domestic garages less than 40 m^2	3	D-s3, d2
Other rooms (including garages)	1	C-s3, d2
Circulation spaces within dwelling houses	1	C-s3, d2

Fire-resisting ceilings

The need for cavity barriers in some concealed floor or roof spaces can be reduced by using a fire resisting ceiling below the cavity. B2 3.6 (V1) B2 6.6 (V2)

Heat alarms

Heat detectors and heat alarms should be:

- designed and installed in accordance with BS 5839-6:2004; B1 1.10
- sited so that the sensor in ceiling-mounted devices is between 25 mm and 150 mm below the ceiling; B1 1.15c
- mains-operated and conform to BS 5446-2:2003; and B1 1.4 (V1)
- have a standby power supply such as a rechargeable (or non rechargeable) battery. B1 1.5 (V2)

Lighting diffusers

Thermoplastic lighting diffusers (i.e. translucent or open structured elements that allow light to pass through) should not be used in fire-protecting or fire-resisting ceilings, unless they have been satisfactorily tested as part of the ceiling system that is to be used to provide the appropriate fire protection. B2 3.12 (V1) B2 6.14 (V2)

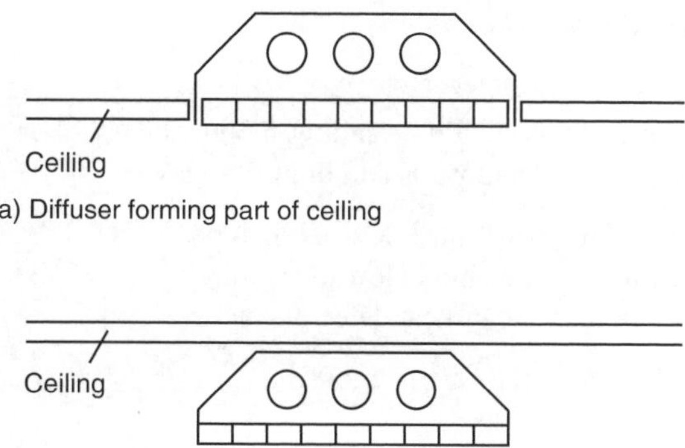

Figure 6.119 Lighting diffuser in relation to ceiling

Rooflights

Rooflights should meet the relevant classification in Table 6.46.	B2 3.7 (V1) B2 6.7 (V2)
Rooflights may be constructed of a thermoplastic material if: • the lower surface has a TPI(a) (rigid) or a TP(b) classification; • the size and disposition of the rooflights accords with the limits in Table 6.47.	B2 3.10 (V1) B2 6.12 (V2)

Table 6.47 Limitations applied to thermoplastic rooflights and lighting diffusers in suspended ceilings and Class 3 plastic rooflights

Minimum classification of lower surface	Use of space below the diffusers or rooflight	Maximum area of each diffuser panel or rooflight	Maximum total area of diffuser panels and rooflights as %age of floor area of the space in which the ceiling is located	Minimum separation distance between diffuser panels or rooflights
TP(a)	Any (except a protected stairway)	No limit	No limit	No limit
Class 3 or TP(b)	Rooms	5	50	3
	Circulation spaces (except protected stairways)	5	15	3

Smoke alarms

Smoke alarm systems should be ceiling-mounted and at least 300 mm from walls and light fittings.	B1 1.15b (V1) B1 1.14b (V2)
The sensor in ceiling-mounted devices shall be between 25 mm and 600 mm below the ceiling (25–150 mm in the case of heat detectors or heat alarms).	B1 1.15c (V1) B1 1.14c (V2)

Suspended ceilings

A suspended, fire protected, ceiling should meet the requirements of Table 6.48.

Table 6.48 Limitations on fire-protecting suspended ceilings

Height of building or separated part (m)	Type of floor	Provision for fire resistance of floor (minutes)
Less than 18	Not compartment	60 or less
	Compartment	Less than 60
18 or more	Any	60 or less
No limit	Any	More than 60

for further details see Part B, Appendix A, Table A3.

Suspended or stretched-skin ceilings

The ceiling of a room may be constructed from panels of a thermoplastic material of the TP(d) flexible classification, provided that it is not part of a fire-resisting ceiling.	B2 3.14 (V1) B2 6.16 (V2)
Each panel should not exceed 5 m^2 in area and should be supported on all of its sides.	

Thermoplastic materials

Thermoplastic materials may be used in windows, rooflights and lighting diffusers in suspended ceilings	B2 3.8 (V1) B2 6.10 (V2)
Flexible thermoplastic material may be used in panels to form a suspended ceiling	B2 3.8 (V1) B2 6.10 (V2)
Flexible membranes covering a structure shall be in accordance with Appendix A of BS 7157:1989.	B2 6.8 (V2)

Venting of heat and smoke from basements

Smoke outlets should be sited at high level, either in the ceiling or in the wall of the space they serve.	B5 18.3 (V2) B5 18.5 (V2) B5 18.7 (V2)

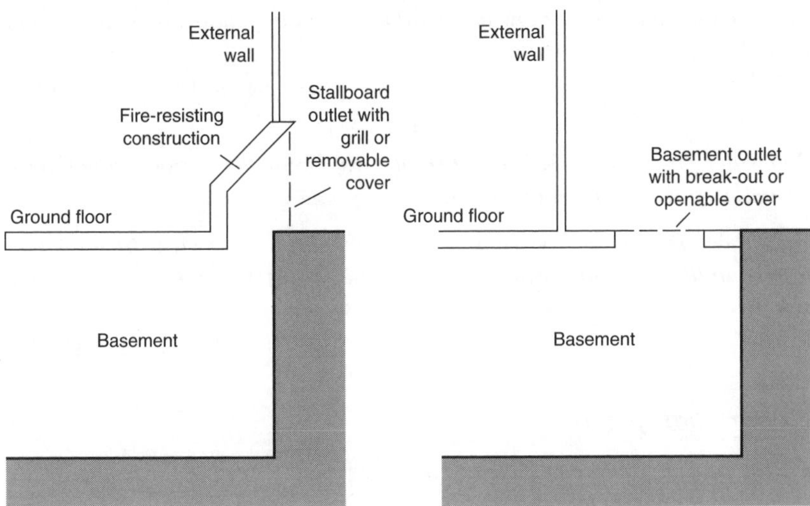

Figure 6.120 Fire-resisting construction for smoke outlet shafts

6.10 Roofs

The roof of a brick-built house is normally an aitched (sloping) roof comprising rafters fixed to a ridge board, braced by purlins, struts and ties and fixed to wall plates bedded on top of the walls. They are then usually clad with slates or tiles to keep the rain out.

Timber-framed houses usually have trussed roofs – prefabricated triangulated frames that combine the rafters and ceiling joists – which are lifted into place and supported by the ails. The trusses are joined together with horizontal and diagonal ties. A ridge board is not fitted, nor are purlins required. Roofing felt battens and tiling are applied in the usual way.

6.10.1 Requirements

Structure

The building shall be constructed so that the combined dead, imposed and wind loads are sustained and transmitted by it to the ground.

(Approved Document A1)

As a fire precaution, all materials used for internal linings of a building should have a low rate of surface flame spread and (in some cases) a low rate of heat release.

(Approved Document B2)

The roof of the building shall be resistant to the penetration of moisture from rain or snow to the inside of the building.

All floors next to the ground, walls and roof shall not be damaged by moisture from the ground, rain or snow and shall not carry that moisture to any part of the building which it would damage.

(Approved Document C2)

Rainwater from roofs shall be carried away from the surface either by a drainage system or by other means.

The rainwater drainage system shall carry the flow of rainwater from the roof to an outfall (e.g. a soakaway, a watercourse, a surface water or a combined sewer).

(Approved Document H3)

External fire spread

- *The roof shall be constructed so that the risk of spread of flame and/or fire penetration from an external fire source is restricted.*
- *The risk of a fire spreading from the building to a building beyond the boundary, or vice versa shall be limited.*

(Approved Document B4)

Internal fire spread (structure)

- *Ideally the building should be sub-divided by elements of fire-resisting construction into compartments.*
- *All openings in fire-separating elements shall be suitably protected in order to maintain the integrity of the continuity of the fire separation.*
- *Any hidden voids in the construction shall be sealed and sub-divided to inhibit the unseen spread of fire and products of combustion, in order to reduce the risk of structural failure, and the spread of fire.*

(Approved Document B3)

Ventilation

There shall be adequate means of ventilation provided for people in the building.

(Approved Document F)

Conservation of fuel and power

Reasonable provision shall be made for the conservation of fuel and power in buildings by:

(a) limiting heat gains and losses:
  - *(i) through thermal elements and other parts of the building fabric; and*
  - *(ii) from pipes, ducts and vessels used for space heating, space cooling and hot water services;*

(b) providing and commissioning energy-efficient fixed building services with effective controls; and

(c) providing to the owner sufficient information about the building, the fixed building services and their maintenance requirements so that the building can be operated in such a manner as to use no more fuel and power than is reasonable in the circumstances.

(Approved Document L1)

Safety

Pedestrian guarding should be provided for any roof to which people have access.

(*Approved Document K2*)

6.10.2 Meeting the requirements

Precipitation

Roofs should:

- resist the penetration of precipitation to the inside of the building; C6.2a
- not be damaged by precipitation; C6.2b
- not carry precipitation to any part of the building which would be damaged by it; C6.2b
- be designed and constructed so that their structural and thermal performance are not adversely affected by interstitial condensation. C6.2c

Resistance to moisture from the outside

Roofs should be designed so as to protect the building from precipitation either by holding the precipitation at the face of the roof or by stopping it from penetrating beyond the back of the roofing system. C6.3

Roofs that are jointless or have sealed joints should be impervious to moisture. C6.4a

Roofs that have overlapping dry joints should be weather resistant and backed by a material (such as roofing felt) to direct any precipitation that does enter the roof towards the outer face. C6.4b

Materials that can deteriorate rapidly without special care should only be used as the weather-resisting part of a roof. C6.5

Weather-resistant parts of a roofing system shall not include paint or include any coating, surfacing or rendering which will not itself provide all the weather resistance. C6.5

Roofing systems may be:

- impervious – such as metal, plastic and bituminous products; C6.6a
- weather-resistant – such as natural stone or slate, cement-based products, fired clay and wood; C6.6b
- moisture-resisting – such as bituminous and plastic products lapped at the joints; C6.6c
- jointless materials and sealed joints – that would allow for structural and thermal movement. C6.6d

Dry joints between roofing sheets should be designed so that precipitation will not pass through them. C6.7

Any precipitation that does enter a joint shall be drained away without penetrating beyond the back of the roofing system. C6.7

Each sheet, tile and section of roof should be fixed in accordance with the guidance contained in BS 8000-6: 1990. C6.8

Resistance to damage from interstitial condensation

Roofs shall be designed and constructed in accordance with Clause 8.4 of BS 5250: 2002 and BS EN ISO 13788: 2001. C6.10

Cold deck roofs (i.e. those roofs where the moisture from the building can permeate the insulation) shall be ventilated. C6.11

Any parts of a roof which have a pitch of 70° or more shall be insulated as though it were a wall. C6.11

All gaps and penetrations for pipes and electrical wiring should be filled and sealed to avoid excessive moisture transfer to roof voids. C6.12

An effective draught seal should be provided to loft hatches to reduce inflow of warm air and moisture. C6.12

Specialist advice should be sought when designing swimming pools and other buildings where interstitial condensation in the walls (caused by high internal temperatures and humidities) can cause high levels of moisture being generated. C6.13

Resistance to surface condensation and mould growth

Roofs shall be designed and constructed so that the:

- thermal transmittance (U-value) does not exceed 0.35 W/m^{2}K at any point; C6.14a
- junctions between elements and the details of openings (such as windows) are in accordance with the recommendations in the report on robust construction details. C6.14b

Building height

For residential buildings, the maximum height of the building measured from the lowest finished ground level adjoining the building to the highest point of any roof should not be greater than 15 m. A1/2 2C4i

General

Roofs shall be constructed so that they: A1/2 1A2d

- provide local support to the walls;
- act as horizontal diaphragms capable of transferring the wind forces to buttressing elements of the building.

Note: A traditional cut timber roof (i.e. using rafters, purlins and ceiling joists) generally has sufficient built-in resistance to instability and wind forces. However, the need for diagonal rafter bracing equivalent to that recommended in BS 5268: Part 3: 1998 or Annex H of BS 8103: Part 3: 1996 for trussed rafter roofs, should be considered especially for single-hipped and non-hipped roofs of greater than 40° pitch to detached houses.

Roofs should:

- act to transfer lateral forces from walls to buttressing walls, piers or chimneys; A1/2 2C33a
- be secured to the supported wall. A1/2 2C33b

The roof shall be braced (in accordance with BS 5268: Part 3): A1/2 2C38(i)h

- at rafter level;
- horizontally at eaves level;
- at the base of any gable by roof decking, rigid sarking or diagonal timber bracing (as appropriate).

Vertical strapping may be omitted if the roof: A1/2 2C36a–d

- has a pitch of 15° or more; and
- is tiled or slated; and
- is of a type known by local experience to be resistant to wind gusts; and
- has main timber members spanning onto the supported wall at not more than 1.2 m centres.

Gable walls should be strapped to roofs as shown in Figure 6.121(a) and (b) by tension straps. A1/2 2C36

Walls shall be tied to the roof structure vertically and horizontally and have a horizontal lateral restraint at roof level. A1/2 2C38(i)i

Wall ties should also be provided, spaced not more than 300 mm apart vertically, within a distance of 225 mm from the vertical edges of all roof verges. A1/2 2C8

Walls shall be tied to the roof structure vertically and horizontally and have a horizontal lateral restraint at roof level. A1/2 2C38(i)i

Walls should be tied horizontally at no more than 2 m centres to the roof structure at eaves level, base of gables and along roof slopes (as shown in Figure 6.122) with straps.	A1/2 2C38(iv)
Isolated columns should also be tied to the roof structure (see Figure 6.122).	A1/2 2C38(iv)
The roof structure of an annexe shall be secured to the structure of the main building at both rafter and eaves level.	A1/2 2C38(1)j
Access to the roof shall only be for the purposes of maintenance and repair.	A1/2 2C38(1)d

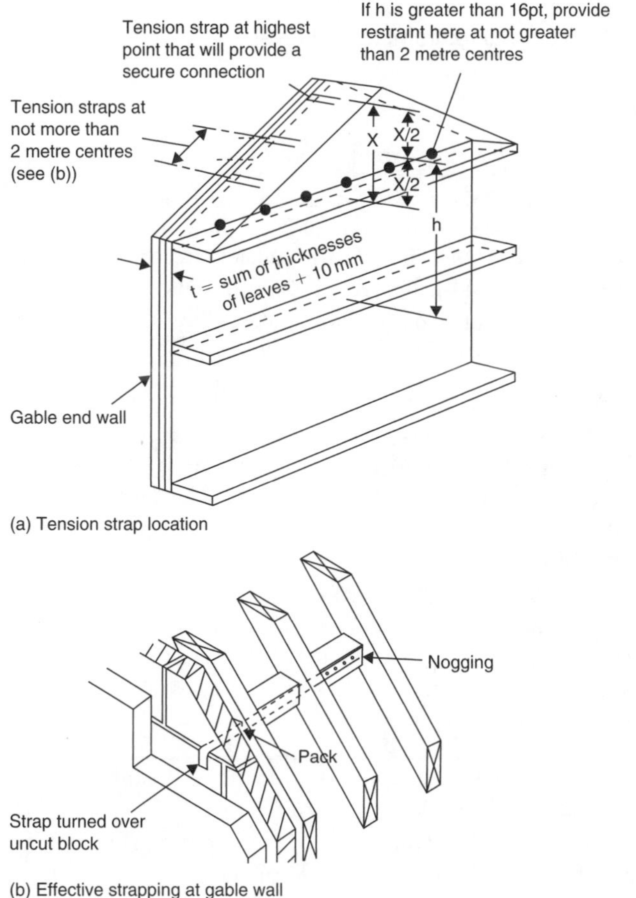

Figure 6.121 Lateral support at roof level

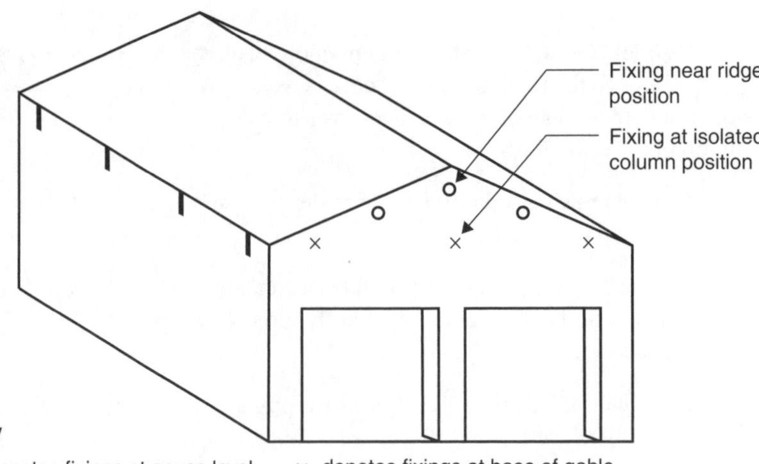

Figure 6.122 Lateral restraint at roof level

Timber

Softwood timber used for roof construction or fixed in the roof space (including ceiling joists within the void spaces of the roof), should be adequately treated to prevent infestation by the house longhorn beetle (*Hylotrupes bajulus* L.), particularly in the following areas: A1/2 2B2

- The Borough of Bracknell Forest, in the parishes of Sandhurst and Crowthorne
- The Borough of Elmbridge
- The District of Hart, in the parishes of Hawley and Yateley
- The District of Runnymede
- The Borough of Spelthorne
- The Borough of Surrey Heath
- The Borough of Rushmoor, in the area of the former district of Farnborough
- The Borough of Woking.

Note: Guidance on suitable preservative treatments is given within the *British Wood Preserving and Damp-Proofing Association's Manual* (2000 revision), available from 1 Gleneagles House, Vernongate, South Street, Derby DE1 1UP.

Note: Guidance on the sizing of roof members is given in BS 5268: Part 2: 2002 & Part 3: 1998 as well as 'Span tables for solid timber members in floors, ceilings and roofs (excluding trussed rafter roofs) for dwellings', published by TRADA (available from Chiltern House, Stocking Lane, Hughenden Valley, High Wycombe, Bucks HP14 4ND).

Openings

Where an opening in a roof for a stairway adjoins a supported wall and interrupts the continuity of lateral support:

- the maximum permitted length of the opening is to be 3 m, measured parallel to the supported wall; A1/2 2C37a
- connections (if provided by means other than by anchor) should be throughout the length of each portion of the wall situated on each side of the opening; A1/2 2C37b
- connections via mild steel anchors should be spaced closer than 2 m on each side of the opening to provide the same number of anchors as if there were no opening; A1/2 2C37c
- there should be no other interruption of lateral support. A1/2 2C37d

Means of escape

A flat roof being used as a means of escape should: C1

- be part of the same building from which escape is being made;
- should lead to a storey exit or external escape route;
- should provide 30 minutes' fire resistance.

Where a balcony or flat roof is provided for escape purposes guarding may be needed (see also Approved Document K, Protection from falling, collision and impact).

Construction of escape stairs

If an escape route is over flat roof:	B1 2.10 (V1)
• the roof should be part of the same building from which escape is being made;	B1 2.7 (V2)
• the route across the roof should lead to a storey exit or external escape route;	B1 5.35 (V2)
• the part of the roof forming the escape route, its supporting structure, together with any opening within 3 m of the escape route, should provide 30 minutes fire resistance (see Appendix A, Table Al of Approved Document B for fire resistance figures for elements of structure etc.);	
• the route should be adequately defined and guarded by walls and/or protective barriers (which meet the provisions of Part K).	B1 2.11 (V1)

Note: Part K (Protection from falling, collision and impact) specifies a minimum guarding height of 800 mm, except in the case of a window in a roof where the bottom of the opening may be 600 mm above the floor.

An escape over a flat roof is permissible if:	B1 2.31 (V2)
• more than one escape route is available from a storey, or part of a building;	
• the route does not serve as an Institutional building;	
• part of a building is intended for use by members of the public;	
• the roof is fire-resistant in accordance with Tables Al and A2 of Appendix A to Part B;	B1 5.3 (V2)
• the exit from a flat etc. is remote from the main entrance door to that flat.	B1 2.17 (V2)

Provision of refuges

Refuges are relatively safe waiting areas for short periods. They are **not** areas where disabled people should be left alone indefinitely until rescued by the fire and rescue service, or until the fire is extinguished.

A refuge such as a flat roof (balcony, podium or similar compartment, protected lobby, protected corridor or protected stairway) should be provided for each protected stairway.	B1 4.8 (V2)

Note: The number of refuge spaces need not necessarily equal the sum of the number of wheelchair users who can be present in the building.

Compartmentation

To prevent the spread of fire within a building, whenever possible, the building should be sub-divided into compartments separated from one another by walls and/or floors of fire-resisting construction.	B3 5.1 (V1) B3 8.1 (V2)
Compartment walls in a top storey beneath a roof should be continued through the roof space.	B3 5.8 (V1) B3 8.24 (V2)
When a compartment wall meets the underside of the roof covering or deck, the wall/roof junction shall maintain continuity of fire resistance'	B3 8.28 (V2)
Double skinned insulated roof sheeting should incorporate a band of material of limited combustibility.	B3 8.29 (V2)

Compartment walls between buildings

If a fire penetrates a roof near a compartment wall there is a risk that it will spread over the roof to the adjoining compartment. To reduce this risk either: • the wall should be extended up through the roof for a height of at least 375 mm above the top surface of the adjoining roof covering; or • a 1500 mm wide zone on either side of the wall should have a suitable covering (see Figure 6.123).	B3 5.12 (V1) B3 8.29 (V2) B3 8.30 (V2)

Concealed spaces (cavities)

Concealed spaces or cavities in roofs will provide an easy route for smoke and flame spread which, because it is obscured, will present a greater danger than would a more obvious weakness in the fabric of the building. For this reason:

• cavity barriers need not be provided between double-skinned corrugated or profiled insulated roof sheeting, if the sheeting is a material of limited combustibility.	B3 6.4 (V1) B3 9.5 (V2)

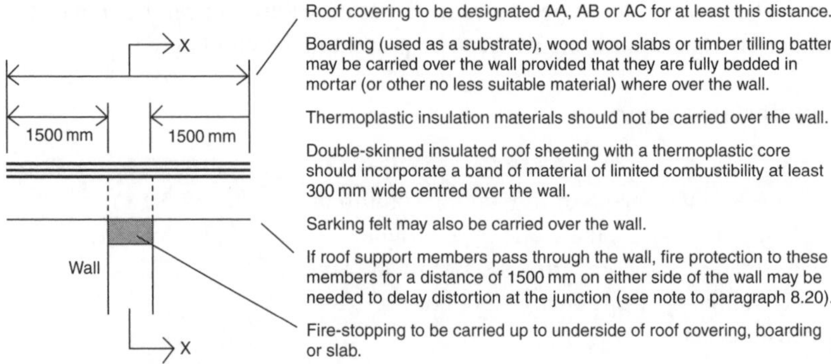

X
1500 mm
1500 mm
Wall
X
Roof covering to be designated AA, AB or AC for at least this distance.
Boarding (used as a substrate), wood wool slabs or timber tilling battens may be carried over the wall provided that they are fully bedded in mortar (or other no less suitable material) where over the wall.
Thermoplastic insulation materials should not be carried over the wall.
Double-skinned insulated roof sheeting with a thermoplastic core should incorporate a band of material of limited combustibility at least 300 mm wide centred over the wall.
Sarking felt may also be carried over the wall.
If roof support members pass through the wall, fire protection to these members for a distance of 1500 mm on either side of the wall may be needed to delay distortion at the junction (see note to paragraph 8.20).
Fire-stopping to be carried up to underside of roof covering, boarding or slab.

Figure 6.123 Junction of compartment wall with roof

Note: Separate rules exist for bedrooms in Institutional and Other Residential buildings (see B3 9.7 (V2)).

Roof covering

The re-covering of roofs is commonly undertaken to extend the useful life of buildings; however, roof structures may be required to carry underdrawing or insulation provided at a time later than their initial construction.

All materials used to cover roofs (including transparent or translucent materials, but excluding windows of glass in residential buildings with roof pitches of not less than 15°) shall be capable of safely withstanding the concentrated imposed loads upon roofs specified in BS 6399 Pt 3. A1/2 4.1

Where the work involves a significant change in the applied loading the structural integrity of the roof structure and the supporting structure should be checked to ensure that upon completion of the work the building is not less compliant. A1/2 4.3

Note: Re-covering roofs is commonly undertaken to extend the useful life of buildings (for example, roof structures may be required to be insulated at a later date).

Where such checking of the existing roof structure indicates that the construction is unable to sustain any proposed increase in loading (e.g. due to overstressed members or unacceptable deflection leading to ponding), appropriate strengthening work or replacement of roofing members should be undertaken. A1/2 4.5

This is classified as a material alteration.

Requirement	Reference
Where work will significantly decrease the roof dead loading, the roof structure and its anchorage to the supporting structure should be checked to ensure that an adequate factor of safety is maintained against uplift of the roof under imposed wind loading. **Note:** A significant change in roof loading is when the loading upon the roof is increased by more than 15%.	A1/2 4.7
Plastic rooflights should have a minimum of class 3 lower surface.	B4 10.6 (V1) B4 14.6 (V2)
When used in rooflights, unwired glass shall be at least 4 mm thick and shall be AA designated (see Table 6.50).	B4 10.8 (V1) B4 14.8 (V2)
Thatch and wood shingles should be regarded as having an AD/BD/CD designation (see Table 6.50).	B4 10.9 (V1) B4 14.9 (V2)
Separation distances (i.e. the minimum distance from the roof, or part of the roof, to the relevant or notional boundary) shall be in accordance with Part B, Volume 1, Table 5 according to the type of roof covering and the size and use of the building.	B4 10.5 (V1) B4 14.5 (V2)
A rigid thermoplastic sheet product made from polycarbonate or from unplasticized PVC may be used in rooflights.	B4 10.7 (V1) B4 14.7 (V2)
Unwired glass at least 4 mm thick may be used in rooflights.	B4 10.8 (V1) B4 14.8 (V2)

Note: See Part B, Volume 1, Table 5 for limitations on roof coverings and BS EN 13501-5:2005 for guidance on roof coverings (which has been updated to incorporate the new European classification system).

Rooflights

Requirement	Reference
Rooflights should meet the relevant classification in Table 6.46).	B2 3.7 (V1) B2 6.7 (V2)
Thermoplastic materials may be used in windows, rooflights and lighting diffusers in suspended ceilings.	B2 3.8 (V1) B2 6.10 (V2)
Rooflights may be constructed of a thermoplastic material if:	B2 3.10 (V1)
• the lower surface has a TPI(a) (rigid) or a TP(b) classification	B2 6.12 (V2)

- the size and disposition of the rooflights accords with the limits in Table 6.49.

Table 6.49 Limitations applied to thermoplastic rooflights and lighting diffusers in suspended ceilings and Class 3 plastic rooflights

Minimum classification of lower surface	Use of space below the diffusers or rooflight	Maximum area of each diffuser panel or rooflight	Maximum total area of diffuser panels and rooflights as %age of floor area of the space in which the ceiling is located	Minimum separation distance between diffuser panels or rooflights
TP(a)	Any (except a protected stairway)	No limit	No limit	No limit
Class 3	Rooms	5	50	3
or TP(b)	Circulation spaces (except protected stairways)	5	15	3

Thatched roofs

In thatched roofs:

- the rafters should be overdrawn with construction having not less than 30 minutes fire resistance; B4 10.9 (V1)
- smoke alarms should be installed in the roof space. B4 14.9 (V2)

Fire resistance

The need for cavity barriers in some concealed roof spaces can be reduced by using a fire resisting ceiling below the cavity. B2 3.6 (V1) B2 6.6 (V2)

Where the conversion of an existing roof space (such as a loft conversion to a two-storey house) means that a new storey is going to be added, then the stairway will need to be protected with fire-resisting doors and partitions. B1 2.20b

Table 6.50 Limitations on roof coverings

Designation of roof covering	Minimum distance from any point on relevant boundary			
	Less than 6 m	At least 6 m	At least 12 m	At least 20 m
AA, AB or AC	Acceptable	Acceptable	Acceptable	Acceptable
BA, BB or BC	Not acceptable	Acceptable	Acceptable	Acceptable
CA, CB or CC	Not acceptable	Acceptable	Acceptable	Acceptable
AD, BD or CD	Not acceptable	Acceptable	Acceptable	Acceptable
DA, DB, DC or DD	Not acceptable	Not acceptable	Not acceptable	Acceptable

Pitched roofs covered with slates or tiles

Covering material	Supporting structure	Designation
1. Natural slates 2. Fibre reinforced cement slates 3. Clay tiles 4. Concrete tiles	Timber rafters with or without underfelt, sarking, boarding, woodwool slabs, compressed straw slabs, plywood, wood chipboard, or fibre insulating board.	AA

Although the table does not include guidance for pitched roofs covered with bitumen felt, it should be noted that there is a wide range of materials on the market and information on specific products is readily available from manufacturers.

Pitched roofs covered with self-supporting sheet

Roof covering material	Construction	Supporting structure	Designation
Profiled sheet of galvanized steel, aluminium, fibre reinforced cement, or pre-painted (coil coated) steel or aluminium with a PVC or PVF2 coating	Single skin without underlay, or with underlay or plasterboard, or woodwool slab	Structure of timber, steel or concrete	AA
Profiled sheet of galvanized steel, aluminium, fibre reinforced cement, or pre-painted (coil coated) steel or aluminium with a PVC or PVF2 coating	Double skin without interlayer, or with interlayer of resin bonded or concrete glass fibre, mineral wool slab, polystyrene, or polyurethane	Structure of timber, steel or concrete	AA

Flat roofs with bitumen felt

A flat roof consisting of bitumen felt should (irrespective of the felt specification) be deemed to be of designation AA if the felt is laid on a deck constructed of 6 mm plywood, 12.5 mm wood chipboard, 16 mm (finished) plain-edged timber boarding, compressed straw slab, screeded woodwool slab, profiled fibre reinforced cement or steel deck (single or double skin) with or without fibre insulating board overlay, profiled aluminium deck (single or double skin) with or without fibre insulating board overlay, or concrete or clay pot slab (in situ or pre-cast), and has a surface finish of:

(a) bitumen-bedded stone chippings covering the whole surface to a depth of at least 12.5 mm;
(b) bitumen-bedded tiles of a non-combustible material;
(c) sand and cement screed; or
(d) tarmacadam.

Pitched or flat roofs covered with fully supported materials

Covering material	Supporting structure	Designation
1. Aluminium sheet 2. Copper sheet 3. Zinc sheet	Timber joists and tongued and grooved boarding, or plain-edged boarding	AA
4. Lead sheet 5. Mastic asphalt 6. Vitreous enamelled steel 7. Lead/tin alloy-coated steel sheet 8. Zinc/aluminium alloy-coated steel sheet	Steel or timber joists with deck of woodwool slabs, compressed straw slab, wood chipboard, fibre insulating board, or 9.5 mm plywood	AA
9. Pre-painted (coil coated) steel sheet including liquid-applied PVC coatings	Concrete or clay pot slab (in situ or pre-cast) or non-combustible deck of steel, aluminium, or fibre cement (with or without insulation)	AA

Lead sheet supported by timber joists and plain edged boarding should be regarded as having a BA designation.

Rating of material and products

Table 6.51 Typical performance ratings of some generic materials and products

Rating	Material or product
Class 0	1. Any non-combustible material or material of limited combustibility. 2. Brickwork, blockwork, concrete and ceramic tiles. 3. Plasterboard (painted or not with a PVC facing not more than 0.5 mm thick) with or without an air gap or fibrous or cellular insulating material behind. 4. Woodwool cement slabs. 5. Mineral fibre tiles or sheets with cement or resin binding.
Class 3	6. Timber or plywood with a density more than 400 kg/ml, painted or unpainted. 7. Wood particle board or hardboard, either untreated or painted. 8. Standard glass reinforced polyesters.

Roof with a pitch of 15° or more

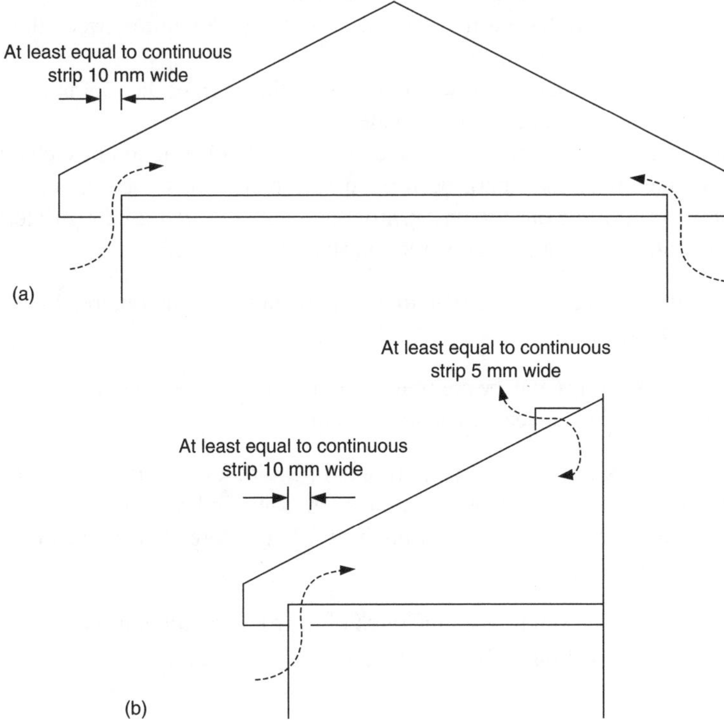

Figure 6.124 Ventilating roof voids. (a) Pitched roof. (b) Lean-to roof

- Pitched roof spaces should have ventilation openings at least 10 mm wide at eaves level to promote cross ventilation.
- A pitched roof that has a single slope and abuts a wall should have ventilation openings at eaves level at least 10 mm wide and at high level (i.e. at the junction of the roof and the wall) at least 5 mm wide.

Roof with a pitch of less than 15°

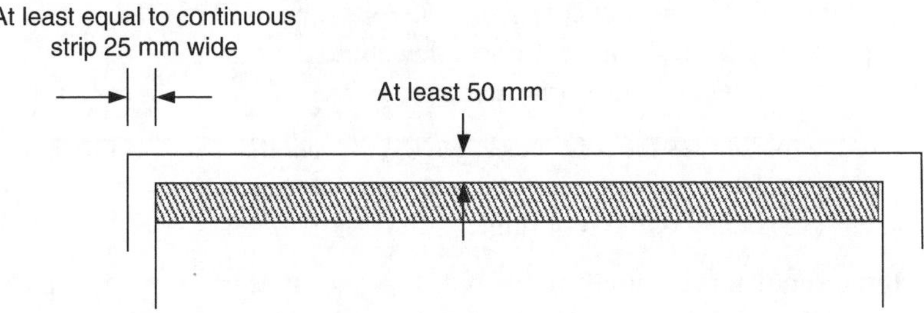

Figure 6.125 Ventilating roof void – flat roof

- Roof spaces should have ventilation openings at least 25 mm wide in two opposite sides to promote cross ventilation.
- The void should have a free air space of at least 50 mm between the roof deck and the insulation.
- Pitched roofs where the insulation follows the pitch of the roof need ventilation at the ridge at least 5 mm wide.
- Where the edges of the roof abut a wall or other obstruction in such a way that free air paths cannot be formed to promote cross ventilation or the movement of air outside any ventilation openings would be restricted, an alternative form of roof construction should be adopted.

Roofs with a span exceeding 10 m may require more ventilation, totalling 0.6% of the roof area.

Ventilation openings may be continuous or distributed along the full length and may be fitted with a screen, facia, baffle, etc.

Where necessary (i.e. for the purposes of health and safety), ventilation to small roofs such as those over porches and bay windows should always be provided and a roof which has a pitch of 70° or more shall be insulated as though it were a wall.

If the ceiling of a room follows the pitch of the roof, ventilation should be provided as if it were a flat roof.

Passive stack ventilation

In roof spaces: • ducts should, ideally, be secured to a wooden strut that is securely fixed at both ends; • flexible ducts should be allowed to curve gently at each end of the strut; • ducts should be insulated with at least 25 mm of a material having a thermal conductivity of 0.04 W/mK.	F App D
For stability, rigid ducts should be used for any outside part of the PSV system that is above the roof slope and to provide stability, it should project down into the roof space far enough to allow firm support.	F App D
If a duct penetrates the roof more than 0.5 m from the roof ridge, then it must extend above the roof slope to at least the height of the roof ridge.	F App D
If tile ventilators are used on the roof slope they must be positioned no more than 0.5 m from the roof ridge.	F App D

If a duct extends above the roof level, then that section of the duct should be insulated or be fitted with a condensation trap just below roof level.	F App D
Ducts should be securely fixed to the roof outlet terminal so that it cannot sag or become detached.	F App D
Separate ducts shall be taken from the ceilings of wet rooms to separate terminals on the roof.	F App D
Terminals should be designed such that any condensation forming inside it cannot run down into the dwelling but will run off onto the roof.	F App D

Note: Placing the outlet terminal at the ridge of the roof is the preferred option as it is not prone to wind gusts and/or certain wind directions.

Ceiling and roof junctions

General

Where a type 1 separating wall is used it should be continuous to the underside of the roof.	E2.55 E2.91 E2.133
The junction between the separating wall and the roof should be filled with a flexible closer which is also suitable as a fire stop.	E2.56 E2.92 E2.134

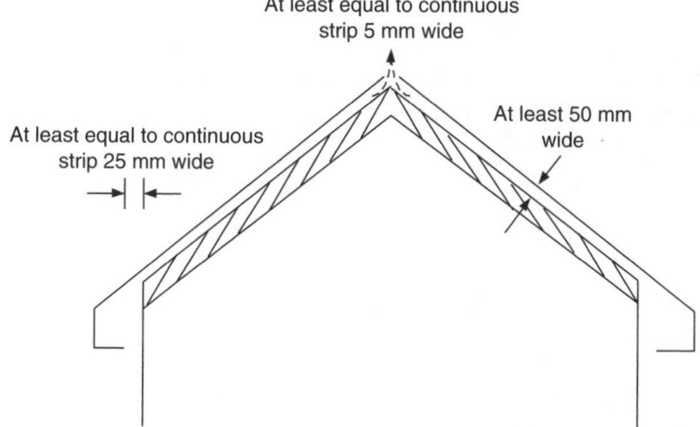

Figure 6.126 Ventilating roof void – ceiling following pitch of roof

Wall type 1 – solid masonry

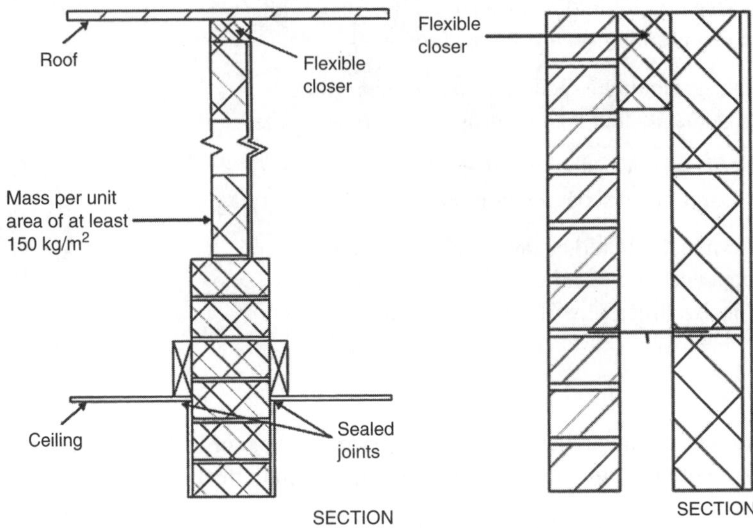

If lightweight aggregate blocks of density less than 1200 kg/m^3 are used above ceiling level, then one side should be sealed with cement, paint or plaster skim.	E2.58 E2.94 E2.138
Where the roof or loft space is not a habitable room (and there is a ceiling with a minimum mass per unit area of 10 kg/m^2 with sealed joints) then the mass per unit area of the separating wall above the ceiling may be reduced to 150 kg/m^2.	E2.57
Where there is an external cavity wall, the cavity should be closed at eaves level with a suitable flexible material (e.g. mineral wool).	E2.59
A rigid connection between the inner and external wall leaves should be avoided.	Ep23
If a rigid material is used, then it should only be rigidly bonded to one leaf.	Ep23

Wall type 2 – cavity masonry

Where a type 2 separating wall is used, it should be continuous to the underside of the roof.	E2.91

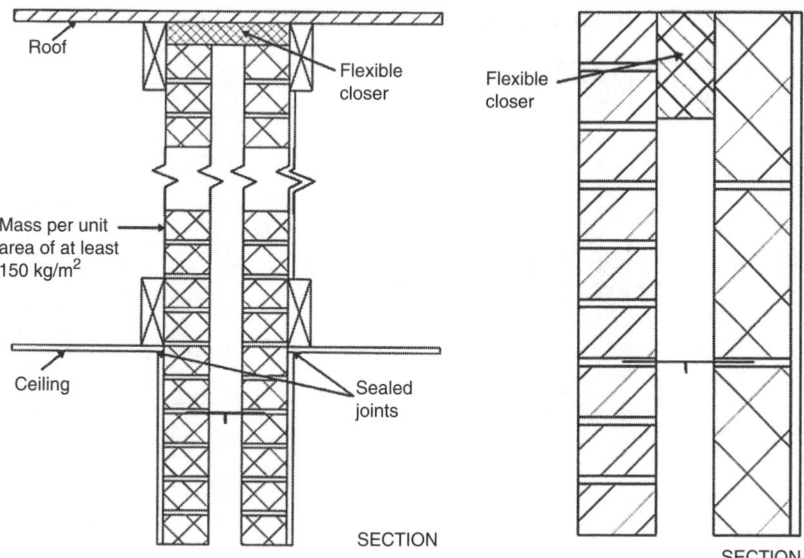

Wall type 2 – ceiling and roof junction

External cavity wall at eaves level

The junction between the separating wall and the roof should be filled with a flexible closer which is also suitable as a fire stop.	E2.92
If lightweight aggregate blocks (with a density less than 1200 kg/m^3 are used) above ceiling level, then one side should be sealed with cement paint or plaster skim.	E2.94
The cavity of an external cavity wall should be closed at eaves level with a suitable flexible material (e.g. mineral wool).	E2.95
A rigid connection between the inner and external wall leaves should be avoided.	E2.95
If a rigid material has to be used, then it should **only** be rigidly bonded to one leaf.	E2.95

Note: If the roof or loft space is not a habitable room (and there is a ceiling with a minimum mass per unit area of 10 kg/m^2 with sealed joints) then the mass per unit area of the separating wall above the ceiling may be reduced to 150 kg/m^2 – **but** it should still be a cavity wall.

Wall type 3 – masonry between independent panels

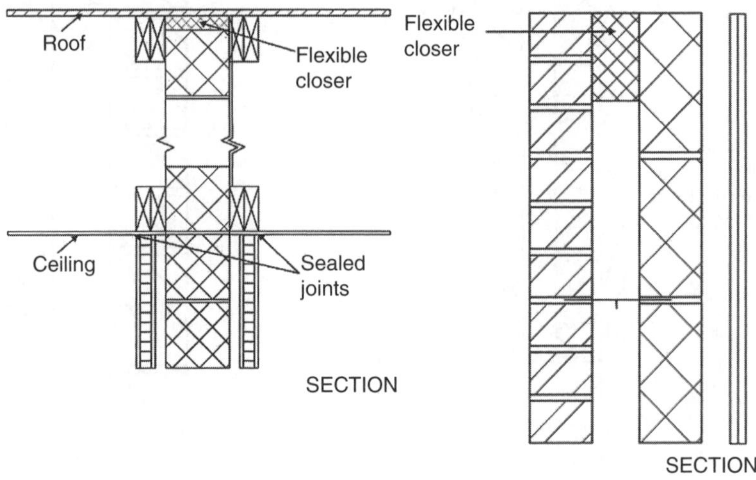

Wall types 3.1 and 3.2 – ceiling and roof junction

External cavity wall at eaves level

Where a type 3 separating wall is used, the masonry core should be continuous to the underside of the roof.	E2.133
The junction between the separating wall and the roof should be filled with a flexible closer which is also suitable as a fire stop.	E2.134
The junction between the ceiling and independent panels should be sealed with tape or caulked with sealant.	E2.135
If there is an external cavity wall, the cavity should be closed at eaves level with a suitable flexible material (e.g. mineral wool).	E2.136
Rigid connections between the inner and external wall leaves should be avoided where possible.	E2.136
If a rigid material is used, then it should only be rigidly bonded to one leaf.	E2.136
For wall types 3.1 and 3.2 (solid masonry core):	
• if the roof or loft space is not a habitable room (and there is a ceiling with a minimum mass per unit area of 10 kg/m^2 and it has of sealed joints) the independent panels may be omitted in the roof space and the mass per unit area of the separating wall above the ceiling may be a minimum of 150 kg/m^2;	E2.137

- if lightweight aggregate blocks with a density less than 1200 kg/m^3 are used above ceiling level, then one side should be sealed with cement paint or plaster skim. E2.138

For wall type 3.3 (cavity masonry core):

- if the roof or loft space is not a habitable room (and there is a ceiling with a minimum mass per unit area of 10 kg/m^2 and it has sealed joints) the independent panels may be omitted in the roof space, but the cavity masonry core should be maintained to the underside of the roof. E2.139

Wall type 4 – framed walls with absorbent material

Where a type 4 separating wall is used, the wall should preferably be continuous to the underside of the roof. E2.160

The junction between the separating wall and the roof should be filled with a flexible closer. E2.161

The junction between the ceiling and the wall linings should be sealed with tape or caulked with sealant. E2.162

If the roof or loft space is not a habitable room (and there is a ceiling with a minimum mass per unit area of 10 kg/m^2 with sealed joints), then: E2.162a

- either the linings on each frame may be reduced to two layers of plasterboard, each sheet with a minimum mass per unit area of 10 kg/m^2; or E2.162a
- the cavity may be closed at ceiling level without connecting the two frames rigidly together. E2.162b

Note: In which case there need only be one frame in the roof space provided there is a lining of two layers of plasterboard, each sheet of minimum mass per unit area 10 kg/m^2, on both sides of the frame.

External wall cavities should be closed at eaves level with a suitable material. E2.163

6.10.3 Conservation of fuel and power

Thermal and opaque elements should have U-values that are no worse than those shown in Table 6.52.	L1B 22c L2B 32c

Table 6.52 Standards for thermal units (W/m^{2}K) for an existing dwelling and/or building

Element Existing building	Existing dwelling	Existing building
Pitched roof – insulation at ceiling level	0.16	
Pitched roof – insulation between rafters	0.20	
Flat roof or roof with integral insulation	0.25	
Roof		0.35
Windows, roof windows, rooflights & doors		3.3

Replacement roof windows, lights and ventilators should have U-values that are no worse than that shown in Table 6.53.	L1B 22c L2B 32c

Table 6.53 Standards for glazed elements in conservatories

Element	Type of building	Replacement fittings (W/m^{2}K)
Windows, roof windows, rooflights & doors	Existing dwelling	2.2 (whole unit)
	Existing building	1.2 (centre pane)
Roof ventilators	Existing building	6.0

Newly constructed thermal elements that are part of an extension (and thermal elements that are constructed/renovated as replacements for existing elements in an existing dwelling) should have U-values that are no worse than that shown in Table 6.54.	L1B 50, 51 and 54 L2B 70a, 86 and 88

Table 6.54 Standards for thermal elements (W/m^{2}K)

Element	Type of building	New elements	Replacement and/or renovated elements
Pitched roof – insulation at ceiling level	Existing dwelling Existing building	0.16	0.16

(*Continued*)

Table 6.54 (*Continued*)

Element	Type of building	New elements	Replacement and/or renovated elements
Pitched roof – insulation between rafters	Existing dwelling Existing building	0.20	0.20
Flat roof or roof with integral insulation	Existing dwelling Existing building	0.20	0.25

> Retained thermal elements whose U-value is worse than the threshold value shall be upgraded to achieve the improved U-value for that element (see Table 6.55). L1B 57

Table 6.55 Standards for replacement elements in an existing building (W/m^{2}K)

Element	Threshold value	Improved value
Pitched roof – insulation at ceiling level	0.35	0.16
Pitched roof – insulation between rafters	0.35	0.20
Flat roof or roof with integral insulation	0.35	0.25

> The area weighted U-value for each element type shall be no worse than the value for similar work being carried out on domestic buildings (see Table 6.56). L2B 29b

Table 6.56 Limiting U-value standards

Element	Area-weighted average U-value
Roof	0.25
Roof windows and rooflights	2.2

Reasonable limits for plane element U-values for building fabric elements are shown in Table 6.57 below.

Note: Display windows and similar glazing are **not** required to meet the standard given for 'Windows and rooflight'.

Table 6.57 Limiting U-value standards (W/m^{2}K)

Element	Area-weighted dwelling average	Worst individual sub-element
Roof	0.25	0.35
Windows, roof windows, rooflights & doors	2.2	3.3
Windows and rooflights	2.2	3.0
Roof ventilators (including smoke vents)	6.0	6.0

Material changes of use (domestic buildings)

When a building is subject to a material change of use, then:

any existing roof window or rooflight which separates a conditioned space from the external environment and which has a U-value that is worse than 3.3 W/m^{2}K, should be replaced. L1B 21

Work on controlled services or fittings (domestic buildings)

When working on a controlled service or fitting (i.e. where the service or fitting is subject to the requirements of Part G, H, J, L or P of Schedule 1), roof windows and rooflights should be provided with draught-proofed units.

Material changes of use (domestic buildings)

When a building is subject to a material change of use, then:

any existing roof window or rooflight which separates a conditioned space from an unconditioned space (or the external environment) and which has a U-value that is worse than 3.3 W/m^{2}K, should be replaced. L1B 21

6.11 Chimneys and fireplaces

6.11.1 The requirement (Building Act 1984 Section 73)

If a person erects or raises a building that is (or is going to be) taller than the chimneys and/or flues from an adjoining building that is either joined by a party wall or less than six feet away from the taller building, then the local authority may:

- if reasonably practical, require that person to build up those chimneys and flues, so that their top is of the same height as the top of the chimneys of the taller building or the top of the taller building, whichever is the higher;

- require the owner or occupier of the adjoining building to allow the person erecting or raising the building, access to the adjacent building so that he can carry out such work as may be necessary to comply with the notice served on him.

The owner or occupier of the adjacent building is entitled to complete the work himself by (within fourteen days) serving a 'counter-notice' that he has elected to carry out the work himself.

The building shall be constructed so that the combined dead, imposed and wind loads are sustained and transmitted by it to the ground

- *safely;*
- *without causing such deflection or deformation of any part of the building (or such movement of the ground) as will impair the stability of any part of another building.*

(Approved Document A1)

Fire precautions (construction)

Any hidden voids in the construction shall be sealed and subdivided to inhibit the unseen spread of fire and products of combustion, in order to reduce the risk of structural failure, and the spread of fire.

(Approved Document B3)

Protection of building

Combustion appliances and fluepipes shall be so installed, and fireplaces and chimneys shall be so constructed and installed, as to reduce to a reasonable level the risk of people suffering burns or the building catching fire in consequence of their use.

(Approved Document J3)

Provision of information

Where a hearth, fireplace, flue or chimney is provided or extended, a durable notice containing information on the performance capabilities of the hearth, fireplace, flue or chimney shall be affixed in a suitable place in the building for the purpose of enabling combustion appliances to be safely installed.

(Approved Document J4)

6.11.2 Meeting the requirement

End restraints

The ends of every wall (except single leaf walls less than 2.5 m in storey height and length) in small single storey non-residential buildings and annexes should be bonded or otherwise securely tied throughout their full height to a buttressing wall, pier or chimney.	A1/2 (2C25)
Long walls may be provided with intermediate support, dividing the wall into distinct lengths; each distinct length is a supported wall for the purposes of this section.	A1/2 (2C25)
The buttressing wall, pier or chimney should provide support from the base to the full height of the wall.	A1/2 (2C25)
The sectional area on plan of chimneys (excluding openings for fireplaces and flues) should be not less than the area required for a pier in the same wall, and the overall thickness should not be less than twice the required thickness of the supported wall (see Figure 6.127).	A1/2 (2C27b)
Floors and roofs should act to transfer lateral forces (see Table 6.30) from walls to buttressing walls, piers or chimneys and be secured to the supported wall as shown.	A1/2 (2C33a)

Masonry chimneys

Where a chimney is not adequately supported by ties or securely restrained in any way, its height H (measured from the highest point of any chimney pot or other flue terminal) should not exceed 4.5 times the width W (the least horizontal dimension of the chimney measured at the same point of intersection) – provided that the density of the masonry is greater than 1500 kg/m^3.	1D1

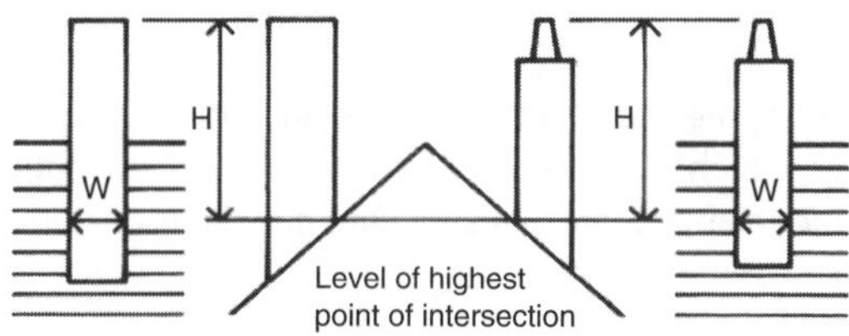

Figure 6.127 Proportions for masonry chimneys

The foundation of piers, buttresses and chimneys should project as indicated in Figure 6.128 and the projection X should never be less than P.

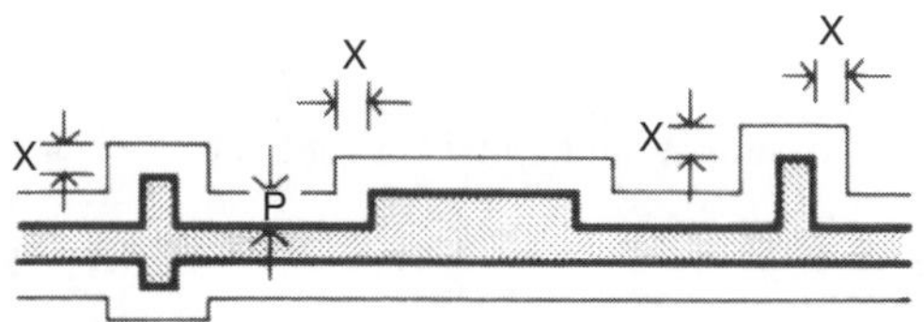

Figure 6.128 Piers and chimneys

Flues, etc.

Flue walls should have a fire resistance of at least one half of that required for the compartment wall or floor and be of non-combustible construction.	B3
If a flue, (or duct containing flues and/or ventilation duct(s)), passes through a compartment wall, or is built into a compartment wall, each wall of the flue or duct should have a fire resistance of at least half that of the wall or floor (see Figure 6.129).	B3 7.11 (V1) B3 10.16 (V2)

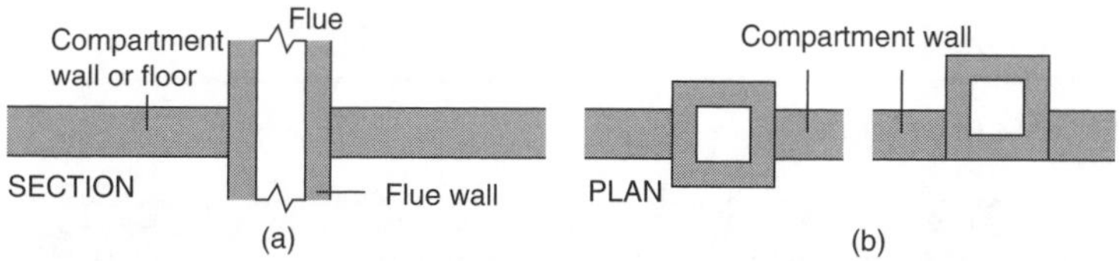

Figure 6.129 Flues penetrating compartment walls or floors. (a) Flue passing through compartment wall or floor. (b) Flue built into compartment wall

Fire stopping

Proprietary fire-stopping and sealing systems (including those designed for service penetrations) which have been shown by test to maintain the fire resistance of the wall or other element, are available and may be used. Other fire-stopping materials include:

- cement mortar,
- gypsum-based plaster,
- cement or gypsum-based vermiculite/perlite mixes,
- glass fibre, crushed rock, blast furnace slag or ceramic based products (with or without resin binders), and
- intumescent mastics (B3 11.14).

Joints between fire separating elements should be fire-stopped.	B3 7.12a (V1) B3 10.17a (V2)
All openings for pipes, ducts, conduits or cables to pass through any part of a fire separating element should be: • kept as few in number as possible; • kept as small as practicable; • fire-stopped (which in the case of a pipe or duct, should allow thermal movement).	B3 7.12b (V1) B3 10.17b (V2)
Cables concealed in floors and walls (in certain circumstances) are required to have an earthed metal covering, be enclosed in steel conduit, or have additional mechanical protection (see BS 7671 for more information).	P App A 2d
To prevent displacement, materials used for fire-stopping should be reinforced with (or supported by) materials of limited combustibility.	B3 7.13 (V1) B3 10.18 (V2)

Chimney construction

Chimneys shall consist of a wall or walls enclosing one or more flues (see Figure 6.130).	J (0.4–7)

In the gas industry, the chimney for a gas appliance is commonly called the flue.

Down-draughts that could interfere with the combustion performance of an open-flued appliance (see Figure 6.131) shall be minimized. J (0.4–11)

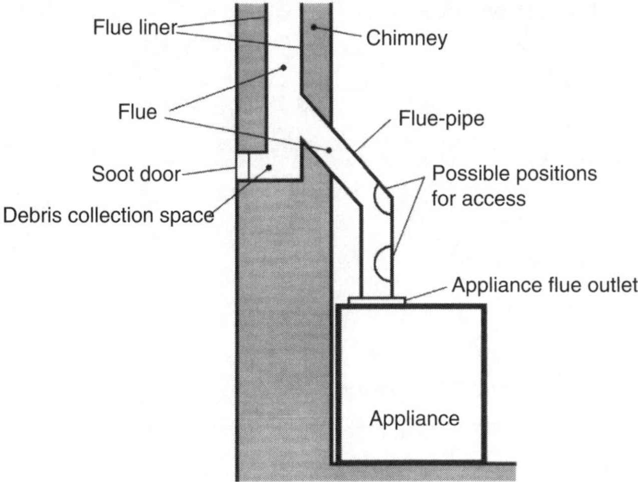

Figure 6.130 Chimneys and flues

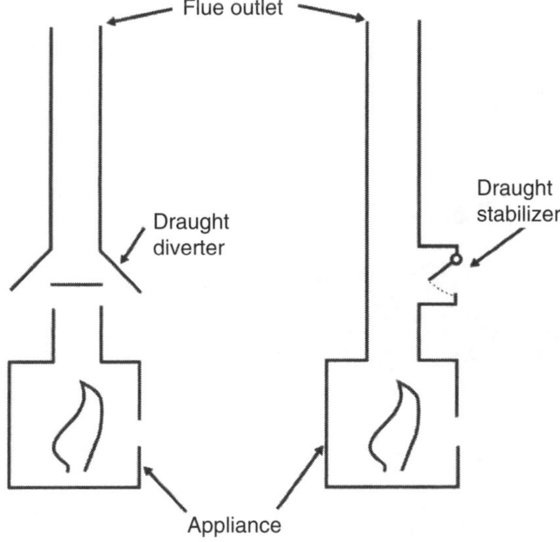

Figure 6.31 Draught diverters and draught stabilizers

Fireplaces

Fireplace recesses (sometimes called a builder's opening) shall be formed in a wall or in a chimney breast, from which a chimney leads and which has a hearth at its base. J (0.4–16)

Simple recesses are suitable for closed appliances such as roomheaters, stoves, cookers or boilers. They are **not** suitable for an open fire without a canopy.

Fireplace recesses are used for accommodating open fires and freestanding fire baskets.

Fireplace recesses are often lined with firebacks to accommodate inset open fires.

Lining components and decorative treatments fitted around openings reduce the opening area. It is the finished fireplace opening area that determines the size of flue required for an open fire in such a recess.

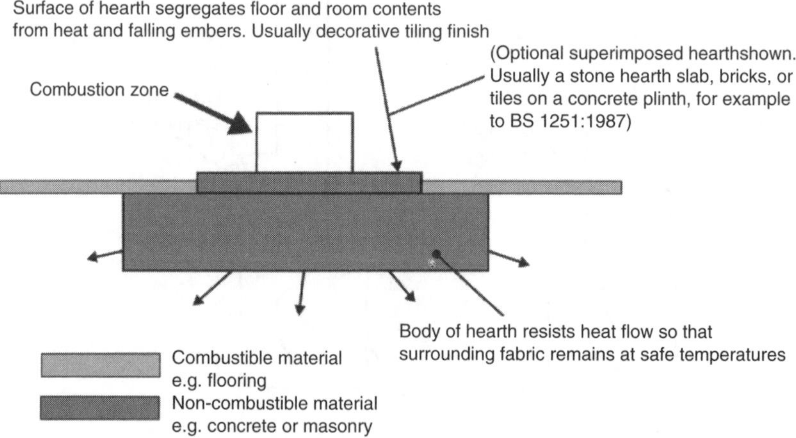
Surface of hearth segregates floor and room contents from heat and falling embers. Usually decorative tiling finish
(Optional superimposed hearthshown. Usually a stone hearth slab, bricks, or tiles on a concrete plinth, for example to BS 1251:1987)
Combustion zone
Body of hearth resists heat flow so that surrounding fabric remains at safe temperatures
Combustible material e.g. flooring
Non-combustible material e.g. concrete or masonry

Figure 6.132 The functions of a hearth

Hearths

A hearth shall safely isolate a combustion appliance from people, combustible parts of the building fabric and soft furnishings.	J (0.4–26)

Flueblock chimneys

Flueblock chimneys should be constructed of factory-made components suitable for the intended application installed in accordance with manufacturer's instructions.	J (1.29)
Joints should be sealed in accordance with the flueblock manufacturer's instructions.	J (1.30)
Bends and offsets should only be formed with matching factory-made components.	J (1.30)

Masonry chimneys (change of use)

Where a building is to be altered for different use (e.g. it is being converted into flats) the fire resistance of walls of existing masonry chimneys may need to be improved.	J (1.31)

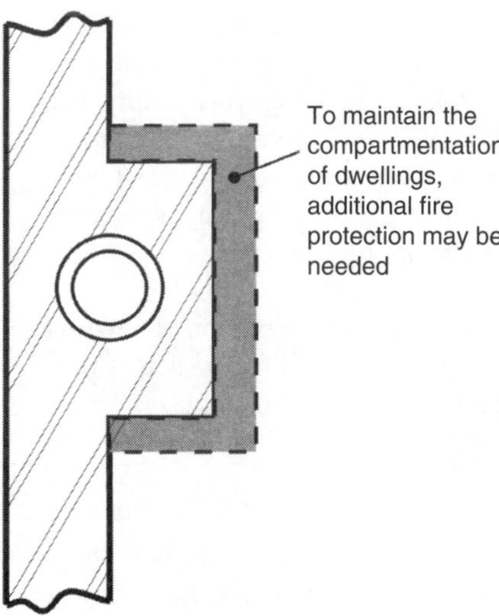
To maintain the compartmentation of dwellings, additional fire protection may be needed

Figure 6.133 Fire protection of chimneys passing through other dwellings

Connecting fluepipes

Whenever possible, fluepipes should be manufactured from: J (1.32)

- cast iron (BS41: 1973 (1998));
- mild steel fluepipes (BS1449, Part 1: 1991, with a flue wall thickness of at least 3 mm);
- stainless steel (BS EN 10088-1: 1995 grades 1.4401, 1.4404, 1.4432 or 1.4436 with a flue wall thickness of at least 1 mm);
- vitreous enamelled steel (BS 6999: 1989 (1996)).

Fluepipes with spigot and socket joints should be fitted with the socket facing upwards to contain moisture and other condensates in the flue. J (1.33)

Joints should be made gas-tight. J (1.33)

Repair of flues

If renovation, refurbishment or repair amounts to or involves the provision of a new or replacement flue liner, it is considered 'building work' within the meaning of Regulation 3 of the Building Regulations and must, therefore, **not** be undertaken without prior notification to the local authority. Examples of work that would need to be notified include: J (1.34–1.35)

- relining work comprising the creation of new flue walls by the insertion of new linings such as rigid or flexible prefabricated components;
- a cast in situ liner that significantly alters the flue's internal dimensions.

If you are in doubt you should consult the building control department of your local authority, or an approved inspector.

Re-use of existing flues

Where it is proposed to bring a flue in an existing chimney back into use (or to re-use a flue with a different type or rating of appliance) the flue and the chimney should be checked and, if necessary, altered to ensure that they satisfy the requirements for the proposed use. J (1.36)

Oversize flues can be unsafe. A flue may, however, be lined to reduce the flue area to suit the intended appliance. J (1.38)

Relining

If a chimney has been relined in the past using a metal lining system and the appliance is being replaced, the metal liner should also be replaced unless the metal liner can be proven to be recently installed and can be seen to be in good condition. J (1.39)

In certain circumstances, relining is considered 'building work' within the meaning of Regulation 3 of the Building Regulations and must, therefore, **not** be undertaken without prior notification to the local authority. If you are in doubt you should consult the building control department of your local authority, or an approved inspector.

Flexible flue liners should only be used to reline a chimney and should not be used as the primary liner of a new chimney. J (1.40)

Plastic fluepipe systems can be acceptable in some cases, for example with condensing boiler installations, where the fluepipes are supplied by or specified by the appliance manufacturer. J (1.41)

Factory-made metal chimneys

Where a factory-made metal chimney passes through a wall, sleeves should be provided to prevent damage to the flue or building through thermal expansion. J (1.43)

To facilitate the checking of gas-tightness, joints between chimney sections should not be concealed within ceiling joist spaces or within the thicknesses of walls. J (1.43)

When installing a factory-made metal chimney, provision should be made to withdraw the appliance without the need to dismantle the chimney. J (1.44)

Factory-made metal chimneys should be kept a suitable distance away from combustible materials. J (1.45)

One way of meeting this requirement is by locating the chimney not less than distance 'X' from combustible material, where 'X' is defined in BS 4543-1: 1990 (1996) as shown in Figure 6.134.

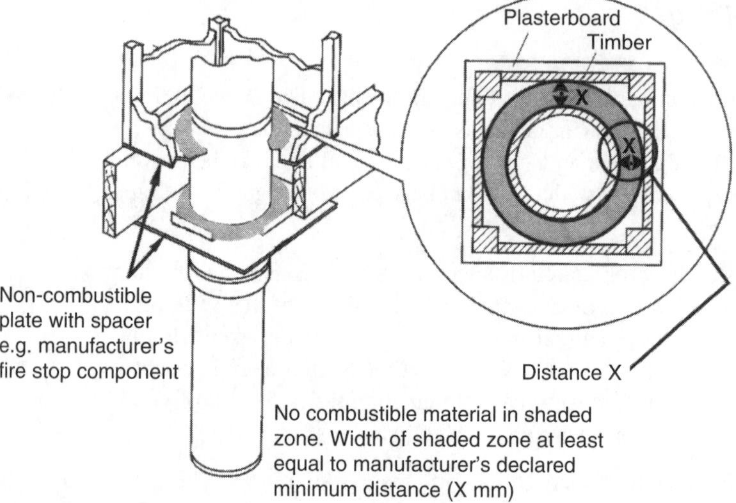

Figure 6.134 The separation of combustible material from a factory-made metal chimney meeting BS 4543, Part 1 (1990)

Flue systems

Flue systems should offer least resistance to the passage of flue gases by minimizing changes in direction or horizontal length. J (1.47)

Wherever possible flues should be built so that they are straight and vertical except for the connections to combustion appliances with rear outlets where the horizontal section should not exceed 150 mm. Where bends are essential, they should be angled at no more than 45° to the vertical. J (1.47)

Provisions should be made to enable flues to be swept and inspected (see Figure 6.135). J (1.48)

A flue should **not** have openings into more than one room or space except for the purposes of: J (1.49)

- inspection or cleaning; or
- fitting an explosion door, draught break, draught stabilizer or draught diverter.

Openings for inspection and cleaning should be formed using purpose factory-made components compatible with the flue system, having an access cover that has the same level of gas-tightness as the flue system and an equal level of thermal insulation. J (1.50)

After the appliance has been installed, it should be possible to sweep the whole flue. J (1.50)

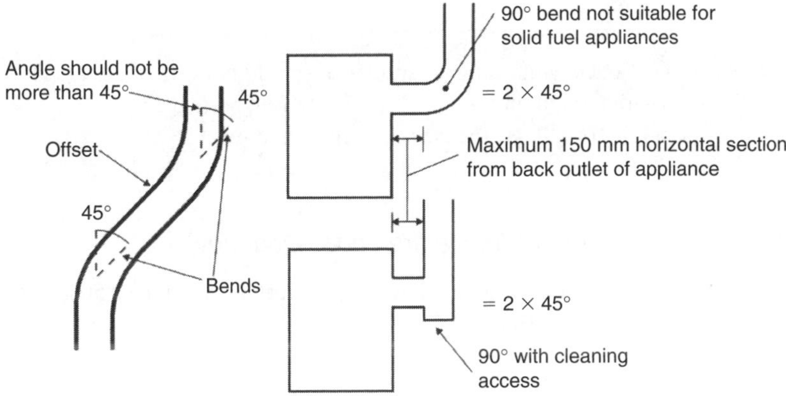

Figure 6.135 Bends in flues

Dry lining around fireplace openings

Where a decorative treatment, such as a fireplace surround, masonry cladding or dry lining is provided around a fireplace opening, any gaps that could allow flue gases to escape from the fireplace opening into the void behind the decorative treatment, should be sealed to prevent such leakage. J (1.52)

The sealing material should be capable of remaining in place despite any relative movement between the decorative treatment and the fireplace recess. J (1.53)

Notice plates for hearths and flues (Requirement J4)

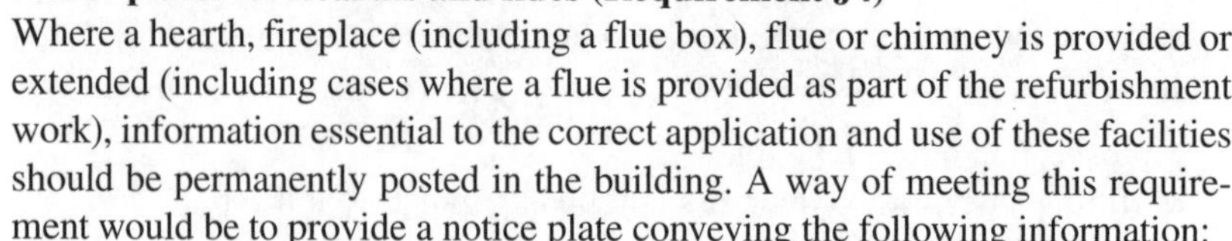

Where a hearth, fireplace (including a flue box), flue or chimney is provided or extended (including cases where a flue is provided as part of the refurbishment work), information essential to the correct application and use of these facilities should be permanently posted in the building. A way of meeting this requirement would be to provide a notice plate conveying the following information:

- the location of the hearth, fireplace (or flue box) or the location of the beginning of the flue;
- the category of the flue and generic types of appliances that can be safely accommodated;
- the type and size of the flue (or its liner if it has been relined) and the manufacturer's name;
- the installation date.

Additional provisions for appliances burning solid fuel (with a rated output up to 50 kW)

Any room or space containing an appliance should burning solid fuel (with a rated output up to 50 kW) have a permanent air vent opening of at least the size shown in Figure 6.136.	J (2.2)

Open fire with no throat (e.g. a fire under a canopy)

Permanently open air vent(s) should have a total free area of at least 50% of the crosssectional area of the flue.

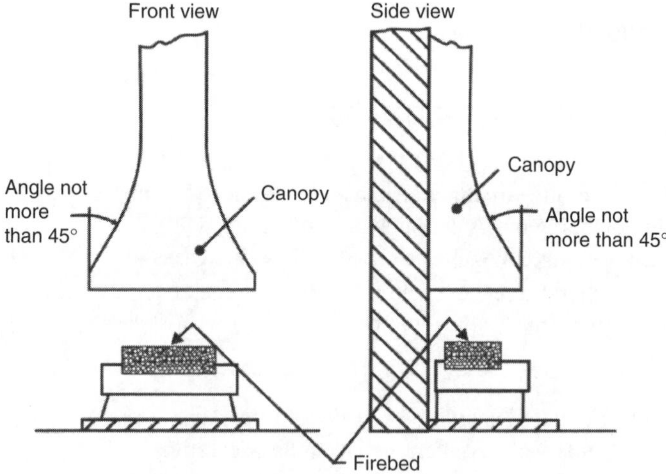

Figure 6.136 Canopy for an open solid fuel fire

Open fire with a throat and gather

Permanently open air vent(s) should have a total free area of at least 50% of the throat opening area.

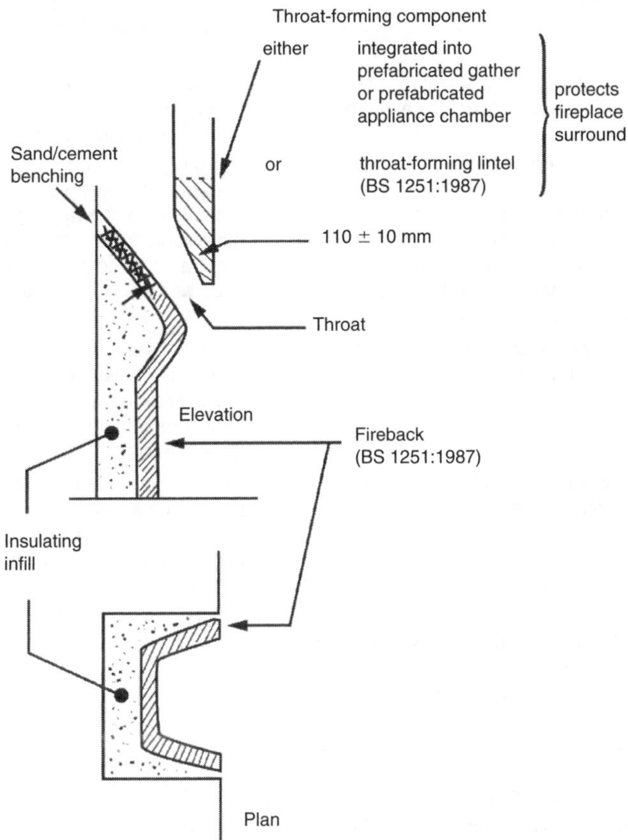

Figure 6.137 Open fireplaces – throat and fireplace components

Other appliance (such as a stove, cooker or boiler)

Permanently open air vent(s).

Size of flues

Fluepipes should have the same diameter or equivalent cross-sectional area as that of the appliance flue outlet.	J (2.4)
Flues should be not less than the size of the appliance flue outlet or that recommended by the appliance manufacturer.	J (2.5)

Flues in chimneys

Table 6.58 Size of flues in chimneys (see Figure 6.138)

Fireplace with an opening up to 500 mm × 550 mm.	200 mm diameter or rectangular/square flues having the same cross-sectional dimension not less than 175 mm.
Fireplace with an opening in excess of 500 mm × 550 mm or a fireplace exposed on two or more sides.	If rectangular/square flues are used the minimum dimension should not be less than 200 mm.
Closed appliance up to 20 kW rated output which: • burns smokeless or low volatile fuel; or • is an appliance that meets the requirements of the Clean Air Act when burning an appropriate bitumous coal (these appliances are known as 'exempted fireplaces').	125 mm diameter or rectangular/square flues having the same cross-sectional area and a minimum dimension not less than 100 mm for straight flues or 125 mm for flues with bends or offsets.
Other closed appliance of up to 35 kW rated output burning any fuel.	150 mm diameter or rectangular/square flues having the same cross-sectional area and a minimum dimension not less than 125 mm.
Closed appliance of up to 30 kW and up to 50 kW rated output burning any fuel.	175 mm diameter or rectangular/square flues having the same cross-sectional area and a minimum dimension not less than 150 mm.
For fireplaces with openings larger than 500 mm × 550 mm or fireplaces exposed on two or more sides (such as a fireplace under a canopy or open on both sides of a central chimney breast) a way of showing compliance would be to provide a flue with a cross-sectional area equal to 15% of the total face area of the fireplace opening(s) using the formula: Fireplace opening area (mm^2) × Total horizontal length of fireplace opening L (mm) × Height of fireplace opening H (mm) **Examples of L and H for large and unusual fireplace openings are shown in Figure 6.138**	J (2.7)

Height of flues (see Figure 6.138)

Flues should be high enough (normally 4.5 m is sufficient) to ensure sufficient draught to clear the products of combustion. J (2.8)

The outlet from a flue should be above the roof of the building in a position where the products of combustion can discharge freely and will not present a fire hazard, whatever the wind conditions (see Figure 6.139 and Table 6.59). J (2.10)

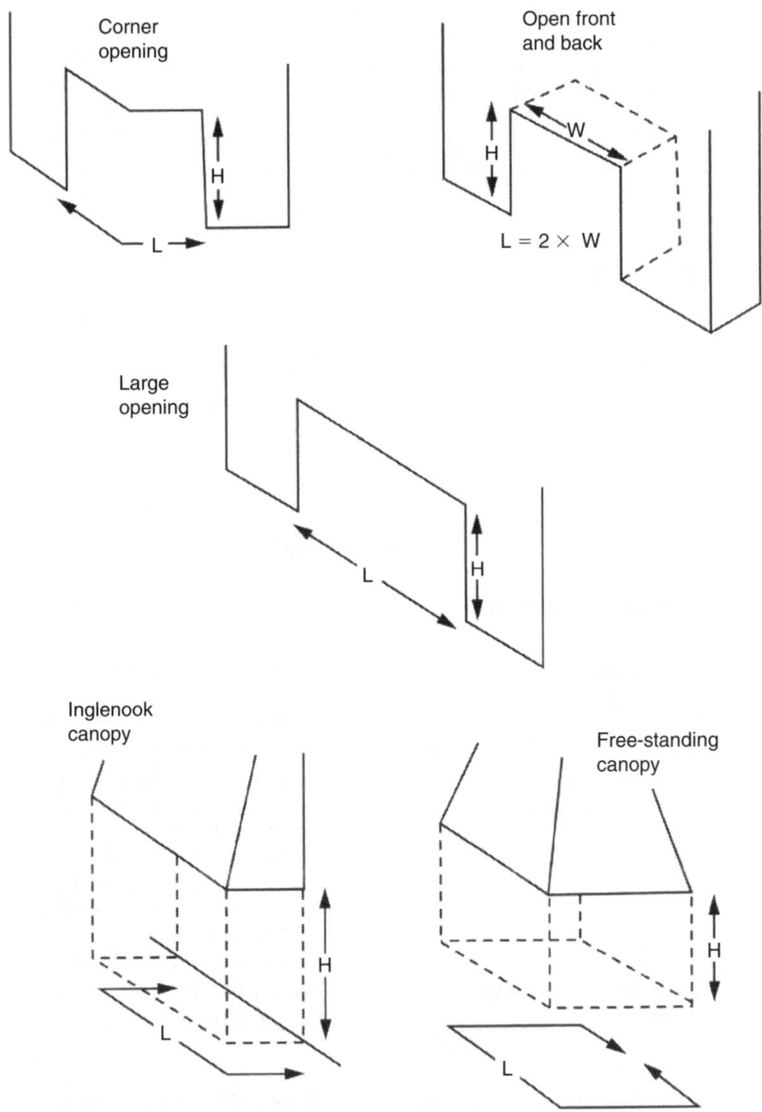

Figure 6.138 Large or unusual fireplace openings

Flue outlet clearances – thatched or shingled roof

The clearances to flue outlets which discharge on, or are in close proximity to, roofs with surfaces which are readily ignitable (e.g. covered in thatch or shingles) should be increased to those shown in Figure 6.140. J (2.12)

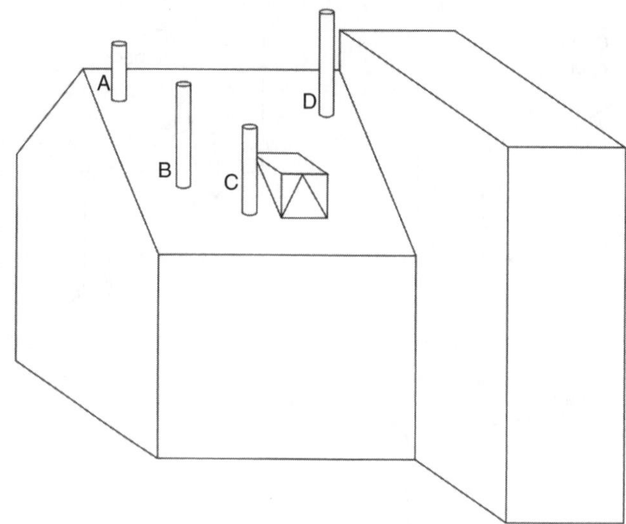

Figure 6.139 Flue outlet positions for solid fuel appliances

Table 6.59 Flue outlet positions

Point where flue passes through weather surfaces (e.g. roof, tiles or external walls)		Clearance to flue outlet
A	At or within 600 mm of the ridge.	At least 600 mm above the ridge.
B	Elsewhere on a roof (whether pitched or flat).	At least 2300 mm horizontally from the nearest point on the weather surface and: • at least 1000 mm above the highest point of intersection of the chimney and the weather surface; or • at least as high as the ridge.
C	Below (on a pitched roof) or within 2300 mm horizontally to an openable rooflight, dormer window or other opening.	At least 1000 mm above the top of the opening.
D	Within 2300 mm of an adjoining or adjacent building whether or not beyond the measurements is the boundary.	At least 600 mm above the adjacent building.

Connecting fluepipes

Connecting fluepipes should not pass through any roof space, partition, internal wall or floor, unless they pass directly into a chimney through J (2.14)

either a wall of the chimney or a floor supporting the chimney.

Connecting fluepipes should be guarded if they are likely to be damaged or if the burn hazard they present to people is not immediately apparent. J (2.14)

Connecting fluepipes should be located so as to avoid igniting combustible material by minimizing horizontal and sloping runs and separation of the fluepipe from combustible material. J (2.15 and 1.45)

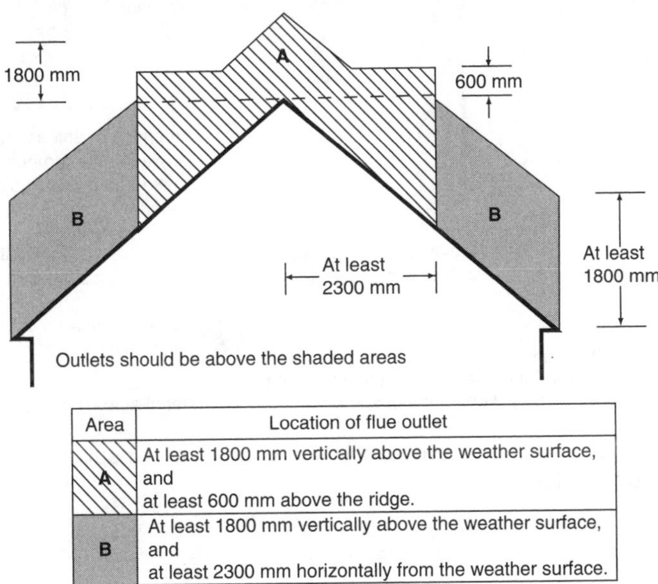

Area	Location of flue outlet
A	At least 1800 mm vertically above the weather surface, and at least 600 mm above the ridge.
B	At least 1800 mm vertically above the weather surface, and at least 2300 mm horizontally from the weather surface.

Figure 6.140 Flue outlet positions for solid fuel appliances – discharging near easily ignited roof coverings

Masonry and flueblock chimneys

The thickness of the walls around the flues, excluding the thickness of any flue liners, should be in accordance with Figure 6.142. J (2.17)

Combustible material should not be located where it could be ignited by the heat dissipating through the walls of fireplaces or flues. J (2.18)

Construction of fireplace gathers

To minimize resistance to the proper working of flues, tapered gathers should be provided in fireplaces for open fires J (2.21), or corbelling of masonry, as

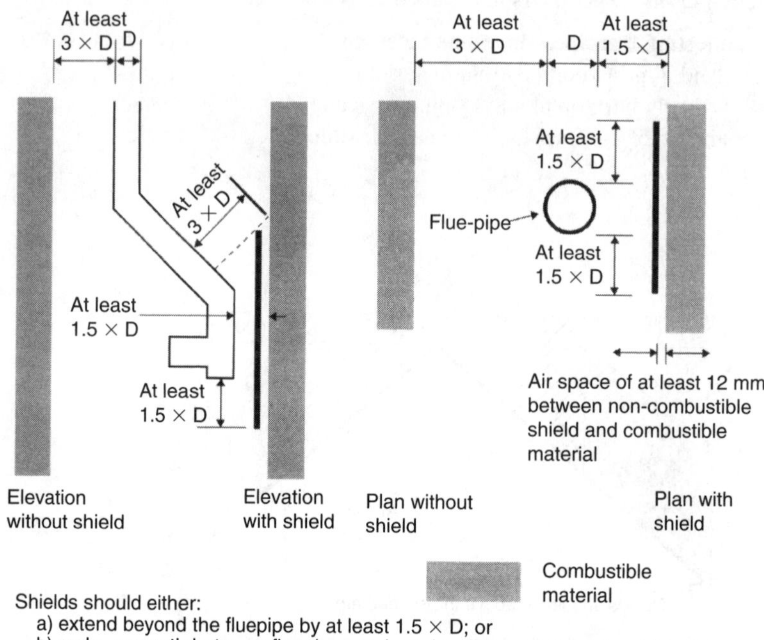

Figure 6.141 Protecting combustible material from uninsulated fluepipes for solid fuel appliances

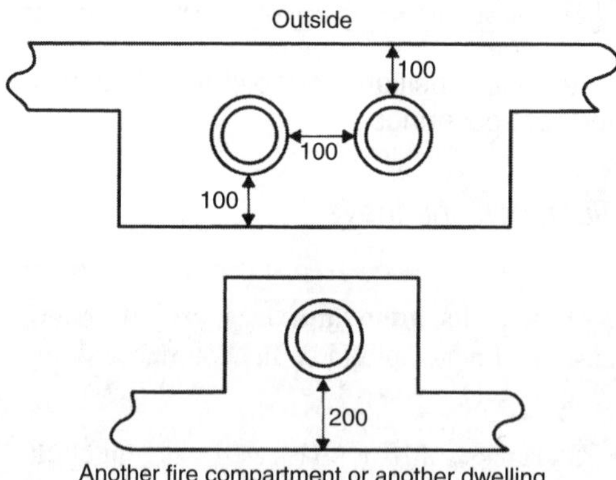

Figure 6.142 Wall thickness for masonry and flueblock chimneys. Dimensions in mm

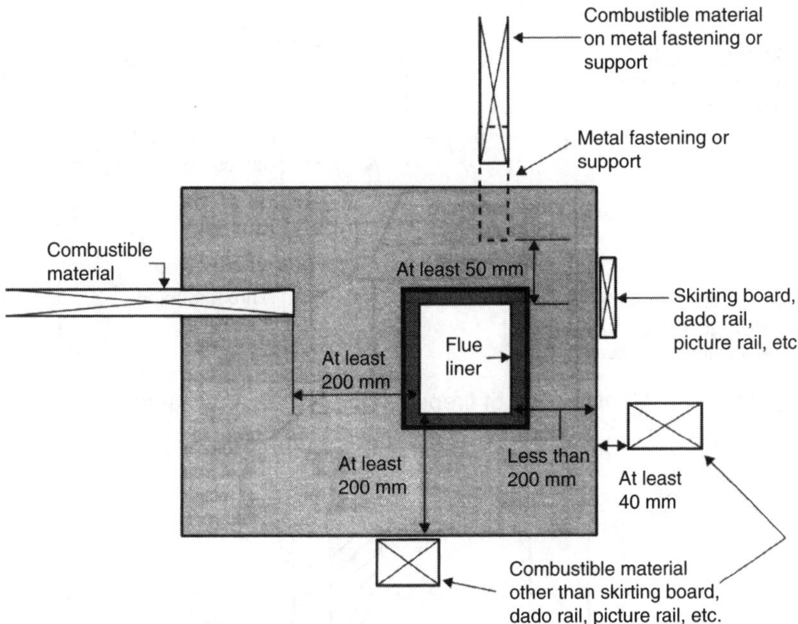

Figure 6.143 Minimum separation distances for combustible material in or near a chimney

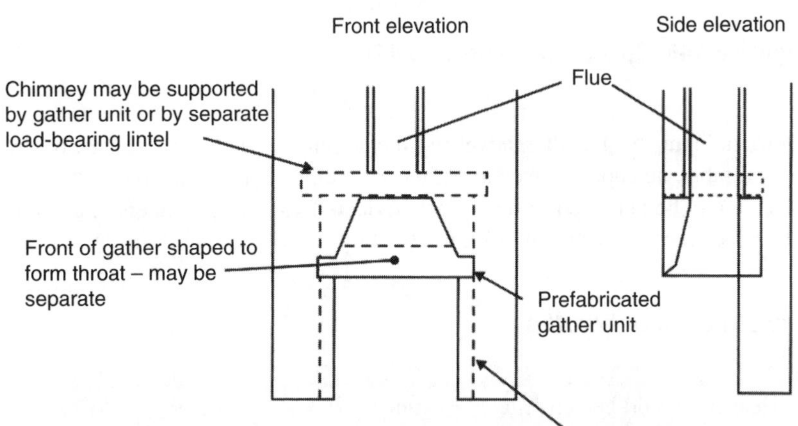

Figure 6.144 Construction of fireplace gathers – using prefabricated components

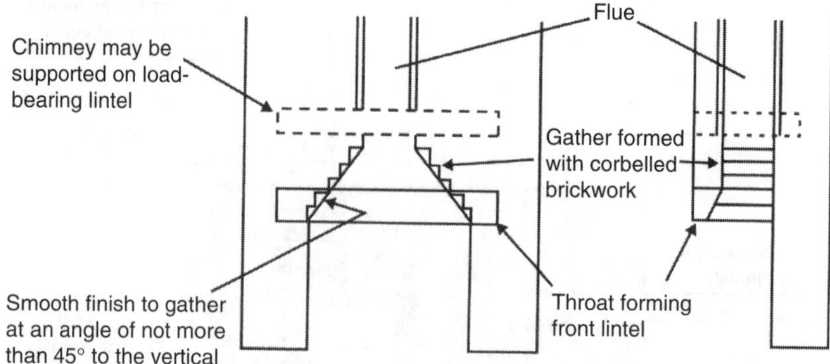

Figure 6.145 Construction of fireplace gathers – using masonry

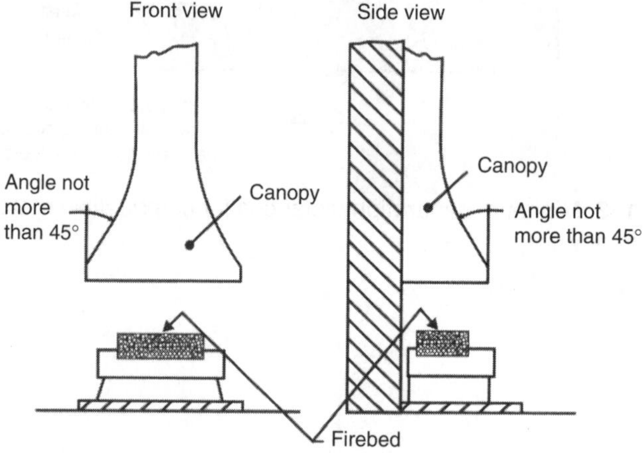

Figure 6.146 Canopy for an open fuel fire

shown in Figure 6.145. Alternatively a suitable canopy (as shown in Figure 6.146) or a prefabricated appliance chamber incorporating a gather may be used.

This can be achieved by using prefabricated gather components built into a fireplace recess, as shown in Figure 6.144.

Construction of hearths

Hearths should be constructed of suitably robust materials and to appropriate dimensions such that, in normal use, they prevent combustion appliances setting fire to the building fabric and furnishings, and they limit the risk of people being accidentally burnt. J (2.22)

The hearth should be able to accommodate the weight of the appliance and its chimney if the chimney is not independently supported. J (2.22)

Appliances should stand wholly above either hearths made of non-combustible board/sheet material, tiles at least 12 mm thick or constructional hearths. J (2.23)

Constructional hearths should have plan dimensions as shown in Figure 6.147. J (2.24a)

Constructional hearths should be made of solid, non-combustible material, such as concrete or masonry, at least 125 mm thick, including the thickness of any non-combustible floor and/or decorative surface. J (2.24b)

Combustible material should not be placed beneath constructional hearths unless there is an air-space of at least 50 mm between the underside of the hearth and the combustible material, or the combustible material is at least 250 mm below the top of the hearth (see Figure 6.148). J (2.25)

An appliance should be located on a hearth so that it is surrounded by a surface free of combustible material (as shown in Figure 6.134) or it may be the surface of a superimposed hearth laid wholly or partly upon a constructional hearth. J (2.26)

The edges of this surface should be marked to provide a warning to the building occupants and to discourage combustible floor finishes such as carpet from being laid too close to the appliance. A way of achieving this would be to provide a change in level. J (2.26)

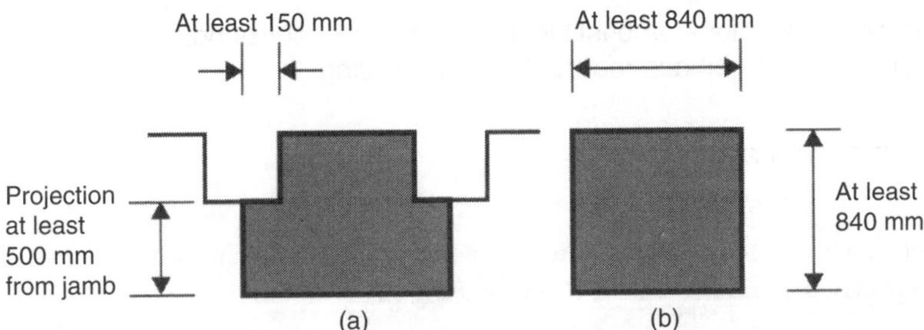

Figure 6.147 Constructional hearth suitable for solid fuel appliances (including open fires) – plan. (a) Fireplace recess. (b) Freestanding

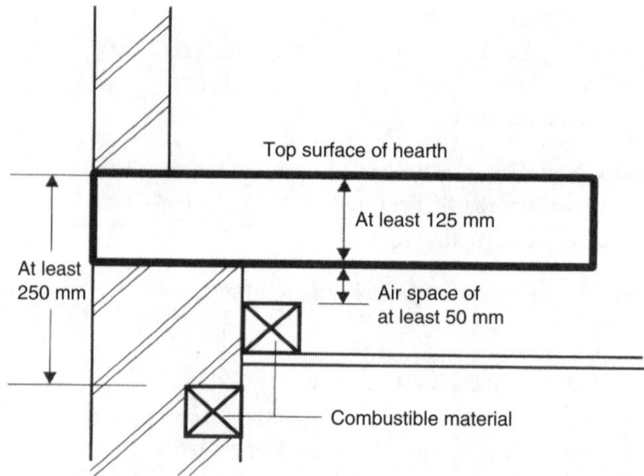

Figure 6.148 Constructional hearth suitable for solid fuel appliances (including open fires) – section

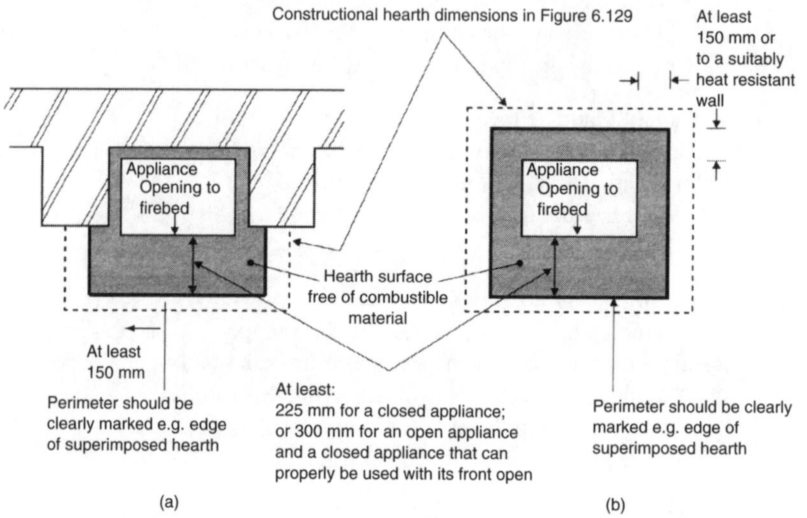

Figure 6.149 Non-combustible hearth surface surrounding a solid fuel appliance. (a) Fireplace recess. (b) Freestanding

Fireplace recesses

Fireplaces need to be constructed such that they adequately protect the building fabric from catching fire.	J (2.29)
Fireplace recesses can be from masonry or concrete as shown in Figure 6.149.	J (2.29a)

Fireplace recesses can also be prefabricated factory-made appliance chambers using components that are made of insulating concrete having a density of between 1200 and 1700 kg/m^3 and with the minimum thickness as shown in Table 6.59.	J (2.29b)

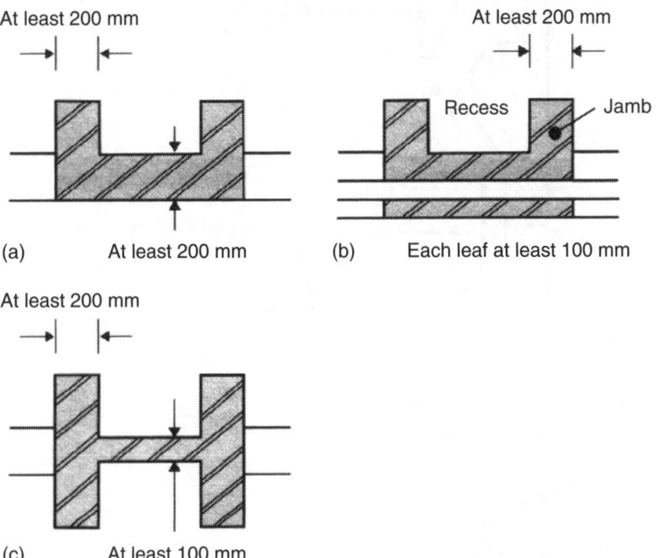

Figure 6.150 Fireplace recesses. (a) Solid wall. (b) Cavity wall. (c) Back-to-back (within the same dwelling)

Fireplace lining components

Fireplace recesses containing inset open fires, need to be heat protected and should either be lined with suitable firebricks or lining components as shown in Table 6.60.	J (2.30)

Table 6.60 Prefabricated appliance chambers: minimum thickness

Component	Minimum thickness (mm)
Base	50
Side section, forming wall on either side of chamber	75
Back section, forming rear of chamber	100
Top slab, lintel or gather, forming top of chamber	100

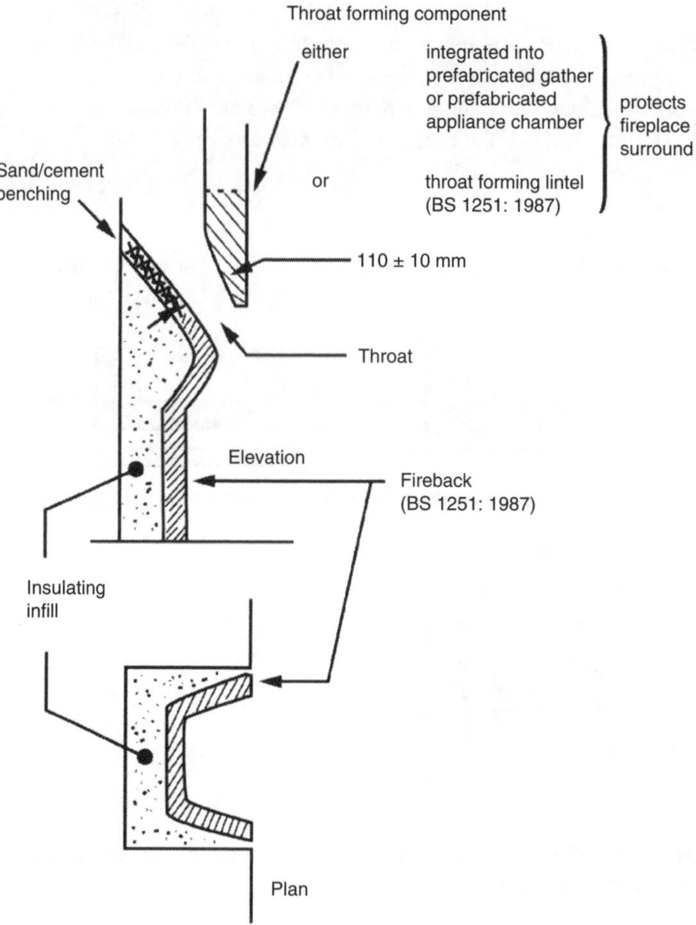

Figure 6.151 Open fireplaces – throat and fireplace components

Walls adjacent to hearths

Walls that are not part of a fireplace recess or a prefabricated appliance chamber but are adjacent to hearths or appliances also need to protect the building from catching fire. A way of achieving the requirement is shown in Figure 6.151. J (2.31)

Additional provisions for gas burning devices

The Gas Safety (Installation and Use) Regulations require that (a) gas fittings, appliances and gas storage vessels must only be installed by a person with the required competence and (b) any person having control to any extent of gas work must ensure that the person carrying out that work has the required

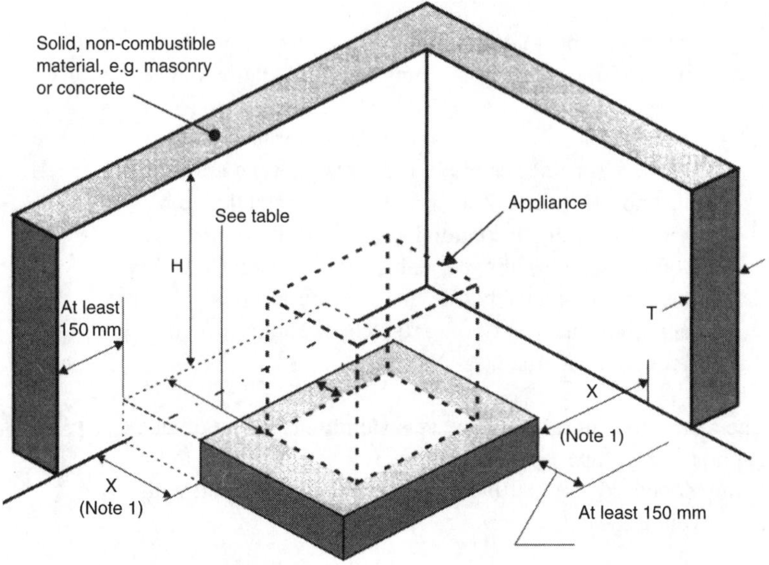

Location of hearth or appliance	Solid, non-combustible material	
	Thickness (T)	Height (H)
Where the hearth abuts a wall and the appliance is not more than 50 mm from the wall	200 mm	At least 300 mm above the appliance and 1.2 m above the hearth
Where the hearth abuts a wall and the appliance is more than 50 mm but not more than 300 mm from the wall	75 mm	At least 300 mm above the appliance and 1.2 m above the hearth
Where the hearth does not abut a wall and is no more than 150 mm from the wall (see Note 1)	75 mm	At least 1.2 m above the hearth
Note 1: There is no requirement for protection of the wall where X is more than 150 mm		

Figure 6.152 Walls adjacent to hearths

competence and (c) any person carrying out gas installation, whether an employee or self-employed, must be a member of a class of persons approved by the HSE; for the time being this means they must be registered with CORGI, the Council for Registered Gas Installers.

Important elements of the Regulations include that:

(a) any appliance installed in a room used or intended to be used as a bath or shower room must be of the room-sealed type; J (3.5a)

(b) a gas fire, other gas space heater or gas water heater of more than 14 kW (gross) heat input (12.7 kW (net) J (3.5b) heat input) must not be installed in a room used or intended to be used as sleeping accommodation unless the appliance is room-sealed;

(c) a gas fire, other space heater or gas water heater of up to 14 kW (gross) heat input (12.7 kW (net) heat input) must not be installed in a room used or intended to be used as sleeping accommodation unless it is room-sealed or equipped with a device designed to shut down the appliance before there is a build-up of a dangerous quantity of the products of combustion in the room concerned. J (3.5c)

The restrictions in (a)–(c) above also apply in respect of any cupboard or compartment within the rooms concerned and to any cupboard, compartment or space adjacent to and with an air vent into such a room. J (3.5d)

Instantaneous water heaters (installed in any room) must be room-sealed or have fitted a safety device to shut down the appliance as in (c) above. J (3.5e)

Precautions must be taken to ensure that all installation pipework, gas fittings, appliances and flues are installed safely. When any gas appliance is installed, checks are required for ensuring compliance with the Regulations, including the effectiveness of the flue, the supply of combustion air, the operating pressure or heat input (or where necessary both), and the operation of the appliance to ensure its safe functioning. J (3.5f)

All flues must be installed in a safe position. J (3.5g)

No alteration is allowed to any premises in which a gas fitting or gas storage vessel is fitted that would adversely affect the safety of that fitting or vessel, causing it no longer to comply with the Regulations. J (3.5h)

LPG storage vessels and LPG-fired appliances fitted with automatic ignition devices or pilot lights must not be installed in cellars or basements. J (3.5i)

Outlets from flues should be situated externally so as to allow the products of combustion to dispel, and, if a balanced flue, the intake of air – see Figure 6.153.

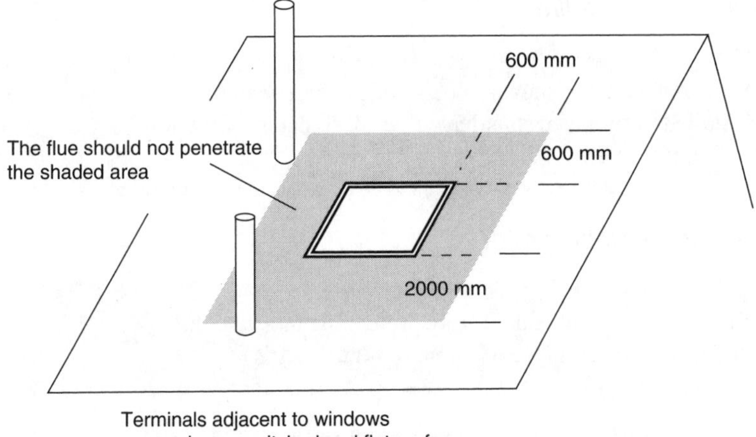

Figure 6.153 Location of outlets near roof windows from flues serving gas appliances

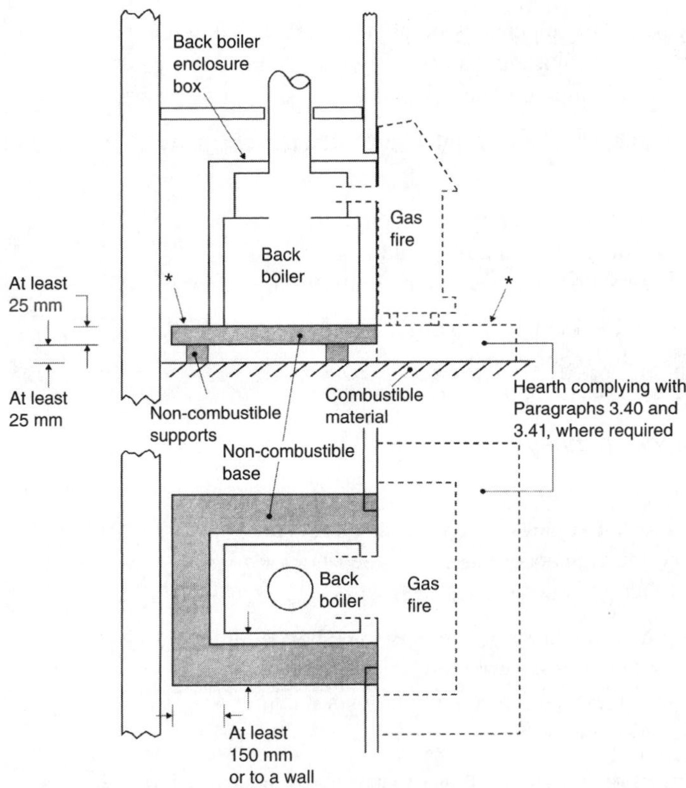

Figure 6.154 Bases for back boilers

Fireplaces – gas fires

Provided it can be shown to be safe, gas fires may be installed in fireplaces that have flues designed to serve solid fuel appliances.	J (3.7)

Bases for back boilers

Back boilers should adequately protect the fabric of the building from heat (see example in Figure 6.154).	J (3.39)

Kerosene and gas oil burning appliances

Kerosene (Class C2) and gas oil (Class D) appliances have the following, additional, requirements:

Open-fired oil appliances should not be installed in rooms such as bedrooms and bathrooms where there is an increased risk of carbon monoxide poisoning. J (4.2)

The outlet from a flue should be so situated externally to ensure: J (4.6)

- the correct operation of a natural draft flue;
- the intake of air if a balanced flue;
- dispersal of the products of combustion.

Figure 6.155 (and Table 6.60) indicates typical positioning to meet this requirement.

Flueblock chimneys

Flueblock chimneys should be installed with sealed joints in accordance with the flueblock manufacturer's installation instructions.	J (4.16)
Flueblocks that are not intended to be bonded into surrounding masonry should be supported and restrained in accordance with the manufacturer's installation instructions.	J (4.16)
Where a fluepipe or chimney penetrates a fire compartment wall or floor, it must not breach the fire separation requirements.	J (4.18) Approved Doc. B

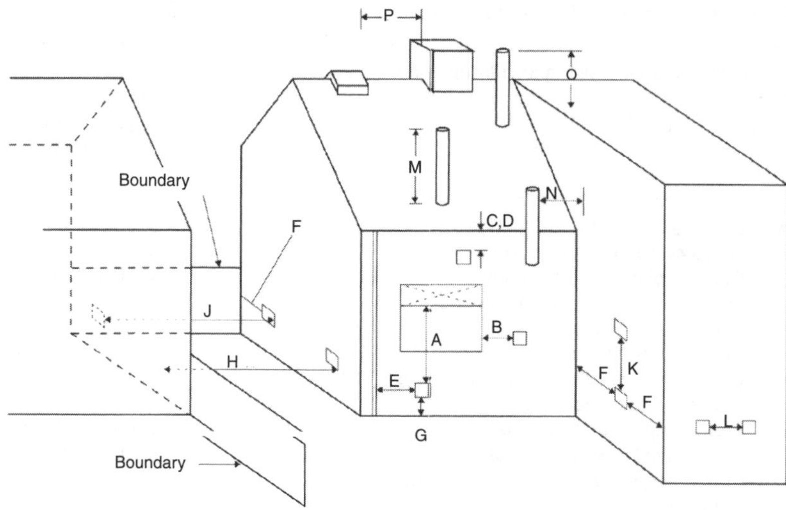

Figure 6.155 Location of outlets from flues serving oil-fired appliances

Relining chimney flues (for oil appliances)

Flexible metal flue liners should be installed in one complete length without joints within the chimney.	J (4.22)
Other than for sealing at the top and the bottom, the space between the chimney and the liner should be left empty (unless this is contrary to the manufacturer's instructions).	J (4.22)
Flues that may be expected to serve appliances burning Class D oil (i.e. gas oil) should be made of materials that are resistant to acids.	J (4.23)

Hearths for oil appliances

Oil appliance hearths are needed to prevent the building catching fire and, whilst it is not a health and safety provision, it is customary to top them with a tray for collecting spilled fuel.	J (4.24)

Table 6.61 Location of outlets from flues serving oil fired appliances

Minimum separation distances for terminals in mm

	Location of outlet[1]	Appliance with pressure jet burner	Appliance with vaporizing burner
A	Below an opening[2,3]	600	should not be used
B	Horizontally to an opening[2,3]	600	should not be used
C	Below a plastic/painted gutter, drainage pipe or eaves if combustible material protected[4]	75	should not be used
D	Below a balcony or a plastic/painted gutter, drainage pipe or eaves without protection to combustible material	600	should not be used
E	From vertical sanitary pipework	300	should not be used
F	From an external or internal corner or from a surface or boundary alongside the terminal	300	should not be used
G	Above ground or balcony level	300	should not be used
H	From a surface or boundary facing the terminal	600	should not be used
J	From a terminal facing the terminal	1200	should not be used
K	Vertically from a terminal on the same wall	1500	should not be used
L	Horizontally from a terminal on the same wall	750	should not be used
M	Above the highest point of an intersection with the roof	600[6]	1000[5]
N	From a vertical structure to the side of the terminal	750[6]	2300
O	Above a vertical structure which is less than 750 mm (pressure jet burner) or 2300 mm (vaporizing burner) horizontally from the side of the terminal	600[6]	1000[5]
P	From a ridge terminal to a vertical structure on the roof	1500	should not be used

Notes:

1. Terminals should only be positioned on walls where appliances have been approved for such configurations when tested in accordance with BS EN 303-1: 1999 or OFTEC standards OFS A100 or OFS A101.
2. An opening means an openable element, such as an openable window, or a permanent opening such as a permanently open air vent.
3. Notwithstanding the dimensions above, a terminal should be at least 300 mm from combustible material, e.g. a window frame.
4. A way of providing protection of combustible material would be to fit a heat shield at least 750 mm wide.
5. Where a terminal is used with a vaporizing burner, the terminal should be at least 2300 mm horizontally from the roof.
6. Outlets for vertical balanced flues in locations M, N and O should be in accordance with manufacturer's instructions.

6.12 Stairs

6.12.1 Requirements

The building shall be constructed so that the combined dead, imposed and wind loads are sustained and transmitted by it to the ground:

- *safely;*
- *without causing such deflection or deformation of any part of the building (or such movement of the ground) as will impair the stability of any part of another building.*

(Approved Document A1)

The building shall be designed and constructed so that there are appropriate provisions for the early warning of fire, and appropriate means of escape in case of fire from the building to a place of safety outside the building capable of being safely and effectively used at all material times.

(Approved Document B)

For a typical one- or two-storey dwelling, the requirement is limited to the provision of smoke alarms and to the provision of openable windows for emergency exit (see B1.i).

Airborne and impact sound

Dwellings shall be designed so that the noise from domestic activity in an adjoining dwelling (or other parts of the building) is kept to a level that:

- *does not affect the health of the occupants of the dwelling;*
- *will allow them to sleep, rest and engage in their normal activities in satisfactory conditions.*

(Approved Document E1)

Dwellings shall be designed so that any domestic noise that is generated internally does not interfere with the occupants' ability to sleep, rest and engage in their normal activities in satisfactory conditions.

(Approved Document E2)

Domestic buildings shall be designed and constructed so as to restrict the transmission of echoes.

(Approved Document E3)

Schools shall be designed and constructed so as to reduce the level of ambient noise (particularly echoing in corridors).

(Approved Document E4)

Stairs, ladders and ramps

All stairs, steps and ladders shall provide reasonable safety between levels in a building.

(Approved Document K1)

In a public building the standard of stair, ladder or ramp may be higher than in a dwelling, to reflect the lesser familiarity and greater number of users.

This requirement only applies to stairs, ladders and ramps that form part of the building.

Pedestrian guarding should be provided for any part of a floor, gallery, balcony, roof, or any other place to which people have access and any light well, basement area or similar sunken area next to a building.

(Approved Document K2)

Requirement K2 (a) applies only to stairs and ramps that form part of the building.

Access and facilities for disabled people

In addition to the requirements of the Disability Discrimination Act 1995 precautions need to be taken to ensure that:

- *new non-domestic buildings and/or dwellings (e.g. houses and flats used for student living accommodation etc.);*
- *extensions to existing non-domestic buildings;*
- *non-domestic buildings that have been subject to a material change of use (e.g. so that they become a hotel, boarding house, institution, public building or shop);*

are capable of allowing people, regardless of their disability, age or gender, to:

- *gain access to buildings;*
- *gain access within buildings;*
- *be able to use the facilities of the buildings (both as visitors and as people who live or work in them);*
- *use sanitary conveniences in the principal storey of any new dwelling.*

(Approved Document M)

Note: See Annex A for guidance on access and facilities for disabled people.

6.12.2 Meeting the requirements

As the upper surfaces of floors and stairs are not significantly involved in a fire until it is well developed, they do not play an important part in fire spread in the early stages of a fire.

Means of escape

Except for kitchens, all habitable rooms in the upper storey(s) of a dwelling house which are served by only one stair should be provided with: • a window (or external door); or • direct access to a protected stairway.	B1 2.4 (V1) B1 2.12 (V2)
If direct escape to a place of safety is impracticable, it should be possible to reach a place of relative safety such as a protected stairway within a reasonable travel distance.	B1 V (b)

Table 6.62 Limitations on distance of travel in common areas of blocks of flats

Maximum distance of travel (m) from flat entrance door to common stair, or to stair lobby	
Escape in one direction only	Escape in more than one direction
7.5 m	30 m

Escape routes should be planned so that people do not have to pass through one stairway enclosure to reach another.	B1 2.23 (V2)
Common corridors should be protected corridors. The wall between each flat and the corridor should be a compartment wall.	B1 2.24 (V2) B1 2.24 (V2)
Means of ventilating common corridors/lobbies (i.e. to control smoke and so protect the common stairs) should be available.	B1 2.25 (V2)
In large buildings, the corridor or lobby adjoining the stair should be provided with a vent that is located as high as practicable and with its top edge at least as high as the top of the door to the stair.	B1 2.26 (V2)

There should also be a vent, with a free area of at least $1.0\,m^2$ from the top storey of the stairway to the outside. B1 2.26 (V2)

In single stair buildings the smoke vents on the fire floor and at the head of the stair should be actuated by means of smoke detectors in the common access space providing access to the flats. B1 2.26 (V2)

In buildings with more than one stair the smoke vents may be actuated manually. B1 2.26 (V2)

Note: Self-closing fire doors should be positioned so that smoke will not affect access to more than one stairway.

Stairs

Where an opening in a floor or roof for a stairway or the like adjoins a supported wall and interrupts the continuity of lateral support:

- the maximum permitted length of the opening is to be 3 m, measured parallel to the supported wall; A1/2 2C37a
- connections (if provided by means other than by anchors) should be throughout the length of each portion of the wall situated on each side of the opening; A1/2 2C37b
- connections via mild steel anchors should be spaced closer than 2 m on each side of the opening to provide the same number of anchors as if there were no opening; A1/2 2C37c
- there should be no other interruption of lateral support. A1/2 2C37d

Stairs that separate a dwelling from another dwelling (or part of the same building) shall resist: E

- the transmission of impact sound (such as footsteps and furniture moving);
- the flow of sound energy through walls and floors;
- the level of airborne sound;
- flanking transmission from stairs connected to the separating wall. E2

All new stairs constructed within a dwelling-house (flat or room used for residential purposes) – whether purpose built or formed by a material change of use – shall meet the laboratory sound insulation values set out in Table 6.63. E0.9

Table 6.63 Dwelling-houses and flats – performance standards for separating floors and stairs that have a separating function

	Airborne sound insulation $D_{nT,w} + C_{tr}$ dB (minimum values)	Impact sound insulation $L'_{nT,w}$ dB (maximum values)
Purpose built rooms for residential purposes	45	62
Purpose built dwelling houses and flats	45	62
Rooms for residential purposes formed by material change of use	43	64
Dwelling-houses and flats formed by material change of use	43	64

Notes:

(1) The sound insulation values in this table include a built-in allowance for 'measurement uncertainty' and so if any these test values are not met, then that particular test will be considered as failed.

(2) Occasionally a higher standard of sound insulation may be required between spaces used for normal domestic purposes and noise generated in and to an adjoining communal or non-domestic space. In these cases it would be best to seek specialist advice before committing yourself.

(3) If the stair is not enclosed, then the potential sound insulation of the internal floor will not be achieved, nevertheless, the internal floor should still satisfy Requirement E2.

(4) In some cases it may be that an existing wall, floor or stair in a building will achieve these performance standards without the need for remedial work, for example if the existing construction was already compliant.

Figure 6.156 illustrates the relevant parts of the building that should be protected from airborne and impact sound in order to satisfy Requirement E2.

Sound insulation testing

The person carrying out the building work should arrange for sound insulation testing to be carried out (by a test body with appropriate third party accreditation) in accordance with the procedure described in Annex B of this Approved Document E.	E0.3 E0.4
Impact sound insulation tests should be carried out without a soft covering (e.g. carpet, foam backed vinyl etc.) on the stair floor.	E1.10

Requirement E2(a) – Internal walls

Requirement E2(b) – Internal floors

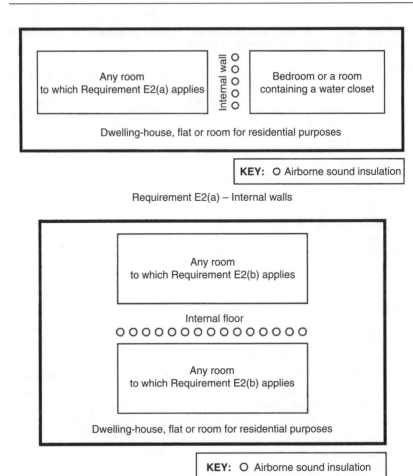

Figure 6.156 Airborne and impact sound requirements

Testing should not be carried out between living spaces, corridors, stairwells or hallways.	E1.8
Test bodies conducting testing should preferably have UKAS accreditation (or a European equivalent) for field measurements.	E0.4

Note: Some properties, for example loft apartments, may be sold before being fitted out with internal walls and other fixtures and fittings. In these cases sound insulation measurements should be made between the available spaces.

If stairs form a separating function then they are subject to the same sound insulation requirements as floors. In this case, the resistance to airborne sound depends mainly on:

- the mass of the stair;
- the mass and isolation of any independent ceiling;
- the air tightness of any cupboard or enclosure under the stairs;
- the stair covering (which reduces impact sound at source).

Stair treatment 1

Stair treatment 1 consists of a stair covering and independent ceiling with absorbent material.

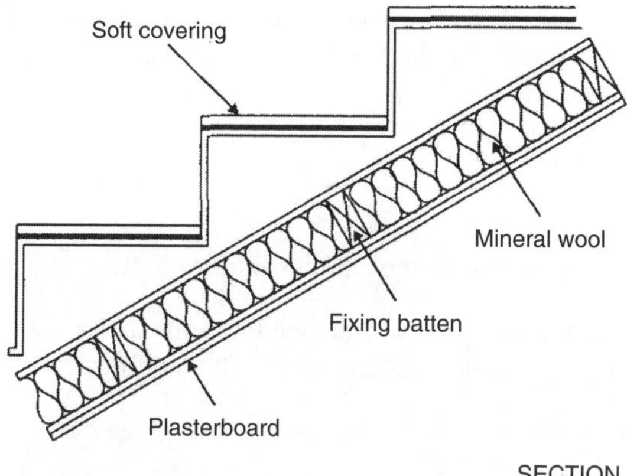

Figure 6.157 Stair covering and independent ceiling with absorbent material

The soft covering should be:

- at least 6 mm thickness; E4.37
- laid over the stair treads;
- be securely fixed (e.g. glued) so it does not become a safety hazard.

If there is a cupboard under all, or part, of the stair: E4.37

- the underside of the stair within the cupboard should be lined with plasterboard (minimum mass per unit area 10 kg/m^2) together with an absorbent layer of mineral wool (minimum density 10 kg/m^3);
- the cupboard walls should be built from two layers of plasterboard (or equivalent), each sheet with a minimum mass per unit area of 10 kg/m^2;
- a small, heavy, well fitted door should be fitted to the cupboard.

If there is no cupboard under the stair, an independent ceiling should be constructed below the stair (see Floor treatment 1).	E4.37
Where a staircase performs a separating function it shall conform to Building Regulation Part B – Fire safety.	E4.38

Reverberation

Requirement E3 requires that '*domestic buildings shall be designed and constructed so as to restrict the transmission of echoes*'. The guidance notes provided in Part E cover two methods (Method A and Method B) which can be used in determining the amount of additional absorption to be used in corridors, hallways, stairwells and entrance halls that give access to flats and rooms for residential purposes. Method A is applicable to stairs and requires the following to be observed:

Method A

Cover the ceiling area with the additional absorption.	E7.10
Cover the underside of intermediate landings, the underside of the other landings, and the ceiling area on the top floor.	E7.11
The absorptive material should be equally distributed between all floor levels.	E7.12
For stairwells (or a stair enclosure), calculate the combined area of the stair treads, the upper surface of the intermediate landings, the upper surface of the landings (excluding ground floor) and the ceiling area on the top floor. Either cover an area equal to this calculated area with a Class D absorber, or cover an area equal to at least 50% of this calculated area with a Class C absorber or better.	E7.11

Note: Method A can generally be satisfied by the use of proprietary acoustic ceilings.

Piped services

Piped services (excluding gas pipes) and ducts that pass through separating floors should be surrounded with sound absorbent material for their full height and enclosed in a duct above and below the floor.

Junctions with floor penetrations (excluding gas pipes)

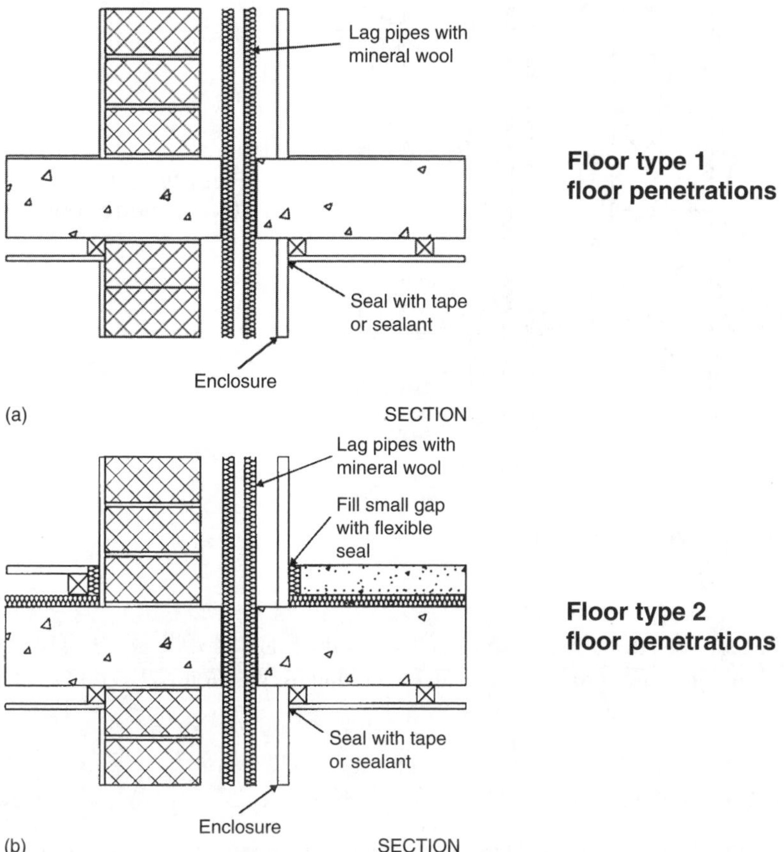

Figure 6.158a,b,c Junctions with floor penetrations (excluding gas pipes)

Pipes and ducts that penetrate a floor separating habitable rooms in different flats should be enclosed for their full height in each flat.	E3.41 E3.79 E3.117
The enclosure should be constructed of material having a mass per unit area of at least 15 kg/m^2.	E3.32 E3.80 E3.118
The enclosure should either be lined or the duct (or pipe) within the enclosure wrapped with 25 mm unfaced mineral fibre.	E3.42 E3.80 E3.118

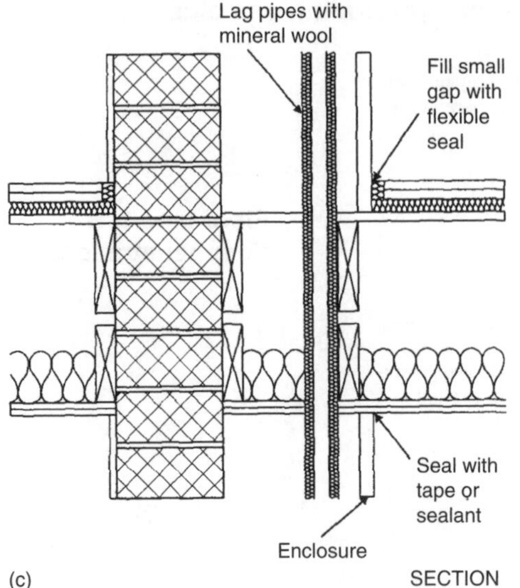

Floor type 3 floor penetrations

Figure 6.158 (*Continued*)

Penetrations through a separating floor by ducts and pipes should have fire protection to satisfy Building Regulation Part B – Fire safety.	E3.43 E3.82 E3.120
Fire stopping should be flexible to prevent a rigid contact between the pipe and the floor.	E3.43 E3.121
A small gap (sealed with sealant or neoprene) of about 5 mm should be left between the enclosure and the floating floor.	E3.81 E3.119
Where floating floor (a) or (b) is used the enclosure may go down to the floor base (provided that the enclosure is isolated from the floating layer).	E3.81 E3.119

Junctions with floor penetrations (including gas pipes)

Gas pipes may be contained in a separate (ventilated) duct or can remain unenclosed.	E3.43 E3.120
If a gas service is installed it shall comply with the Gas Safety (Installation and Use) Regulations 1998, SI 1998 No 2451.	E3.43 E3.120

In the Gas Safety Regulations there are requirements for ventilation of ducts at each floor where they contain gas pipes. Gas pipes may be contained in a separate ventilated duct or they can remain unducted.

Stairs, ladders and ramps

The rise of a stair shall be between 155 mm and 220 mm with any going between 245 mm and 260 mm and a maximum pitch of 42°. K1 (1.1–1.4)

The normal relationship between the dimensions of the rise and going is that twice the rise plus the going (2R + G) should be between 550 mm and 700 mm.

Stairs with open risers that are likely to be used by children under 5 years should be constructed so that a 100 mm diameter sphere cannot pass through the open risers. K1 (1.9)

Stairs which have more than 36 risers in consecutive flights should make at least one change of direction, between flights, of at least 30°. K1 (1.14)

If a stair has straight and tapered treads, then the going of the tapered treads should not be less than the going of the straight tread. K1 (1.20)

The going of tapered treads should measure at least 50 mm at the narrow end. K1 (1.18)

The going should be uniform for consecutive tapered treads. K1 (1.19)
K1 (1.22–1.24)

Stairs should have a handrail on both sides if they are wider than 1 m and on at least one side if they are less than 1 m wide. K1 (1.27)

Handrail heights should be between 900 mm and 1000 mm measured to the top of the handrail from the pitch line or floor. K1 (1.27)

Spiral and helical stairs should be designed in accordance with BS 5395. K1 (1.21)

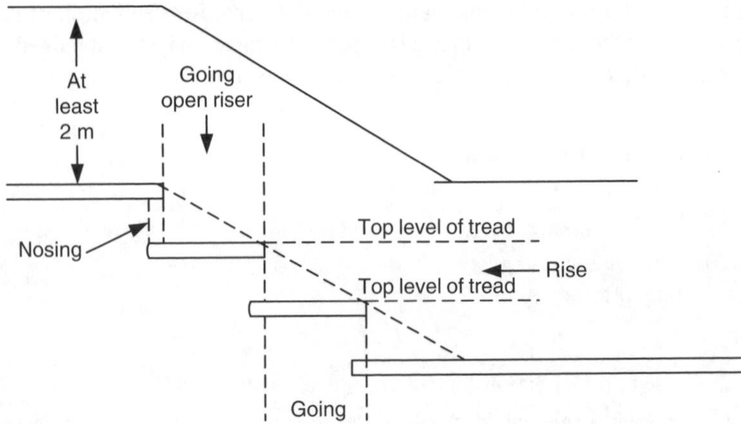

Figure 6.159 Rise and going plus headroom

Steps

Steps should have level treads.	K1 (1.8)
Steps may have open risers, but treads should then overlap each other by at least 16 mm.	K1 (1.8)
Steps should be uniform with parallel nosings, the stair should have handrails on both sides and the treads should have slip resistant surfaces.	
The headroom on the access between levels should be no less than 2 m.	K1 (1.10)
Landings should be provided at the top and bottom of every flight.	K1 (1.15)
The width and length of every landing should be the same (or greater than) the smallest width of the flight.	K1 (1.15)
Landings should be clear of any permanent obstruction.	K1 (1.16)
Landings should be level.	K1 (1.17)
Any door (entrance, cupboard or duct) that swings across a landing at the top or bottom of a flight of stairs must leave a clear space of at least 400 mm across the full width of the flight.	K1 (1.16)

Flights and landings should be guarded at the sides when there is a drop of more than 600 mm.	K1 (1.28–1.29)

For stairs that are likely to be used by children under 5 years the construction of the guarding shall be such that a 100 mm sphere cannot pass through any openings in the guarding and children will not easily climb the guarding.

• For loft conversions, a fixed ladder should have fixed handrails on both sides.	K1 (1.25)

Whilst there are no recommendations for minimum stair widths, designers should bear in mind the requirements of Approved Documents B (means of escape) and M (access and facilities for disabled people).

Ramps

All ramps shall provide reasonable safety between levels in a building (where the difference in level is more than 600 mm) and other buildings where the change of level is more than 380 mm.

Ramps should be clear of permanent obstructions.	K1 (2.4)
The slope of a ramp shall be no more than 1:12.	K1 (2.1)
Ramps should have a handrail on both sides if they are wider than 1 m and on at least one side if they are less than 1 m wide.	K1 (2.5) M
Handrail heights should be between 900 mm and 1000 mm measured to the top of the handrail from the pitch line or floor.	K1 (2.5) M
All ramps should have landings.	K1 (2.6)
All ramps (and associated landings) should have a clear headroom throughout of at least 2 m.	K1 (2.2)
Ramps and landings should be guarded at the sides when there is a drop of more than 600 mm.	K1 (2.7)

For stairs that are likely to be used by children under 5 years the construction of the guarding shall be such that a 100 mm sphere cannot pass through any openings in the guarding and children will not easily climb the guarding.

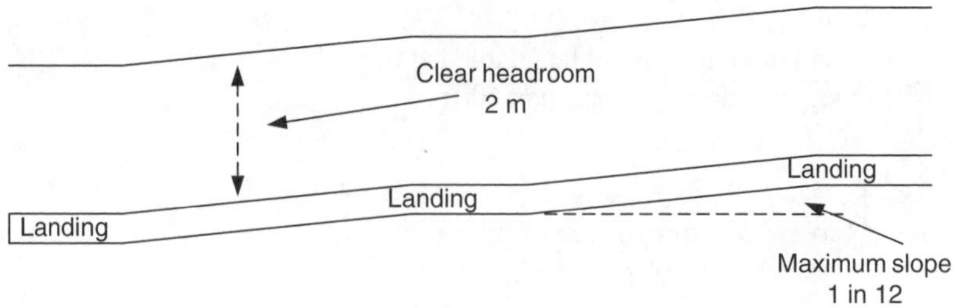

Figure 6.160 The recommended design of a ramp

Protection from falling

All stairs, landings, ramps and edges of internal floors shall have a wall, parapet, balustrade or similar guard at least 900 mm high.	K3 (3.2)
All guarding should be capable of resisting at least the horizontal force given in BS 6399: Part 1: 1996.	K3 (3.2)
If glazing is used as (or part of) the pedestrian guarding, see Approved Document N: Glazing – safety in relation to impact, opening and cleaning.	N
If a building is likely to be used by children under 5 years, the guarding should not have horizontal rails, should stop children from easily climbing it, and the construction should prevent a 100 mm sphere being able to pass through any opening of that guarding.	K3 (3.3)
All external balconies and edges of roofs shall have a wall, parapet, balustrade or similar guard at least 1100 mm high.	K3 (3.2)

Wall cladding

Where wall cladding is required to function as pedestrian guarding to stairs, ramps, vertical drops of 600 mm or greater or as a vehicle barrier, account should be taken of the additional imposed loading as stipulated in Part K.	A1/2 3.5

Where wall cladding is required to safely withstand lateral pressures from crowds, an appropriate design loading is given in BS 6399 Part 1 and the *Guide to Safety at Sports Grounds* (4th Edition, 1997).

Escape routes

Any storey which has more than one escape stair should be planned so that it is not necessary to pass through one stairway to reach another. B1 3.13 (V2)

If an escape stair forms part of the only escape route from an upper storey of a large building it should not be continued down to serve any basement storey. B1 2.44 (V2)

The basement should be served by a separate stair.

If there is more than one escape stair from an upper storey of a building, only one of the stairs serving the upper storeys of the building need be terminated at ground level. B1 2.45 (V2)

Note: Other stairs may connect with the basement storey(s) if there is a protected lobby or a protected corridor between the stair(s) and accommodation at each basement level.

Doors on escape routes

Unless escape stairways and corridors are protected by a pressurization system complying with BS EN 12101-6:2005, every dead-end corridor exceeding 4.5 m in length should be separated by self-closing fire doors (together with any necessary associated screens) from any part of the corridor which: B1 3.27 (V2)

- provides two directions of escape;
- continues past one storey exit to another.

A door that opens towards a corridor or a stairway should be sufficiently recessed to prevent its swing from encroaching on the effective width of the stairway or corridor. B1 5.16 (V2)

Vision panels shall be provided where doors on escape routes sub-divide corridors, or where any doors are hung to swing both ways. B1 5.17 (V2)

see also Parts M and N.

Revolving doors, automatic doors and turnstiles should not be placed across escape routes. B1 5.18 (V2)

Escape stairs

An external escape stair may be used, provided that:

• there is at least one internal escape stair from every part of each storey (excluding plant areas);	B1 2.49 (V2)
• in the case of an Assembly and Recreation building, the route is not intended for use by members of the public; or	B1 4.44 (V2)
• in the case of an Institutional building, the route serves only office or residential staff accommodation.	
• all doors giving access to the stair are fire-resisting and self-closing;	B1 2.15 (a, b and c) (V1)
• any part of the external envelope of the building within 1800 mm of (and 9 m vertically below) the flights and landings of an external escape stair is of fire resisting construction (see Figure 6.161);	B1 5.25 (V2)
• there is protection by fire-resisting construction for any part of the building within 1800 mm of the escape route from the stair to a place of safety;	
• glazing should also be fire resistant and fixed shut.	

Note: Glazing in any fire-resisting construction should be fire-resisting and fixed shut.

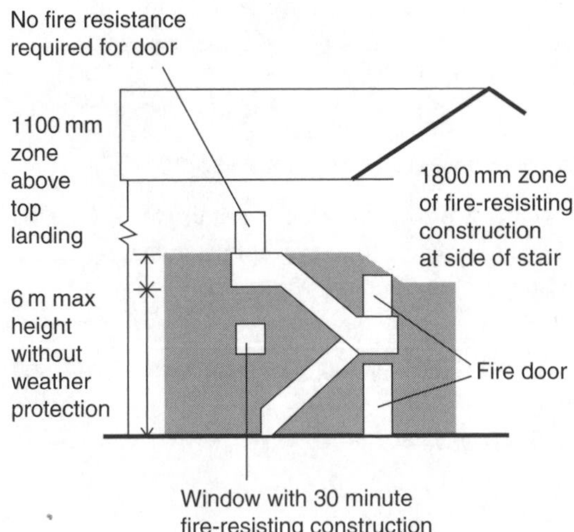

Figure 6.161 Fire resistance of areas adjacent to external stairs

Escape stairs shall have a protected lobby or protected corridor at all levels (except the top storey, all basement levels and when the stair is a firefighting stair) if the:	B1 4.34 (V2)
• stair is the only one serving a building which has more than one storey above or below the ground storey; • stair serves a storey that is higher than 18 m; or • building is designed for phased evacuation;	
• the stairway is near (or potentially next to) a place of special fire hazard.	B1 4.35 (V2)
External escape stairs greater than 6 m in vertical extent shall be protected from the effects of adverse weather conditions.	B1 2.15d
If the building (or part of the building) is served by a single access stair, that stair may be external if it: • serves a floor not more than 6 m above the ground level; and • meets the provisions in paragraph 5.25.	B1 2.48 (V2)

Protection of escape stairs

Escape stairs need to have a satisfactory standard of fire protection.	B1 4.31 (V2)
Internal escape stairs should be a protected stairway within a fire-resisting enclosure.	B1 4.32 (V2)
Except for bars and restaurants, an escape stair may be open provided that: • it does not connect more than two storeys and reaches the ground storey not more than 3 m from the final exit; and • the storey is also served by a protected stairway; or • it is a single stair in a small premises with the floor area in any storey not exceeding 90 m^2.	B1 4.33 (V2)
A dwelling house with more than one floor 4.5 m above ground level may either have a protected stairway which: • extends to the final exit (see Figure 6.162); or • gives access to at least two escape routes at ground level, each delivering to final exists and separated	B1 2.6 (V1)

from each other by fire-resisting construction and fire doors (see Figure 6.162(b)); or

- the top floor can be separated from the lower storeys by fire-resisting construction and be provided with its own alternative escape route leading to its own final exit (see Figure 6.162).

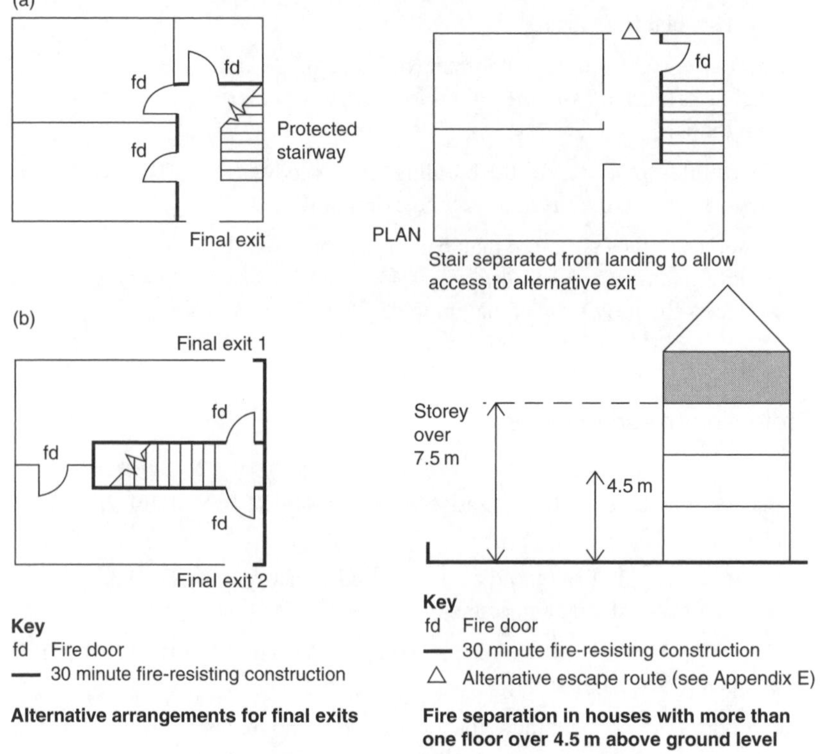

Figure 6.162 Alternative arrangements for final exits

Construction of escape stairs

The flights and landings of every escape stair should be constructed using materials of limited combustibility particularly if it is: B1 5.19 (V2)

- the only stair serving the building;
- within a basement storey;

- serves any storey having a floor level more than 18 m above ground or access level;
- external;
- a firefighting stair.

Single steps may **only** be used on an escape route if they are prominently marked. B1 5.21 (V2)

Helical and spiral stairs forming part of an escape route should be: B1 5.22a (V2)

- designed in accordance with BS 5395-2:1984;
- be type B (Public stair) if they are intended to serve members of the public.

Fixed ladders should not be used as a means of escape for members of the public. B1 5.22b (V2)

Note: See Part K for guidance on the design of helical and spiral stairs and fixed ladders.

If a protected stairway projects beyond, or is recessed from, or is in an internal angle adjoining external wall of the building, then the distance between any unprotected area in the external enclosures to the building and any unprotected area in the enclosure to the stairway should be at least 1800 mm (see Figure 6.163). B1 5.24 (V2)

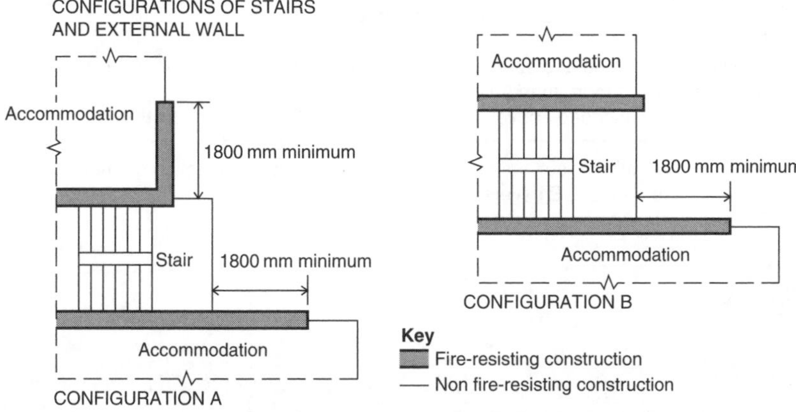

Figure 6.163 External protection to protected stairways

The width of escape stairs should: B1 4.15 (V2)

- Not be less than the width of any exit(s);
- Not be less than the minimum widths given in Table 6.64;
- not exceed 1400 mm if their vertical extent is more than 30 m, **unless** it is provided with a central handrail;
- not reduce in width at any point on the way to a final exit.

In public buildings, if the width of the stair is more than 1800 mm, the stair should have a central handrail. B1 4.16 (V2)

Every escape stair should be wide enough to accommodate the number of persons needing to use it in an emergency. B1 4.18 (V2)

Note: For further guidance and worked examples see Appendix C to Part B and Sections 4.18 (V2) to 4.25 (V2).

Table 6.64 Minimum widths of escape stairs

	Stair situation	Maximum number of people served	Minimum stair width
1a	In an Institutional building (unless the stair is only used by staff)	150	1000 mm
1b	In an Assembly building and serving an area used for assembly purposes (unless the area is less than 100 m^2)	220	1100 mm
1c	In any other building and serving an area with an occupancy of more than 50 people	Over 2200	1000–1800 mm*
2	Any stair not described above	50	800 mm

*Depending on whether the stairs are used for simultaneous evacuation or phased evacuation (see Table 7 of Part B1 (Version 2)).

Lighting

Lighting to escape stairs should be on a separate circuit from that supplying any other part of the escape route. B1 5.36 (V2)

The installation of an escape lighting system shall be in accordance with BS 5266-1:2005 and BS 7671 (latest edition).

Number of escape stairs

The number of escape stairs required in a building (or part of a building) will be determined by: • the constraints imposed by the design of horizontal escape routes; • whether independent stairs are required in mixed occupancy buildings; • whether a single stair is acceptable; and • the width for escape (and the possibility that a stair may have to be discounted because of fire or smoke).	B1 4.2 (V2)
Provided that independent escape routes are not necessary from areas in different purpose groups, single escape stairs may be used from: • small premises (other than bars or restaurants); • office buildings comprising not more than five storeys above the ground storey; • factories comprising not more than one storey above the ground storey if the building is of normal risk (two storeys if the building is of low risk); or • process plant buildings with an occupant capacity of not more than 10 people.	B1 4.6 (V2)

Protected shafts

Protected shafts (i.e. spaces that connect compartments, such as stairways and service shafts) shall be protected to restrict fire spread between the compartments.	B2 8.7 (V2)
The uses of protected shafts should be restricted to stairs, lifts, escalators, chutes, ducts and pipes.	B2 8.36 (V2)
Any stairway or other shaft passing directly from one compartment to another should be enclosed in a protected shaft so as to delay or prevent the spread of fire between compartments.	B2 8.35 (V2)

Protected shafts provide for the movement of people (e.g. stairs, lifts), or for passage of goods, air or services such as pipes or cables between different compartments. The elements enclosing the shaft (unless formed by adjacent external walls) are compartment walls and floors. Figure 6.164 shows three common examples which illustrate the principles.

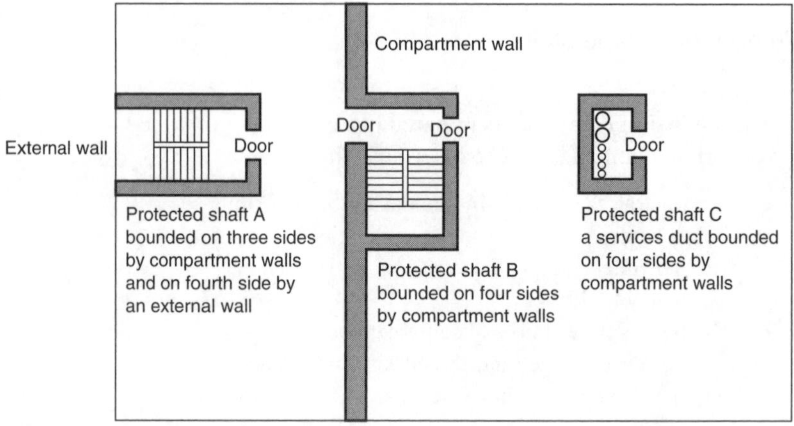

Figure 6.164 Protected shafts

An uninsulated glazed screen may be incorporated in the enclosure to a protected shaft between a stair and a lobby or corridor which is entered from the stair provided that: B2 8.38 (V2)

- the fire resistance for the stair enclosure is not more than 60 minutes; and
- the glazed screen has at least 30 minutes fire resistance; and
- the lobby or corridor is enclosed to at least a 30 minute standard (see Figure 6.165).

If a protected shaft contains a stair and/or a lift, it should **not** also contain: B2 8.40 (V2)

- a pipe conveying oil (other than in the mechanism of a hydraulic lift); or
- a ventilating duct (other than a duct provided solely for ventilating the stairway).

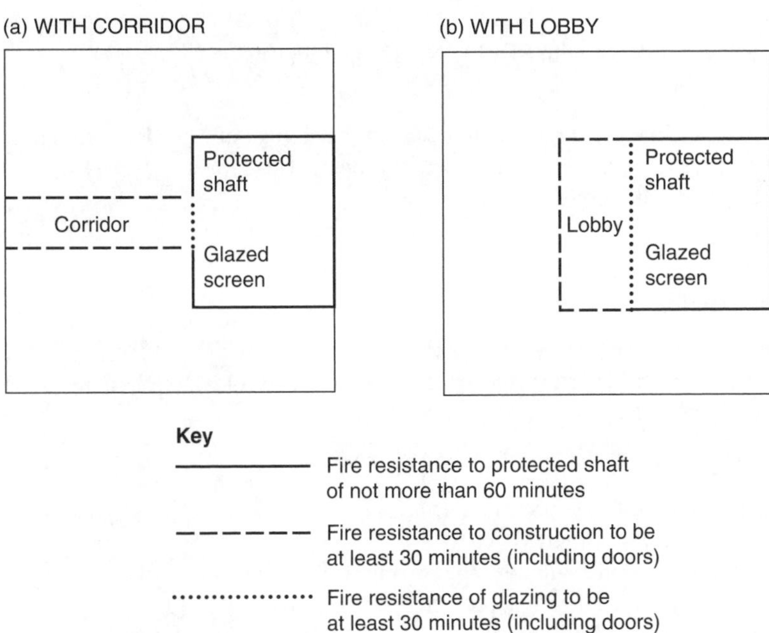

Figure 6.165 Uninsulated glazed screen separating protected shaft from lobby or corridor

Protected stairways

Protected stairways should discharge: B1 2.38 (V2)

- directly to a final exit; or
- via a protected exit passageway to a final exit.

Where two protected stairways (or exit passageways leading to different final exits) are adjacent, they should be separated by an imperforate enclosure. B1 2.39 (V2)

A protected stairway should be relatively free of potential sources of fire. B1 2.40 (V2)

In single stair buildings, meters located within the stairway should be enclosed within a secure cupboard which is separated from the escape route with fire-resisting construction. B1 2.40 (V2)

Gas service and installation pipes (together with their associated meters etc.) should **not** be incorporated within a protected stairway unless the gas installation B1 4.40 (V2) B1 2.42 (V2)

is in accordance with SI 1996 No 825 and SI 1998 No 2451.	
Refuse chutes and rooms provided for the storage of refuse should **not** be located within protected stairways or protected lobbies.	B1 5.55 and 5.56 (V2)

Ventilation

Separate ventilation systems should be provided for each protected stairway.	B1 5.47 (V2)
Air circulation systems shall ensure that:	B1 2.17 (V1)
• smoke or fire is prevented from spreading into a protected stairway;	B3 7.10 (V1)
• transfer grilles are not fitted in any wall, door, floor or ceiling enclosing a protected stairway;	B1 2.18 (V2)
• any duct passing through the enclosure to a protected stairway is of rigid steel construction and all joints between the ductwork and the enclosure are fire-stopped;	B3 10.2 (V2)
• ventilation ducts supplying or extracting air directly to or from a protected stairway do not serve other areas as well;	
• any system of mechanical ventilation which recirculates air, and which serves both the stairway and other areas, is designed to shut down on the detection of smoke within the system.	

Passenger lifts

Passenger lifts which serve floors more than 4.5 m above ground level should either be located in the enclosure to the protected stairway or be contained in a fire-resisting lift shaft.	B1 2.18
Lift wells should be either:	B1 5.42 (V2)
• contained within the enclosures of a protected stairway; or	
• enclosed throughout their height with fire-resisting construction.	

In basements and enclosed (i.e. non open-sided) car parks the lift should be approached only by a protected lobby or protected corridor (unless it is within the enclosure of a protected stairway).	B1 5.43 (V2)
Lift shafts should not be continued down to serve a basement storey if it is within the enclosure to an escape stair which is terminated at ground level.	B1 5.44 (V2)

Common stairs

All common stairs should be situated within a fire-resisting enclosure (i.e. it should be a protected stairway), to reduce the risk of smoke and heat making use of the stair hazardous.	B1 2.36 (V2)
A single common stair can be acceptable in some cases, but otherwise there should be access to more than one common stair for escape purposes.	B1 2.32 (V2)
Where a common stair forms part of the only escape route from a flat, it should **not** also serve any covered car park, boiler room, fuel storage space or other ancillary accommodation of similar fire risk.	B1 2.46 (V2)
Common stairs which do not form part of the only escape route from a fiat may also serve ancillary accommodation if they are separated from the ancillary accommodation by a protected lobby or a protected corridor.	B1 2.47 (V2)
If the stair serves an enclosed (non open-sided) car park, or place of special fire hazard, the lobby or corridor should have not less than $0.4\,m^2$ permanent ventilation or be protected by a mechanical smoke control system.	B1 2.47 (V2)

Fire protected stairways

Fire protected stairways shall, as far as reasonably possible:

• exclude all flames, smoke and gases shall be designed to provide effective 'fire sterile' areas that lead to places of safety outside the building;	B1 1.viii

• consist of fire-resistant material and fire resistant doors and have an appropriate form of smoke control system;	B1 1.viii
• contain a fire main outlet;	B5 17.10 (V2)
• have cavity barriers above the enclosures	B1 2.14
• be free of potential sources of fire;	B1 4.38 (V2)
• discharge: – directly to a final exit; or – by way of a protected exit passageway to a final exit.	B1 4.36 (V2)

Note: Any such protected exit passageway should have the same standard of fire resistance and lobby protection as the stairway it serves.

If two protected stairways are adjacent, they (and any protected exit passageways linking them to final exits) shall be separated by an imperforate enclosure.	B1 4.37 (V2)

Firefighting stairs

Any stair used as a firefighting stair should:	B1 2.33 (V2)
• be at least 1100 mm wide (see Part B V2, Appendix C for measurement of width);	
• not have a protected lobby or protected corridor;	B1 4.34 (V2)
• be constructed using materials of limited combustibility;	B1 5.19 (V2)
• be approached from the accommodation, through a firefighting lobby (unless it is in blocks of flats)	B5 17.11 (V2)
All firefighting shafts should be equipped with fire mains having outlet connections and valves at every storey.	B5 17.12 (V2)

Refuges

Refuges are relatively safe waiting areas for short periods. They are **not** areas where disabled people should be left alone indefinitely until rescued by the fire and rescue service, or until the fire is extinguished.

A refuge (i.e. an enclosure such as a compartment, protected lobby, protected corridor, protected stairway (see Figure 6.166) or an area in the open air such as a flat roof, balcony, podium or similar) should be provided for each protected stairway. B1 4.8 (V2)

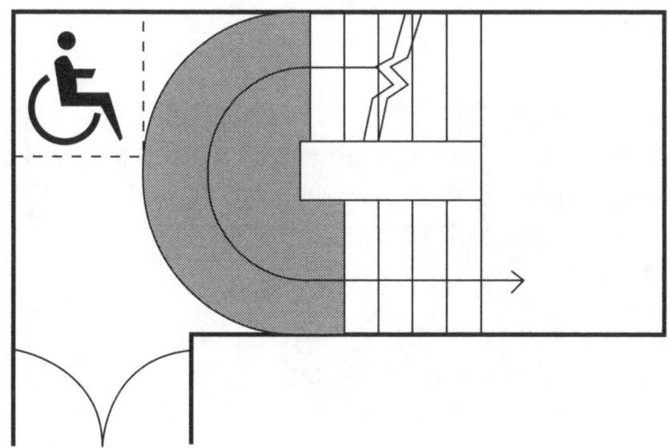

Provision where access to the wheelchair space is counter to the access flow within the stairway

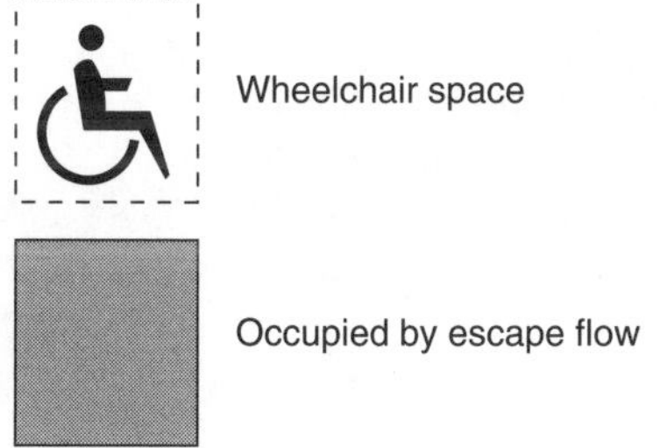

Figure 6.166 Refuge formed in a protected stairway

Note: The number of refuge spaces need not necessarily equal the sum of the number of wheelchair users who can be present in the building.

Refuges and evacuation lifts should be clearly identified by appropriate fire safety signs. B1 4.10 (V2)

If a refuge is in a lobby or stairway the sign should be accompanied by a blue mandatory sign worded ***'Refuge – keep clear'.***

Smoke alarms

Smoke alarms should **not** be fixed over a stair or any other opening between floors.	B1 1.16 (V1) B1 1.15 (V2)

Flats

Except for kitchens:

all habitable rooms in the upper storey(s) of a multi-storey flat which are served by only one stair should (depending on the height of the top storey) be provided with: • a window (or external door); or • direct access to a protected stairway.	B1 2.12 (V2)

Where the vertical distance between the floor of the entrance storey and the floors above and below it does not exceed 7.5 m, multi-storey flats are required to:

• provide a protected stairway **plus** additional smoke alarms in all habitable rooms and a heat alarm in any kitchen; or	B1 2.16c (V2)
• provide a protected stairway **plus** a sprinkler system and smoke alarms.	B1 2.16d (V2)

An alternative exit from a flat should:

• be remote from the main entrance door to the flat; and • lead to a defined exit or common stair by way of: – a door onto an access corridor, access lobby or common balcony; or – an internal private stair leading to an access corridor, access lobby or common balcony at another level; or – a door into a common stair; or – a door onto an external stair; or – a door onto an escape route over a flat roof.	B1 2.17 (V2)

Flats in mixed use buildings

The stairs of buildings which are no more than three storeys above the ground storey may serve both flats and other occupancies, **provided** that the stairs are separated from each occupancy by protected lobbies at all levels. B1 2.50 (V2)

The stairs of buildings that are more than three storeys above the ground storey may serve both flats and other occupancies, **provided** that: B1 2.51 (V2)

- the flat is ancillary to the main use of the building and is provided with an independent alternative escape route;
- the stair is separated from any other occupancies on the lower storeys by protected lobbies (at those storey levels);
- any automatic fire detection and alarm system with which the main part of the building is fitted also covers the flat;
- any security measures should not prevent escape at all material times.

Galleries

Any cooking facilities within a room containing a gallery should either: B1 2.12 (V1)

- be enclosed with fire-resisting construction; or B1 2.8 (V2)
- be remote from the stair to the gallery.

Loft conversions

Where the conversion of an existing roof space (such as a loft conversion to a two-storey house) means that a new storey is going to be added, then the stairway will need to be protected with fire-resisting doors and partitions. B1 2.20b

Basements

Because of their situation, basement stairways are more likely to be filled with smoke and heat than stairs located in ground and upper storeys. Special measures are, therefore, required in order to prevent a basement fire endangering upper storeys.

If an escape stair forms part of the only escape route from an upper storey of a building it should **not** be continued down to serve a basement storey (i.e. the basement should be served by a separate stair).	B1 4.42 (V2)
If there is more than one escape stair from an upper storey of a building then only **one** of the stairs serving the upper storeys of the building need be terminated at ground level.	B1 4.43 (V2)

Owing to the possibility of a single stairway becoming blocked by smoke from a fire in the basement or ground storey:

• basement storeys in a dwelling house that contain a habitable room shall be provided with either:	B1 2.13 (V1)
– a protected stairway leading from the basement to a final exit; or	B1 2.6 (V2)
– an external door or window suitable for egress from the basement.	

Access and facilities for disabled people

Internal steps, stairs and ramps

Stepped access

A stepped access should:	
• have a level landing at the top and bottom of each flight;	M (3.51a)
• be 1200 mm long each landing and be unobstructed.	
Doors should not swing across landings.	M (3.51a)
The surface width of flights between enclosing walls, strings or upstands should not be less than 1.2 m.	M (3.51a)
There should be no single steps.	M (3.51a)
Nosings for the tread and the riser should be 55 mm wide and of a contrasting material.	M (3.51a)
Step nosings should not project over the tread below by more than 25 mm (see Figure 6.167).	M (3.51a)

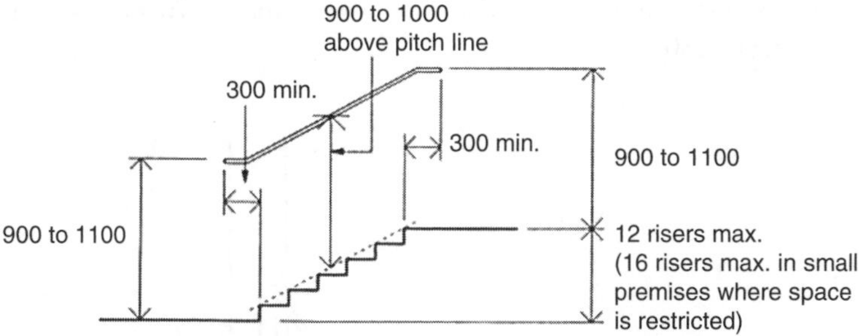

Figure 6.167 Internal stairs – key dimensions

The rise and going of each step should be consistent throughout a flight. M (3.51a)

The rise of each step should be between 150 mm and 170 mm. M (3.51a)

The going of each step should be between 280 mm and 425 mm. M (3.51a)

Rises should not be open. M (3.51a)

There should be a continuous handrail on each side of a flight and landings. M (3.51a)

If additional handrails are used to divide the flight into channels, then they should not be less than 1 m wide or more than 1.8 m wide. M (3.51a)

Flights between landings should contain no more than 12 risers. M (3.51b)

The rise of each step should be between 150 mm and 170 mm. M (3.51c)

The going of each step should be at least 250 mm. M (3.51d)

For mobility-impaired people, a going of at least 300 mm is preferred.

Materials for treads should not present a slip hazard. M (3.50)

Areas below stairs or ramps with a soffit less than 2.1 m above ground level should be protected by guarding and low level cane detection. M (3.51e)

Any feature projecting more than 100 mm onto an access route should be protected by guarding that includes a kerb (or other solid barrier) that can be detected using a cane (see Figure 6.168). M (3.51e)

Note: For school buildings, the rise should not exceed 170 mm, with a preferred going of 280 mm.

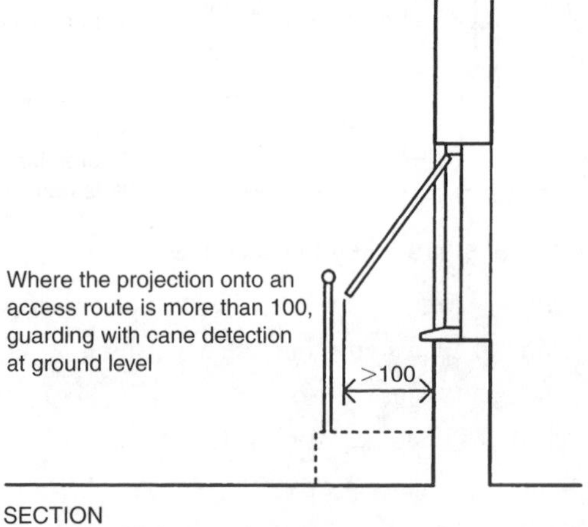

Figure 6.168 Avoiding hazards on access routes

Internal ramps

Where an internal ramp is provided: M (3.53)

- the approach should be clearly signposted;
- the going should be no greater than 10 m;
- the rise should be no more than 500 mm;
- if the total rise is greater than 2 m then an alternative means of access (e.g. a lift) should be provided for wheelchair users;
- the ramp surface should be slip resistant;
- the ramp surface should be of a contrasting colour with that of the landings;
- frictional characteristics of ramp and landing surfaces should be similar;
- landings at the foot and head of a ramp should be at least 1.2 m long and clear of any obstructions;
- intermediate landings should be at least 1.5 m long and clear of obstructions;
- all landings should be:
  - level;
  - have a maximum gradient of 1:60 along their length;
  - have a maximum cross fall gradient of 1:40;
- there should be a handrail on both sides;

- in addition to the guarding requirements of Park K, there should be a visually contrasting kerb on the open side of the ramp (or landing) at least 100 mm high.

Where the change in level is 300 mm or more, two or more clearly signposted steps should be provided (i.e. in addition to the ramp).	M (3.53b)
If the change in level is no greater than 300 mm, a ramp should be provided instead of a single step.	M (3.53c)
All landings should be level and a maximum gradient of 1:60 along their entire length.	M (3.53d)
Areas below stairs or ramps with a soffit less than 2.1 m above ground level should be protected by guarding and low level cane detection.	M (3.53e)
Any feature projecting more than 100 mm onto an access route should be protected by guarding that includes a kerb (or other solid barrier) that can be detected using a cane (see Figure 6.169).	M (3.53e)
Gradients should be as shallow as practicable.	M (3.52)

Handrails to internal steps, stairs and ramps

Handrails to external stepped or ramped access should be positioned as per Figure 6.169.	M (1.37a)

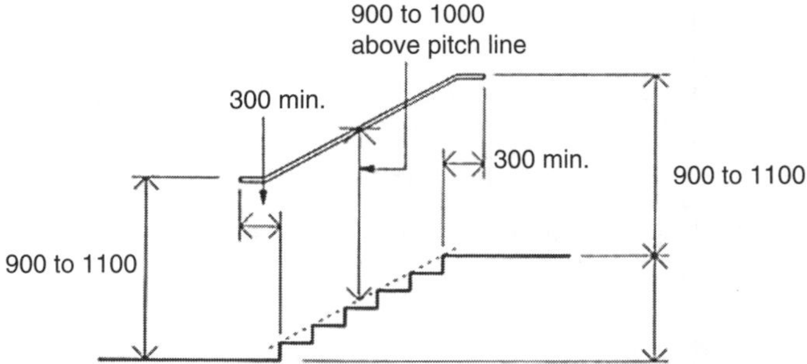

Figure 6.169 Handrails to internal steps, stairs and ramps – key dimensions

Handrails to internal steps, stairs and ramps should: M (3.55)

- be continuous across flights and landings;
- extend at least 300 mm horizontally beyond the top and bottom of a ramped access;
- not project into an access route;
- contrast visually with the background;
- have a slip resistant surface which is not cold to the touch;
- terminate in such a way that reduces the risk of clothing being caught;
- either be circular (with a diameter of between 40 and 45 mm) or oval with a width of 50 mm (see Figure 6.170).

Handrails to external stepped or ramped access should: M (3.55)

- not protrude more than 100 mm into the surface width of the ramped or stepped access where this would impinge on the stair width requirement of Part B1;
- have a clearance of between 60 and 75 mm between the handrail and any adjacent wall surface;
- have a clearance of at least 50 mm between a cranked support and the underside of the handrail;
- ensure that its inner face is located no more than 50 mm beyond the surface width of the ramped or stepped access;
- should be spaced away from the wall and rigidly supported in a way that avoids impeding finger grip;
- should be set at heights that are convenient for all users of the building.

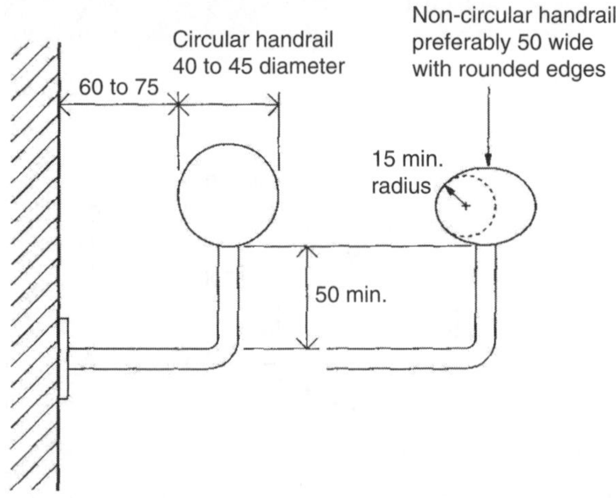

Figure 6.170 Handrail designs

Common stairs in blocks flats

The aim for all buildings containing flats should be to make reasonable provision for disabled people to visit occupants who live on any storey of the building, via a common staircase or a lift.

Common stairs

If there is no passenger lift to provide access between storeys, a stair (designed to suit the needs of ambulant disabled people, people with impaired sight and people with sensory impairments) should be provided.	M (9.3 and 9.4)
If a passenger lift is not installed, a common stair should be provided which has:	
• step nosings with contrasting brightness;	M (9.5a)
• top and bottom landings whose lengths are in accordance with Part K1;	M (9.5b)
• steps with suitable tread nosing profiles (see Figure 6.171) with a uniform rise not more than 170 mm;	M (9.5c)
• a uniform going of each step not less than 250 mm;	M (9.5d)
• risers which are not open;	M (9.5e)
• a continuous handrail on each side of flights and landings (if the rise of the stair comprises two or more rises).	M (9.5f)

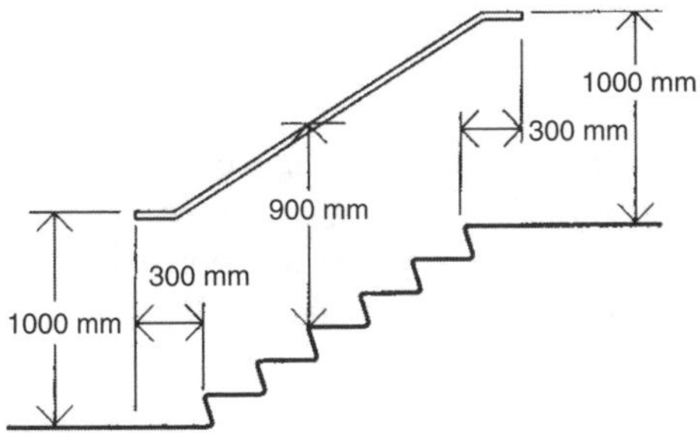

Figure 6.171 Common stairs in blocks of flats

A single common stair can be acceptable in some cases, but otherwise there should be access to more than one common stair for escape purposes.	B1 2.32 (V2)

All common stairs should be situated within a fire-resisting enclosure (i.e. it should be a protected stairway), to reduce the risk of smoke and heat making use of the stair hazardous.	B1 2.36 (V2)
Where a common stair forms part of the only escape route from a flat, it should **not** also serve any covered car park, boiler room, fuel storage space or other ancillary accommodation of similar fire risk.	B1 2.46 (V2)
Common stairs which do not form part of the only escape route from a flat may also serve ancillary accommodation if they are separated from the ancillary accommodation by a protected lobby or a protected corridor.	B1 2.47 (V2)
If the stair serves an enclosed (non open-sided) car park, or place of special fire hazard, the lobby or corridor should have not less than 0.4 m^2 permanent ventilation or be protected by a mechanical smoke control system.	B1 2.47 (V2)

6.13 Windows

6.13.1 Requirements

Ventilation

There shall be adequate means of ventilation provided for people in the building.

(Approved Document F)

Protection from falling

Pedestrian guarding should be provided for any part of a floor (including the edge below an opening window) gallery, balcony, roof (including rooflight and other openings), any other place to which people have access and any light well, basement area or similar sunken area next to a building.

(Approved Document K2)

Conservation of fuel and power

Reasonable provision shall be made for the conservation of fuel and power in buildings by:

(a) limiting heat gains and losses:
  (i) through thermal elements and other parts of the building fabric; and
  (ii) from pipes, ducts and vessels used for space heating, space cooling and hot water services;

(b) providing and commissioning energy-efficient fixed building services with effective controls; and
(c) providing to the owner sufficient information about the building, the fixed building services and their maintenance requirements so that the building can be operated in such a manner as to use no more fuel and power than is reasonable in the circumstances.

(Approved Document L1)

Responsibility for achieving compliance with the requirements of Part L rests with the person carrying out the work. That person may be, for example, a developer, a main (or sub-) contractor, or a specialist firm directly engaged by a private client.

The person responsible for achieving compliance should either themselves provide a certificate, or obtain a certificate from the sub-contractor, that commissioning has been successfully carried out. The certificate should be made available to the client and the building control body.

Protection against impact

Glazing with which people are likely to come into contact whilst moving in or about the building, shall:

- *if broken on impact, break in a way which is unlikely to cause injury; or*
- *resist impact without breaking; or*
- *be shielded or protected from impact.*

(Approved Document N)

Means of escape

In an emergency, the occupants of any part of the building shall be able to escape without any external assistance.

(Approved Document B1)

6.13.2 Meeting the requirement

Emergency egress windows

Except for kitchens, all habitable rooms on the ground floor (and the upper storey(s) of a dwelling house which are served by only one stair) should be provided with an emergency egress window.	B1 2.3 and 2.4 (V1) B1 2.11 and 2.12 (V2)

Note: There are some other alternatives if this is not possible, such as having access to a protected route, so it is best to see the Regulations if you need confirmation of this point.

The window should be at least 450 mm high and 450 mm wide and have an unobstructed openable area of at least 0.33 m^2.	B1 2.8 (V1)
The bottom of the openable area should be not more than 1100 m above the floor.	B1 2.9 (V2)
The window should enable the person escaping to reach a place free from danger of fire (e.g. a courtyard or back garden which is at least as deep as the dwelling house is high – see Figure 6.172).	

Notes:

1. Approved Document K (Protection from falling, collision and impact) specifies a minimum guarding height of 800 mm, except in the case of a window in a roof where the bottom of the opening may be 600 mm above the floor.
2. Locks (with or without removable keys) and stays may be fitted to egress windows, provided that the stay is fitted with a child resistant release catch.
3. Windows should be designed so that they remain in the open position without needing to be held open by the person making their escape.

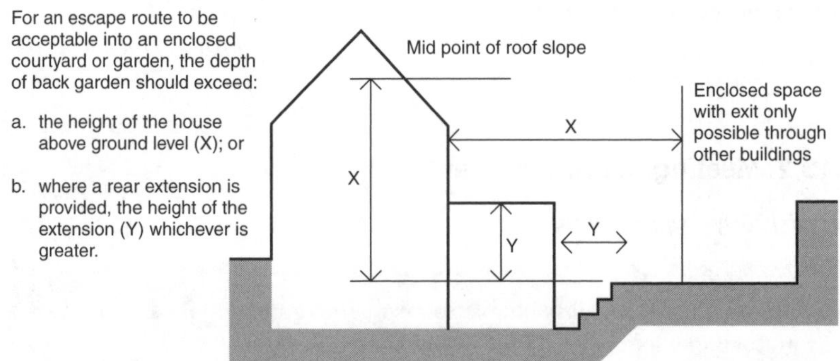
For an escape route to be acceptable into an enclosed courtyard or garden, the depth of back garden should exceed:
a. the height of the house above ground level (X); or
b. where a rear extension is provided, the height of the extension (Y) whichever is greater.
Mid point of roof slope
X
X
Y
Y
Enclosed space with exit only possible through other buildings

Figure 6.172 Ground or basement storey exit into an enclosed space

Any inner room that is a kitchen, laundry or utility room, dressing room, bathroom, WC or shower or situated not more than 4.5 m above ground level and whose only escape route is through another room, shall be provided with an emergency egress window.	B1 2.9 (V1) B1 2.5 (V2)
All galleries shall be provided with an alternative exit or, where the gallery floor is not more than 4.5 m above ground level, an emergency egress window.	B1 2.12 (V1) B1 2.8 (V2)
All basement storeys in a dwelling house that contain a habitable room shall be provided with either an external door or window suitable for egress from the basement.	B1 2.13 (V1) B1 2.6 (V2)

Note: There are certain alternatives. See Regulations for details

Dimensions

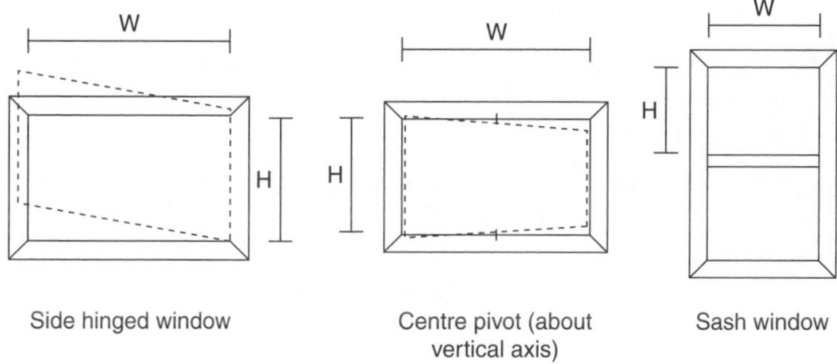

Figure 6.173 Window dimensions

The height times width of the opening part of hinged or pivot windows that are designed to open **more** than 30 and/or sliding sash windows, should be at least 1/20 of the floor area of the room.	F App B
The height times width of the opening Approved Document of hinged or pivot windows designed to open **less** than 30 should be at least 1/10 of the floor area of the room.	F App B

If a room contains more than one openable window, then the areas of **all** the opening parts may be added together to achieve the required floor area.

Ventilation

Habitable rooms **without** openable windows may be either ventilated through another habitable room (i.e. an internal room) provided that the other room has: F 1.12 F 1.13

- purge ventilation; and
- an 8000 mm^2 background ventilator; and
- there is a permanent opening between the two rooms.

Habitable rooms **without** openable windows may be either ventilated through a conservatory provided that that conservatory has: F 1.12 F 1.14

- purge ventilation; and
- an 8000 mm^2 background ventilator; and
- there is a closable opening between the room and the conservatory that is equipped with:
  - purge ventilation; and
  - an 8000 mm^2 background ventilator.

Windows with night latches should **not** be used as they are more liable to draughts as well as being a potential security risk. F 0.17

If a fan is installed in an internal room or an office **without** an openable window, then the fan should have a 15 minute overrun. F Table 1.5 F Table 2.2c

Extensions

If the additional room is connected to an existing habitable room which now has no windows opening to outside (or if it still has windows opening to outside, but with a total background ventilator equivalent area less than 5000 mm^2), then the ventilation opening (or openings) shall be greater than 8000 mm^2 equivalent area. F 3.8ai F 3.8aii

If the additional room is connected to an existing habitable room which still has windows opening to outside (but with a total background ventilator equivalent area of at least 5000 mm^2 equivalent area) then there should be: F 3.8aiii

- background ventilators of at least 8000 mm^2 equivalent area between the two rooms; and
- background ventilators of at least 8000 mm^2 equivalent area between the additional room and outside.

Replacement windows

Where windows are to be replaced the replacement work should comply with the requirements of Parts L and N.	B1 2.19
If a window is currently located where, in the case of a new dwelling house, an escape window would be necessary, then the replacement window opening should be sized to provide at least the same potential for escape as the window it replaces.	B1 2.19

Note: If the original window is larger than necessary (i.e. for escape purposes) the window opening **could** be reduced down to a minimum of 450 mm high and 450 mm wide provided that it has an unobstructed openable area of at least 0.33 m^2 and the openable area is not more than 1100 m above the floor.

Protection from falling

All stairs, landings, ramps and edges of internal floors shall have a wall, parapet, balustrade or similar guard at least 900 mm high.	K3 (3.2)
All guarding should be capable of resisting at least the horizontal force given in BS 6399: Part 1: 1996.	K3 (3.2)
If glazing is used as (or part of) the pedestrian guarding, see Approved Document N: Glazing – safety in relation to impact, opening and cleaning.	N
If a building is likely to be used by children under 5 years, the guarding should not have horizontal rails, should stop children from easily climbing it, and the construction should prevent a 100 mm sphere being able to pass through any opening of that guarding.	K3 (3.3)
All external balconies and edges of roofs shall have a wall, parapet, balustrade or similar guard at least 1100 mm high.	K3 (3.2)
All windows, skylights, and ventilators shall be capable of being left open without danger of people colliding with them by: • installing windows, etc. so that projecting parts are kept away from people moving in and around the building; or	K

- installing features that guide people moving in or about the building away from any open window, skylight or ventilator.

Parts of windows (skylights and ventilators) that project either internally or externally more than about 1000 mm horizontally into spaces used by people moving in or about the building should not present a safety hazard. K

16.13.3 Conservation of fuel and power

Domestic buildings

When working on a controlled service or fitting (i.e. where the service or fitting is subject to the requirements of Part G, H, J, L or P of Schedule 1):

all windows, roof windows, rooflights and/or doors should be provided with draught-proofed units. L1B 32

the area-weighted average performance of draught-proofed units for: L1B 32

- new fittings that are provided as part of the construction of an extension shall be no worse than 1.8 W/m^2 K; L1B 32
- new and/or replacement fittings in an existing dwelling shall be no worse than 2.0 W/m^2 K (whole unit) or 1.2 W/m^2 K (centre pane).

U-values shall be calculated (using the methods and conventions set out in BR 443) and should not exceed the limits shown in Table 6.65. L1B 11

Note: Display windows and similar glazing are **not** required to meet the standard given for 'Windows and rooflight'.

Table 6.65 Limiting U-value standards (W/m^{2}K)

Element	Area-weighted dwelling average	Worst individual sub-element
Windows, roof windows, rooflights & doors	2.2	3.3
Windows and rooflights	2.2	3.0
Pedestrian doors	2.2	3.0
Vehicle access & similar large doors	1.5	4.0
High usage entrance doors	6.0	6.0
Roof ventilators (including smoke vents)	6.0	6.0

High internal temperatures caused by solar gains should be minimized by a combination of: L1A 46 L2A 63

- window size and orientation;
- shading;
- ventilation; and
- high thermal capacity.

The building fabric should be constructed so that there are no reasonably avoidable thermal bridges in the insulation layers caused by gaps within the various elements, at the joints between elements, and at the edges of windows and door openings. L1A 51 L2A 68

The area of windows roof windows and doors in extensions should not exceed the sum of: L1B 15

- 25% of the floor area of the extension **plus**
- the area of any windows or doors, which, as a result of the extension works, no longer exist or are no longer exposed.

If the total floor area of the proposed extension **does exceed** these limits, then the work should be regarded as a **new** building, in which case the area of windows and rooflights in the extension should not exceed the values given in Table 6.66 (L2B 27).

Table 6.66 Opening areas in an extension

Building type	Windows and personnel doors as a %age of the exposed wall	Rooflights as a %age of the roof area
Residential buildings (where people temporarily or permanently reside)	30	20
Offices, shops and places where people assemble	40	20
Industrial and storage buildings	15	20
Vehicle access doors and display windows and similar glazing	As required	N/A

If the extension is part of an existing building, and is a conservatory, then:

- thermal and opaque elements should have U-values that are no worse than 3.3 W/m^2K; L1B 22c L2B 32c

- the windows between the building and the extension should be insulated and weather-stripped to at least the same extent as in the existing building; L1B 22a L2B 32a
- replacement glazed elements should be no worse than 2.2 W/m^{2}K (whole unit) or 1.2 W/m^{2}K (centre pane) L1B 22c L2B 32c
- the building fabric should be constructed so that any thermal bridges in the insulation layers around windows (caused by gaps and joints between the various elements) are avoided. L1B 52 L2B 71

Note: Conservatories with a floor area no greater than 30 m^2 are exempt from the Building Regulations.

Non-domestic buildings

Non-domestic buildings should be constructed and equipped so that there are no unreasonable thermal bridges in the insulation layers caused by gaps within the various elements, at the joints between elements, and at the edges of windows and door openings. L2A 68

In occupied spaces that are not served by a comfort cooling system, the combined solar and internal casual gains (people, lighting and equipment) per unit floor area averaged over the period of daily occupancy should not be greater than 35 W/m^2 calculated over a perimeter area not more than 6 m from the window wall and averaged during the period 06.30–16.30 hrs GMT. L2A 64a

Windows, roof windows, rooflights and/or doors should be provided with draught-proofed units. L2B 75

The area-weighted average performance of draught-proofed units for new fittings in extensions and replacement fittings in an existing dwelling shall be no worse than given in Table 6.67. L2B 75

Note: The U-value should be determined with the window in the vertical position (see SAP 2005). Table 6.67 gives values for different window configurations that can be used in the absence of test data or calculated values.

Table 6.67 Standards for controlled fittings

Element	New fittings in an extension (W/m^2K)	Replacement fittings in an existing dwelling (W/m^2K)
Windows, roof windows and glazed rooflights	1.8	2.2 (whole unit) 1.2 (centre pane)

Material changes of use

When a building is subject to a material change of use, then that part of the building affected shall comply with the requirements of Part L and:

any existing window (including roof window or rooflight) or door which separates a conditioned space from an unconditioned space (or the external environment), and which has a U-value that is worse than 3.3 W/mK, should be replaced.	L2B 27c

Material alterations

If a building is subject to a material alteration and an existing element becomes part of the thermal element of a building (where previously it did not) and it has a U-value worse than 3.3 W/m^2K, it shall be replaced (unless it is a display window or high usage door).	L2B 39c

Consequential improvements

If a building has a total useful floor area greater than 1000 m^2 and the proposed building work includes:

- an extension; or
- the initial provision of any fixed building services; or
- an increase to the installed capacity of any fixed building services;

then the energy efficiency of the whole building should consequentially improved and:

All existing windows (less display windows), roof windows, rooflights or doors (excluding high usage entrance doors) that are within the area served by the fixed building service and which have a U-value worse than 3.3 Wm^2K, should be replaced.	L2B 18-7

All replacement windows should include trickle ventilators or have an equivalent background ventilation opening in the same room.	F 3.4
Ventilation openings should not be smaller than the original opening and should be controllable.	F 3.6
Where there was no previous ventilation opening, or where the size of the original ventilation opening is not known, the replacement window(s) shall be greater than the minimum requirements shown in Tables 6.68 and 6.69.	F 3.6

Table 6.68 Equivalent areas for replacement windows – dwellings

Type of room	Equivalent area
Habitable rooms	5000 mm^2
Kitchen	2500 mm^2
Utility room	2500 mm^2
Bathroom (without a WC)	2500 mm^2

Table 6.69 Equivalent areas for replacement windows – buildings other than dwellings

Type of room	Equivalent area
Occupiable rooms with floor areas $<10\,m^2$	2500 mm^2
Occupiable rooms with floor areas $>10\,m^2$	250 mm^2 per m^2 of floor area
Kitchens (domestic type)	2500 mm^2
Bathrooms and shower rooms	2500 mm^2 per bath or shower
Sanitary accommodation (and/or washing facilities)	2500 mm^2 per WC

What about glazing?

Although the installation of replacement windows or glazing (e.g. by way of repair), is not considered as building work under Regulation 3 of the Building Regulations, on the other hand, glazing that:

- is installed in a location where there was none previously;
- is installed as part of an erection;
- is installed as part extension or material alteration of a building;

is subject to these requirements.

The existence of large uninterrupted areas of transparent glazing represents a significant risk of injury through collision. The risk is at its most severe between areas of a building or its surroundings that are essentially at the same level and where a person might reasonably assume direct access between locations that are separated by glazing.

The most likely places where people can sustain injuries are due to impacts with doors, door side panels (especially between waist and shoulder level) when initial impact can be followed by a fall through the glazing resulting in additional injury to the face and body. Hands, wrists and arms are particularly vulnerable.

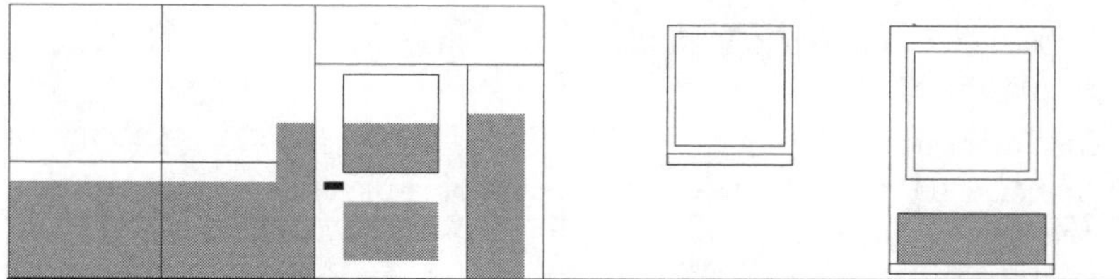

Figure 6.174 Shaded areas show critical locations in internal and external walls

Apart from doors, walls and partitions are a low-level, high-risk area, particularly where children are concerned.

The existence of large uninterrupted areas of transparent glazing represents a significant risk of injury through collision. The risk is at its most severe between areas of a building or its surroundings that are essentially at the same level and where a person might reasonably assume direct access between locations that are separated by glazing.

Approved Document B: Fire safety includes guidance on fire-resisting glazing and the reaction of glass to fire.

Approved Document K: Protection from falling, collision and impact covers glazing that forms part of the protection from falling from one level to another, and that needs to ensure containment as well as limiting the risk of sustaining injury through contact.

Some glazing materials, such as annealed glass, gain strength through thickness; others such as polycarbonates or glass blocks are inherently strong. Some annealed glass is considered suitable for use in large areas forming fronts to shops, showrooms, offices, factories, and public buildings.

Provision of cavity barriers

Cavity barriers should be provided around window openings.	B3 6.3 (V1) B3 9.3 (V2)

Protection against impact

Requirement	Reference
Measures shall be taken to limit the risk of sustaining cutting and piercing injuries.	N1 (0.1)
In critical locations, if glazing is damaged the breakage should only result in small, relatively harmless particles.	N1 (0.2)
Glazing should be sufficiently robust to ensure that the risk of breakage is low.	N1 (0.4)
Steps should be taken to limit the risk of contact with the glazing.	N1 (0.5)
Glazing in critical locations should either be permanently protected, be in small panes or if it breaks, break safely (see BS 6206).	N1 (1.2)
Small panes should not exceed 250 mm and an area of 0.5 m^2.	N1 (1.6)

Transparent glazing

Transparent glazing with which people are likely to come into contact while moving in or about the building shall incorporate features that make it apparent.

Requirement	Reference
The presence of glazing should be made more apparent or visible to people using the building.	N2 (0.8)
The presence of large uninterrupted areas of transparent glazing should be clearly indicated.	N2 (0.6, 2.1, 2.2)
In critical locations (i.e. large areas where the glazing forms part of internal or external walls and doors of shops, showrooms, transoms, offices, factories, public or other non-domestic buildings) the presence of large uninterrupted areas of transparent glazing should be clearly indicated by the use of broken or solid lines, patterns or company logos at appropriate heights and intervals.	N2 (2.4–2.5)

Thermoplastic materials

Requirement	Reference
Thermoplastic materials may be used in windows, rooflights and lighting diffusers in suspended ceilings.	B2 3.8 (V1) B2 6.10 (V2)
External windows to rooms (other than to circulation spaces) may be glazed with thermoplastic materials.	B2 3.9 (V1) B2 6.11 (V2)

Safe opening and closing of windows

Windows, skylights and ventilators that can be opened by people should be capable of being opened, closed or adjusted safely.

Where controls can be reached without leaning over an obstruction they should not be more than 1.9 m above the floor. Where there is an obstruction, the control should be lower (e.g. not more than 1.7 m where there is a 600 mm deep obstruction).	N3 (3.2)
Where controls cannot be positioned within safe reach from a permanent stable surface, a safe means of remote operation, such as a manual or electrical system should be provided.	N3 (3.2)
Where there is a danger of the operator or other person falling through a window above ground floor level, suitable opening limiters should be fitted or guarding provided.	N3 (3.3)

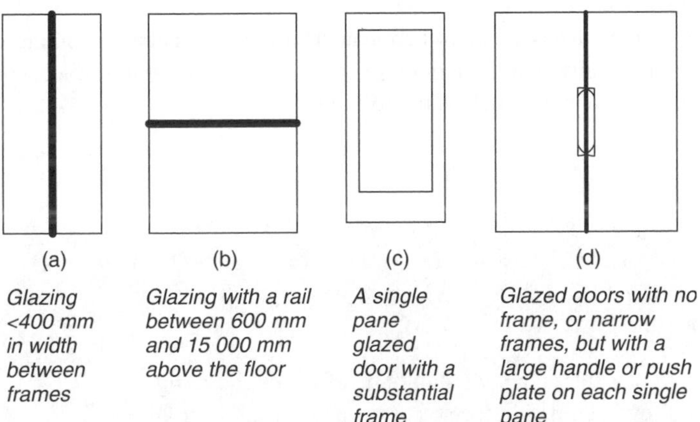

(a)
Glazing <400 mm in width between frames
(b)
Glazing with a rail between 600 mm and 15 000 mm above the floor
(c)
A single pane glazed door with a substantial frame
(d)
Glazed doors with no frame, or narrow frames, but with a large handle or push plate on each single pane

Figure 6.175 Examples of door height glazing not requiring identification further

Safe access for cleaning

All windows, skylights (or any transparent or translucent walls, ceilings or roofs) of a dwelling should be safely accessible for cleaning.	N

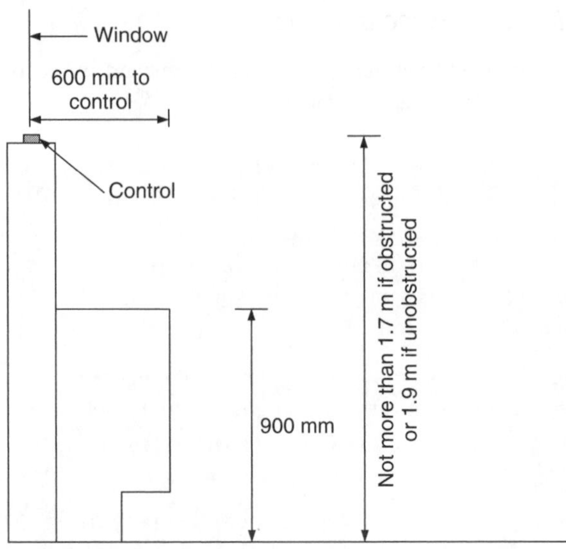

Figure 6.176 Height of controls

Where glazed surfaces cannot be cleaned safely by a person standing on the ground, the requirement for a floor, or other permanent stable surface, could be satisfied by provisions such as the following:

Safe means of access shall be provided for cleaning both sides of glazed surfaces where there is danger of falling more than 2 m.	N4
Where possible windows should be of a size and design that allow the outside surface to be cleaned safely from inside the building.	N4 (4.2)
Windows that reverse for cleaning should be fitted with a mechanism that holds the window in the reversed position (see BS 8213).	N4 (4.2)
For large buildings (e.g. office blocks) a firm, level surface shall be provided to enable portable ladders (not more than 9 m long) to be used and the use of suspended cradles, travelling ladders, abseiling equipment should also be considered.	N4 (4.2)

6.14 Doors

6.14.1 Requirements

Conservation of fuel and power

Energy efficiency measures shall be provided that limit the heat loss through the doors, etc. by suitable means of insulation.

Responsibility for achieving compliance with the requirements of Part L rests with the person carrying out the work. That person may be, for example, a developer, a main (or sub-) contractor, or a specialist firm directly engaged by a private client.

The person responsible for achieving compliance should either themselves provide a certificate, or obtain a certificate from the sub-contractor, that commissioning has been successfully carried out. The certificate should be made available to the client and the building control body.

Ventilation

There shall be adequate means of ventilation provided for people in the building.

(Approved Document F)

Conservation of fuel and power

Reasonable provision shall be made for the conservation of fuel and power in buildings by:

(a) limiting heat gains and losses:
  (i) through thermal elements and other parts of the building fabric; and
  (ii) from pipes, ducts and vessels used for space heating, space cooling and hot water services;

(b) providing and commissioning energy-efficient fixed building services with effective controls; and

(c) providing to the owner sufficient information about the building, the fixed building services and their maintenance requirements so that the building can be operated in such a manner as to use no more fuel and power than is reasonable in the circumstances.

(Approved Document L1)

Fire safety

- *There shall be sufficient escape routes that are suitably located to enable persons to evacuate the building in the event of a fire.*
- *Safety routes shall be protected from the effects of fire.*
- *In an emergency, the occupants of any part of the building shall be able to escape without any external assistance.*

(Approved Document B1)

Access and facilities for disabled people

In addition to the requirements of the Disability Discrimination Act 1995 precautions need to be taken to ensure that:

- *new non-domestic buildings and/or dwellings (e.g. houses and flats used for student living accommodation etc.);*
- *extensions to existing non-domestic buildings;*
- *non-domestic buildings that have been subject to a material change of use (e.g. so that they become a hotel, boarding house, institution, public building or shop);*

are capable of allowing people, regardless of their disability, age or gender, to:

- *gain access to buildings;*
- *gain access within buildings;*
- *be able to use the facilities of the buildings (both as visitors and as people who live or work in them);*
- *use sanitary conveniences in the principal storey of any new dwelling.*

(Approved Document M)

Note: See Appendix A for guidance on access and facilities for disabled people.

6.14.2 Meeting the requirement

Ventilation

To ensure good transfer of air throughout the dwelling, there shall be an undercut of 7600 mm^2 (minimum) in all internal doors above the floor finish (equivalent to an undercut of 10 mm for a standard 760 mm width door).	F Table 1.4 F App E
The height times width of an external door (including patio doors) should be at least 1/20 of the floor area of the room.	F App B
If a room contains more than one external door (or a combination of at least one external door and at least one openable window) then the areas of **all** the opening parts may be added together to achieve the required floor area.	

Access and facilities for disabled people

Doors and gates on main traffic routes (and those that can be pushed open from either side) should have vision panels unless they are low enough (e.g. 900 mm) to see over.	K5 (5.2a)

Sliding doors and gates should have a retaining rail to prevent them falling should the suspension system fail or the rollers leave the track.	K5 (5.2b)
Upward opening doors and gates should be fitted with a device to stop them falling in a way that could cause injury.	K5 (5.2c)
Power operated doors and gates should have safety features to prevent injury to people who are struck or trapped (such as a pressure sensitive door edge that operates the power switch).	K5 (5.2d)
Power operated doors and gates should have a readily identifiable and accessible stop switch.	K5 (5.2d)
Power operated doors and gates should be provided with a manual or automatic opening device in the event of a power failure where and when necessary for health or safety.	K5 (5.2d)

Conservation of fuel and power

Extensions

In a dwelling, the area of doors in extensions should not exceed the sum of — L1B 15

- 25% of the floor area of the extension **plus**
- the area of any windows or doors, which, as a result of the extension works, no longer exist or are no longer exposed.

Note: if the total floor area of the proposed extension exceeds these limits, then the work should be regarded as a **new** building and the requirements of Approved Document L2A and L2B should be used.

The U-value of thermal and/or opaque doors should not exceed 3.3 W/m^2K.	L2B 32c
Doors between the building and an extension should be insulated and weather-stripped to at least the same extent as in the existing building.	L1B 22a L2B 32a
Glazed elements shall comply with the standards given in Table 6.70.	L1B 22c L2B 32c

Table 6.70 Standards for glazed elements in conservatories

Element	Type of building	Replacement fittings (W/m^{2}K)
Doors with more than 50% of their internal face glazed	Existing dwelling	2.2 (whole unit) 1.2 (centre pane)
Doors with less than 50% of their internal face glazed	Existing dwelling	3.0
High usage entrance doors	Existing building	6.0
Vehicle access and large doors	Existing building	1.5

The building fabric should be constructed so that thermal bridges in the insulation layers around door openings (caused by gaps and joints between the various elements) are avoided. L1B 52 L2B 71

U-values

U-values for building fabric elements (calculated using the methods and conventions set out in BR 443) shall not exceed that shown in Table 6.71. L1B 10 L2B 29b L2B 29c

Table 6.71 Limiting U-value standards (W/m^{2}K)

Element	Area-weighted dwelling average	Worst individual sub-element
Doors	2.2	3.3
Pedestrian doors	2.2	3.0
Vehicle access & similar large doors	1.5	4.0
High usage entrance doors	6.0	6.0

Controlled fittings (non-domestic buildings)

Windows, roof windows, rooflights and/or doors should be provided with draught-proofed units. L2B 75

The area-weighted average performance of draught-proofed units for new fittings in extensions and replacement fittings in an existing dwelling shall be no worse than given in Table 6.72. L2B 75

Table 6.72 Standards for controlled fittings

Element	New fittings in an extension (W/m^2K)	Replacement fittings in an existing dwelling (W/m^2K)
Pedestrian doors having more than 50% of their internal face area glazed	2.2	2.2
High usage entrance doors	6.0	6.0
Vehicle access and large doors	1.5	1.5

Consequential improvements (non-domestic buildings)

If a building has a total useful floor area greater than $1000\,m^2$ and the proposed building work includes:

- an extension; or
- the initial provision of any fixed building services; or
- an increase to the installed capacity of any fixed building services;

then consequential improvements should be made to improve the energy efficiency of the whole building by replacing:

all existing doors (excluding high usage entrance doors) within the area served by the fixed building service that have a U-value worse than $3.3\,W/m^2K$ with doors whose U-value is less than $2.2\,W/m^2K$.	L2B 18-7

Material changes of use (domestic buildings)

When a building is subject to a material change of use, then:

any thermal element that is being retained should be upgraded;	L1B 27d
any existing door which separates a conditioned space from an unconditioned space (or the external environment) and which has a U-value that is worse than $3.3\,W/m^2K$, should be replaced by a door whose U-value is less than $2.2\,W/m^2K$.	L1B 27e

Work on controlled services or fittings (domestic buildings)

When working on a controlled service or fitting (i.e. where the service or fitting is subject to the requirements of Part G, H, J, L or P of Schedule 1) and where windows, roof windows, rooflights and/or doors are to be provided:

doors should be provided with draught-proofed units;	L1B 32
the area-weighted average performance of draught-proofed units for new and replacement fittings that are provided as part of the construction of an extension shall be no worse than given in Table 6.73.	L1B 32

Table 6.73 Standards for thermal elements for new fittings in an extension

Element	New fittings in an extension (W/m^2K)	Replacement fittings in an existing dwelling (W/m^2K)
Doors	1.8	2.0 (whole unit) 1.2 (centre pane)
Doors with more than 50% of their internal face glazed	2.2	2.2
Other doors	3.0	3.0

Dwellings should be constructed and equipped so that there are no reasonably avoidable thermal bridges in the insulation layers caused by gaps within the various elements, at the joints between elements, and at the edges of door openings.	L2A 68

Material changes of use (domestic buildings)

When a building is subject to a material change of use, then:

any thermal element that is being retained should be upgraded;	L1B 21
any existing door which separates a conditioned space from an unconditioned space (or the external environment) and which has a U-value that is worse than 3.3 W/m^2K, should be replaced.	L1B 21

Material alterations

If a building is subject to a material alteration then:

if an existing element becomes part of the thermal element of a building (where previously it did not) and it has a U-value worse than 3.3 W/m^2K it should be replaced (unless they are high usage doors).	L2B 39c

Access and facilities for disabled people

Internal doors

The opening force for a manual operating door should not exceed 20 N.	M (3.10a)
The effective clear width through a single leaf door (or one leaf of a double leaf door) should be in accordance with Table 6.74 and Figure 6.177.	M (3.10b)

Table 6.74 Minimum effective clear widths of doors

Direction and width of approach	New buildings (mm)	Existing buildings (mm)
Straight-on (without a turn or oblique approach)	800	750
At right angles to an access route at least 1500 mm wide	800	750
At right angles to an access route at least 1200 mm wide	825	775
External doors to buildings used by the general public	1000	775

There should be an unobstructed space of at least 300 mm on the pull side of the door between the leading edge of the door and any return wall (unless the door is a powered entrance door) (see Figure 6.177).	M (3.10c)
A space alongside the leading edge of a door should be provided to enable a wheelchair user to reach and grip the door handle.	

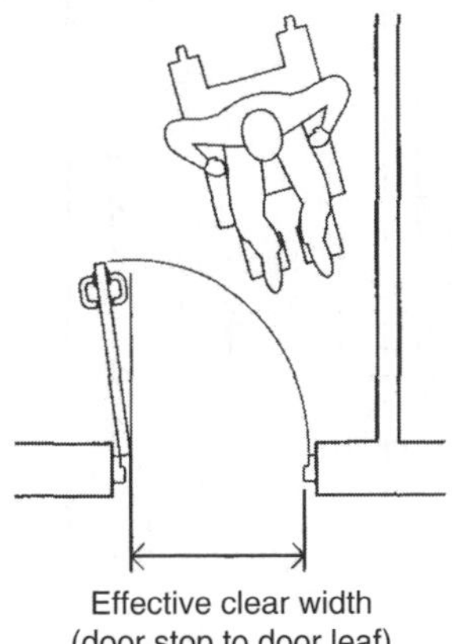

Figure 6.177 Effective clear width and visibility requirements of doors

Door opening furniture should:

- be easy to operate by people with limited manual dexterity;
- be capable of being operated with one hand using a closed fist (e.g. a lever handle); M (3.10d)
- contrast visually with the surface of the door. M (3.10e)

Door frames should contrast visually with the surrounding wall. M (3.10f)

The surface of the leading edge of a non self-closing door should contrast visually with the other door surfaces and its surroundings. M (3.10g)

Door leaves or side panels should be wider than 450 mm. M (3.10h)

Vision panels towards the leading edge of the door should include a visibility zone (or zones) between 500 mm and 1500 mm from the floor. If interrupted (e.g. to accommodate an intermediate horizontal rail – see Figure 6.178) then this should be 800 mm and 1150 mm above the floor. M (3.10h)

Glass entrance doors and glazed screen should be clearly marked (i.e. with a logo or sign) on the glass at two levels, 850 to 1000 mm and 1400 to 1600 mm above the floor. M (3.10i)

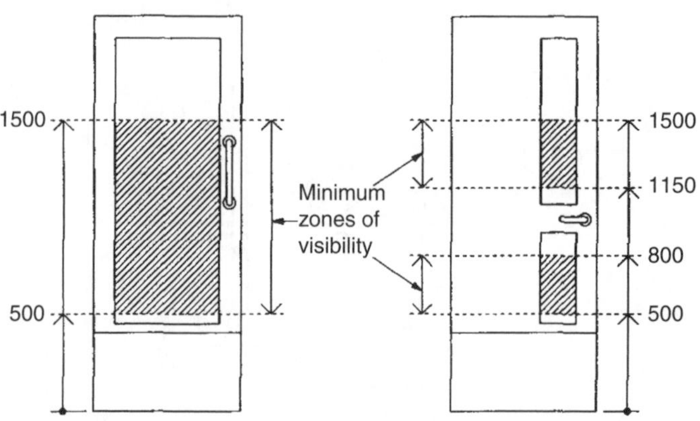

Figure 6.178 Door vision panels

It should be possible to tell between a fully glazed door and any adjacent glazed wall/partition by providing a high-contrast strip at the top and on both sides.	M (3.10j)
Fire doors (particularly those in corridors) should be held open with an electromagnetic device that is capable of self closing when: • the power supply fails; • activated by smoke detectors; • activated by a hand-operated switch.	M (3.10k)
Fire doors (particularly to individual rooms) should be fitted with swing-free devices that close when: • activated by smoke detectors; • the building's fire alarm system is activated; • when the power supply fails.	M (3.10l)
Low energy powered door systems may be used in locations not subject to frequent use or heavy traffic.	M (3.7)
Low energy powered swing door systems should be capable of being operated: • in manual mode; • in powered mode; or • in power-assisted mode.	M (3.10m)
The use of self-closing devices should be minimized as they disadvantage many people (e.g. those pushing prams or carrying heavy objects).	M (3.7)
If closing devices are needed for fire control: • they should be electrically powered hold-open devices or swing-free closing devices; • their closing mechanism should only be activated in case of emergency.	M (3.7)
The presence of doors, whether open or closed, should be apparent to visually impaired people.	M (3.8)

Note: See BS 8300 for guidance on:

- electrically powered hold-open devices;
- swing-free systems;
- low energy powered door systems.

Fire safety

Cavity barriers

Cavity barriers should be provided for all door openings.	B3 6.3 (V1) B3 9.3 (V2)
Openings in a cavity barrier should be limited to those for doors which have at least 30 minutes fire resistance.	B3 6.8 (V1) B3 9.13 (V2)

Note: Detailed guidance on door openings and fire doors is given in Part B Appendix B.

Emergency egress doors

The door should enable the person escaping to reach a place free from danger of fire (e.g. a courtyard or back garden which is at least as deep as the dwelling house is high – see Figure 6.172).	
In a gallery, if the floor is not provided with an alternative exit or escape window:	B1 2.12 (V1) B1 2.8 (V2)
• the distance between the foot of the access stair to the gallery and the door to the room containing the gallery should not exceed 3 m.	
Where an external escape stair is provided:	
• all doors giving access to the stair should be fire-resisting; • doors within 1800 mm of the escape route shall be protected by fire resisting construction.	B1 2.15 (a, b and c)

Note: Glazing in any fire-resisting construction should be fire-resisting and fixed shut.

Basements

Owing to the risk that a single stairway may be blocked by smoke from a fire in the basement or ground storey:

basement storeys in a dwelling house that contain a habitable room shall be provided with either an external door or window suitable for egress from the basement.	B1 2.13 (V1) B1 2.6 (V2)

Fire alarm systems

When fitted, manual call points for fire alarm systems shall be adjacent to exit doors.	B1 1.29 (V2)

Fire doors

Fire doors now **only** need to not be provided with self closing devices, if they are between a dwelling house and an integral garage.

Self-closing fire doors should be positioned so that smoke will not affect access to more than one stairway.

Doors that need to be fire-resisting should meet the requirements given in Table B1 of Appendix B to Part B.	B1 5.6 (V2)
Doors on escape routes should be hung to open not less than 90 degrees.	B1 5.15 (V2)
Doors giving access to an external escape should be fire-resisting and self-closing.	B1 5.25 (V2)
Door-closing devices for fire doors: • shall be in accordance with BS EN 1155:1997; • should take account of the needs of residents; • should have free-swing door closers in bedrooms; • should have hold-open devices in circulation spaces.	B1 3.51 (V2)
Doors on escape routes (both within and from the building) should be readily openable.	B1 5.10 (V2)
Doors on escape routes (whether or not the doors are fire doors), should either not be fitted with lock, latch or bolt fastenings, or they should only be fitted with simple fastenings that can be readily operated from the side approached by people making an escape, without the use of a key and without having to manipulate more than one mechanism.	B1 5.11 (V2)
Doors that open towards a corridor or a stairway should be sufficiently recessed to prevent swing from encroaching on the effective width of the stairway or corridor.	B1 5.16 (V2)
Fire doors should be capable of performing in accordance with Table 6.75 of Appendix B (to Volume 1 of Part B).	App B 1 (V1)

Fire doors: should be fitted with a self-closing device except for: • fire doors to cupboards and to service ducts which are normally kept locked shut; and • fire doors within flats.	App B 2 (V2)
Fire doors should be marked with the appropriate fire safety sign complying with BS 5499-5:2002 according to whether the door is: • to be kept closed when not in use (***Fire door keep shut***); • to be kept locked when not in use (***Fire door keep locked shut***); or • held open by an automatic release mechanism or free swing device (***Automatic fire door keep clear***).	App B 8 (V2)
Fire doors (excet those to cupboards and to service ducts) should be marked on **both** sides.	
Fire doors serving an attached or integral garage should be fitted with a self-closing device.	App B 2 (V1)
Electrically powered locks should return to the unlocked position: • on operation of the fire alarm system; • on loss of power or system error; • on activation of a manual door release unit.	B1 5.11 (V2)
In assembly places, shops and commercial buildings, doors on escape routes from rooms with an occupant capacity of more than 60 should either not be fitted with lock, latch or bolt fastenings, or be fitted with panic fastenings in accordance with BS EN 1125:1997.	B1 5.12 (V2)
see also Appendix B for guidance about door closing and 'hold open' devices for fire doors.	
No more than 25% of the length of a compartment wall should consist of door openings.	App B 5 (V2)
Revolving doors, automatic doors and turnstiles should not be placed across escape routes.	B1 5.18 (V2)
Secure doors that need to be operated by a code, combination, swipe or proximity card, biometric data or similar means, should also be capable of being	B1 5.11 (V2)

overridden from the side approached by people making their escape.	
Self-closing fire doors may be held open by • a fusible link; or • an automatic release mechanism actuated by an automatic fire detection and alarm system; or • a door closer delay device.	App B 3 (V2)
The door of any doorway or exit should be hung to open in the direction of escape.	B1 5.14 (V2)
The essential components of any hinge on which a fire door is hung should be made entirely from materials having a melting point of at least 800°C.	App B 3 (V1) App B 7 (V2)
Two fire doors may be fitted in the same opening so that the total fire resistance is the sum of their individual fire resistances, provided that each door is capable of closing the opening.	App B 4 (V2)
Vision panels shall be provided where doors on escape routes sub-divide corridors, or where any doors are hung to swing both ways.	B1 5.17 (V2)
see also Parts M and N.	

Fire doors in dwellings

Table 6.75 Provision for fire doors (dwellings)

Position of door	Minimum fire resistance of door in terms of integrity (minutes) when tested to BS 476-22:1987	Minimum fire resistance of door in terms of integrity (minutes) when tested to the relevant BSEN 1634 European standard
Any door		
With a cavity barrier	FD 30	E30
Between a dwelling house and a garage	FD 30s	E30Sa
Forming part of the enclosure to a protected stairway in a single family dwelling house	FD 20	E20
Within any other fire resisting construction in a dwelling house not described elsewhere in the table	Fd 20	E 20

Notes:

1. Minimum fire resistance for doors in buildings other than dwellings as is given in Table B1 (page 134) of Part B Volume 2.
2. BS 8214:1 990 gives recommendations for the specification, design, construction, installation and maintenance of fire doors constructed with non-metallic door leaves.
3. Guidance on timber fire-resisting doorsets may be found in *'Timber fire-resisting doorsets: maintaining performance under he new European test standard'* published by TRADA.
4. In flats where the habitable rooms do not have direct access to the entrance hall, bedrooms should be separated from the living accommodation by fire-resisting construction and fire doors (B1 2.14(V2)).

Fire protected stairways

Fire protected stairways shall (as far as is reasonably possible) consist of fire-resistant material and fire resistant doors and have an appropriate form of smoke control system.	B1 1.viii

Fire service (access and facilities)

Door(s) should be provided such that there is no more than 60 m between each door and/or the end of that elevation (e.g. a 150 m elevation would need at least 2 doors).	B5 16.5 (V2)
Every elevation to which vehicle access is provided should have a suitable door(s), not less than 750 mm wide, giving access to the interior of the building.	B5 11.3 (V2) B5 16.5 (V2)

Garages

Fire doors now only need to not be provided with self closing devices if they are between a dwelling house and an integral garage.

If a door is provided between a dwelling house and the garage, the floor of the garage should: • be laid to fall to allow fuel spills to flow away from the door to the outside; or • the door opening should be positioned at least 100 mm above garage floor level.	B3 5.5 (V1)

Loft conversions

Where a loft conversion means that a new storey is going to be added, then the stairway will need to be protected with fire-resisting doors and partitions.	B1 2.20b

Means of escape

Common corridors that connect two or more storey exits should be sub-divided by a self-closing fire door with, if necessary, an associated fire-resisting screen.	B1 2.28 (V2)
Every corridor more than 12 m long which connects two or more storey exits, should be sub-divided by self-closing fire doors positioned approximately mid-way between the two storey exits.	B1 3.26 (V2)
Except doorways, all escape routes should have a clear headroom of not less than 2 m.	B1 3.17 (V2)
Except for kitchens, all habitable rooms shall be provided with suitable means for emergency egress from each storey via doors or windows.	B1 2.1–2.4 (V1) B1 2.11–2.12 (V2)
If the only escape route from an inner room is through another room then there should be a vision panel not less 0.1 m^2 located in the door or walls of the inner room.	B1 3.10 (V2)
Unless escape stairways and corridors are protected by a pressurization system complying with BS EN 12101-6:2005, every dead-end corridor exceeding 4.5 m in length should be separated by self-closing fire doors.	B1 3.27 (V2)
In residential Care Homes, bedrooms should be enclosed in fire-resisting construction with fire resisting doors and every corridor serving bedrooms should be a protected corridor.	B1 3.48 (V2)
The dead-end portion of any common corridor should be separated from the rest of the corridor by a self-closing fire door (see Figure 6.181).	B1 2.29 (V2)

Note: Generally, in residential care homes for the elderly it is reasonable to assume that at least a proportion of the residents will need some assistance to evacuate.

Protected stairways

Dwelling houses with one floor more than 4.5 m above ground level should have a protected stairway which separated by fire doors.	B1 2.6 (V1)
Transfer grilles belonging to air circulation systems shall not be fitted in any door leading to a protected stairway.	B1 2.17

Sprinkler systems

Where a sprinkler system is provided fire doors to bedrooms need not be fitted with self closing devices.	B1 3.52 (V2)

6.15 Vertical circulation within the building

6.15.1 The requirement

In addition to the requirements of the Disability Discrimination Act 1995 precautions need to be taken to ensure that:

- *new non-domestic buildings and/or dwellings (e.g. houses and flats used for student living accommodation etc.);*
- *extensions to existing non-domestic buildings;*
- *non-domestic buildings that have been subject to a material change of use (e.g. so that they become a hotel, boarding house, institution, public building or shop);*

are capable of allowing people, regardless of their disability, age or gender, to:

- *gain access to buildings;*
- *gain access within buildings;*
- *be able to use the facilities of the buildings (both as visitors and as people who live or work in them);*
- *use sanitary conveniences in the principal storey of any new dwelling.*

(Approved Document M)

Note: See Annex A for guidance on access and facilities for disabled people.

Fire safety

- *There shall be sufficient escape routes that are suitably located to enable persons to evacuate the building in the event of a fire.*
- *Safety routes shall be protected from the effects of fire.*
- *In an emergency, the occupants of any part of the building shall be able to escape without any external assistance.*

(Approved Document B1)

6.15.2 Meeting the requirement

Lifting devices

For all buildings, a passenger lift is considered the most suitable form of access for people moving from one storey to another.

Wherever possible a lifting device (e.g. a passenger lift or a lifting platform) serving all storeys should be provided in: • new developments; • existing buildings.	M (3.17, 3.24a to d)
Note: In exceptional circumstances (e.g. a listed building, or an infill site in a historic town centre) where a passenger lift cannot be accommodated, a wheelchair platform stair lift serving an intermediate level or a single storey may be used.	
The location of lifting devices that are accessible by mobility impaired people should be clearly visible from the building entrance.	M (3.18)
Signs should be available at each landing to identify the floor reached by the lifting device.	M (3.18)
In addition to the lifting device, internal stairs (designed to suit ambulant disabled people and those with impaired sight) should always be provided.	M (3.19)

General requirements for lifting devices

There should be an unobstructed manoeuvring space of 1500 mm × 1500 mm, or a straight access route 900 mm wide, in front of each lifting device.	M (3.28a)

The landing call buttons should be located between 900 mm and 1100 mm from the floor and at least 500 mm from any return wall.	M (3.28b)
Landing call buttons and lifting device control button symbols: • should contrast visually with the surrounding face plate; • should be raised to facilitate tactile reading; • should be accessible by wheelchair users.	M (3.248c and d) M (3.27)
The floor of the lifting device should not be of a dark colour.	M (3.24e)
A handrail (at 900 mm nominal) should be provided on at least one wall of the lifting device.	M (3.24f)
A suitable emergency communication system should be fitted.	M (3.24g)

Note: See also:

- Lift Regulations 1997 SI 1997/831;
- Lifting Operations and Lifting Equipment Regulations 1998 SI 1998/2307;
- Provision and Use of Work Equipment Regulations 1998 SI 1998/2306;
- Management of Health and Safety at Work Regulations 1999 SI 1999/3242;
- BS 8300.

Lifts

Lifts should **not** be used when there is a fire in the building – unless it is a fire-fighting lift.

Lifts (except suitably designed and installed evacuation lifts, are **not** considered acceptable as a means of escape.	B1 vi
Lift entrances should be separated from the floor area on every storey by a protected lobby.	B1 5.42 (V2)
Lift shafts should not be continued down to serve any basement storey if it is: • in a building served by only one escape; • within the enclosure to an escape stair which is terminated at ground level.	B1 5.44 (V2)

If a protected shaft contains a stair and/or a lift, it should **not** also contain: • a pipe conveying oil (other than in the mechanism of a hydraulic lift); or • a ventilating duct (other than a duct provided solely for ventilating the stairway).	B2 8.40 (V2)
Lift wells should be either: • contained within the enclosures of a protected stairway; or • enclosed throughout their height with fire-resisting construction.	B1 5.42 (V2)
A lift well connecting different compartments should form a protected shaft.	B1 5.42 (V2)
Lift machine rooms should be sited over the lift well whenever possible.	B1 5.45 (V2)
In basements and enclosed (i.e. non open-sided) car parks the lift should be approached only by a protected lobby or protected corridor (unless it is within the enclosure of a protected stairway).	B1 5.43 (V2)
Evacuation lifts and refuges should be clearly identified by appropriate fire safety signs.	B1 4.10 (V2)
If a refuge is in a lobby or stairway the sign should be accompanied by a blue mandatory sign worded ***'Refuge – keep clear'.***	

Firefighting lifts

Buildings with a floor at more than 18 m above (or with a basement at more than 10 m below) fire and rescue service vehicle access level, should be provided with at least two firefighting shafts containing firefighting lifts.	B5 17.2 (V2) B5 17.8 (V2)

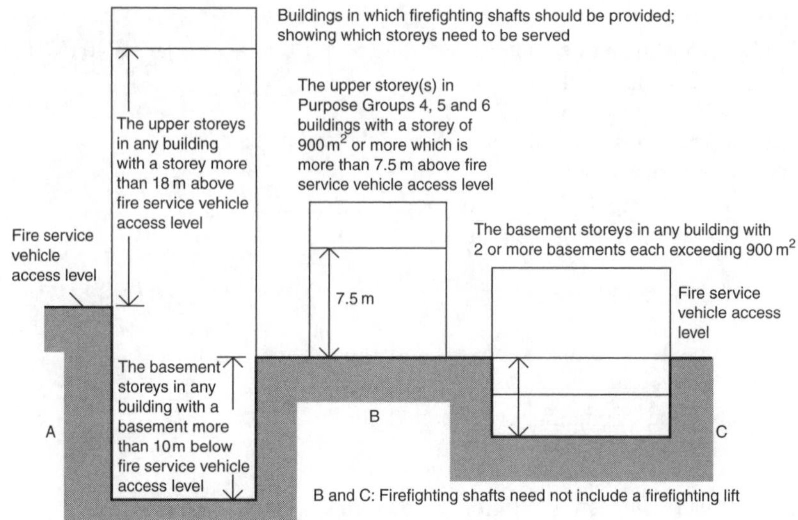

Figure 6.179 Provision of firefighting shafts

Other than in blocks of flats, all firefighting lifts should be approached from the accommodation, through a firefighting lobby.	B5 17.11 (V2)

Passenger lifts

Lift sizes should be chosen to suit the anticipated density of use of the building and the needs of disabled people.

Passenger lifts should conform to the requirements of:

- the Lift Regulations 1997, SI 1997/831;
- the relevant British Standards, EN 81 series of standards; — M (3.34a)
- BS EN 81-70: 2003 (*Safety rules for the construction and installation of lifts*).

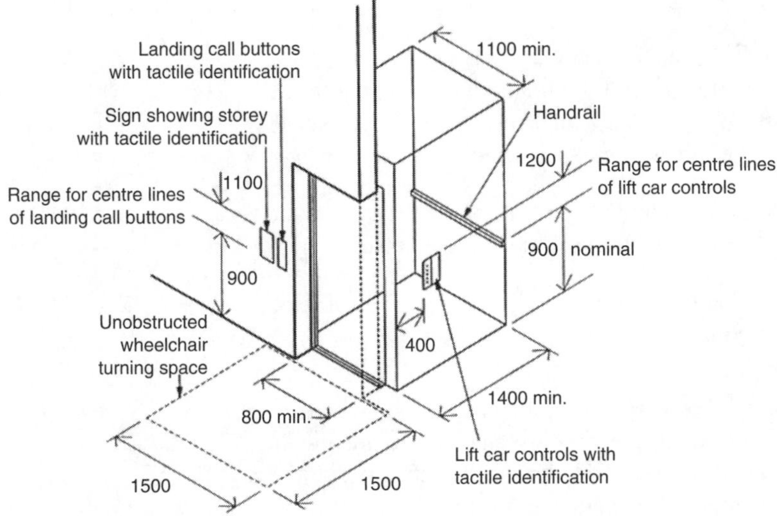

Figure 6.180 Key dimensions associated with passenger lifts

Passenger lifts should:

• be accessible from the remainder of the storey;	M (3.34b)
• have power-operated horizontal sliding doors which provide an effective clear width of at least 800 mm (nominal);	M (3.34e)
• have doors fitted with timing devices (and re-opening activators) to allow enough time for people and any assistance dogs to enter or leave;	M (3.34f)
• locate the car controls between 900 mm and 1200 mm (preferably 1100 mm) from the car floor and at least 400 mm from any return wall;	M (3.34g)
• locate all landing call buttons between 900 mm and 1100 mm from the floor of the landing and at least 500 mm from any return wall;	M (3.34h)
• be fitted with lift landing and car doors that are visually distinguishable from adjoining walls;	M (3.34i)
• include (in the lift car and the lift lobby) audible and visual information to tell them that a lift has arrived, which floor it has reached and where in a bank of lifts it is located;	M (3.31 and 3.34j)
• ensure that all glass areas can be easily identified by people with impaired vision;	M (3.34k)
• conform to BS 5588-8 if the lift is to be used to evacuate disabled people in an emergency.	M (3.34l)

If the lift is not large enough to allow a wheelchair user to turn around within the lift car, then a mirror should be provided that enables the wheelchair user to see behind the wheelchair.	M (3.34d)
The minimum dimensions of the lift cars should be 1100 mm wide and 1400 mm deep (see Figure 6.180);	M (3.34c)
Visually and acoustically reflective wall surfaces should not be used.	M (3.32)
Where possible, lift cars (used for access between two levels only) may be provided with opposing doors to allow a wheelchair user to leave without reversing out.	M (3.33)

Passenger lifts in dwelling houses (which serve any floor more than 4.5 m above ground level) should either be located in the enclosure to a protected stairway or a fire-resisting lift shaft.	B1 2.18

Lifting platforms

A lifting platform should only be provided to transfer wheelchair users, people with impaired mobility and their companions vertically between levels or storeys (M3.35).

Lifting platforms should:

• conform to the requirements of the Supply of Machinery (Safety) Regulations 1992, SI 1992/3073, the relevant British Standards and the EN81 series of standards;	M (3.43a)
• restrict the vertical travel distance to: – not more than 2 m if there is no liftway enclosure and/or floor penetration; – more than 2 m, where there is a liftway enclosure;	M (3.43b)
• restrict the rated speed of the platform so that it does not exceed 0.15 m/s;	M (3.43c)

- locate their controls between 80 mm and 1100 mm from the floor of the lifting platform and at least 400 mm from any return wall; M (3.43d)
- locate all landing call buttons between 900 mm and 1100 mm from the floor of the landing and at least 500 mm from any return wall; M (3.43f)
- have continuous pressure controls (e.g. push buttons); M (3.43e)
- have doors with an effective clear width of at least: M (3.43h)
  - 900 mm for an 1100 mm wide and 1400 mm deep lifting platform;
  - 800 mm in other cases;
- be fitted with clear instructions for use; M (3.43i)
- have their entrances accessible from the remainder of the storey; M (3.43j)
- should have doors visually distinguishable from adjoining walls; M (3.43k)
- have an audible and visual announcement of platform arrival and level reached; M (3.43l)
- have areas of glass that are identifiable by people with impaired vision. M (3.43m)

The minimum clear dimensions of the platform should be: M (3.43g)

- 800 mm wide by 1250 mm deep (where the lifting platform is not enclosed and provision has been made for an unaccompanied wheelchair user);
- 900 mm wide by 1400 mm deep (where the lifting platform is enclosed and provision has been made for an unaccompanied wheelchair user);
- 1100 mm wide by 1400 mm deep (where two doors are located at 90° relative to each other, the lifting platform is enclosed or where provision is being made for an accompanied wheelchair user).

All users including wheelchair users should be able to reach and use the controls that summon and direct the lifting platform. M (3.36)

Where possible, lifting platforms (used for access between two levels only) may be provided with opposing doors to allow a wheelchair user to leave without reversing out. M (3.41)

Visually and acoustically reflective wall surfaces should not be used. M (3.42)

Wheelchair platform stairlifts

Wheelchair platform stairlifts are only intended for the transportation of wheelchair users and should only be considered for conversions and alterations where it is not practicable to install a conventional passenger lift or a lifting platform (3.44).

Wheelchair platform stairlifts should conform to the requirements of the Supply of Machinery (Safety) Regulations 1992, SI 1992/3073 the relevant British Standards, EN81 series of standards.	M (3.49a)
Buildings with single stairways shall maintain the required clear width.	M (3.49b)
The speed of the platform should not exceed 0.15 m/s.	M (3.49c)
Continuous pressure controls (e.g. joystick) should be provided.	M (3.47 and 3.49d)
The platform should have minimum clear dimensions of 800 mm wide and 1250 mm deep.	M (3.49e)
Wheelchair platform stairlifts should:	
• be fitted with clear instructions for use;	M (3.49f)
• provide an effective clear width of at least 800 mm;	M (3.49g)
• not be installed where their operation restricts the safe use of the stair by other people.	M (3.44)

Passenger lifts in blocks of flats

If a passenger lift access is installed, it should:	
• be suitable for an unaccompanied wheelchair user;	M (9.4)
• have a minimum load capacity of 400 kg;	M (9.6)
• have a clear landing at least 1500 mm wide and 1500 mm long in front of its entrance;	M (9.7a)
• have a door (or doors) with a clear opening width of at least 800 mm;	M (9.7b)
• have a car at least 900 mm wide and 1250 mm long;	M (9.7c)
• have landing and car controls between 900 mm and 1200 mm above the landing and the car floor and at least 400 mm from the front wall;	M (9.7d)
• have suitable tactile indication (on the landing and adjacent to the lift call button) to identify the storey;	M (9.7e)
• have suitable tactile indication (on or adjacent to lift buttons within the car) to confirm the floor selected;	M (9.7f)

• incorporate a signalling system that provides visual notification that the lift is answering a landing call;	M (9.7g)
• have a 'dwell time' of five seconds before its doors begin to close after they are fully open;	M (9.7g)
• provide a visual and audible indication of the floor reached (when the lift serves more than three storeys).	M (9.7h)

Vertical circulation within the entrance storey of a dwelling

A stair providing vertical circulation within the entrance/principal storey of a dwelling should have: • flights with clear widths of at least 900 mm; • a suitable continuous handrail on each side of the flight (and any intermediate landings where the rise of the flight comprises three or more rises); • the rise and going in accordance with the guidance in Approved Document K for private stairs in dwelling.	M (7.7a to c)

6.16 Corridors and passageways

6.16.1 The requirement

Fire safety

- *There shall be sufficient escape routes that are suitably located to enable persons to evacuate the building in the event of a fire.*
- *Safety routes shall be protected from the effects of fire.*
- *In an emergency, the occupants of any part of the building shall be able to escape without any external assistance.*

(Approved Document B1)

6.16.2 Meeting the requirement

General requirements for corridors

In large buildings, the corridor adjoining the stair should be provided with a vent that is located as high as practicable and with its top edge is at least as high as the top of the door to the stair.	B1 2.26 (V2)

In buildings containing flats, the wall between each flat and the corridor should be a compartment wall. B1 2.24 (V2)

Stores and other ancillary accommodation should **not** be located within, or entered from, a protected corridor forming part of the only common escape route from a flat on the same storey as that ancillary accommodation. B1 2.30 (V2)

Every corridor more than 12 m long which connects two or more storey exits, should be sub-divided by self-closing fire doors positioned approximately mid-way between the two storey exits B1 3.26 (V2)

Unless escape stairways and corridors are protected by a pressurization system complying with BS EN 12101-6:2005, every dead-end corridor exceeding 4.5 m in length should be separated by self-closing fire doors (together with any necessary associated screens) from any part of the corridor which: B1 3.27 (V2)

- provides two directions of escape;
- continues past one storey exit to another.

Corridors associated with a common stair serving an enclosed (non open-sided) car park, or place of special fire hazard shall have not less than 0.4 m^2 permanent ventilation or be protected by a mechanical smoke control system. B1 2.47 (V2)

Protected corridors

Protected corridors shall be installed for: B1 3.24 (V2)

- corridors serving bedrooms;
- dead-end corridors (excluding recesses not exceeding 2 m deep);
- any corridor that is common to more than one different occupancies.

Escape stairs shall have a protected lobby or protected corridor at all levels (except the top storey, all basement levels and when the stair is a firefighting stair) if: B1 4.34 (V2)

- the stair is the only one serving a building which has more than one storey above or below the ground storey;
- where the stair serves any storey at a height greater than 18 m; or
- where the building is designed for phased evacuation.

Corridors serving bedrooms in Residential Care Homes should be protected corridors.	B1 3.48 (V2)

Fire safety

Common corridors should be protected corridors.	B1 2.24 (V2)
Means of ventilating common corridors should be available.	B1 2.25 (V2)
Common corridors that connect two or more storey exits should be sub-divided by a self-closing fire door with, if necessary, an associated fire-resisting screen (see Figure 6.181a).	B1 2.28 (V2)
The dead-end portion of any common corridor should be separated from the rest of the corridor by a self-closing fire door (see Figure 6.181b).	B1 2.29 (V2)

Note: Self-closing fire doors should be positioned so that smoke will not affect access to more than one stairway.

Disabled access

Corridors and passageways should:

• be wide enough to allow wheelchair users, people with buggies, people carrying cases and/or people on crutches to pass others on the access route;	M (3.11)
• **not** have projecting elements such as columns, radiators and fire hoses;	M (3.14a)
• have an unobstructed width of at least 1200 mm.	M (3.14b)

For school buildings, the preferred corridor width dimension is 2700 mm where there are lockers within the corridor.

• have passing places at least 1800 mm long and at least 1800 mm wide at corridor junctions;	M (3.14c)
• have a floor level no steeper than 1:60;	M (3.14d)
• have an internal ramp in accordance with Table 6.76 and Figure 6.182 for floors with a gradient of 1:20 or steeper.	M (3.14d)

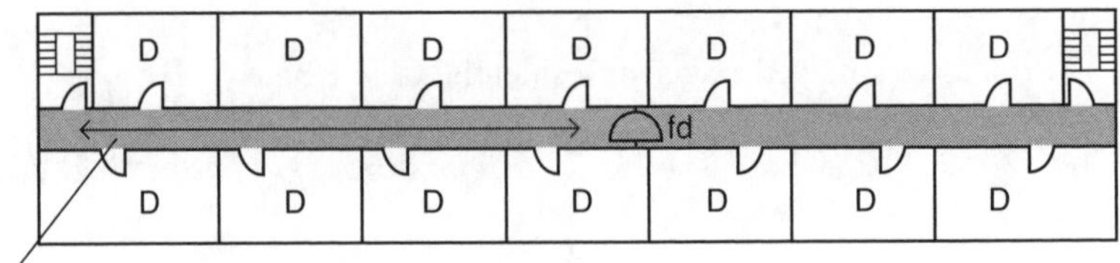

30 m max.

(a) Corridor access without dead ends

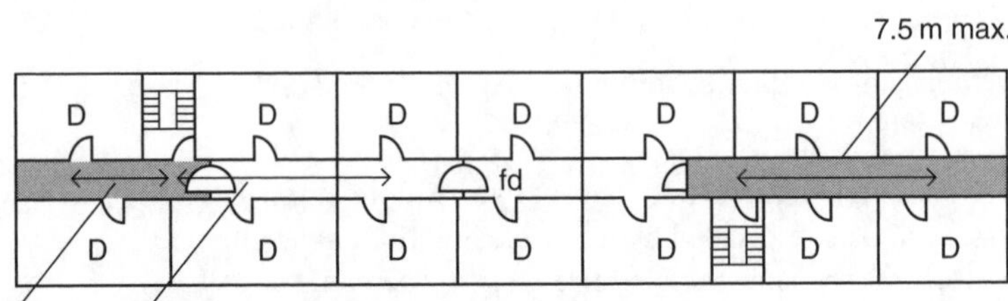

7.5 m max. 30 m max.

(b) Corridor access with dead ends
The central door may be omitted if maximum travel distance is not more than 15 m

Note:

The arrangements shown also apply to the top storey.

Key

D Dwelling
fd Fire door
[shaded] Shaded area indicates zone where ventilation should be provided in accordance with paragraph 2:26
(An external wall vent or smoke shaft located anywhere in the shaded area)

Figure 6.181 Flats served by more than one common stair

Table 6.76 Limits for ramp gradients

Going of a flight (m)	Maximum gradient	Maximum rise (mm)
10	1:20	500
5	1:15	333
2	1:12	16

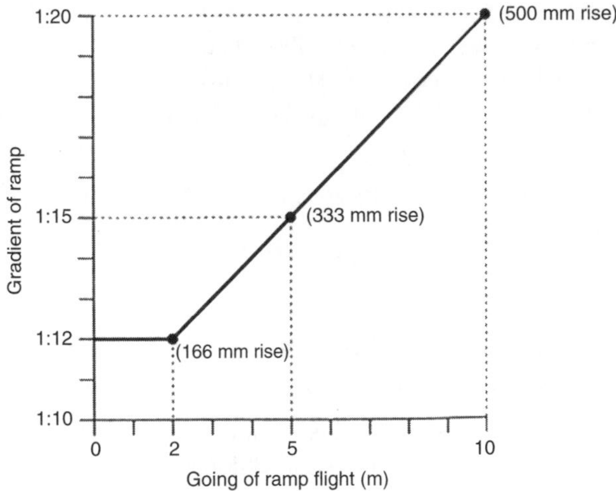

Figure 6.182 Relationship for ramp gradients

Corridors should be at least 1800 mm wide where doors from a unisex wheelchair-accessible toilet project open into that corridor.	M (3.14h)
If a section of the floor has a gradient steeper than 1:60 (but less than 1:20) it should rise no more than 500 mm without a level rest area at least 1500 mm long.	M (3.14e)
Sloping sections should extend the full width of the corridor or have the exposed edge protected by guarding.	M (3.14f)
Doors opening towards a corridor, which is a major access route or an escape route, should be recessed.	M (3.14g)
On a major access route (or an escape route) the wider leaves of double doors should be on the same side of the corridor.	M (3.14i)
The use of floor surface finishes which have patterns that could be mistaken for steps or changes of level should be avoided.	M (3.14j)
Floor finishes should be slip resistant.	M (3.14k)

Glazed screens alongside a corridor should be clearly marked (i.e. with a logo or sign) on the glass at two levels, 850 to 1000 mm and 1400 to 1600 mm above the floor.	M (3.14l)
The acoustic design should be neither too reverberant nor too absorbent.	M (3.13)

Corridors, passageways and internal doors within the entrance storey of a dwelling

The objective is to make it easy for wheelchair users, ambulant disabled people, people of either sex with babies and small children, or people with luggage to gain access to an entrance and/or principal storey of a dwelling, into habitable rooms and/or a room containing a WC on that level.

Corridors and passageways in the entrance storey should be sufficiently wide enough for a wheelchair user to circumnavigate.	M (7.2)
Internal doors should be wide enough for wheelchairs to go through with ease.	M (7.4)
Permanent obstructions in a corridor (such as a radiator) should be no longer than 2 m (provided that the width of the corridor is not less than 750 mm and the obstruction is not opposite a door to a room).	M (7.2 and 7.5b)
Corridors and/or other access routes in the entrance storey should have an unobstructed width in accordance with Table 6.77 and Figure 6.183.	M (7.5a)
Doors to habitable rooms and/or rooms containing a WC should have minimum clear opening widths shown in Table 6.77 and Figure 6.182.	M (7.5c)

Table 6.77 Minimum widths of corridors and passageways for a range of doorway widths

Doorway clear opening width (mm)	Corridor/passageway width (mm)
750 or wider	900 (when approached head on)
750	1200 (when not approached head on)
775	1050 (when not approached head on)
800	900 (when not approached head on)

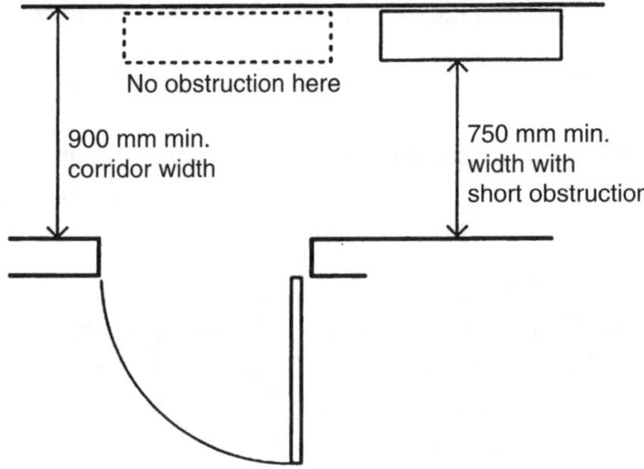

Figure 6.183 Corridors, passages and internal doors

6.17 Facilities in buildings other than dwellings

6.17.1 The requirement

Fire safety

- *There shall be sufficient escape routes that are suitably located to enable persons to evacuate the building in the event of a fire.*
- *Safety routes shall be protected from the effects of fire.*
- *In an emergency, the occupants of any part of the building shall be able to escape without any external assistance.*

(Approved Document B1)

In addition to the requirements of the Disability Discrimination Act 1995 precautions need to be taken to ensure that:

- *new non-domestic buildings and/or dwellings (e.g. houses and flats used for student living accommodation etc.);*
- *extensions to existing non-domestic buildings;*
- *non-domestic buildings that have been subject to a material change of use (e.g. so that they become a hotel, boarding house, institution, public building or shop);*

are capable of allowing people, regardless of their disability, age or gender, to:

- *gain access to buildings;*
- *gain access within buildings;*
- *be able to use the facilities of the buildings (both as visitors and as people who live or work in them);*
- *use sanitary conveniences in the principal storey of any new dwelling.*

(Approved Document M)

Note: See Appendix A for guidance on access and facilities for disabled people.

Ventilation

There shall be adequate means of ventilation provided for people in the building.
(Approved Document F)

Fresh air supplies should be protected from contaminants that would be injurious to health.	F 2.3

Offices

All office sanitary accommodation, washrooms and food and beverage preparation areas shall be provided with intermittent air extract ventilation capable of meeting the requirements of Table 6.78.	F 2.11
Extract fans that are located in an internal room which does not have an openable window, should have a 15 minute over-run.	F Table 2.2c
Extract ventilators should be located as high as practicable and preferably not less than 400 mm below the ceiling level.	F Table 2.2b
Passive stack ventilators (PSVs) should be located in the ceiling of the room.	F Table 2.2b
Purge ventilation shall be provided in each office.	F 2.13
Purged air should be taken directly to outside and should not be recirculated to any other part of the building.	F 2.13
PSV can be used as an alternative to a mechanical extract fan for office sanitary and washrooms and food preparation areas.	F Table 2.2a
PSV controls can be either manual or automatic.	F Table 2.2c
The controls for extract fans can be either manual or automatic.	F Table 2.2c
Printers and photocopiers that are being used in large numbers and which are in almost	F 2.11

constant use (i.e. greater than 30 minutes per hour) shall:

- be located in a separate room;
- have extract facilities capable of providing an extract rate greater than 20 l/s per machine, during use (see Table 6.78).

The whole building ventilation rate for the supply of air to the offices should be greater than 10 l/s per person (see Table 6.78). F 2.12

The following air flow rates can mainly be provided by natural ventilation

Table 6.78 Extract ventilation rate

Room	Air extract rate
Rooms containing printers and photocopiers in substantial use (greater than 30 minutes per hour)	20 l/s per machine during use
Office sanitary accommodation and washrooms	15 l/s per shower/bath 6 l/s per WC/urinal
Food and beverage preparation areas (not commercial kitchens)	15 l/s with microwave and beverages only 30 l/s adjacent to the hob with cooker(s) 60 l/s elsewhere with cooker(s)
Specialist buildings and spaces (e.g. commercial kitchens, fitness rooms)	See Table 2.3 of Approved Document F

The outdoor air supply rates shown below for offices are based on controlling body odours with low levels of other pollutants.

Table 6.79 Whole building ventilation rate for air supply to offices

Offices	Air supply rate
Total outdoor air supply rate for offices (no-smoking and no significant pollutant sources)	10 l/s per person

Other types of buildings

The ventilation requirements for other buildings (such as assembly halls, broadcasting studios, computer rooms, factories, hospitals, hotels, museums, schools, sports centres and warehouses etc.) are listed in Table 2.3 of Approved Document F, which also provides a link to the relevant controlling Acts of Parliament, Statutory Instruments, BS, CIBSE and HSE standards, practices and recommendations.

Car parks

Where a car park is well ventilated, there is a low probability of fire spread from one storey to another. Ventilation, however, is an important factor and, as heat and smoke cannot be dissipated so readily from a car park that is not open-sided, fewer concessions are made. For more guidance see Section 11 of Part B3 (Volume 2).

Non combustible materials should be used in the construction of any 'open sided' car park.

Underground car parks, enclosed car parks and multi-storey car parks designed to limit the concentration of carbon monoxide to not more than 30 parts per million averaged over an eight-hour period and peak concentrations.	F 2.19
Naturally ventilated car parks shall have openings at each car parking level: • at least 1/20th of the floor area at that level; • with a minimum of 25% on each of two opposing walls.	F 2.21a
Ramps and exits shall not go above 90 parts per million for periods not exceeding 15 minutes.	F 2.19
Car parks should have a separate and independent extraction system and the extracted air should not be recirculated.	B1 5.50 (V2)
Mechanically ventilated car parks can either have natural ventilation openings that are not less than 1/40th of the floor area or a mechanical ventilation system capable of at least three air changes per hour (ach).	F 2.21b
Mechanically ventilated basement car parks shall be capable of at least six air changes per hour (ach).	F 2.21b
Mechanically ventilated exits and ramps (i.e. where cars queue inside the building with engines running) shall be capable of at least ten air changes per hour (ach).	F 2.21b
Where a common stair forms part of the only escape route from a flat, it should **not** also serve any covered car park.	B1 2.46 (V2)

In enclosed (i.e. non open-sided) car parks, lifts should be approached by a protected lobby or protected corridor (unless it is within the enclosure of a protected stairway).	B1 5.43 (V2)
Car parks are not normally expected to be fitted with sprinklers.	B5 18.13 (V2)

Hospitals

HTM 05 'Firecode' should be used for used for the design of hospitals and similar health care premises.

Offices

In small premises:

• floor areas should be generally undivided (except for ancillary offices and stores) to ensure that exits are clearly visible from all parts of the floor areas;	B1 3.34 (V2)
• clear glazed areas should be provided in any partitioning separating an office from the open floor area to enable any person within the office to obtain early visual warning of an outbreak of fire.	B1 3.36 (V2)

Disabled facilities

The overall aim should be that **all** people can have access to (and be able to use) **all** of the facilities provided within a building. Everyone (no matter their disability) should be able to fully participate in lecture/conference facilities as well as be able to enjoy entertainment, leisure and social venues – not just as spectators, but also as participants and/or staff. To achieve these aims:

All floor areas (even when located at different levels) should be accessible.	M (4.3)
In hotels, motels and student accommodation:	M (4.4)
• a proportion of the sleeping accommodation should be designed for wheelchair users;	

- the remainder should include facilities suitable for people with sensory, dexterity or learning difficulties.

If there is a reception point: • it should be easily accessible and convenient to use; • information about the building should be clearly available from notice boards and signs; • the floor surface should be slip resistant.	M (3.2 to 3.5)
Disabled people should be able to have: • a choice of seating location at spectator events; • a clear view of the activity taking place (whilst not obstructing the view of others).	M (4.2)
Bars and counters in refreshment areas should be at a suitable level for wheelchair users.	M (4.3)

6.17.2 Audience and spectator facilities

Audience and spectator facilities fall primarily into three categories:

- lecture/conference facilities;
- entertainment facilities (e.g. theatres/cinemas);
- sports facilities (e.g. stadia).

Audience facilities generally

Wheelchair users (as well as those with mobility and/or sensory problems) may need to see or listen from a particular side, or sit at the front to lip read or read sign interpreters.

For this reason they should be provided with a selection of spaces into which they can manoeuvre easily and which offer them a clear view of an event – taking particular care that these do not become segregated into 'special areas'.

For audience seating generally

The route to wheelchair spaces should be accessible to users.	M (12a)
Stepped access routes to audience seating should be provided with fixed handrails.	M (12b)
Handrails to external stepped or ramped access should be positioned as per Figure 6.184.	M (4.12a)

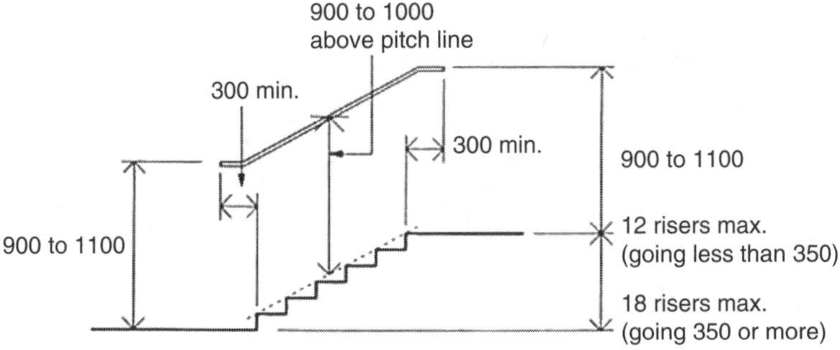

Figure 6.184 Handrails to external stepped and ramped access – key dimensions

Handrails to external stepped or ramped access should:

- be continuous across flights and landings; M (4.12b)
- extend at least 300 mm horizontally beyond the top and bottom of a ramped access;
- not project into an access route;
- contrast visually with the background;
- have a slip resistant surface which is not cold to the touch;
- terminate in such a way that reduces the risk of clothing being caught;
- either be circular (with a diameter of between 40 and 45 mm) or oval with a width of 50 mm (see Figure 6.185).

Handrails to external stepped or ramped access should:

- not protrude more than 100 mm into the surface width of the ramped or stepped access where this would impinge on the stair width requirement of Part B1; M (4.12b)
- have a clearance of between 60 and 75 mm between the handrail and any adjacent wall surface;
- have a clearance of at least 50 mm between a cranked support and the underside of the handrail;
- ensure that its inner face is located no more than 50 mm beyond the surface width of the ramped or stepped access;
- be spaced away from the wall and rigidly supported in a way that avoids impeding finger grip;
- be set at heights that are convenient for all users of the building.

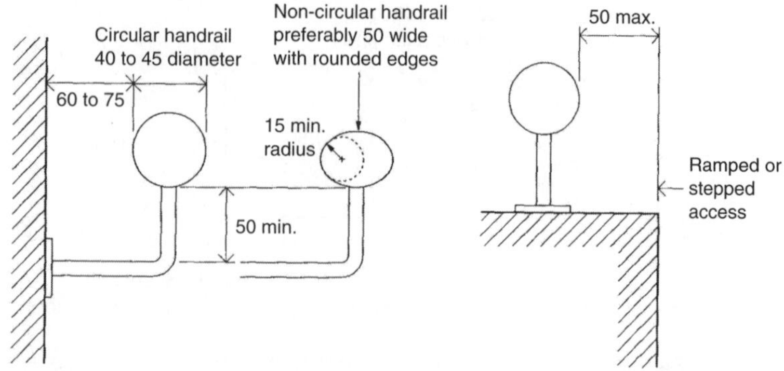

Figure 6.185 Handrail – key dimensions

Seating

The minimum number of permanent and removable spaces provided for wheelchair users is as per Table 6.80.	M (4.12c)

Table 6.80 Provision of wheelchair spaces for audience seating

Seating capacity	Minimum provision of spaces for wheelchairs	
	Permanent	Removable
Up to 600	1% of total seating capacity (rounded up)	Remainder to make a total of 6
Over 600 but less than 10,000	1% of total seating capacity (rounded up)	Additional provision if desired

Some wheelchair spaces should be provided in pairs, with standard seating on at least one side (see Figure 6.186). M (4.12d)

If more than two wheelchair spaces are provided, they should be located so as to give a range of views of the event. M (4.12e)

The minimum clear space for:

- access to wheelchair spaces should be 900 mm; M (4.12f)
- wheelchair spaces in a parked position should be 900 mm wide by 1400 mm deep. M (4.12g)

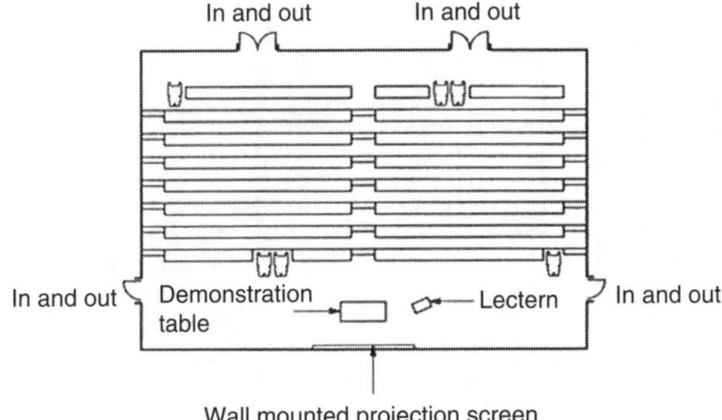

Figure 6.186 An example of wheelchair spaces in a lecture theatre

The floor of each wheelchair space should be horizontal.	M (4.12h)
Seats at the ends of rows **and** next to wheelchair spaces should have detachable (or lift-up) arms.	M (4.12j)

For seating on a stepped terraced floor

Wheelchair spaces at the back of a stepped terraced floor should be in accordance with Figure 6.187.	M (4.12k)

Lecture/conference facilities

All people should be able to use presentation facilities.	M (4.9)
People with hearing impairments should be able to participate fully in conferences, committee meetings and study groups.	M (4.9)
The design of the acoustic environment should ensure that audible information can be heard clearly.	M (4.9)
Artificial lighting should be designed to give good colour rendering of all surfaces.	M (4.9)
Glare and reflections from shiny surfaces, and large repeating patterns, should be avoided in spaces where visual accuracy is critical.	M (4.9)
Uplighters mounted at low or floor level should be avoided as they can disorientate some visually impaired people.	M (4.9)

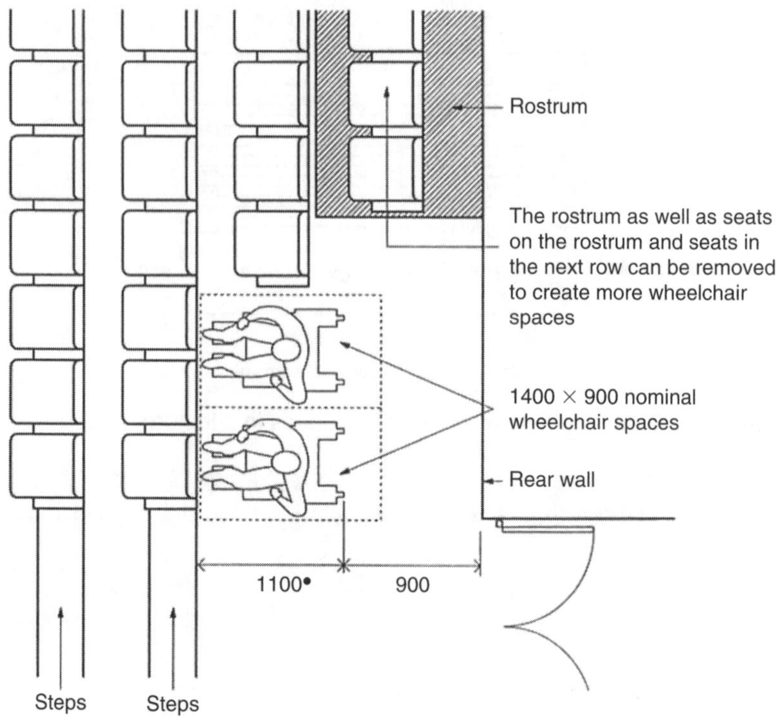

Figure 6.187 Possible location of wheelchair spaces in front of a rear isle

For lecture/conference facilities

Where a podium or stage is provided, wheelchair users should have access to it by means of a ramp or lifting platform.	M (4.12.1)
A clearly audible public address system should be supplemented by visual information.	
Hearing enhancement systems should be installed: • in rooms and spaces designed for meetings, lectures, classes, performances, spectator sports or films; • at service or reception counters (especially when situated in noisy areas or behind glazed screens).	M (4.12.1)
The availability of an induction loop or infrared hearing enhancement system should be indicated by the standard symbol.	M (4.12.1)

Telephones suitable for hearing aid users should: M (4.12.1)

- be clearly indicated by the standard ear symbol;
- incorporate an inductive coupler and volume control.

Text telephones for deaf and hard of hearing people should be clearly indicated by the standard symbol. M (4.12.1)

Artificial lighting should be designed to be compatible with other electronic and radio frequency installations. M (4.12.1)

Toilets are available that have been adapted for people with mobility impairments

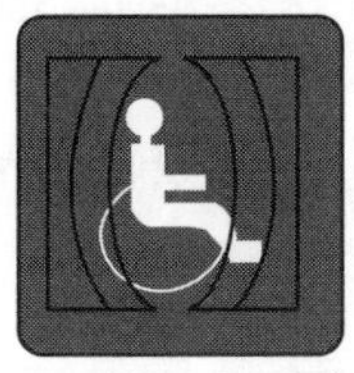

Changing rooms are available for people with mobility impairments

Assistance dogs are welcome on the premises

Figure 6.188 Examples of facility signs

Entertainment, leisure and social facilities

In theatres and cinemas (where seating is normally closely packed together) special care should be given to the design and location of wheelchair spaces. M (4.10)

Sports facilities

 **Note:** See *Guide to Safety at Sports Grounds.*

Refreshment facilities

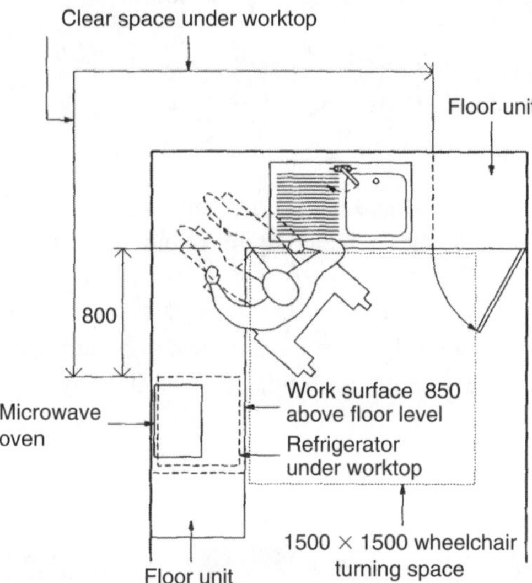

Figure 6.189 An example of a typical shared refreshment facility

Restaurants and bars should be designed so that they can be reached and used by all people independently or with companions.

All people should have access to:

- all parts of the facility; M (4.13)
- staff areas; M (4.13)
- public areas (e.g. lavatory accommodation, public telephones and external terraces); M (4.14)
- self-service facilities (when provided). M (4.14)

Changes of floor level are permitted, provided that all of the different levels are accessible and raised thresholds are avoided.

Part of the working surface of a bar or serving counter should:

- be permanently accessible to wheelchair users; M (4.16b)
- be at a level of not more than 850 mm above the floor. M (4.16b)

Note: If unavoidable, then the total height should not be more than 15 mm, with a minimum number of upstands and slopes and with any upstands higher than 5 mm chamfered or rounded.

In addition:

- the worktop of a shared refreshment facility (e.g. for tea making) should be 850 mm above the floor with a clear space beneath at least 700 mm above the floor (see Figure 6.187); M (4.16c)
- basin taps should either be controlled automatically or capable of being operated using a closed fist, e.g. by lever action; M (4.16c)
- all terminal fittings should comply with Guidance Note G18.5 of the Guidance Document relating to Schedule 2: Requirements for Water Fittings, of the Water Supply (Water Fittings) Regulations 1999, SI 1999/1148. M (4.16c)

6.17.3 Sleeping accommodation

Sleeping accommodation in hotels, motels and student accommodation, should be convenient for all types of people. M (4.17)

A proportion of rooms should be available for wheelchair users. M (4.17)

Wheelchair users should be able to:

- reach all the facilities available within the building; M (4.18)
- manoeuvre around and use the facilities in the room, and operate switches and controls. M (4.19)

En-suite sanitary facilities are the preferred option for wheelchair-accessible bedrooms. M (4.19)

There should be at least as many en-suite shower rooms as en-suite bathrooms (i.e. some mobility-impaired people may find it easier to use a shower than a bath). M (4.19)

In all bedrooms, built-in wardrobes and shelving should be accessible and convenient to use. M (4.20)

Bedrooms not designed for independent use by a person in a wheelchair should have an outer door wide enough to be accessible to a wheelchair user.	M (4.21)

For all bedrooms:

- built-in wardrobe swing doors should open through 180°; M (4.24b)
- handles on hinged and sliding doors: M (4.24c)
  - should be easy to grip and operate;
  - should contrast visually with the surface of the door;
- windows and window controls should be: M (4.24d)
  - located between 800 and 1000 mm above the floor;
  - easy to operate without using both hands simultaneously;
- a visual fire alarm signal should be provided in addition to the requirements of Part B; M (4.24e)
- room numbers should be embossed characters; M (4.24f)
- the effective clear width of the door from the access corridor should comply with Table 6.81. M (4.24a)

Table 6.81 Minimum effective clear widths of doors

Direction and width of approach	New buildings (mm)	Existing buildings (mm)
Straight-on (without a turn or oblique approach)	800	750
At right angles to an access route at least 1500 mm wide	800	750
At right angles to an access route at least 1200 mm wide	825	775
External doors to buildings used by the general public	1000	775

For wheelchair-accessible bedrooms at least one wheelchair-accessible bedroom should be provided for every 20 bedrooms – M (4.24 g).

Wheelchair-accessible bedrooms should:

- be located on accessible routes; M (4.24h)
- be designed to provide a choice of location; M (4.24i)
- have a standard of amenity equivalent to that of other bedrooms; M (4.24i)
- (for en-suite bathroom and shower room doors) have an effective clear width complying with Table 6.81; M (4.24k)

- be large enough (see Figure 6.190) to enable a wheelchair user to manoeuvre with ease. M (4.24l)

In wheelchair-accessible bedrooms:

- if wide angle viewers are provided in the entrance door, they should be located at 1050 mm and 1500 mm above floor level; M (4.24n)
- if a balcony is provided, it should: M (4.24o)
  - comply with Table 6.81;
  - have a level threshold;
  - have no horizontal transoms between 900 mm and 1200 mm above the floor;
- there should be no permanent obstructions in a zone 1500 mm back from any balcony doors; M (4.24p)

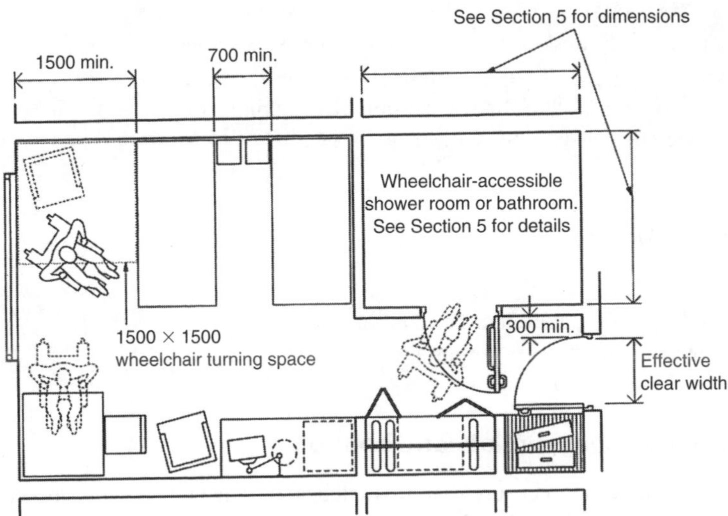

Figure 6.190 Example of a wheelchair accessible hotel bedroom

- emergency assistance alarms should be provided; M (4.24q)
- the door from the access corridor to a wheelchair-accessible bedroom should: M (4.24j)
  - not require more than 20 N opening force;
  - have an effective clear width through a single leaf door (or one leaf of a double leaf door) in accordance with Table 6.81 and Figure 6.191;
  - have an unobstructed space of at least 300 mm on the pull side of the door – see Figure 6.191.

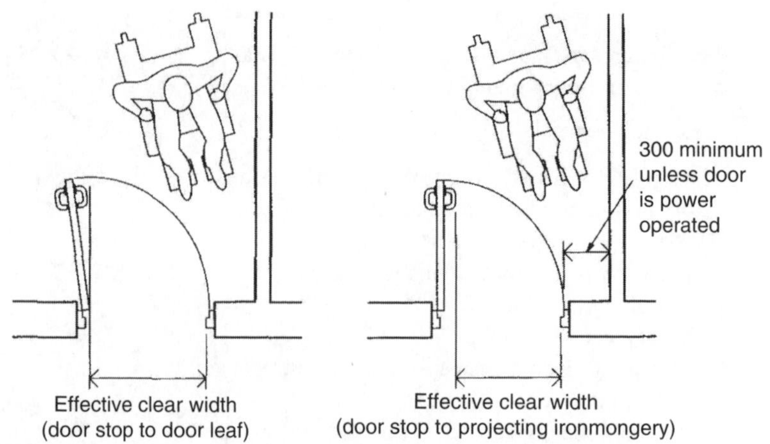

Figure 6.191 Effective clear width of doors

Sanitary facilities, en-suite to a wheelchair-accessible bedroom, should comply with the provisions of M5.15 to M5.21 for 'Wheelchair-accessible bathrooms' or 'Wheelchair-accessible shower facilities'.	M (4.24m)

Smoke alarms

Smoke alarms should not be fixed in bathrooms, showers, cooking areas or garages.	B1 1.17 (V1) B1 1.16 (V2)

6.17.4 Switches, outlets and controls

The aim should be to ensure that all switches, outlets and controls are easy to operate, visible and are free from obstruction.

Light switches should: • have large push pads; • align horizontally with door handles; • be within 900 to 1100 mm from the entrance door opening.	M (4.30h and I)
Switches and controls should be located between 750 mm and 1200 mm above the floor.	M (4.30c and d)
The operation of all switches, outlets and controls should not require the simultaneous use of both hands (unless necessary for safety reasons).	M (4.30j)

Switched socket outlets should indicate whether they are 'on'.	M (4.30k)
Mains and circuit isolator switches should clearly indicate whether they are 'on' or 'off'.	M (4.30l)
Individual switches on panels and on multiple socket outlets should be well separated.	M (4.29)
All socket outlets should be wall mounted.	M (4.30a and b)
All telephone points and TV sockets should be located between 400 mm and 1000 mm above the floor (or 400 mm and 1200 mm above the floor for permanently wired appliances).	M (4.30a and b)
Socket outlets should be located no nearer than 350 mm from room corners.	M (4.30g)
Controls that need close vision (e.g. thermostats) should be located between 1200 mm and 1400 mm above the floor.	M (4.30f)
Emergency alarm pull cords should be: • coloured red; • located as close to a wall as possible; • have two red 50 mm diameter bangles.	M (4.30e)
Front plates should contrast visually with their backgrounds.	M (4.30m)
The colours red and green should **not** be used in combination as indicators of 'on' and 'off' for switches and controls.	M (4.28)

6.17.5 Aids to communication

The design of the acoustic environment should ensure that audible information can be heard clearly.	M (4.33)
A clearly audible public address system should be supplemented by visual information.	
Note: To assist people with impaired hearing to fully participate in public discussions (meetings and performances) may require an advanced sound level system (e.g. induction loop, infrared or radio system) to be installed.	M (4.36a)
Hearing enhancement systems should be installed: • in all rooms and spaces designed for meetings, lectures, classes, performances, spectator sport or films; • at service or reception counters (especially when they are are situated in noisy areas or they are behind glazed screens.	M (4.36b)

The availability of an induction loop or infrared hearing enhancement system should be indicated by the standard symbol.	M (4.36c)
Telephones suitable for hearing aid users should: • be clearly indicated by the standard ear and 'T' symbol; • incorporate an inductive coupler and volume control.	M (4.36d)
Text telephones for deaf and hard of hearing people should be clearly indicated by the standard symbol.	M (4.36e)
Artificial lighting should be designed to be compatible with other electronic and radio frequency installations.	M (4.36f)
Artificial lighting should be designed to give good colour rendering of all surfaces.	M (4.34)
Glare and reflections from shiny surfaces, and large repeating patterns, should be avoided in spaces where visual acuity is critical.	M (4.32)
Uplighters mounted at low or floor level should be avoided as they can disorientate some visually impaired people.	M (4.34)

A hearing system is available in certain locations

Mini-com and/or text phone facility available

Staff have received disability awareness training

Figure 6.192 Examples of facility signs

Note: Detailed guidance on surface finishes, visual, audible and tactile signs, as well as the characteristics and appropriate choice and use of hearing enhancement systems, is available in BS 8300.

6.18 Water (and earth) closets, bathrooms and showers

6.18.1 The requirement

Sanitary conveniences (provided in separate rooms or in bathrooms) shall be:

- *separated from places where food is prepared;*

- *provided with washbasins plumbed with hot and cold water;*
- *be designed and installed so as to allow effective cleaning.*

(Building Act 1984 Section 26)

All plans for buildings must include at least one (or more) water or earth closets **unless** the local authority are satisfied in the case of a particular building that one is not required (for example in a large garage separated from the house).

If you propose using an earth closet, the local authority cannot reject the plans unless they consider that there is insufficient water supply to that earth closet.

Ventilation

There shall be adequate means of ventilation provided for people in the building.

(Approved Document F)

Sanitary conveniences

***All** dwellings (houses, flats or maisonettes) should have at least one closet and one washbasin which should:*

- *be separated by a door from any space used for food preparation or where washing-up is done in washbasins;*
- *ideally, be located in the room containing the closet;*
- *have smooth, non-absorbent surfaces and be capable of being easily cleaned;*
- *be capable of being flushed effectively;*
- *only be connected to a flush pipe or discharge pipe;*
- *washbasins should have a supply of hot and cold water.*

(Approved Document G1)

Access and facilities for disabled people

In addition to the requirements of the Disability Discrimination Act 1995 precautions need to be taken to ensure that:

- *new non-domestic buildings and/or dwellings (e.g. houses and flats used for student living accommodation etc.);*
- *extensions to existing non-domestic buildings;*
- *non-domestic buildings that have been subject to a material change of use (e.g. so that they become a hotel, boarding house, institution, public building or shop);*

are capable of allowing people, regardless of their disability, age or gender, to:

- *gain access to buildings;*
- *gain access within buildings;*
- *be able to use the facilities of the buildings (both as visitors and as people who live or work in them);*

- *use sanitary conveniences in the principal storey of any new dwelling.*

(Approved Document M)

Note: See Annex A for guidance on access and facilities for disabled people.

If the proposed building is going to be used as a workplace or a factory in which persons of both sexes are going to be employed, then separate closet accommodation **must** be provided unless the local authority approve otherwise.

The Building Act 1984 Sections 64–68

Under existing regulations, all buildings (except factories and buildings used as workplaces) shall be provided with sufficient closet accommodation (or privy) according to the intended use of that building and the amount of people using that building. The only exceptions are if the building (in the view of the local authority) has an insufficient water supply and a sewer is not available.

If a building already has a sufficient water supply and sewer available, the local authority have the authority to insist that the owner of the property replaces any other closet (e.g. an earth closet) with a water closet. In these cases the owner is entitled to claim 50% of the expense of doing this off the local authority.

If the local authority completes the work, then they are entitled to claim 50% back from the owner.

The owner of the property has **no** right of appeal in these cases.

In the Greater London area, a 'water closet' can **also** be taken to mean a urinal.

Business premises

There may be some additional requirements regarding numbers, types and siting of appliances in business premises. If this applies to you then you will need to look at:

- the Offices, Shops and Railway Premises Act 1963,
- the Factories Act 1961, or
- the Food Hygiene (General) Regulations 1970.

6.18.2 Meeting the requirement

All dwellings (houses, flats or maisonettes) should have at least one closet and one washbasin.	G1
Closets (and/or urinals) should be separated by a door from any space used for food preparation or where washing-up is done.	G1

Washbasins should, ideally, be located in the room containing the closet, or, if not, adjacent to the WC.	G1
The surfaces of a closet, urinal or washbasin should be smooth, non-absorbent and capable of being easily cleaned.	G1
Closets (and/or urinals) should be capable of being flushed effectively.	G1
Closets (and/or urinals) should only be connected to a flush pipe or discharge pipe.	G1
Washbasins should have a supply of hot and cold water.	G1
Closets and/or urinals fitted with flushing apparatus should discharge through a trap and discharge pipe into a discharge stack or a drain.	G1
Closets fitted with macerators and pump can be connected to a discharge pipe discharging to a discharge stack if the macerator and pump system is approved under the current European Technical Approval system.	G1
Washbasins should discharge through a grating, a trap and a branch discharge pipe to a discharge stack or (if it is a ground floor location) into a gully or directly into a drain.	G1
If there is no suitable water supply or means of disposing foul water, closets (and/or urinals) can use chemical treatment.	G1

Although the above are the minimum requirements for meeting sanitary conveniences and washing facilities, other local authority regulations may apply and it is worth seeking the advice of the local planning officer before proceeding.

Workplace conveniences

If the building is a workplace used by both sexes, then sufficient and satisfactory accommodation is required for persons of each sex.

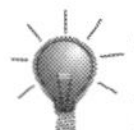

This requirement does not apply to premises to which the Offices, Shops and Railway Premises Act of 1963 applies.

Loan of temporary sanitary conveniences

If the local authority is maintaining, improving or repairing drainage systems and this requires the disconnection of existing buildings from these sanitary conveniences, then, on request from the occupier of the building, the local authority are required to supply (on temporary loan and at no charge) sanitary conveniences:

- if the disconnection is caused by a defect in a public sewer;
- if the local authority has order the replacement of earth closets (see above);
- for the first seven days of any disconnection.

Erection of public conveniences

You are not allowed to erect a public sanitary convenience in (or on) any location that is accessible from a street, without the consent of the local authority. Any person who contravenes this requirement is liable to a fine and can be made (at his own expense) to remove or permanently close it.

This requirement does not apply to sanitary conveniences erected by a railway company within their railway station, yard or approaches and by dock undertakers on land belonging to them.

Protection of openings for pipes

Pipes which pass through a compartment wall or compartment floor (unless the pipe is in a protected shaft), or through a cavity barrier, should conform to one of the following alternatives:

Proprietary seals (any pipe diameter) that maintain the fire resistance of the wall, floor or cavity barrier.	B3 (11.5–11.6)
Pipes with a restricted diameter where fire-stopping is used around the pipe, keeping the opening as small as possible.	B3 (11.5 and 11.7)
Sleeving – a pipe of lead, aluminium, aluminium alloy, fibre-cement or UPVC, with a maximum nominal internal diameter of 160 mm, may be used with a sleeving of non-combustible pipe as shown in Figure 6.193.	B3 (11.5 and 11.8)

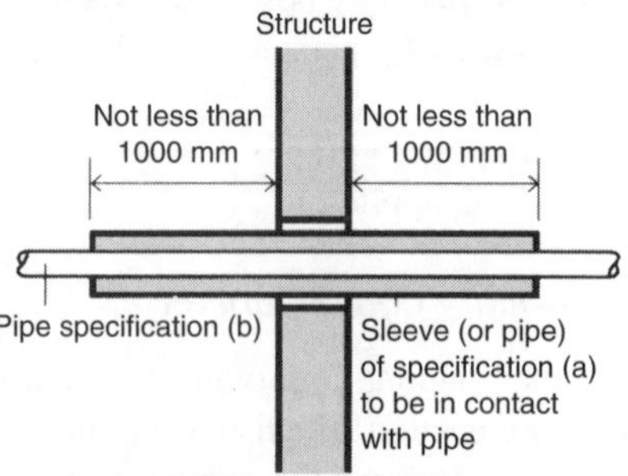

Figure 6.193 Pipes penetrating a structure

Make the opening in the structure as small as possible and provide fire-stopping between pipe and structure.

Joints between fire-separating elements should be fire-stopped and all openings for pipes that pass through any part of a fire-separating element should be: • kept as few in number as possible; and • kept as small as practicable; and • fire-stopped (which in the case of a pipe or duct should allow thermal movement).	B3 7.12 (V1) B3 10.17 (V2)

Ventilation

Meeting the requirement

Extract ventilation concerns the removal of air directly from a space or spaces to outside. Extract ventilation may be by natural means such as Passive Stack Ventilation (PSV) or by mechanical means (e.g. by an extract fan or central system).

All sanitary accommodation shall be provided with extract ventilation to the outside which is capable of operating either intermittently or continuously with a minimum extract rate of 6 l/s.	F 1.5
Ventilation devices designed to work continuously shall **not** have automatic controls such as a humidity control when used for sanitary accommodation.	F Table 1.5
As odour is the main pollutant, humidity controls should **not** be used for intermittent extract in sanitary accommodation.	F Table 1.5
PSV can be used as an alternative to a mechanical extract fan for office sanitary and washrooms.	F Table 2.2a
PSV devices **shall** have a minimum internal duct diameter of 80 mm and a minimum internal cross-sectional area of 5000 mm^2. **Note:** Purge ventilation may be used **provided** that security is not an issue.	F
Common outlet terminals and/or branched ducts shall **not** be used for wet rooms such as WCs.	F App D
Where there was no previous ventilation opening, or where the size of the original ventilation opening	F 3.6

is not known, replacement window(s) shall have an equivalent area greater than 2500 mm^2 per WC.	
In buildings other than dwellings, fresh air supplies should be protected from contaminants that could be injurious to health.	F 2.3
All office sanitary accommodation and washrooms shall be provided with intermittent air extract ventilation.	F 2.11

Disabled access

The aim of the amended Approved Document M is that suitable sanitary accommodation should be available to *everybody*, including sanitary accommodation specifically designed for wheelchair users, ambulant disabled people, people of either sex with babies and small children and/or people with luggage.

Provision of toilet accommodation

Where sanitary facilities are provided in a building, at least one wheelchair-accessible unisex toilet should be available.	M (5.7b)
If there is only space for one toilet in a building, then it should be a wheelchair-accessible unisex type.	M (5.7a)
In separate-sex toilet accommodation, at least one WC cubicle should be provided for ambulant disabled people.	M (5.7c)
In separate-sex toilet accommodation having four or more WC cubicles, at least one should be an enlarged cubicle for use by people who need extra space.	M (5.7d)
Wheelchair-accessible unisex toilets should always be provided in addition to any wheelchair-accessible accommodation in separate-sex toilet washrooms.	M (5.5)
If there is only space for **one** toilet in a building:	
• then it should be a wheelchair-accessible unisex type;	M (5.7a)
• its width should be increased from 1.5 m to 2 m;	M (5.7e)
• it should include a standing height wash basin, in addition to the finger rinse basin associated with the WC.	M (5.7e)

For specific guidance on the provision of sanitary accommodation in sports buildings, refer to 'Access for Disabled People'.

Sanitary accommodation generally

Sanitary accommodation and washrooms may be included in protected shafts.

Doors

Doors to WC cubicles and wheelchair-accessible unisex toilets should: M (5.3)

- (ideally) open outwards;
- be operable by people with limited strength or manual dexterity;
- be capable of being opened if a person has collapsed against them while inside the cubicle.

Doors to wheelchair-accessible unisex toilets, changing rooms or shower rooms should:

- be fitted with light action privacy bolts; M (5.4d)
- be capable of being opened using a force no greater than 20 N; M (5.4d)
- have an emergency release mechanism so that they are capable of being opened outwards (from the outside) in case of emergency. M (5.4e)

Door opening furniture should: M (5.4c)

- be easy to operate by people with limited manual dexterity;
- be easy to operate with one hand using a closed fist (e.g. a lever handle);
- contrast visually with the surface of the door.

Doors when open should not obstruct emergency escape routes. M (5.4f)

Sanitary fittings

The surface finish of sanitary fittings and grab bars should contrast visually with background wall and floor finishes. M (5.4k)

Taps should be operable by people with limited strength and/or manual dexterity.	M (5.3)
Bath and wash basin taps should either be controlled automatically or capable of being operated using a closed fist, e.g. by lever action.	M (5.4a)
All terminal fittings should comply with Guidance Note G18.5 of the Guidance Document relating to Schedule 2: Requirements for Water Fittings, of the Water Supply (Water Fittings) Regulations 1999, SI 1999/1148.	M (5.4b)

Outlets, controls and switches

The aim should be to ensure that all controls and switches should be easy to operate, visible and free from obstruction: M (5.4i)

- they should be located between 750 mm and 1200 mm above the floor;
- they should not require the simultaneous use of both hands (unless necessary for safety reasons) to operate;
- light switches should:
  - have large push pads;
  - align horizontally with door handles;
  - be within 900 to 1100 mm from the entrance door opening;
- switched socket outlets should indicate whether they are 'on';
- mains and circuit isolator switches should clearly indicate whether they are 'on' or 'off';
- individual switches on panels and on multiple socket outlets should be well separated;
- controls that need close vision (e.g. thermostats) should be located between 1200 mm and 1400 mm above the floor;
- emergency alarm pull cords should be:
  - coloured red;
  - located as close to a wall as possible;
  - have two red 50 mm diameter bangles;

- front plates should contrast visually with their backgrounds.

Heat emitters should either be screened or have their exposed surfaces kept at a temperature below 43°C.	M (5.4j)
Where possible, light switches with large push pads should be used in preference to pull cords.	M (5.3)
The colours red and green should not be used in combination as indicators of 'on' and 'off' for switches and controls.	M (5.3)

Smoke alarms

Smoke alarms should be positioned in the circulation spaces between sleeping spaces and places where fires are most likely to start (e.g. in kitchens and living rooms).	B1 1.11

Wheelchair-accessible unisex toilets

General

Where sanitary facilities are provided in a building, at least one wheelchair-accessible unisex toilet should be available.	M (5.7b)
Wheelchair-accessible unisex toilets should **not** be used for baby changing.	M (5.5)
Wheelchair-accessible unisex toilets should:	
• be located as close as possible to the entrance and/or waiting area of the building;	M (5.10a)
• not be located in a way that compromises the privacy of users;	M (5.10b)
• be located in a similar position on each floor of a multi-storey building;	M (5.10c)
• allow for right- and left-hand transfer on alternate floors;	M (5.10c and d)
• be located on accessible routes that are direct and obstruction-free;	M (5.10f)

- always be provided in addition to any wheelchair-accessible accommodation in separate-sex toilet washrooms; M (5.5)
- not be used for baby changing. M (5.5)

The minimum overall dimensions of, and arrangement of fittings within, a wheelchair-accessible unisex toilet should comply with Figure 6.194. M (5.10i)

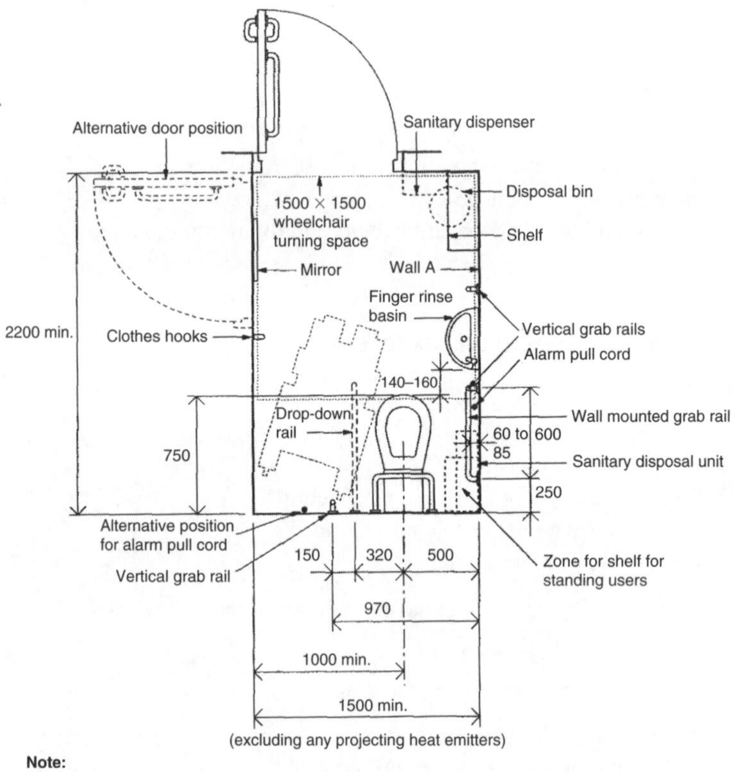

Figure 6.194 Example of a unisex wheelchair-accessible toilet with a corner WC

Accessibility

The approach to a unisex toilet should be separate to other sanitary accommodation. M (5.9)

Wheelchair users should:	
• not have to travel more than 40 m on the same floor to reach a unisex toilet;	M (5.9 and 5.10h)
• not have to travel more than combined horizontal distance where the unisex toilet accommodation is on another floor of the building (accessible by passenger lift);	M (5.9 and 5.10h)
• be able to approach, transfer to, and use the sanitary facilities provided within a building.	M (5.8)

Heights and arrangements

The heights and arrangement of fittings in a wheelchair-accessible unisex toilet should comply with Figure 6.194 and (as appropriate) Figure 6.195.	M (5.10)
The space provided for manoeuvring should enable wheelchair users to adopt various transfer techniques that allow independent or assisted use.	M (5.8)
The transfer space alongside the WC should be kept clear to the back wall.	M (5.8)
The relationship of the WC to the finger rinse basin and other accessories should allow a person to wash and dry hands while seated on the WC.	M (5.8)
Heat emitters (if located) should not restrict:	
• the minimum clear wheelchair manoeuvring space; • the space beside the WC used for transfer from the wheelchair to the WC.	M (5.10p)

Doors

Doors should:	
• preferably open outwards; • be fitted with a horizontal closing bar fixed to the inside face.	M (5.10g)

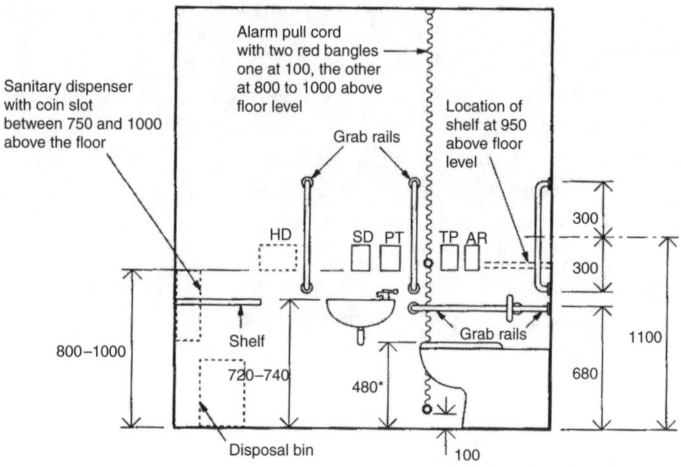

* Height subject to manufacturing tolerance of WC pan

HD: Possible position for automatic hand dryer
SD: Soap dispenser
PT: Paper towel dispenser
AR: Alarm reset button
TP: Toilet paper dispenser

* Height of drop-down rails to be the same as the other horizontal grab rails

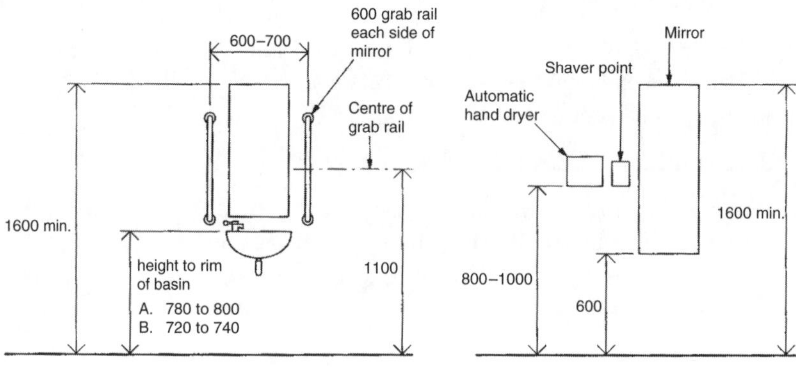

Height of independent wash basin and location of associated fittings, for wheelchair users and standing people

A. For people standing
B. For use from WC

Mirror located away from wash basin suitable for seated and standing people (Mirror and associated fittings used within a WC compartment or serving a range of compartments)

Figure 6.195 Typical heights of arrangement of fittings in a unisex wheelchair-accessible toilet

Support rails

The rail on the open side can be a drop-down rail, but on the wall side, it can either be a wall-mounted grab rail or, alternatively, a second drop-down rail in addition to the wall-mounted grab rail.	M (5.8)
If the horizontal support rail (on the wall adjacent to the WC) is set at the minimum spacing from the wall, an additional drop-down rail should be provided on the wall side, 320 mm from the centre line of the WC.	M (5.10j)
If the horizontal support rail (on the wall adjacent to the WC) is set so that its centre line is 400 mm from the centre line of the WC, there is no additional drop-down rail.	M (5.10k)

Emergency assistance

An emergency assistance alarm system should be provided that has:	
• an outside emergency assistance call signal that can be easily seen and heard by those able to give assistance;	M (5.10n)
• visual and audible indicators to confirm that an emergency call has been received;	M (5.10m)
• a reset control reachable from a wheelchair, WC, or from a shower/changing seat;	M (5.10m)
• a signal that is distinguishable visually and audibly from the fire alarms provided.	M (5.10m)
Emergency assistance pull cords should be:	M (5.10o)
• easily identifiable; • reachable from the WC and from the floor close to the WC; • coloured red; • located as close to a wall as possible; • have two red 50 mm diameter bangles.	

Fire safety

There shall be an early warning fire alarm system for persons in the building.
(Approved Document B1)

Alarms

Fire alarms should emit an audio and visual signal to warn occupants with hearing or visual impairments. M (5.4g)

Emergency assistance alarm systems should have: M (5.4h)

- visual and audible indicators to confirm that an emergency call has been received;
- a reset control reachable from a wheelchair, WC, or from a shower/changing seat;
- a signal that is distinguishable visually and audibly from the fire alarm.

WC pans

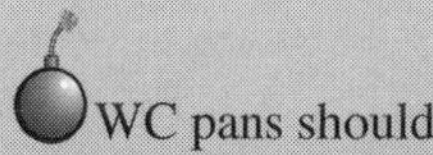

WC pans should:

- conform to BS 5503-3 or BS 5504-4; M (5.10q)
- be able to accept a variable height toilet seat riser; M (5.10q)
- have a flushing mechanism positioned on the open or transfer side of the space – irrespective of handing. M (5.10r)

See BS 8300 for more detailed guidance on the various techniques used to transfer from a wheelchair to a WC, as well as appropriate sanitary and other fittings.

Toilets in separate-sex washrooms

General

There should be at least the same number of WCs (for women) as urinals (for men). M (5.13)

Ambulant disabled people should have the opportunity to use a WC compartment within any separate-sex toilet washroom. M (5.11)

A wheelchair-accessible compartment (where provided) shall have the same layout and fittings as the unisex toilet. M (5.14f)

Where a separate-sex toilet washroom can be accessed by wheelchair users, it should be possible for them to use both a urinal (where appropriate) and a washbasin at a lower height than is provided for other users. M (5.13)

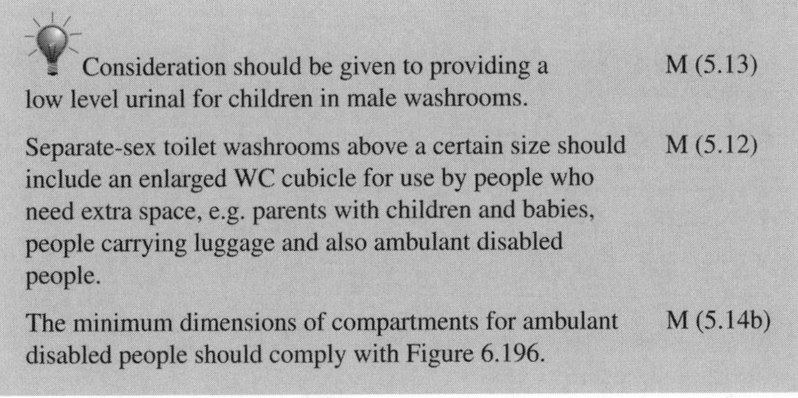

Consideration should be given to providing a low level urinal for children in male washrooms. M (5.13)

Separate-sex toilet washrooms above a certain size should include an enlarged WC cubicle for use by people who need extra space, e.g. parents with children and babies, people carrying luggage and also ambulant disabled people. M (5.12)

The minimum dimensions of compartments for ambulant disabled people should comply with Figure 6.196. M (5.14b)

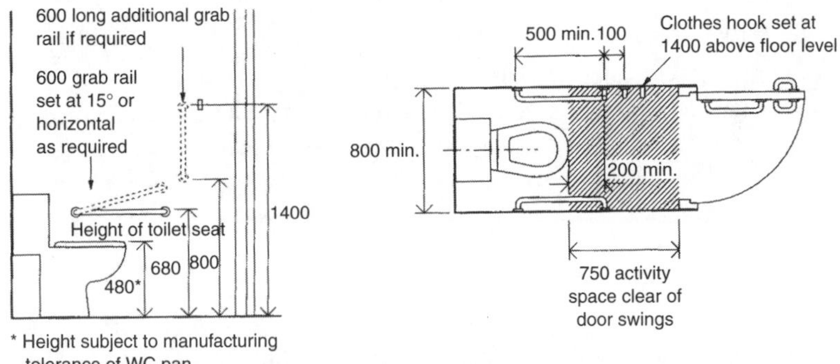

Figure 6.196 Example of a WC cubicle for an ambulant disabled person

Accessibility

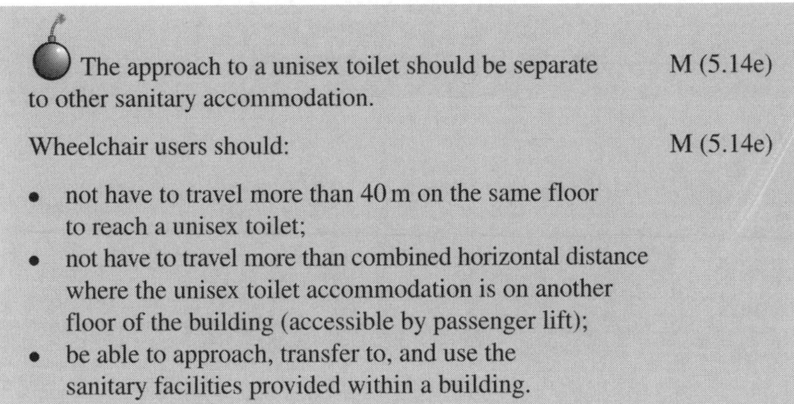

The approach to a unisex toilet should be separate to other sanitary accommodation. M (5.14e)

Wheelchair users should: M (5.14e)

- not have to travel more than 40 m on the same floor to reach a unisex toilet;
- not have to travel more than combined horizontal distance where the unisex toilet accommodation is on another floor of the building (accessible by passenger lift);
- be able to approach, transfer to, and use the sanitary facilities provided within a building.

Height and arrangement

Compartments used for ambulant disabled people should: • be 1200 mm wide; • include a horizontal grab bar adjacent to the WC; • include a vertical grab bar on the rear wall; • include space for a shelf and fold-down changing table.	M (5.14d)
A wheelchair-accessible washroom (where provided) shall have: • at least one washbasin with its rim set at 720 to 740 mm above the floor; • for men, at least one urinal with its rim set at 380 mm above the floor, with two 600 mm long vertical grab bars with their centre lines at 1100 mm above the floor, positioned either side of the urinal.	M (5.14g)
The compartment should: • be fitted with support rails; • include a minimum activity space to accommodate people who use crutches, or otherwise have impaired leg movements.	M (5.11)

Doors

Doors to compartments for ambulant disabled people should: • preferably open outwards; • be fitted with a horizontal closing bar fixed to the inside face.	M (5.14d)
The swing of any inward opening doors to standard WC compartments should enable a 450 mm diameter manoeuvring space to be maintained between the swing of the door, the WC pan and the side wall of the compartment.	M (5.14a)

WC pans

WC pans should:

• conform to BS 5503-3 or BS 5504-4;	M (5.14e)
• accommodate the use of a variable height toilet seat riser.	M (5.14e)

More detailed guidance on appropriate sanitary and other fittings is given in BS 8300.

Wheelchair-accessible changing and shower facilities

A choice of shower layout together with correctly located shower controls and fittings will enable disabled people to independently make use of the facilities – or be assisted by others where necessary.

General

In large building complexes (e.g. retail parks and large sports centres) there should be one wheelchair-accessible unisex toilet capable of including an adult changing table.	M (5.17)
The dimensions of the self-contained compartment should allow sufficient space for a helper.	M (5.16)
A combined facility should be divided into distinct 'wet' and 'dry' areas.	M (5.16)

For changing and shower facilities

A choice of layouts suitable for left-hand and right-hand transfer should be provided when more than one individual changing compartment or shower compartment is available.	M (5.18a)
Wall mounted drop-down support rails and wall mounted slip-resistant tip-up seats (not spring-loaded) should be provided.	M (5.18b)
Subdivisions (with the same configuration of space and equipment as for self-contained facilities) should be provided for communal shower facilities and changing facilities.	M (5.18c)
In addition to communal separate-sex facilities individual self-contained shower and changing facilities should be available in sports amenities.	M (5.18d)
An emergency assistance alarm system should be provided and which should have: • visual and audible indicators to confirm that an emergency call has been received; • a reset control reachable from a wheelchair, WC, or from a shower/changing seat; • a signal that is distinguishable visually and audibly from the fire alarm.	M (5.18f)
An emergency assistance pull cord should be provided which should: • be easily identifiable and reachable from the wall mounted tip-up seat (or from the floor);	M (5.18e)

- be located as close to a wall as possible;
- be coloured red;
- have two red 50 mm diameter bangles.

Facilities for limb storage should be included for the benefit of amputees. M (5.18)

For changing facilities

The floor of a changing area should be level and slip resistant when dry or when wet – particularly when associated with shower facilities. M (5.18i)

There should be a manoeuvring space 1500 mm deep in front of lockers in self-contained and/or communal changing areas. M (5.18j)

The minimum overall dimensions of (and the arrangement of equipment and controls within) individual self-contained changing facilities should comply with Figure 6.197. M (5.18h)

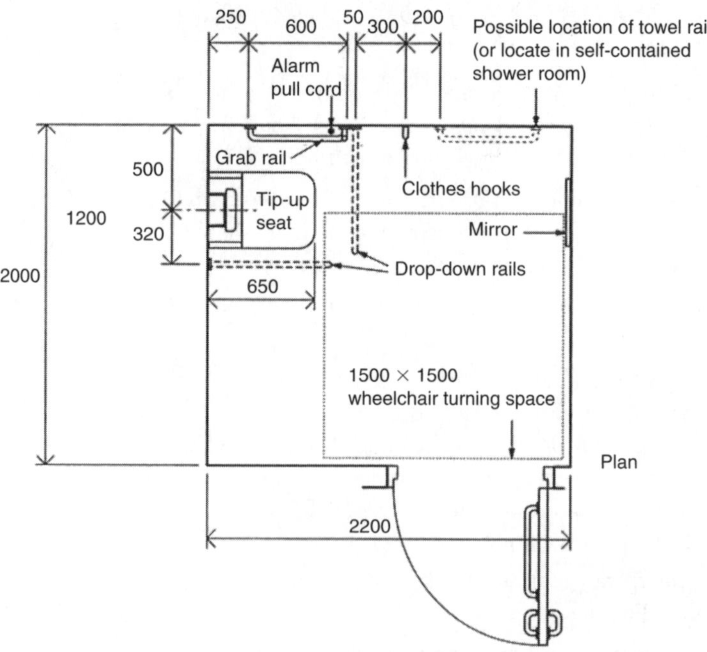

Figure 6.197 Example of a self-contained changing room for individual use

For shower facilities

A shower curtain (enclosing the seat and rails and which can be operated from the shower seat) should be provided.	M (5.18m)
A shelf (that can be reached from the shower seat and/or from the wheelchair) should be provided for toiletries.	M (5.18n)
The floor of the shower and shower area should be slip resistant and self-draining.	M (5.18o)
The shower controls should be positioned between 750 and 1000 mm above the floor in all communal area wheelchair-accessible shower facilities.	M (5.18q)
If showers are provided in commercial developments for the benefit of staff, at least one wheelchair-accessible shower compartment (complying with Figure 6.182) should be made available.	M (5.18l)
Individual self-contained shower facilities should comply with Figure 6.198.	M (5.18k)

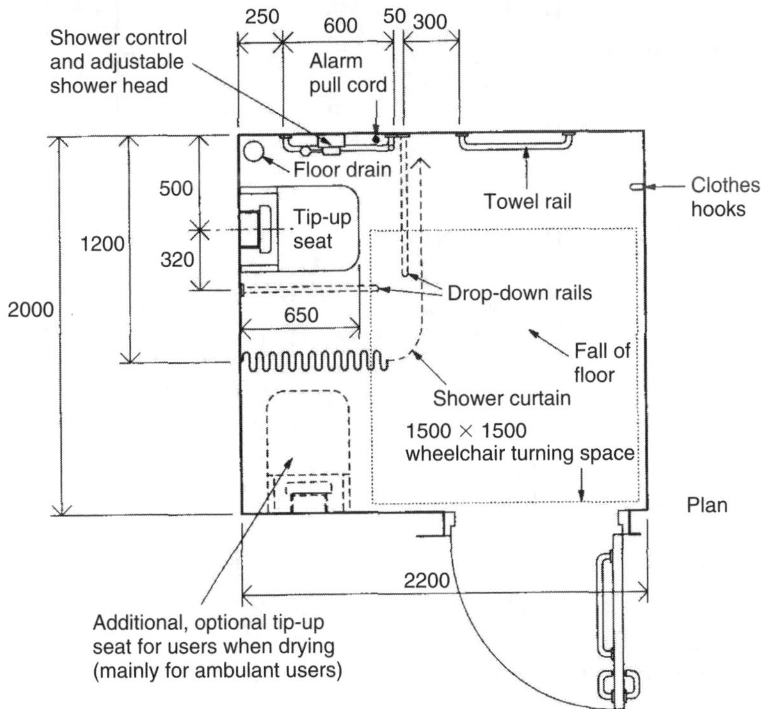

Figure 6.198 Example of a self-contained shower room for individual use

For shower facilities incorporating a WC

A choice of left-hand and right-hand transfer layouts should be available when more than one shower area includes a corner WC. M (5.18s)

The minimum overall dimensions of (and the arrangement of fittings within) an individual self-contained shower area incorporating a corner WC – e.g. in a sports building – should comply with Figure 6.199. M (5.18r)

More detailed guidance on appropriate sanitary and other fittings is given in BS 8300.

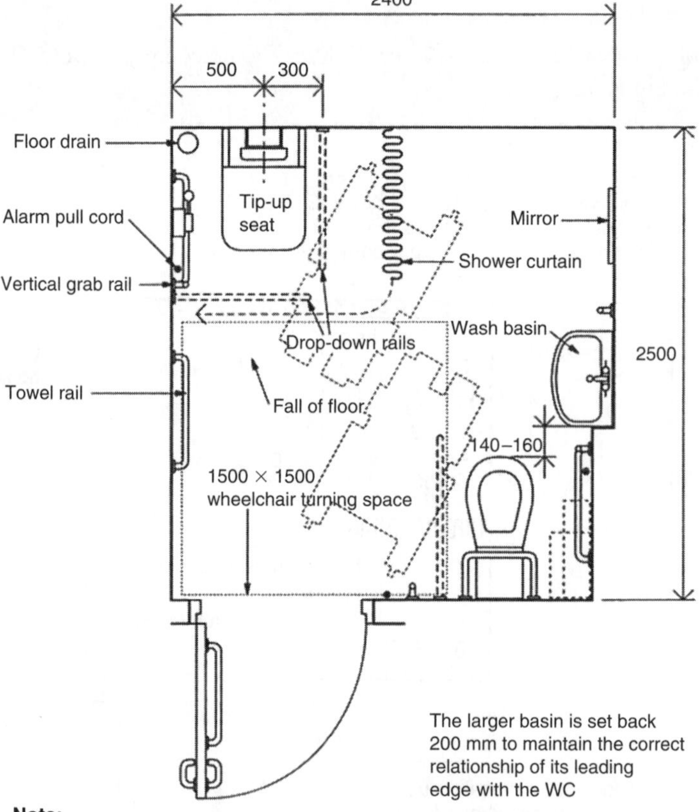

Figure 6.199 Example of a shower room incorporating a corner WC for individual use

Wheelchair-accessible bathrooms

Wheelchair users and ambulant disabled people (in hotels, motels, relatives' accommodation in hospitals and student accommodation and sports facilities) should be able to wash or bathe either independently or with assistance from others.

The minimum overall dimensions of (and the arrangement of fittings within) a bathroom for individual use incorporating a corner WC should comply with Figure 6.200. M (5.21a)

A choice of layouts suitable for left-hand and right-hand transfer should be provided when more than one bathroom is available. M (5.21b)

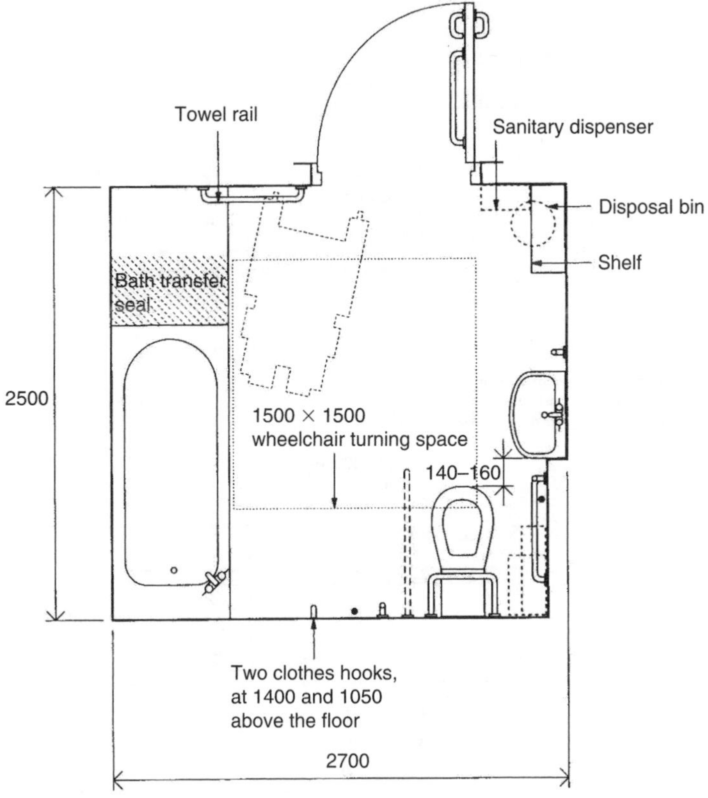

Note:
Layout shown for right-hand transfer to bath and WC.

Figure 6.200 Example of a bathroom containing a WC

The floor of a bathroom should be slip resistant when dry or when wet. M (5.21c)

The bath should be provided with a transfer seat that is 400 mm deep and equal to the width of the bath. M (5.21d)

Outward opening doors, fitted with a horizontal closing bar fixed to the inside face, should be provided. M (5.21e)

An emergency assistance pull cord should be provided which should: M (5.21f)

- be easily identifiable and reachable from the wall mounted tip-up seat (or from the floor);
- be located as close to a wall as possible;
- be coloured red;
- have two red 50 mm diameter bangles.

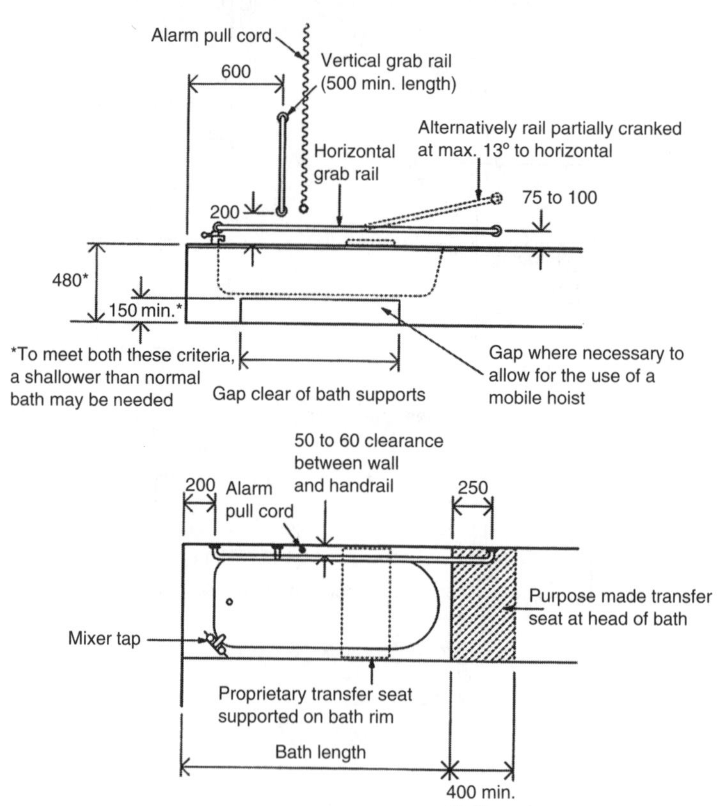

Figure 6.201 Grab rails and fitting associated with a bath

Note:

(1) More detailed guidance on appropriate sanitary and other fittings, including facilities for the use of mobile and fixed hoists, is given in BS 8300.

(2) Guidance on the slip resistance of floor surfaces is given in Annex C of BS 8300:2001.

WC provision in the entrance storey of the dwelling

Whenever possible, a WC should be provided in the entrance storey of the dwelling so that there is no need to negotiate a stair to reach it from the habitable rooms in that storey.

If there is a bathroom in the principal storey, then a WC may be collocated with it. M (10.2)

The door to the WC compartment should: M (10.3b)

- open outwards;
- be positioned so as to allow wheelchair users access to it;
- have a clear opening width in accordance with Table 6.82.

Table 6.82 Minimum widths of corridors and passageways for a range of doorway widths

Doorway clear opening width (mm)	Corridor/passageway width (mm)
750 or wider	900 (when approached head on)
750	1200 (when not approached head on)
775	1050 (when not approached head on)
800	900 (when not approached head on)

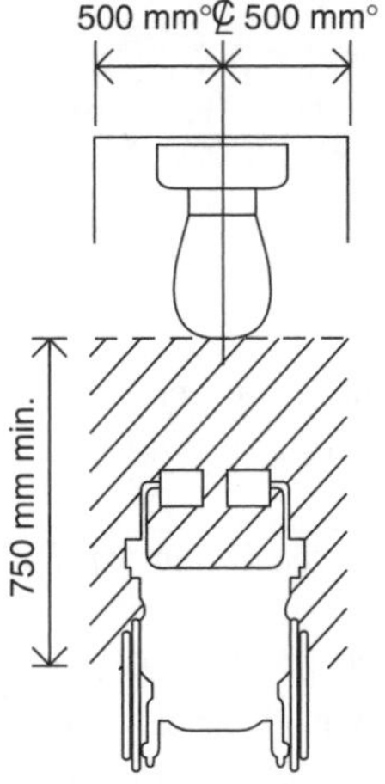

Clear space for frontal access to WC

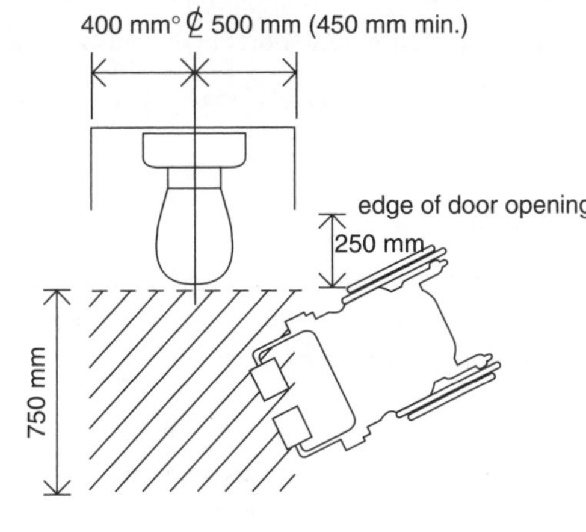

Clear space for oblique access to WC

The WC compartment should:	M (10.3c)
• provide a clear space for wheelchair users to access the WC; • position the washbasin so that it does not impede access.	

Note: For further information see the Disability Rights Commission's website at http://www.drc-gb.org.

6.19 Electrical safety

6.19.1 The requirement

Although:

- Part E (Resistance to the passage of sound);
- Part J (Combustion appliances and fuel storage systems); and
- Part K (Protection from falling, collision and impact);

all have a number of requirements concerning electrical safety and electrical installations (see below for details), the main requirements are contained in:

- Part P (Electrical safety) together with:
  - Part B(Fire safety)
  - Part L (Conservation of fuel and power) and, where necessary,
  - Part M (Access and use of buildings).

A visual and audible fire alarm signal should be provided in buildings where it is anticipated that one or more persons with impaired hearing may be in relative isolation (e.g. hotel bedrooms and sanitary accommodation).	B1 1.34 (V2)

Electrical work

Reasonable provision shall be made in the design and installation of electrical installations in order to protect persons operating, maintaining or altering the installations from fire or injury

(Approved Document P1)

Note: Although Part P makes requirements for the safety of fixed electrical installations, this does not cover system functionality (such as electrically powered fire alarm systems, fans and pumps) which are covered in other Parts of the Building Regulations and other legislation.

Conservation of fuel and power

Reasonable provision shall be made for the conservation of fuel and power in buildings by:

(a) limiting heat gains and losses:
  (i) through thermal elements and other parts of the building fabric; and
  (ii) from pipes, ducts and vessels used for space heating, space cooling and hot water services;
(b) providing and commissioning energy-efficient fixed building services with effective controls; and
(c) providing to the owner sufficient information about the building, the fixed building services and their maintenance requirements so that the building can be operated in such a manner as to use no more fuel and power than is reasonable in the circumstances.

(Approved Document L1)

Responsibility for achieving compliance with the requirements of Part L rests with the person carrying out the work. That 'person' may be, for example, a developer, a main (or sub) contractor, or a specialist firm directly engaged by a private client.

Note: The person responsible for achieving compliance should either themselves provide a certificate, or obtain a certificate from the sub-contractor, that commissioning has been successfully carried out. The certificate should be made available to the client and the building control body.

Access and facilities for disabled people

In addition to the requirements of the Disability Discrimination Act 1995 precautions need to be taken to ensure that:

- *new non-domestic buildings and/or dwellings (e.g. houses and flats used for student living accommodation etc.);*
- *extensions to existing non-domestic buildings;*
- *non-domestic buildings that have been subject to a material change of use (e.g. so that they become a hotel, boarding house, institution, public building or shop);*

are capable of allowing people, regardless of their disability, age or gender to:

- *gain access to buildings;*
- *gain access within buildings;*
- *be able to use the facilities of the buildings (both as visitors and as people who live or work in them);*
- *use sanitary conveniences in the principal storey of any new dwelling.*

(Approved Document M)

Note: See Annex A for guidance on access and facilities for disabled people.

6.19.2 Meeting the requirement

Electrical installations

All electrical installations shall comply with Approved Document P (Electrical safety).	B1 1.20

Electrical installations must be inspected and tested during, at the end of installation and before they are taken into service to verify that they:

- comply with Part P (and any other relevant Parts) of the Building Regulations; — P 1.7
- are safe to use, maintain and alter;
- meet the relevant equipment and installation standards; — P 0.1b
- meet the requirements of the Building Regulations. — P 3.1

Any proposal for a new mains supply installation (or where significant alterations are going to be made to an existing mains supply) **must** be agreed with the electricity distributor. — P 1.2

Design

Where it is critical for electrical circuits to be able to continue to function during a fire, protected circuits (meeting the requirements of BS EN 50200:2006) shall be installed.	B1 5.38 (V2)

Electrically powered locks should return to the unlocked position:

- on operation of the fire alarm system; — B1 5.11 (V2)
- on loss of power or system error;
- on activation of a manual door release unit.

Electrical installations must be designed and installed (suitably enclosed and appropriately separated) so that they:

- comply with Part P (and any other relevant Parts) of the Building Regulations; — P 1.7 and 3.1

- comply with the relevant equipment and installation standards; P 0.1b
- do not present an electric shock or fire hazard to people; P 0.1a
- provide mechanical and thermal protection; P 1.3
- provide adequate protection against mechanical and thermal damage; P 0.1a
- provide adequate protection for persons against the risks of electric shock, burn or fire injuries; P1.3
- are safe to use, maintain and alter. P 1.7

When downlighters, loudspeakers and other electrical accessories are installed, additional protection may be required to maintain the integrity of a wall or floor.

Transfer grills of air circulation systems should **not** be fitted in any wall, door, floor or ceiling enclosing a protected stairway. B1 2.17

Electrical installations must be designed and installed (suitably enclosed and appropriately separated) so that they comply with the requirements of BS 7671, as amended. P 1.4

Note: See Appendix A of Part P to the Building Regulations for details of some of the types of electrical services normally found in dwellings, some of the ways they can be connected and the complexity of wiring and protective systems that can be used to supply them.

Extensions, material alterations and material changes of use

In accordance with Regulation 4(2) the **whole** of an existing installation **does not** have to be upgraded to current standards, but only to the extent necessary for the new work to meet current standards except where upgrading is required by the energy efficiency requirements of the Building Regulations.

Where any electrical installation work is classified as an extension, a material alteration or a material change of use, the work must consider and include:

- confirmation that the mains supply equipment is suitable and can carry the additional loads envisaged; P 2.1b–2.2

• the earthing and bonding systems are satisfactory and meet the requirements;	P 2.1a–2.2c
• the necessary additions and alterations to the circuits which feed them;	P 2.1a
• the protective measures required to meet the requirements;	P 2.1a–2.2b
• the rating and the condition of existing equipment (belonging to both the consumer and the electricity distributor) is sufficient.	P 2.2a

See Figure 6.202 for details of some of the types of electrical services normally found in dwellings, some of the ways they can be connected and the complexity of wiring and protective systems that can be used to supply them.

Note: Appendix C to Part P of the Building Regulations offers guidance on some of the older types of installations that might be encountered during alteration work and Appendix D provides guidance on the application of the now harmonized European cable identification system.

New habitable rooms that are the result of a material alteration and which are above ground floor level (or at ground floor level where no final exit has been provided) shall be equipped with:	B1 1.8 (V1) and B1 1.6 (V2)
• a fire detection and fire alarm system;	
• smoke alarms (in accordance with paragraphs 1.10 to 1.18) in the circulation spaces.	

Electricity distributor's responsibilities

The electricity distributor is responsible for:	
• evaluating and agreeing proposals for new installations or significant alterations to existing ones;	P 1.2
• installing the cut-out and meter in a safe location;	P 1.5
• ensuring that it is mechanically protected and can be safely maintained;	
• taking into consideration the possible risk of flooding.	P 1.5

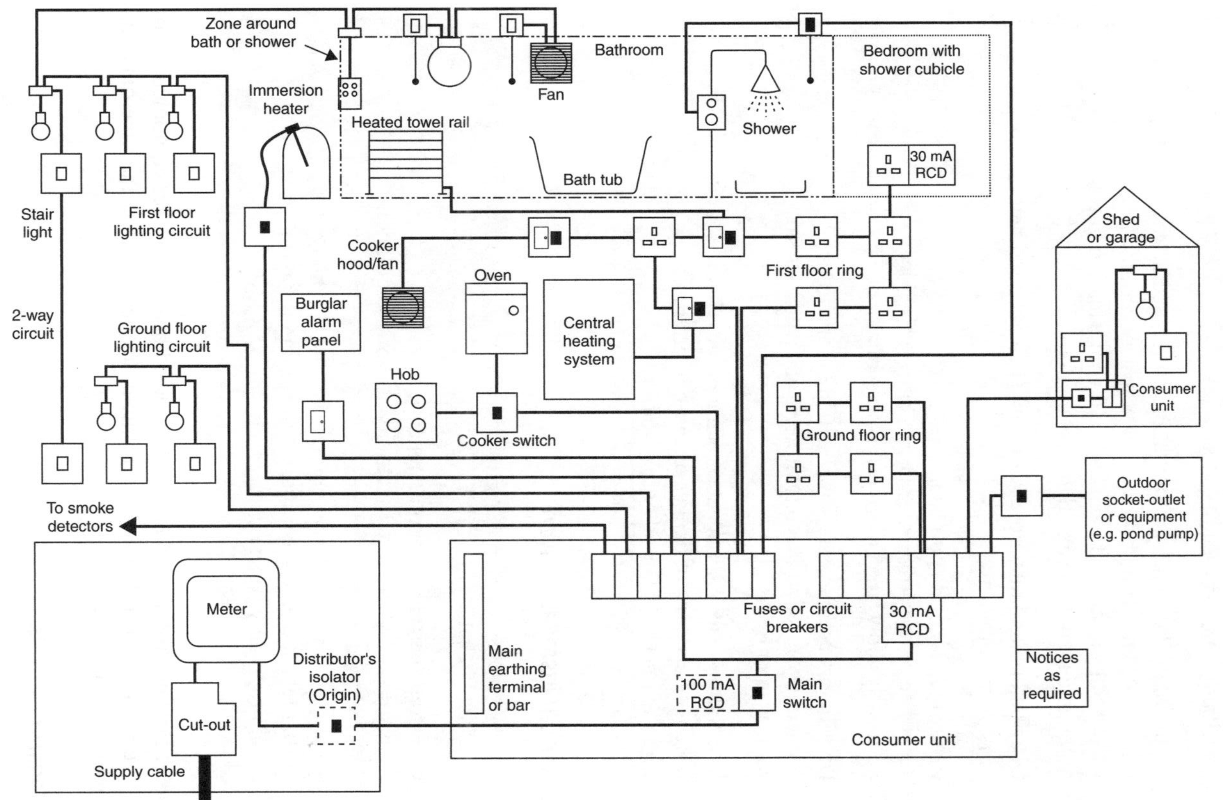

Figure 6.202 Typical fixed installations that might be encountered in new (or upgraded) existing dwellings

Note: See DGLC publication 'Preparing for flooding' which can be downloaded from www.dglc.gov.uk.

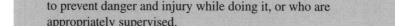

Distributors are also required to:

- maintain the supply within defined tolerance limits; P 3.8
- provide an earthing facility for new connections;
- provide certain technical and safety information to consumers to enable them to design their installations.

Distributors (and meter operators) must ensure that their equipment on consumers' premises:

- is suitable for its purpose; P 3.9
- is safe in its particular environment;
- clearly shows the polarity of the conductors.

Distributors are prevented by the Regulations from connecting installations to their networks which do not comply with BS 7671. P 3.12

Distributors may disconnect consumers' installations which are a source of danger or cause interference with their networks or other installations. P 3.12

Electrical installation work

Electrical installation work:

- is to be carried out professionally; P 1.1
- is to comply with the Electricity at Work Regulations 1989 as amended;
- may only be carried out by persons that are competent to prevent danger and injury while doing it, or who are appropriately supervised. P 3.4a

Note: Persons installing domestic combined heat and power equipment **must** advise the local distributor of their intentions before (or at the time of) commissioning the source.

Consumer units

Accessible consumer units should be fitted with a child-proof cover or installed in a lockable cupboard. P 1.6

Earthing

All electrical installations shall be properly earthed.	P AppC

Note: The most usual type is an electricity distributor's earthing terminal, provided for this purpose near the electricity meter.

All lighting circuits shall include a circuit protective conductor.	P AppC
All socket outlets which have a rating of 32 A or less, and which may be used to supply portable equipment for use outdoors, shall be protected by a Residual Current Device (RCD).	P AppC
Distributors are required to provide an earthing facility for new connections.	P 3.8
It is not permitted to use a gas, water or other metal service pipe as a means of earthing for an electrical installation (this does not rule out, however, equipotential bonding conductors being connected to these pipes).	P AppC
New or replacement, non-metallic light fittings, switches and/or other components must not require earthing (e.g. non-metallic varieties) unless new circuit protective (earthing) conductors are provided.	P AppC
Socket-outlets that will accept unearthed (2-pin) plugs must **not** use supply equipment that needs to be earthed.	P AppC
Where electrical installation work is classified as an extension, a material alteration or a material change of use, the work must consider and include earthing and bonding systems that are satisfactory and meet the requirements.	P 2.1a–2.2c

See Figure 6.203 for details of some earth and bonding conductors that might be part of an electrical installation.

Equipotential bonding conductors

Main equipotential bonding conductors are required to all water service pipes, gas installation pipes, oil supply pipes	P AppC

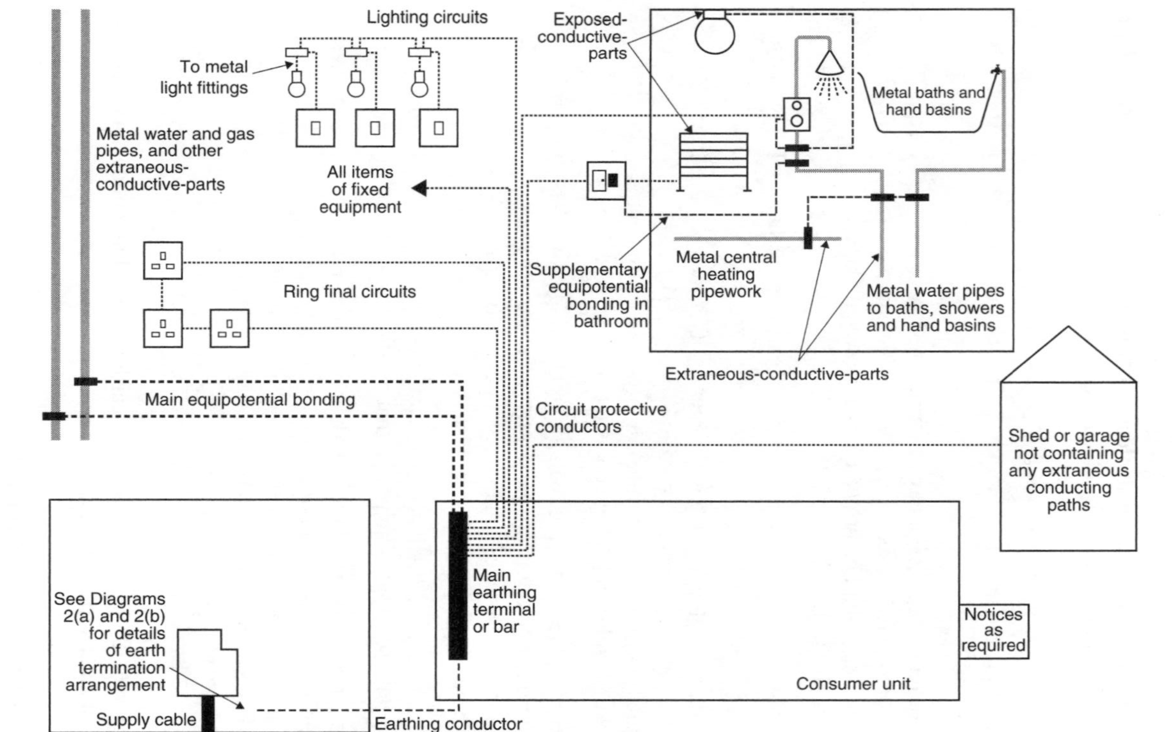

Figure 6.203 Typical earth and bonding conductors that might be part of the electrical installation shown in Figure 6.189

plus certain other 'earthy' metalwork that may be present on the premises.	
The installation of supplementary equipotential bonding conductors is required for installations and locations where there is an increased risk of electric shock (e.g. such as bathrooms and shower rooms).	P AppC
The minimum size of supplementary equipotential bonding conductors (without mechanical protection) is 4 mm^2.	P AppC

Types of wiring or wiring system

Cables concealed in floors (and walls in certain circumstances) are required to have an earthed metal covering, be enclosed in steel conduit, or have additional mechanical protection (see BS 7671 for more information).	P AppC
Cables to an outside building (e.g. garage or shed) if run underground, should be routed and positioned so as to give protection against electric shock and fire as a result of mechanical damage to a cable.	P AppC
Heat-resisting flexible cables are required for the final connections to certain equipment (see makers instructions).	P AppC

PVC insulated and sheathed cables are likely to be suitable for much of the wiring in a typical dwelling.

Electrical components

All electrical work should be inspected (during installation as well as on completion) to verify that the components have:	
• been made in compliance with appropriate British Standards or harmonized European Standards;	P 1.11a i
• been selected and installed in accordance with BS 7671;	P 1.11a ii
• been evaluated against external influences (such as the presence of moisture);	P 1.11a ii
• been tested as per sections 71–74 of BS 7671:2001;	P 1.9

• been tested using appropriate and accurate instruments;	P 1.12
• been tested to check satisfactory performance with respect to continuity of conductors, insulation resistance, separation of circuits, polarity, earthing and bonding arrangements, earth fault loop impedance and functionality of all protective devices including residual current devices;	P 1.11b
• not been visibly damaged (or are defective) so as to be unsafe; had their test results recorded using the models in Appendix 6 of BS 7671 (see Annex D, Appendices 1–4 for examples).	P 1.11a iii P 1.12

Note: Inspections and testing of DIY work should **also** meet the above requirements.

Socket outlets

Where necessary, socket outlets should comply with the requirements of Part M.	P 1.6
Socket outlets should be wall-mounted.	M 4.30a and b
Socket outlets that will accept unearthed (2-pin) plugs must not be used to supply equipment that needs to be earthed.	P AppC
Socket outlets which have a rating of 32 A or less and which may be used to supply portable equipment for use outdoors shall be protected by an RCD.	P AppC
Socket outlets should be located no nearer than 350 mm from room corners.	M 4.30
Switched socket outlets should indicate whether they are 'ON'.	M 4.30k
Front plates should contrast visually with their backgrounds.	M 4.30m
Mains and circuit isolator switches should clearly indicate whether they are 'on' or 'off'.	M 4.30l
Older types of socket outlet designed non-fused plugs must not be connected to a ring circuit.	P AppC
The colours red and green should **not** be used in combination as indicators of 'on' and 'off' for switches and controls.	M 4.28

Wall sockets

Wall sockets shall meet the following requirements:

Table 6.83 Building Regulations requirements for wall sockets

Type of wall	Requirement	Section
Timber framed	Power points may be set in the linings provided there is a similar thickness of cladding behind the socket box.	E p14
	Power points should not be placed back to back across the wall.	E p14
Solid masonry	Deep sockets and chases should **not** be used in separating walls.	E 2.32
	The position of sockets on opposite sides of the separating wall should be staggered.	E 2.32f
Cavity masonry	The position of sockets on opposite sides of the separating wall should be staggered.	E 2.65e
	Deep sockets and chases should **not** be used in a separating wall.	E 2.65d2
	Deep sockets and chases in a separating wall should **not** be placed back to back.	E 2.65d2
Framed walls with absorbent material	Sockets should:	
	• be positioned on opposite sides of a separating wall;	E 2.146b
	• not be connected back to back;	E 2.146b2
	• be staggered a minimum of 150 mm edge to edge.	E 2.146b2

Wall-mounted switches and socket outlets

Where necessary, wall-mounted switches and socket outlets should comply with the requirements of Part M.	P 1.6
A cable or stud detector shall be used when attempting to drill into walls, floors or ceilings.	P AppC
Individual switches on panels and on multiple socket outlets should be well separated.	M 4.29
Switches and socket outlets for lighting and other equipment should be located so that they are easily reachable.	M 8.2
Switches and socket outlets for lighting and other equipment in habitable rooms should be located between 450 mm and 1200 mm from finished floor level (see Figure 6.204).	M 8.3

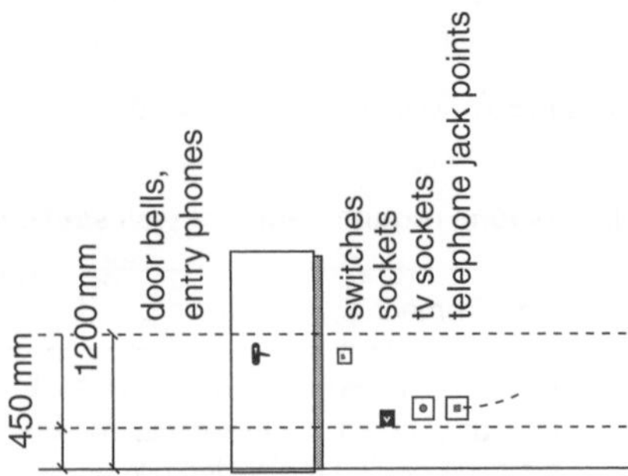

Figure 6.204 Heights of switches and sockets etc

The aim is to help people with limited reach (e.g. seated in a wheelchair) access a dwelling's wall-mounted switches and socket outlets.

Outlets, controls and switches

The aim should be to ensure that all controls and switches are easy to operate, visible, free from obstruction and: • are located between 750 mm and 1200 mm above the floor; • do not require the simultaneous use of both hands (unless necessary for safety reasons) to operate; • switched socket outlets indicate whether they are 'on'; • mains and circuit isolator switches clearly indicate whether they are 'on' or 'off'; • individual switches on panels and on multiple socket outlets are well separated; • controls that need close vision (e.g. thermostats) are located between 1200 mm and 1400 mm above the floor; • front plates contrast visually with their backgrounds.	M 5.4i
The operation of all switches, outlets and controls should not require the simultaneous use of both hands (unless necessary for safety reasons).	M 4.30j
Where possible, light switches with large push pads should be used in preference to pull cords.	M 5.3
The colours red and green should not be used in combination as indicators of 'on' and 'off' for switches and controls.	M 5.3

Light switches

Light switches should: • have large push pads; • align horizontally with door handles; • be within 900–1100 mm from the entrance door opening.	M 4.30h and l
Switches and controls should be located between 750 mm and 1200 mm above the floor.	M 4.30c and d
Where possible, light switches with large push pads should be used in preference to pull cords.	M 5.3
The colours red and green should not be used in combination as indicators of 'on' and 'off' for switches and controls.	M 5.3

Heat emitters

In toilets and bathrooms designed for disabled people, heat emitters (if located) should not restrict: • the minimum clear wheelchair manoeuvring space; • the space beside a WC used to transfer from the wheelchair to the WC.	M 5.10p
Heat emitters should either be screened or have their exposed surfaces kept at a temperature below 43°C.	M 5.4j

Telephone points and TV sockets

All telephone points and TV sockets should be located between 400 mm and 1000 mm above the floor (or 400 mm and 1200 mm above the floor for permanently wired appliances).	M 4.30a and b

Thermostats

Controls that need close vision (e.g. thermostats) should be located between 1200 mm and 1400 mm above the floor.	M 4.30f

Lighting circuits

All lighting circuits shall include a circuit protective conductor. P App C

Light fittings

New or replacement, non-metallic light fittings, switches or other components must not require earthing unless new circuit protective (earthing) conductors are provided. App C

Fixed internal lighting (domestic buildings)

In order that dwelling occupiers may benefit from the installation of efficient electric lighting, whenever L1B 43

- a dwelling is extended, or
- a new dwelling is created from a material change of use, or
- an existing lighting system has been replaced as part of re-wiring works,

then the re-wiring works **must** comply with Part P.

Lighting fittings (including lamp, control gear, housing, reflector, shade, diffuser or other device for controlling the output light) should only take lamps with a luminous efficiency greater than 40 lumens per circuit-watt. L1A 42

Note: Light fittings in less-frequented areas such as cupboards and other storage areas do not count).

Fixed external lighting (i.e. lighting that is fixed to an external surface of the dwelling and which is powered from the dwelling's electrical system) should either: L1A 45

- have a lamp capacity not exceeding 150 W per light fitting that automatically switches off:
  - when there is enough daylight and
  - when it is not required at night or
- include sockets that can only be used with lamps which have an efficiency greater than 40 lumens per circuit watt.

Fixed energy-efficient light fittings (one per 25 m^2 dwelling floor area (excluding garages) and one per four fixed light fittings) should be installed in the most frequented locations in the dwelling.	L1B 44

See GIL 20, Low energy domestic lighting, BRECSU, 1995 for further guidance.

Office, industrial and storage areas (non-domestic buildings)

All areas that involve predominantly desk-based tasks (i.e. such as classrooms, seminar and conference rooms – **including** those in schools) shall have an average efficiency of not less than 45 luminaire-lumens/circuit-watt (averaged over the whole area).	L2B 56

Note: In other spaces (i.e. other than office or storage spaces) less efficient lamps may be used **provided** that the installed lighting has an average initial (100 hour) lamp plus ballast efficacy of not less than 50 lamp lumens per circuit-watt.

Lighting controls (non-domestic buildings)

The distance between the local switch and the luminaire it controls should generally be not more than six metres, or twice the height of the luminaire above the floor if this is greater.	L2A 57
Local switches should be:	L2B 60
• located in easily accessible positions within each working area (or at boundaries between working areas and general circulation routes);	
• operated by the deliberate action of the occupants (referred to as occupant control), either manually or remotely.	
• located within six metres (or twice the height of the light fitting above the floor if this is greater) of any luminaire it controls.	L2B 61
Occupant control of local switching may be supplemented by automatic systems which:	L2B 62
• switch the lighting off when they sense the absence of occupants; or	

- dim (or switch) the lighting off when there is sufficient daylight.

Automatically switched lighting systems should be subject to a risk assessment. L2A 54

Lighting controls should be provided to switch off the lighting during daylight hours and when the area is unoccupied. L2A 54

If the space is daylit space served by side windows, the perimeter row of luminaires should be separately switched. L2A 57

Manually operated local switches should be in easily accessible positions within each working area, at boundaries between working areas, and at general circulation routes. L2A 55

Local (manual) switching can be supplemented by automatic controls which: L2A 58

- switch the lighting off when they sense the absence of occupants; or
- dim (or switch off) the lighting when there is sufficient daylight.

Display lighting in all types of space (non-domestic buildings)

Display lighting should have an average initial (100 hour) efficiency of not less than 15 lamp-lumens per circuit-watt. L2A 60 L2B 64 and 65

Note: In spaces where it would be reasonable to expect cleaning and restocking outside public access hours, general lighting should also be provided.

Where possible, display lighting should be connected in dedicated circuits that can be switched off at times when people will not be inspecting exhibits or merchandise or attending entertainment events. L2A 60 L2B 66

Note: In a retail store, for example, this could include timers that switch off the display lighting outside store opening hours.

Emergency escape lighting (non-domestic buildings)

Emergency escape lighting, specialist process lighting and vertical transportation systems are not subject to the requirements of Part L. L2B 68

General lighting efficiency in all other types of space (non-domestic buildings)

Lighting (over the whole of these areas) should have an average initial efficacy of not less than 45 luminaire-lumens/circuit-watt.	L2A 51
Lighting systems serving other types of space may use lower powered and less efficient lamps.	L2A 53

Limiting the effects of solar gains in summer – buildings other than dwellings

In occupied spaces that are not served by a comfort cooling system, the combined solar and internal casual gains (people, lighting and equipment) per unit floor area averaged over the period of daily occupancy should not be greater than 35 W/m^2 calculated over a perimeter area not more than 6 m from the window wall and averaged during the period 06.30–16.30 hrs GMT.	L2A 64a

Consequential improvements (non-domestic buildings)

If a building has a total useful floor area greater than 1000 m^2 and the proposed building work includes:

- an extension; or
- the initial provision of any fixed building services; or
- an increase to the installed capacity of any fixed building services, then:

any general lighting system serving an area greater than 100 m^2 which has an average lamp efficacy of less than 40 lamp-lumens per circuit-watt, should be upgraded with new luminaires or improved controls.	L2B 18 4

Controlled services (non-domestic buildings)

Where the work involves the provision of a controlled service:

new lighting systems should be provided with controls that achieve reasonable standards of energy efficiency.	L2B 41c

Inspection and commissioning of the building services systems (non-domestic buildings)

When building services systems are commissioned, the systems and their controls shall be left in their intended working order and are capable of operating efficiently regarding the conservation of fuel and power.	L2A 77–78
Systems should be provided with meters to efficiently manage energy use and to ensure that at least 90% of the estimated annual energy consumption of each fuel is assigned to the various end-use categories (heating, lighting etc.).	L2A 43 L2B 67
Whenever a cooling plant (i.e. a chiller) is being replaced, cooling loads should (if practical and cost effective) be improved through solar control and/or more efficient lighting.	L2B 45 and L2B 46
Air-handling systems should be capable of achieving a specific fan power at 25% of design flow rate.	L2A 47

Commissioning (non-domestic buildings)

Building services systems should be commissioned so that at completion, the system(s) and their controls are left in working order and operate efficiently (i.e. for the purposes of the conservation of fuel and power.	L2B 70

External lighting fixed to the building

External lighting (including lighting in porches, but not lighting in garages and carports) should:

Automatically extinguish when there is enough daylight, and when not required at night.	L1 45a
Have sockets that can only be used with lamps having an efficacy greater than 40 lumens per circuit-watt (such as fluorescent or compact fluorescent lamp types, and **not** GLS tungsten lamps with bayonet cap or Edison screw bases).	L1 45b

Emergency alarms

Emergency assistance alarm systems should have: • visual and audible indicators to confirm that an emergency call has been received; • a reset control reachable from a wheelchair, WC, or from a shower/changing seat; • a signal that is distinguishable visually and audibly from the fire alarm.	M 5.4h

Emergency alarm pull cords should be: • coloured red; • located as close to a wall as possible; • have two red 50 mm diameter bangles.	M 4.30e
Front plates should contrast visually with their backgrounds.	M 4.30m
The colours red and green should **not** be used in combination as indicators of 'on' and 'off' for switches and controls.	M 4.28

Smoke alarms

With the introduction of Part B 2007:

- smoke alarms now need to be installed in accordance with BS 5839-6:2004;
- all smoke alarms should have a standby power supply;
- the provision of smoke alarms shall be based on an assessment of the risk to the occupants in the event of fire.

Note: If a dwelling house is extended, then smoke alarms should be provided in all circulation spaces.

Smoke alarms should:	B1 1.4
• be mains-operated and conform to BS 5446-1:2000;	
• have a standby power supply such as a rechargeable (or non-rechargeable) battery;	
• be designed and installed in accordance with BS 5839-6:2004;	B1 1.10 (V1) B1 1.9 (V2)
• be positioned in the circulation spaces between sleeping spaces and places where fire are most like to start (e.g. in kitchens and living rooms);	B1 1.11 (V1) B1 1.10 (V2)
• be installed on every storey of a dwelling house.	B1 1.12 (V1)

Note: Where the kitchen area is not separated from the stairway or circulation space by a door, there should be a compatible interlinked heat detector or heat alarm in the kitchen, in addition to whatever smoke alarms are needed in the circulation space(s).	B1 1.11 (V2) B1 1.13 (V1) B1 1.12 (V2)
If more than one alarm is installed they should be linked so that the detection of smoke or heat by one unit operates the alarm signal in all of them.	B1 1.14 (V1) B1 1.13 (V2)

Smoke alarms/detectors should be sited so that:	B1 1.15a (V1)
• there is a smoke alarm in the circulation space within 7.5 m of the door to every habitable room;	B1 1.14a (V2)
• they are ceiling-mounted and at least 300 mm from walls and light fittings;	B1 1.15b (V1) B1 1.14b (V2)
• the sensor in ceiling-mounted devices is between 25 mm and 600 mm below the ceiling (25–150 mm in the case of heat detectors or heat alarms).	B1 1.15c (V1) B1 1.14c (V2)

Smoke alarms should **not** be fixed:	B1 1.16 (V1)
• over a stair or any other opening between floors;	B1 1.15 (V2)
• next to or directly above heaters or air-conditioning outlets;	B1 1.17 (V1) B1 1.16 (V2)
• in bathrooms, showers, cooking areas or garages;	B1 1.17 (V1) B1 1.16 (V2)
• in any place where steam, condensation or fumes could give false alarms;	B1 1.17 (V1) B1 1.16 (V2)
• in places that get very hot (such as a boiler room);	B1 1.18 (V1) B1 1.17 (V2)
• in places that get very cold (such as an unheated porch);	B1 1.18 (V1) B1 1.17 (V2)
• to surfaces which are normally much warmer or colder than the rest of the space.	B1 1.18 (V1) B1 1.17 (V2)

Smoke detectors – power supplies

The power supply for a smoke alarm system should: B1 1.19

- be derived from the dwelling-house's mains electricity supply;
- comprise a single independent circuit at the dwelling-house's main distribution board (consumer unit or a single regularly used local lighting circuit.

It should be possible to isolate the power to the smoke alarms without isolating the lighting. B1 1.19

The electrical installation should comply with Approved Document P (Electrical safety). B1 1.20

Any cable suitable for domestic wiring may be used for the power supply and interconnection to smoke alarm systems (except in large buildings where the cable needs to be fire resistant (see BS 5839-6:2004)). B1 1.21

Conductors use to interconnect alarms (e.g. signalling) should be colour coded so as to distinguish them from those supplying mains power. B1 1.21

Mains powered smoke alarms may be interconnected using radio-links, provided that this does not reduce the lifetime or duration of any standby power supply below 72 hours. B1 1.21

Where the vertical distance between the floor of the entrance storey and the floors above and below it does not exceed 7.5 m, multi-storey flats are required to:

- provide a protected stairway plus additional smoke alarms in all habitable rooms and a heat alarm in any kitchen; or B1 2.16c (V2)
- provide a protected stairway plus a sprinkler system and smoke alarms. B1 2.16d (V2)

Smoke alarms in thatched roofs

In thatched roofs: B4 10.9 (V1)

- the rafters should be overdrawn with construction having not less than 30 minutes fire resistance; B4 14.9 (V2)
- a smoke alarm should be installed in the roof space.

Fire alarms

Fire alarms should emit an audio and visual signal to warn occupants with hearing or visual impairments.	M 5.4g

All fire detection and fire-warning systems shall be properly designed, installed and maintained.

All buildings should have arrangements for detecting fire.	B1 1.27 (V2)
All buildings should have the means of raising an alarm in case of fire (e.g. rotary gongs, handbells or shouting 'fire') or be fitted with a suitable electrically operated fire warning system (in compliance with BS 5839).	B1 1.28 (V2)
The fire warning signal should be distinct from other signals that may be in general use.	B1 1.32 (V2)
In premises used by the general public, e.g. large shops and places of assembly, a staff alarm system (complying with BS 5839) may be used.	B1 1.33 (V2)
In small buildings raising the alarm may be a comparatively simple matter, but when this does not apply, the building should be provided with an electrically operated fire warning system with manual call points adjacent to exit doors which shall comply with BS 5839-1:2002.	B1 1.29 and 1.30 (V2)
Call points for electrical alarm systems should be installed in accordance with BS 5839-1 and comply with either BS 5839-2:1983, or Type A of BS EN 54-11:2001.	B1 1.31 (V2)

Type B call points should **only** used with the approval of the Building Control Body.

Where it is critical for electrical circuits to be able to continue to function during a fire, protected circuits (meeting the requirements of BS EN 50200:2006) shall be installed.	B1 5.38 (V2)

Electrically powered locks

Electrically powered locks should:	B1 5.11 (V2)
• return to the unlocked position – on operation of the fire alarm system; – on loss of power or system error; – on activation of a manual door release unit.	
• comply with Part P (and any other relevant Parts) of the Building Regulations;	P 1.7 and 3.1
• comply with the relevant equipment and installation standards;	P 0.1b
• not present an electric shock or fire hazard to people;	P 0.1a
• provide mechanical and thermal protection;	P 1.3
• provide adequate protection against mechanical and thermal damage;	P 0.1a
• provide adequate protection for persons against the risks of electric shock, burn or fire injuries;	P1.3
• be safe to use, maintain and alter.	P 1.7

Power-operated doors

Power-operated doors and gates should:

• be provided with a manual or automatic opening device in the event of a power failure where and when necessary for health or safety;	K5 5.2d
• have safety features to prevent injury to people who are struck or trapped (such as a pressure-sensitive door edge which operates the power switch);	
• have a readily identifiable and accessible stop switch.	

Power-operated entrance doors

Doors to accessible entrances shall be provided with a power-operated door opening and closing system if a force greater than 20 N is required to open or shut a door.	M 2.13a

Once open, all doors to accessible entrances should be wide enough to allow unrestricted passage for a variety of users, including wheelchair users, people carrying luggage, people with assistance dogs, and parents with pushchairs and small children.

The effective clear width through a single leaf door (or one leaf of a double leaf door) should be in accordance with Table 6.84. M 2.13b

Table 6.84 Minimum effective clear widths of doors

Direction and width of approach	New buildings (mm)	Existing buildings (mm)
Straight-on (without a turn or oblique approach)	800	750
At right angles access route at least 1500 mm wide	800	750
At right angles to an access route at least 1200 mm wide	825	775
External doors to buildings used by the general public	1000	775

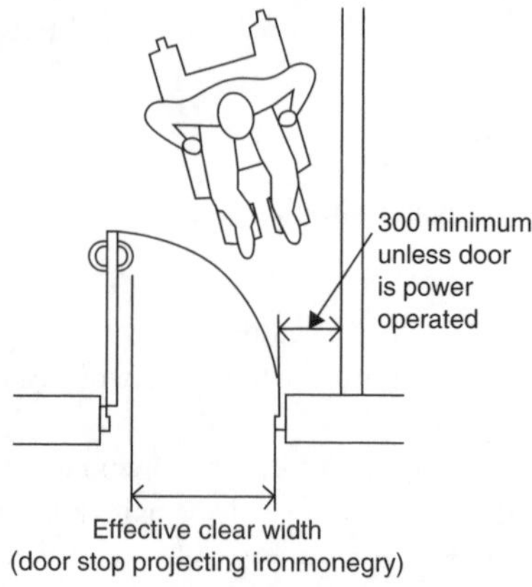

Figure 6.205 Effective clear width and visibility requirements of doors

Power-operated entrance doors should have a sliding, swinging or folding action controlled manually (by a push pad, card swipe, coded entry, or remote control) or automatically controlled by a motion sensor or proximity sensor such as contact mat.

Power-operated entrance doors should:

- open towards people approaching the doors; M 2.21a
- provide visual and audible warnings that they are operating (or about to operate); M 2.21c

• incorporate automatic sensors to ensure that they open early enough (and stay open long enough) to permit safe entry and exit;	M 2.21c
• incorporate a safety stop that is activated if the doors begin to close when a person is passing through;	M 2.21b
• revert to manual control (or fail safe) in the open position in the event of a power failure;	M 2.21d
• when open, should **not** project into any adjacent access route	M 2.21e
• ensure that its manual controls: – are located between 750 mm and 1000 mm above floor level; – are operable with a closed fist;	M 2.21f
• be set back 1400 mm from the leading edge of the door when fully open;	M 2.21g
• be clearly distinguishable against the background;	M 2.21g
• contrast visually with the background.	M 2.19 and 2.21g

Note: Revolving doors are **not** considered 'accessible' as they create particular difficulties (and possible injury) for people who are visually impaired, people with assistance dogs or mobility problems and for parents with children and/or pushchairs.

Cellars or basements

LPG storage vessels and LPG fired appliances fitted with automatic ignition devices or pilot lights must not be installed in cellars or basements.	J 3.5i

Lecture/conference facilities

Artificial lighting should be designed to:

• give good colour rendering of all surfaces;	M 4.9 and 4.34
• be compatible with other electronic and radio-frequency installations;	M 4.12.1
• be compatible with other electronic and radio-frequency installations.	M 4.36f

Swimming pools and saunas

Swimming pools and saunas are subject to special requirements specified in Part 6 of BS 7671:2001.	P AppA

Inspection and testing

Electrical installations must be inspected and tested:	
• during, at the end of installation and before they are taken into service to verify that they are reasonably safe **and** that they comply with BS 7671:2001;	P 1.6
• to verify that they meet the relevant equipment and installation standards.	P 0.1b
All electrical work should be inspected (during installation as well as on completion) to verify that the components have:	
• been selected and installed in accordance with BS 7671;	P 1.8a (ii)
• been made in compliance with appropriate British Standards or harmonized European Standards;	P 1.8a (i)
• been evaluated against external influences (such as the presence of moisture);	P 1.8a (ii)
• not been visibly damaged (or are defective) so as to be unsafe;	P 1.8a (iii)
• been tested to check satisfactory performance with respect to continuity of conductors, insulation resistance, separation of circuits, polarity, earthing and bonding arrangements, earth fault loop impedance and functionality of all protective devices including residual current devices;	P 1.8b
• been inspected for conformance with section 712 of BS 7671:2001;	P 1.9
• been tested as per section 713 of BS 7671:2001;	P 1.10
• been tested using appropriate and accurate instruments;	P 1.10
• had their test results recorded using the model in Appendix 6 of BS 7671;	P 1.10
• had their test results compared with the relevant performance criteria to confirm compliance.	P 1.10

Note: Inspections and testing of DIY work should **also** meet the above requirements.

Inspection and testing of non-notifiable work

Although it is not necessary for non-notifiable electrical installation work to be checked by a building control body, it nevertheless **must** be carried out in accordance with the requirements of BS 7671:2001.	P 1.30
Installers who are qualified to complete BS 7671 installation certificates and who carry out non-notifiable work should issue the appropriate electrical installation certificate for all but the simplest of like-for-like replacements.	P 1.32

Certification

BS 7671 Installation certificates

Compliance with Part P can be demonstrated by the issue of the appropriate BS 7671 electrical installation certificate.	P 1.8
An electrical installation certificate may **only** be issued by the installer responsible for the installation work.	P 1.28
Inspection and testing should be carried out in compliance with BS 7671:2001.	P 1.9
Section 712 of BS 7671:2001 provides a list of all the inspections that may be necessary whilst Section 713 provides a list of tests.	
Tests should be carried out using appropriate and accurate instruments under the conditions given in BS 7671, and the results compared with the relevant performance criteria to confirm compliance.	P 1.15
A copy of the installation certificate should be supplied to the person ordering the work.	P 1.9
Certificates may only be made out and signed by someone with the appropriate qualifications, knowledge and experience to carry out the inspection and test procedures.	P 1.9

Certificates should show that the electrical installation work has been: • inspected during erection as well as on completion and that components have been: – made in compliance with appropriate British Standards or harmonized European Standards; – selected and installed in accordance with BS 7671:2001 (including consideration of external influences such as the presence of moisture); – not visibly damaged or defective so as to be unsafe.	P 1.11a
• tested for continuity of conductors, insulation resistance, separation of circuits, polarity, earthing and bonding arrangements, earth fault loop impedance and functionality of all protective devices (including residual current devices).	P 1.11b
A full electrical installation certificate should be used for the replacement of consumer units.	P 1.13

Minor works certificate

Appropriate tests (according to the nature of the work) should be carried out.	P 1.16
A minor works certificate should be issued whenever inspection and testing has been carried out, irrespective of the extent of the work undertaken.	P 1.13
A minor works certificate shall **not** be used for the replacement of consumer units or similar items.	P 1.13

Building Regulations compliance certificates

The following are additional certificates that are issued on completion of notifiable works as evidence of compliance with the Building Regulations: • A Building Regulations Compliance Certificate (issued by Part P competent person scheme installers) – see Appendix 4 to Annex D for an example); • completion certificates (issued by local authorities); • final notices (issued by approved inspectors).	P 1.17

These documents are **different** documents to the BS 7671 installation certificate and are used to attest compliance with **all** relevant requirements of the Building Regulations – not just to Part P.

Certification of notifiable work

Only installers registered with a Part P competent person self-certification scheme are qualified to complete BS 7671 installation certificates and should do so in respect of every job they undertake.	P 1.18
A copy of the certificate should always be given to the person ordering the electrical installation work.	P 1.18
A Building Regulations Compliance Certificate must be issued to the occupant (and the building control body) either by the installer or the installer's registration body within 30 days of the work being completed.	P 1.19
If notifiable electrical installation work is going to be carried out by a person **not** registered with a Part P competent person self-certification the work should be notified to a building control body (the local authority or an approved inspector) before work starts.	P 1.21
Note: The building control body then becomes responsible for making sure the work is safe and complies with all relevant requirements of the Building Regulations.	
On satisfactory completion of all work, the building control body will issue a Building Regulation Completion Certificate (if they are the local authority) or a final certificate (if they are an approved inspector).	P 1.23
If notifiable electrical installation work is going to be carried out by installers who are not qualified to issue BS 7671 completion certificates (e.g. sub-contractors or DIYers), then the building control body must be notified before the work starts.	P 1.24
Note: The building control body then becomes responsible for making sure that the work is safe and complies with all relevant requirements of the Building Regulations – **but** not at the householder's expense!	P 1.26

Third party certification

Unregistered installers should not themselves arrange for a third party to carry out final inspection and testing.	P 1.28
A third party may only sign a BS 7671:2001 Periodic Inspection Report (or similar) to indicate that electrical safety tests had been carried out on the installation which met BS 7671:2001 criteria.	P 1.29
Third parties are not entitled to verify that the installation complies, fully, with BS 7671 requirements – for example with regard to routing of hidden cables.	

Provision of information

Sufficient information should be left with the occupant to ensure that persons wishing to operate, maintain or alter an electrical installation can do so with reasonable safety.	P 1.33
This information should include: • all items called for by BS 7671:2001; • an electrical installation certificate (describing the installation and giving details of work carried out); • permanent labels (e.g. on earth connections and bonds and on items of electrical equipment such as consumer units and RCDs); • operating instructions and log books; • detailed plans (but only for unusually large or complex installations).	P 1.34

6.19.3 Where can I get more information?

Further guidance concerning the requirements for electrical safety is available from:

- The IET (Institution of Engineering Technology) at www.theiet.org;
- The NICEIC (National Inspection Council for Electrical Installation Contracting) at www.niceic.org.uk;
- the ECA (Electrical Contractors' Association) at www.niceic.org.uk or www.eca.co.uk.

To download .pdf copies of Part P, go to http://www.planningportal.gov.uk/england/professionals/4000000001253.html

For details of fixed wire colour changes, go to http://www.niceic.org.uk/downloads/WiringSupp.pdf.

6.20 Combustion appliances

6.20.1 The requirement

Protection of building

Combustion appliances and fluepipes shall be so installed, and fireplaces and chimneys shall be so constructed and installed, as to reduce to a reasonable level the risk of people suffering burns or the building catching fire in consequence of their use.

(Approved Document J3)

Fire safety

- *The spread of flame over the internal linings of the building shall be restricted.*
- *The heat released from the internal linings shall be restricted.*

(Approved Document B2)

Provision of information

Where a hearth, fireplace, flue or chimney is provided or extended, a durable notice containing information on the performance capabilities of the hearth, fireplace, flue or chimney shall be affixed in a suitable place in the building for the purpose of enabling combustion appliances to be safely installed.

(Approved Document J4)

6.20.2 Meeting the requirement

Air supplies for combustion installations

A room containing an open-flued appliance may need permanently open air vents (Figure 6.206(a) and (c)).	J (1.4)
Appliance compartments that enclose open-flued combustion appliances should be provided with vents large enough to admit all of the air required by the appliance for combustion and proper flue operation, whether the compartment draws its air from a room directly from outside (Figure 6.206(b) and (c)).	J (1.5)
Where appliances require cooling air, appliance compartments should be large enough to enable air to circulate and high and low level vents should be provided (Figure 6.207).	J (1.6)

In a flueless situation, air for combustion (and to carry away its products) can be achieved as shown in Figure 6.208.

Where appliances are to be installed within balanced compartments, special provisions will be necessary.	J (1.7)
If an appliance is room-sealed but takes its combustion air from another space in the building (such as the roof void) or if a flue has a permanent opening to another space in the building (such as where it feeds a secondary flue in the roof void), that space should have ventilation openings directly to outside.	J (1.8)

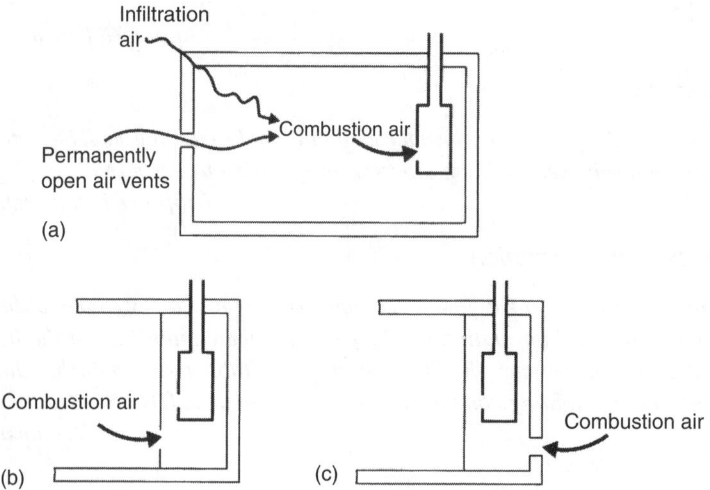

Figure 6.206 Air for combustion and operation of the flue (open flued). (a) Appliance in room. (b) Appliance in appliance compartment with internal vent. (c) Appliance in appliance compartment with external vent

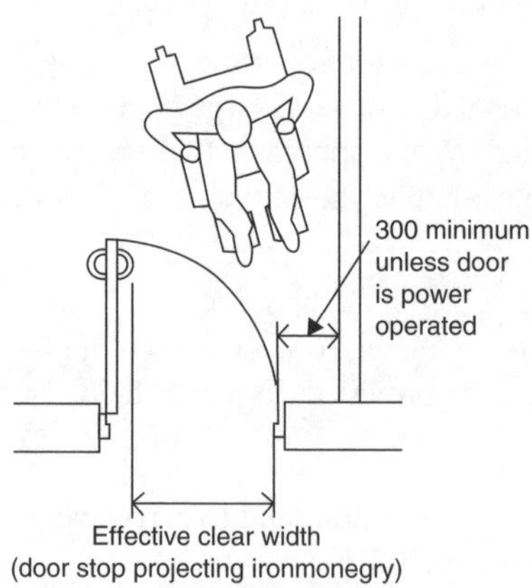

Figure 6.207 Combustion and operation requiring air cooling

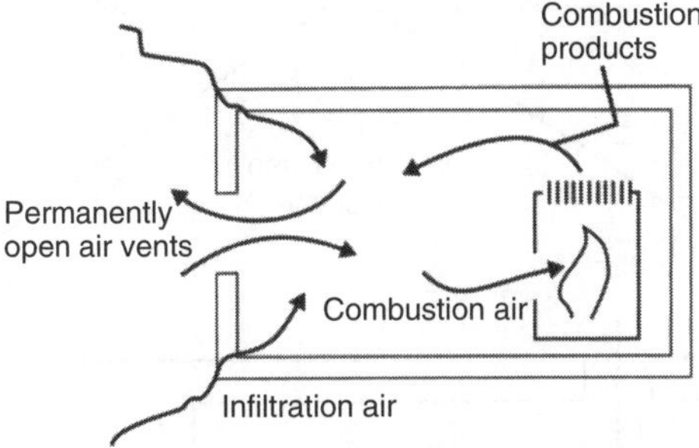

Figure 6.208 Air for combustion and operation of the flue (flueless)

Where flued appliances are supplied with combustion air through air vents which open into adjoining rooms or spaces, the adjoining rooms or spaces should have air vent openings of at least the same size direct to the outside. Air vents for flueless appliances however, should open directly to the outside air. J (1.9)

Any hidden voids in the construction shall be sealed and sub-divided to inhibit the unseen spread of fire and products of combustion. B3

Air vents

Permanently open air vents should be non-adjustable, sized to admit sufficient air for the purpose intended and positioned where they are unlikely to become blocked. J (1.10)

Air vents should be sufficient for the appliances to be installed (taking account where necessary of obstructions such as grilles and anti-vermin mesh). J (1.11)

Air vents should be sited outside fireplace recesses and beyond the hearths of open fires so that dust or ash will not be disturbed by draughts. J (1.11a)

Air vents should be sited in a location unlikely to cause discomfort from cold draughts. J (1.11b)

Grilles or meshes protecting air vents from the entry of animals or birds should have aperture dimensions no smaller than 5 mm. J (1.15)

In noisy areas, it may be necessary to install proprietary noise attenuated ventilators to limit the entry of noise into the building. J (1.16)

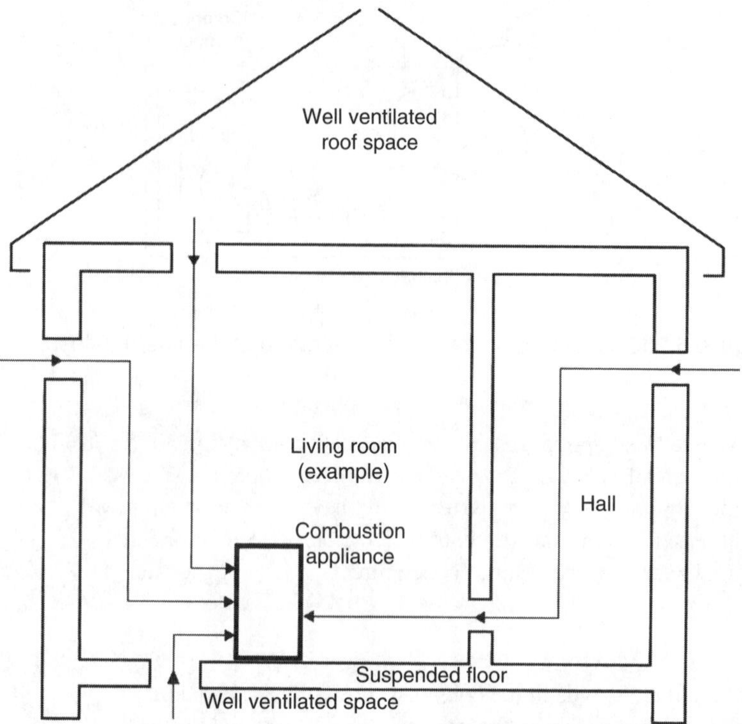

Figure 6.209 Locating permanent air vent openings (examples)

Discomfort from cold draughts can be avoided by placing vents close to appliances (for example by using floor vents), by drawing air from intermediate spaces such as hallways or by ensuring good mixing of incoming cold air by placing air vents close to ceilings.

In buildings where it is intended to install open-flued combustion appliances and extract fans, the combustion appliances should be able to operate safely whether or not the fans are running.	J (1.20)
For gas appliances where a kitchen contains an open-flued appliance, the extract rate of the kitchen extract fan should not exceed 20 litres/second (72 m^3/hour).	J (1.20a)
When installing ventilation for solid fuel appliances avoid installing extract fans in the same room.	J (1.20c)

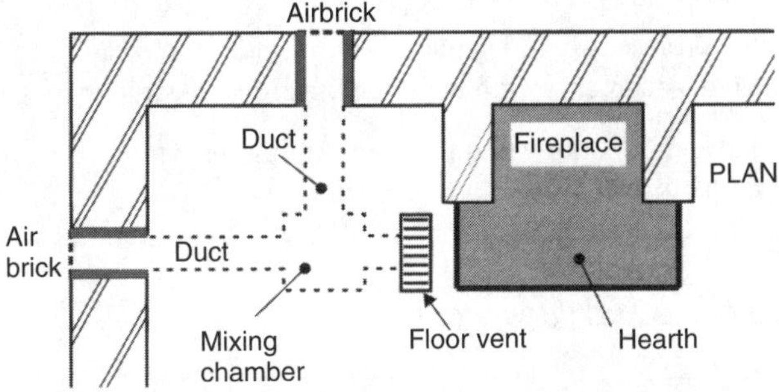

Figure 6.210 Air vent openings in a solid floor

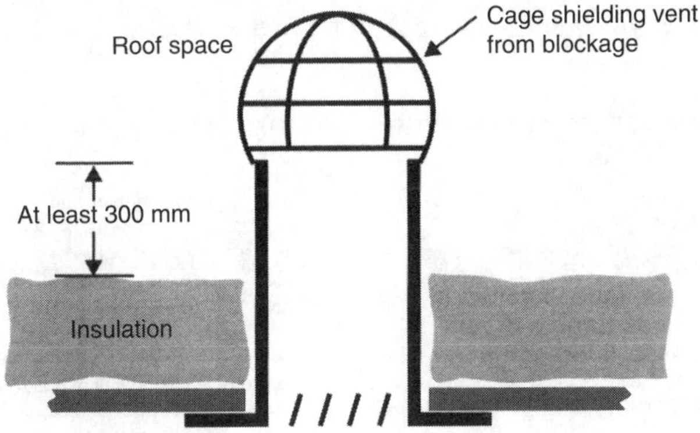

Figure 6.211 Ventilator used in a roof space (e.g. a loft)

Flues

Appliances other than flueless appliances should incorporate or be connected to suitable flues which discharge to the outside air.	J (1.24)
Chimneys and flues should provide satisfactory control of water condensation.	J (1.26)
New chimneys should be constructed with flue liners (clay, concrete or pre-manufactured) and masonry (bricks, medium weight concrete blocks or stone) suitable for the intended application.	J (1.27)
Liners should be installed in accordance with their manufacturer's instructions.	J (1.28)
Liners need to be placed with the sockets or rebate ends uppermost to contain moisture and other condensates in the flue.	J (1.28)

Joints should be sealed with fire cement, refractory mortar or installed in accordance with their manufacturer's instructions.	J (1.28)
Spaces between the lining and the surrounding masonry should not be filled with ordinary mortar.	J (1.28)

Ventilation ducts and flues etc.

If a flue passes through a compartment wall or compartment floor, or is built into a compartment wall, each wall of the flue should have a fire resistance of at least half that of the wall or floor (see Figure 212).	B1 7.11 (V1) B3 10.16 (V2)

(a) FLUE PASSING THROUGH COMPARTMENT WALL OR FLOOR

Flue

Compartment wall or floor

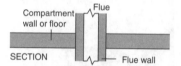

SECTION

Flue wall

Flue walls should have a fire resistance of at least one half of that required for the compartment wall or floor, and be of non-combustible construction.

(b) FLUE BUILT INTO COMPARTMENT WALL

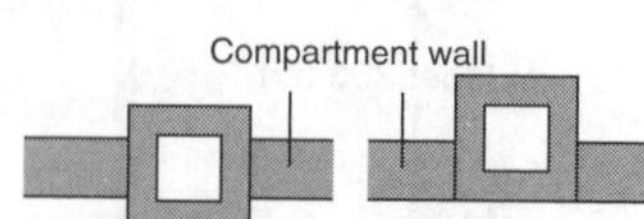

In each case flue walls should have a fire resistance at least one half of that required for the compartment wall and be of non-combustible construction.

Figure 6.212 Flues penetrating compartment walls or floors

6.21 Hot water storage

6.21.1 The requirement

Hot water storage

The hot water system shall:

- *be installed by a competent person;*
- *not exceed 100°C;*
- *discharge safely;*
- *not cause danger to persons in or about the building.*

(Approved Document G3)

Conservation of fuel and power

Reasonable provision shall be made for the conservation of fuel and power in buildings by:

(a) limiting heat gains and losses:

(i) through thermal elements and other parts of the building fabric; and

(ii) from pipes, ducts and vessels used for space heating, space cooling and hot water services;

(b) providing and commissioning energy-efficient fixed building services with effective controls; and

(c) providing to the owner sufficient information about the building, the fixed building services and their maintenance requirements so that the building can be operated in such a manner as to use no more fuel and power than is reasonable in the circumstances.

(Approved Document L1)

6.21.2 Meeting the requirement

Controlled services (non-domestic buildings)

New HVAC systems should be provided with controls that are capable of achieving a reasonable standard of energy efficiency.	L2B 41b
Fixed building services systems should be sub-divided into separate control zones for each area of the building with a significantly different solar exposure, occupancy period or type of use.	L2B 41b(i)
Separate control zones should be capable of independent switching and control set-point.	L2B 41b(ii)
If both heating and cooling are provided, then they should not operate simultaneously.	L2B 41b(iii)
The central plant serving zone-based systems should: • only operate as and when required; • have a default condition that is 'off'.	L2B 41b(iv)

Heating and hot water systems

When heating and hot water systems are commissioned:

• the performance of the building fabric and the heating and hot water systems should be no worse than the design limits;	L2A 9 (Criteria 2)
• the heating and hot water system(s) should be commissioned so that at completion, the system(s) and their controls are left in working order and can operate efficiently for the purposes of the conservation of fuel and power;	L1A 69 L1B 36

• the person carrying out the work shall provide the local authority with a notice confirming that *all fixed building services have been properly commissioned in accordance with a procedure approved by the Secretary of State*;	L1A 69 L1B 38
• independent temperature and on/off controls to all heating appliances shall be provided;	L2B 32b
• the heating system should use heat raising appliances with an efficiency not less than that recommended in the Heating Compliance Guide ICCM;	L1B 35 L2A 44 L2B 32b L2B 43
• if both heating and cooling are provided, they should be controlled so as to not operate simultaneously;	L2A 41c
• energy meters should enable building occupants to assign at least 90% of the estimated annual energy consumption of each fuel used for heating and lighting etc.;	L2A 43 L2B 67
• meters should be provided to enable installed LZC systems to be separately monitored;	L2A 43a L2B 69a
• automatic meter reading and data collection should be provided in all buildings with a total useful floor area that is greater than 1000 m^2.	L2B 69b

Consequential improvements (non-domestic buildings)

Any heating system more than fifteen years old should either be replaced or be equipped with improved controls.	L2B 18 1

Insulation of pipes, ducts and vessels

For buildings other than dwellings:

• insulation should not be less than those shown in the *Non-Domestic Heating, Cooling and Ventilation Compliance Guide*;	L1A 49 L1B 39
• hot and chilled water pipework, storage vessels, refrigerantpipework and ventilation ductwork should be insulated so as to conserve energy and	L2B 52

- to maintain the temperature of the heating or cooling service (see TIMSA HVAC);
- the performance of systems should be better than those described in GPG 2689 (*Energy efficient ventilation in housing*) and their fan powers and heat recovery efficiency should be no not worse than those listed in Table 6.85. L1A 40

Table 6.85 Limits on design flexibility f or mechanical ventilation

System type	Performance
Specific fan power (SFP) for continuous supply only and continuous extract only	0.8 litre/s.W
SFP for balanced systems	2.0 litre/s.W
Heat recovery efficiency	66%

Mechanical cooling

In new dwellings, fixed household air conditioners shall have an energy efficiency classification equal to or better than class C in Schedule 3 of the labelling scheme adopted under The Energy Information (Household Air Conditioners) (No. 2) Regulations 20051. L1B 42 L1A 41

Cooling plant

The requirements for cooling plant in buildings other than dwelling are that:

cooling systems should have a: L2A 45

- suitably efficient cooling plant, and
- an effective control system.

Note: For compliance refer to the checklists contained in the *Non-domestic Air Conditioning Compliance Guide.*

Whenever a chiller is being replaced, cooling loads should (if practical and cost effective) be improved through solar control and/or more efficient lighting (see BR 364 for guidance) using controls that meet the minimum requirements as given in the *Non-domestic Air Conditioning Compliance Guide.* L2B 45 and L2B 46

Mechanical ventilation

In dwellings

Systems should perform not less than that shown in GPG 2689 *Energy efficient ventilation in housing*.	L1B 40
Mechanical ventilation systems **must** satisfy the requirements in Part F.	L1B 33

Air handling plant

For buildings other than dwellings:

• the air-handling plant should be an efficient and effective control system;	L2A 46
• the system should be capable of achieving a specific fan power at 25% of design flow rate;	L2A 47 L2B 49
• ventilation system fans rated at more than 1100 W should be equipped with variable speed drives;	L2A 48
• ventilation ductwork should be made and assembled so as to be reasonably airtight (see HVCA DW144);	L2A 48
• replacement and new air handling plants should be efficient and energy efficient;	L2B 48
• fans that are rated at more than 1100 W and which form part of the environmental control system should be equipped with variable speed drives.	L2B 50
Note: Smoke control fans and similar therefore fall outside this guidance.	
• Ventilation ductwork should be constructed and assembledso as to be reasonably airtight (see HVCA DW144).	L2B 51

A room thermostat for a ducted warm air heating system should be mounted in the living room, at a height between 1370 mm and 1830 mm, and its maximum setting should not exceed 27°C.	B1 2.17 (V1) B1 2.18 (V2) B3 7.10 (V1) B3 10.2 (V2)

6.22 Liquid fuel

6.22.1 Meeting the requirement

Storage and supply

Oil and LPG fuel storage installations (including the pipework connecting them to the combustion appliances in the buildings they serve) shall:

- be located and constructed so that they are reasonably protected from fires that may occur in buildings or beyond boundaries; J (5.1a)
- be reasonably resistant to physical damage and corrosion; J (5.1a)
- be designed and installed so as to minimize the risk of oil escaping during the filling or maintenance of the tank; J (5.1bi)
- incorporate secondary containment when there is a significant risk of pollution; J (5.1bii)
- contain labelled information on how to respond to a leak. J (5.1biii)

Oil pollution

The Control of Pollution (Oil Storage) (England) Regulations 2001 (SI 2001/2954) came into force on 1 March 2002. They apply to a wide range of oil storage installations in England, but they do not apply to the storage of oil on any premises used wholly or mainly as one or more private dwellings, if the capacity of the tank is 3500 litres or less.

Table 6.86 Fire protection for oil storage tanks

Location of tank	Protection usually satisfactory
Within a building	Locate tanks in a place of special fire hazard, which should be directly ventilated to outside. Without prejudice to the need for compliance with all the requirements in Schedule 1, the need to comply with Part B should particularly be taken into account.
Less than 1800 mm from any part of a building	(a) Make building walls imperforate:[1] (1) within 1800 mm of tanks with at least 30 minutes' fire resistance;[2] (2) to internal fire and construct eaves within 1800 mm of tanks and extending 300 mm beyond each side of tanks with at least 30 minutes' fire resistance to external fire and with non-combustible cladding;

(*Continued*)

Table 6.86 (*Continued*)

Location of tank	Protection usually satisfactory
	or (b) Provide a firewall[3] between the tank and any part of the building within 1800 mm of the tank and construct eaves as in (a) above. The firewall should extend at least 300 mm higher and wider than the affected parts of the tank.
Less than 760 mm from a boundary	Provide a firewall between the tank and the boundary or a boundary wall having at least 30 minutes' fire resistance on either side. The firewall or the boundary wall should extend at least 300 mm higher and wider than the top and sides of the tank.
At least 1800 mm from the building and at least 760 mm from a boundary	No further provisions necessary.

Notes:

1. Excluding small openings such as air bricks etc.
2. Fire resistance in terms of insulation, integrity and stability.
3. Firewalls are imperforate non-combustible walls or screens, such as masonry walls or steel screens.

LPG storage

LPG installations are controlled by legislation enforced by the HSE which includes the following requirements applicable to dwellings:

The LPG tank should be installed outdoors and not within an open pit.	J (5.15)
The tank should be adequately separated from buildings, the boundary and any fixed sources of ignition to enable safe dispersal in the event of venting or leaks and in the event of fire to reduce the risk of fire spreading (see Figure 6.213).	J (5.15)
Firewalls may be free-standing built between the tank and the building, boundary and fixed source of ignition (see Figure 6.213(b)) or a part of the building or a fire resistance (insulation, integrity and stability) boundary wall belonging to the property.	J (5.16)

Where a firewall is part of the building or a boundary wall, it should be located in accordance with Figure 6.213(c).	J (5.16)
If the firewall is part of the building then it should be constructed as shown in Figure 6.213(d).	J (5.16)
Firewalls should be imperforate and of solid masonry, concrete or similar construction.	J (5.17)
Firewalls should have a fire resistance (insulation, integrity and stability) of at least 30 minutes.	J (5.17)
If firewalls are part of the building as shown in Figure 6.213(d), they should have a fire resistance (insulation, integrity and stability) of at least 60 minutes.	J (5.17)
To ensure good ventilation, firewalls should not normally be built on more than one side of a tank.	J (5.17)
A firewall should be at least as high as the pressure relief valve.	J (5.18)

Any pipe carrying natural gas or LPG should be: • of screwed steel or of all welded steel construction; • installed in accordance with 1996, SI 1996 No 825 and 1998, SI 1998 No 2451.	B2 8.40 (V2)
A protected shaft conveying piped flammable gas should be adequately ventilated direct to the outside air by ventilation openings at high and low level in the shaft.	B2 8.41 (V2)
If a door is provided between a dwelling house and the garage, the floor of the garage should: • be laid so as to allow fuel spills to flow away from the door to the outside; or • the door opening should be positioned at least 100 mm above garage floor level.	B3 5.5 (V1)

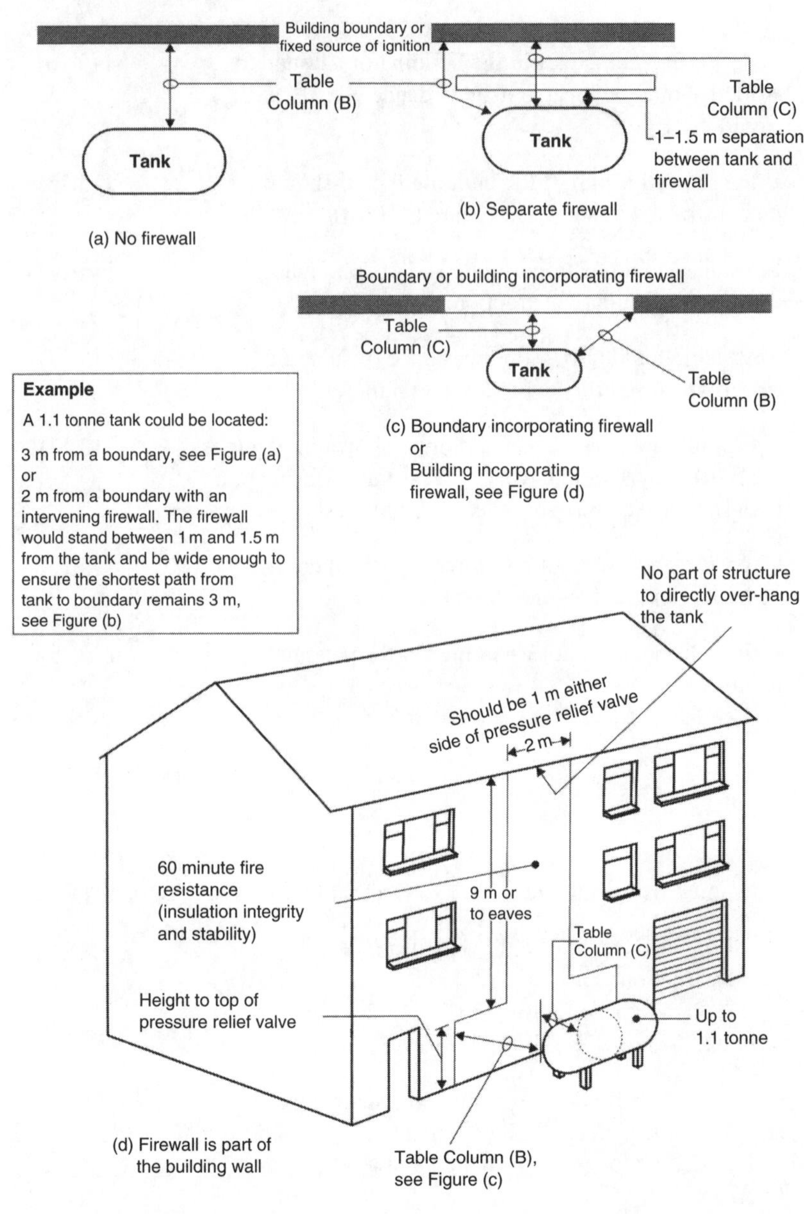

A Capacity of tank (tonnes)	Minimum separation	
	B tank with no firewall	**C** tank shielded by firewall
0.25	2.5	0.3
1.1	3.0	1.5

Figure 6.213 Separation or shielding of LPG tanks from building, boundaries and fixed sources of ignition

Where an LPG storage installation consists of a set of cylinders, a way of meeting the requirements is as shown in Figure 6.214.

Cylinders should stand upright and be secured by straps (or chains) against a wall outside the building, in a well ventilated position at ground level.	J (5.20)
Cylinders should be provided with a firm, level base such as concrete at least 50 mm thick or paving slabs bedded on mortar.	J (5.20)

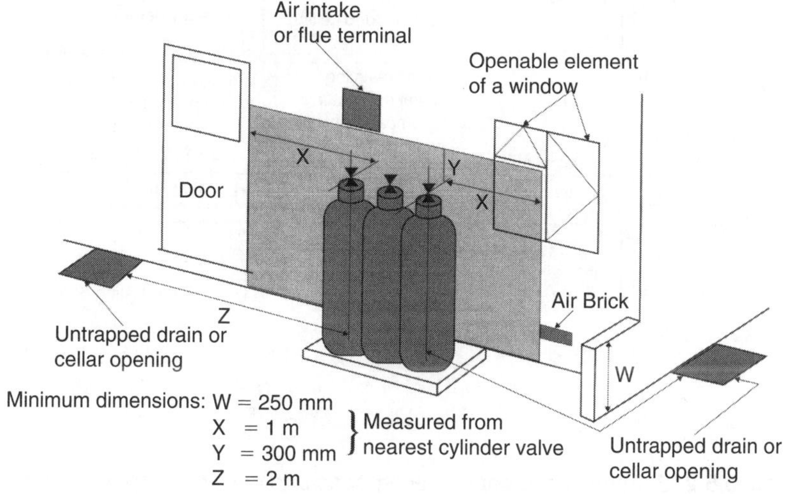

Figure 6.214 Location of LPG cylinders

6.23 Cavities and concealed spaces

6.23.1 The requirement

Internal fire spread (structure)

- *The building should be sub-divided by elements of fire-resisting construction into compartments.*
- *Any hidden voids in the construction shall be sealed and sub-divided to inhibit the unseen spread of fire and products of combustion, in order to reduce the risk of structural failure, and the spread of fire.*

(Approved Document B3)

6.23.2 Meeting the requirement

Concealed spaces or cavities in walls, floors, ceilings and roofs will provide an easy route for smoke and flame spread which, because it is obscured, will present a greater danger than would be more obvious from a weakness in the fabric of the building. To overcome this danger, buildings shall be designed and constructed so that the unseen spread of fire and smoke within concealed spaces in its structure and fabric is inhibited.

With the introduction of Part B 2007, window and door frames are now only suitable for use as cavity barriers if they are constructed of steel or timber of an appropriate thickness.

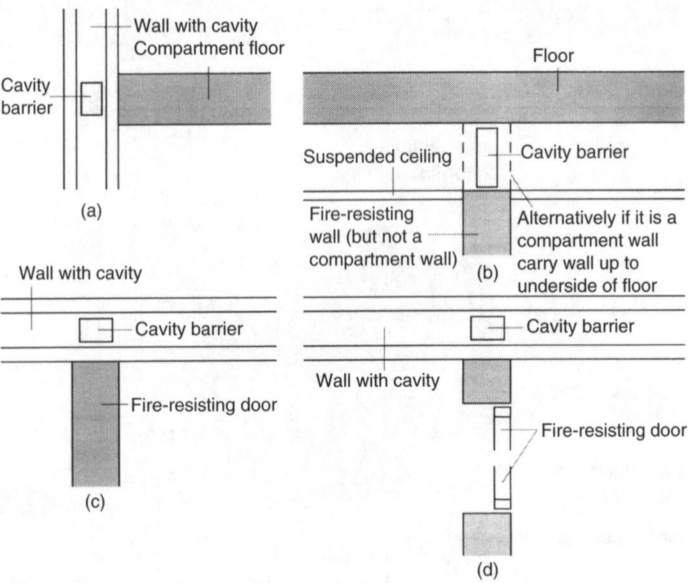

Figure 6.215 Interrupting concealed spaces and cavities. (a), (b) Sections. (c), (d) Plans

Provision of cavity barriers

Cavity barriers should be provided:	B3 6.3 (V1)
• at the edges of cavities;	B3 9.3 (V2)
• around window and door openings;	B3 6.3 (V1) B3 9.3 (V2)
• at the junction between an external cavity wall and a compartment wall;	B3 6.3 (V1) B3 9.3 (V2)
• at the junction between an external cavity wall and a compartment floor;	B3 9.3 (V2)

Table 6.82 Provision of cavity barriers

Cavity barriers to be provided	Dwelling houses	Flats or maisonettes	Other residential or institutional	Non-residential (e.g. office, shop, storage)
At the junction between an external cavity wall and a compartment wall that separates buildings; and at the top of such an external cavity wall.	X	X	X	X
Above the enclosures to a protected stairway in a house with a floor more than 4.5 m above ground level.	X			
At the junction between an external cavity wall and every compartment floor and compartment wall.		X	X	X
At the junction between a cavity wall and every compartment floor, compartment wall, or other wall or door assembly that forms a fire-resisting barrier.		X	X	X
In a protected escape route, above and below any fire-resisting construction that is not carried full storey height, or (in the case of a top storey) to the underside of the roof covering.		X	X	X
Where the corridor should be sub-divided to prevent fire or smoke affecting two alternative escape routes simultaneously above any such corridor enclosures that are not carried full storey height, or (in the case of the top storey) to the underside of the roof covering.			X	X
Above any bedroom partitions that are not carried full storey height, or (in the case of the top storey) to the underside of the roof covering.			X	
To sub-divide any cavity (including any roof space but excluding any underfloor service void).			X	X
Within the void behind the external face of rainscreen cladding at every floor level, and on the line of compartment walls abutting the external wall, of buildings that have a floor 18 m or more above ground level.		X	X	
At the edges of cavities (including around openings).	X	X	X	X

• at the top of such an external cavity wall;	B3 6.3 (V1) B3 9.3 (V2)
• at the junction between an internal cavity wall and any assembly which forms a fire resisting barrier;	B3 9.3 (V2)
• above the enclosures to a protected stairway in a dwelling house with a floor more than 4.5 m above ground level.	B1 2.14
Cavity barriers need not be provided between double-skinned corrugated or profiled insulated roof sheeting, if the sheeting is a material of limited combustibility.	B3 6.4 (V1) B3 9.5 (V2)

Note: Separate rules exist for bedrooms in Institutional and Other Residential buildings (see B3 9.7 (V2)).

Construction and fixings for cavity barriers

Every cavity barrier should be constructed to provide at least 30 minutes fire resistance.	B3 6.5 (V1) B3 9.13 (V2)
A cavity barrier should, wherever possible, be tightly fitted to a rigid construction and mechanically fixed in position.	B3 6.6 (V1) B3 9.14 (V2)
Cavity barriers should be fixed so that their performance is unlikely to be made ineffective by: • movement of the building due to subsidence, shrinkage or temperature change; • movement of the external envelope due to wind; • collapse in a fire of any services penetrating them; • failure in a fire of their fixings; • failure in a fire of any material or construction which they abut.	B3 6.7 (V1) B3 9.15 (V2)
Cavity barriers in a stud wall or partition (or provided around openings) may be made of: • steel at least 0.5 mm thick; or • timber at least 38 mm thick; or • polythene-sleeved mineral wool, or mineral wool slab (in either case under compression);	

- calcium silicate, cement-based or gypsum based boards at least 12 mm thick.

Extensive cavities in floor voids should be subdivided with cavity barriers. B3 (V2)

The need for cavity barriers in some concealed floor or roof spaces can be reduced by using a fire resisting ceiling below the cavity. B2 3.6 (V1) B2 6.6 (V2)

Openings in cavity barriers

Openings in a cavity barrier should be limited to those for: B3 6.8 (V1)

- doors which have at least 30 minutes fire resistance; B3 9.13 (V2)
- the passage of pipes which meet the provisions in Part P Section 7;
- the passage of cables or conduits containing one or more cables;
- openings fitted with a suitably mounted automatic fire damper;
- ducts which are fire-resisting or are fitted with a suitably mounted automatic fire damper where they pass through the cavity barrier.

6.24 Kitchens and utility rooms

6.24.1 The requirement

Fire safety

- *There shall be an early warning fire alarm system for persons in the building.*
- *There shall be sufficient escape routes that are suitably located to enable persons to evacuate the building in the event of a fire.*
- *Safety routes shall be protected from the effects of fire.*
- *In an emergency, the occupants of any part of the building shall be able to escape without any external assistance.*

(Approved Document B1)

- *The spread of flame over the internal linings of the building shall be restricted.*
- *The heat released from the internal linings shall be restricted.*

(Approved Document B2)

Ventilation

There shall be adequate means of ventilation provided for people in the building.

(Approved Document F)

6.24.2 Meeting the requirement

Fire safety

In small premises:	
• store rooms should be enclosed with fire-resisting construction;	B1 3.35 (V2)
• clear glazed areas should be provided in any partitioning separating a kitchen from the open floor area to enable any person within the kitchen to obtain early visual warning of an outbreak of fire.	B1 3.36 (V2)

Smoke alarms

Smoke alarms should be positioned in the circulation spaces **between** sleeping spaces and places where fires are most likely to start (e.g. in kitchens and living rooms).	B1 1.11 (V1) B1 1.10 (V2)
Where the kitchen area is not separated from the stairway or circulation space by a door, there should be a compatible interlinked heat detector or heat alarm in the kitchen, in addition to whatever smoke alarms are needed in the circulation space(s).	B1 1.13 (V1) B1 1.12 (V2)

Inner rooms

Any inner room that is a kitchen or utility room that is situated not more than 4.5 m above ground level, whose only escape route is through another room, shall be provided with an emergency egress window.	B1 2.9 (V1) B1 2.5 (V2)

Ancillary accommodation

Ancillary accommodation such as: • kitchens; • staff changing and locker rooms; and • store rooms; should be enclosed by fire-resisting construction.	B1 3.50 (V2)
Non-domestic kitchens: • should have separate and independent extraction systems; and • extracted air should not be recirculated.	B1 5.50 (V2)

Ventilation

Extract ventilation concerns the removal of air directly from a space or spaces to outside. Extract ventilation may be by natural means (e.g. by passive stack ventilation) or by mechanical means (e.g. by an extract fan or central system).

All kitchens and utility rooms shall be provided with extract ventilation to the outside which is capable of operating either: • intermittently at a minimum extract rate of 30 l/s (adjacent to hob) and 60 l/s elsewhere, or • continuously with a minimum extract rate of 13 l/s.	F 1.5
Common outlet terminals and/or branched ducts shall **not** be used for wet rooms such as a kitchen or a utility room.	F App D
PSV devices shall have a minimum: • internal duct diameter of 125 mm for kitchens, 100 mm for utility rooms; and • a cross-sectional area of 12 000 mm^2 for kitchens, 8000 mm^2 for utility rooms.	F
Where there was no previous ventilation opening, or where the size of the original ventilation opening is not known, replacement window(s) shall have an equivalent area greater than 2500 mm^2.	F 3.6

In buildings other than dwellings, fresh air supplies should be protected from contaminants that would be injurious to health.	F 2.3
Wall/ceiling-mounted centrifugal fans (which are fitted with a 100 mm diameter flexible duct or rectangular duct and which are designed to achieve 60 l/s for kitchens) should not be ducted further than 3 metres and should have no more than one 90° bend.	F App E
Automatic controls for ventilators that are designed to work continuously in kitchens, must be capable of providing sufficient flow during cooking with fossil fuels (e.g. gas) so as to avoid the build-up of combustion products.	F Table 1.5
All office sanitary accommodation, washrooms and food and beverage preparation areas shall be provided with intermittent air extract ventilation.	F 2.11

6.25 Storage of food

6.25.1 The requirement (Building Act 1984 Sections 28 and 70)

All houses or buildings that have been converted into houses must provide sufficient and suitable accommodation for storing food or '*sufficient and suitable space for the provision of such accommodation by the occupier*'.

This could prove to be a problem when submitting plans for approval and you would be wise to consider the possibilities.

Table 6.88 Ventilation of rooms containing openable windows (i.e. located on an external wall)

Room	Rapid ventilation (e.g. opening windows)	Background ventilation (mm^2)	Extract ventilation fan rates or passive stack (PSV)
Habitable room	1/20th of floor area	8000	
Kitchen	Opening window (no minimum size)	4000	30 litres/second adjacent to a hob or 60 litres/second elsewhere or PSV

(*Continued*)

Table 6.88 (*Continued*)

Room	Rapid ventilation (e.g. opening windows)	Background ventilation (mm^2)	Extract ventilation fan rates or passive stack (PSV)
Utility room	Opening window (no minimum size)	4000	30 litres/second or PSV
Bathroom (with or without WC)	Opening window (no minimum size)	4000	15 litres/second or PSV
Sanitary accommodation (separate from bathroom)	1/20th of floor area or mechanical extract at 6 litres/second	4000	

6.26 Refuse facilities

6.26.1 The requirement (Building Act 1984 Section 23)

Probably due to new EU agreements, local authorities have now become far stricter in seeing that the requirements contained in the Building Act for storage and collection of refuse are applied. This means that you have to ensure that the building is equipped with a satisfactory method for storing refuse (with a house this would normally be a simple dustbin; with a block of flats or a factory, however, the system for storage would have to be more sophisticated). The local council will also need to be able to collect and remove this refuse easily and so this should be borne in mind when siting the refuse collection point.

Under the Building Act 1984 it is **unlawful** for any person (except with the consent of the local authority) to close or obstruct the means of access by which refuse or faecal matter is removed from a building.

Fire safety

- *Buildings shall be sub-divided by elements of fire-resisting construction into compartments.*
- *Openings in fire-separating elements shall be suitably protected in order to maintain the integrity of the element (i.e. the continuity of the fire separation).*

(Approved Document B3)

Airborne and impact sound

Dwellings shall be designed so that the noise from domestic activity in an adjoining dwelling (or other parts of the building) is kept to a level that:

- *does not affect the health of the occupants of the dwelling;*

- *will allow them to sleep, rest and engage in their normal activities in satisfactory conditions.*

(Approved Document E1)

Dwellings shall be designed so that any domestic noise that is generated internally does not interfere with the occupants' ability to sleep, rest and engage in their normal activities in satisfactory conditions.

(Approved Document E2)

Domestic buildings shall be designed and constructed so as to restrict the transmission of echoes.

(Approved Document E3)

Schools shall be designed and constructed so as to reduce the level of ambient noise (particularly echoing in corridors).

(Approved Document E4)

6.26.2 Meeting the requirement

Refuse chutes and storage

Refuse storage chambers, refuse chutes and refuse hoppers should be sited and constructed in accordance with BS 5906.	B1 5.54 (V2)
In buildings containing flats, walls that enclose a refuse storage chamber should be constructed as a compartment wall.	B2 8.13 (V2)
Refuse chutes and rooms provided for the storage of refuse should: • be approached either directly from the open air or by way of a protected lobby; • be separated from other parts of the building by fire-resisting construction; • **not** be located within protected stairways or protected lobbies.	B1 5.55 and 5.56 (V2)
The access to refuse storage chambers should not be sited adjacent to escape routes/final exits, or near to windows of flats.	B1 5.57 (V2)

A wall separating a habitable room or kitchen and a refuse chute should have mass (including any finishes) of at least 1320 kg/m^2.	E (2.28)
A wall separating a non-habitable room, which is in a dwelling, from a refuse chute should have a mass (including any finishes) of at least 220 kg/m^2.	E (2.28)

6.27 Fire resistance

The overall aim of fire safety precautions is to ensure that:

- all corridor doors shall meet the requirements for fire safety as described in Building Regulations Part B – Fire safety (see Part E2);
- all doors shall satisfy the Requirements of Building Regulation Part B – Fire safety (see Part E4);
- a satisfactory means of giving an alarm of fire is available;
- a satisfactory means of escape for persons in the event of fire in a building is available (see Part B1);
- external walls and roofs have adequate resistance to the spread of fire over the external envelope;
- fire stopping should be flexible to prevent a rigid contact between the pipe and the floor (see Parts E3 and E4);
- if a fire stop is required in the cavity between frames, then it should either be flexible or only be fixed to one frame (see Part E2);
- if there is an existing lath and plaster ceiling it should be retained as long as it satisfies Building Regulation Part B – Fire safety (see Part E);
- penetrations through a separating floor by ducts and pipes should have fire protection to satisfy Building Regulation Part B – Fire safety (see Parts E3 and E4);
- that fire spread over the internal linings of buildings is inhibited (see Part B2);
- the ceiling void and roof space detail can only be used where the Requirements of Building Regulation Part B – Fire safety can also be satisfied (see Part E).
- the junction between the separating wall and the roof should be filled with a flexible closer which is also suitable as a fire stop (see Part E2);
- the spread of fire from one building to another is restricted (see Part B4);
- the unseen spread of fire and smoke in concealed spaces in buildings is inhibited (see Part B3);
- the stability of buildings is ensured in the event of fire;
- there are facilities in buildings to assist fire fighters in the saving of life of people in and around buildings (see Part B5);
- there is satisfactory access for fire appliances to buildings;

- there is a sufficient degree of fire separation within buildings and between adjoining buildings;
- where a staircase performs a separating function it shall conform to Building Regulation Part B – Fire safety (see Part E4).

Many of the requirements are, of course, closely interlinked. For example, there is a close link between the provisions for means of escape (B1) and those for the control of fire growth (B2), fire containment (B3) and facilities for the fire service (B5). Similarly there are links between B3 and the provisions for controlling external fire spread (B4), and between (B3) and (B5). Interaction between these different requirements should be recognized where variations in the standard of provision are being considered.

Factors that should be taken into account include:

- the ability of a structure to resist the spread of fire and smoke;
- the anticipated probability of a fire occurring;
- the anticipated fire severity;
- the consequential danger to people in and around the building.

Measures that could be incorporated include:

- availability of powers to require staff training in fire safety and fire routines, e.g. under the Fire Precautions Act 1971, the Fire Precautions (Workplace) Regulations 1997, or registration or licensing procedures;
- control of the rate of growth of a fire;
- consideration of the availability of any continuing control under other legislation that could ensure continued maintenance of such systems;
- early fire warning by an automatic detection and warning system;
- facilities to assist the fire service;
- provision of smoke control;
- the adequacy of means to prevent fire;
- the adequacy of the structure to resist the effects of a fire;
- the degree of fire containment;
- fire separation between buildings or parts of buildings;
- the standard of active measures for fire extinguishment or control;
- the standard of means of escape;
- management.

The design of fire safety in hospitals is covered by Health Technical Memorandum (HTM) 81 Fire precautions in new hospitals (revised 1996).

Building Regulations are intended to ensure that a reasonable standard of life safety is provided, in case of fire. The protection of property, including the building itself, may require additional measures, and insurers will in general seek their own higher standards, before accepting the insurance risk. Guidance is given in the LPC Design guide for the fire protection of buildings.

Guidance for assisting protection in Civil and Defence Estates is given in the Crown Fire Standards published by the Property Advisers to the Civil Estate (PACE).

Fire safety engineering

Fire safety engineering can provide an alternative approach to fire safety and in certain circumstances it could be the only practical way to achieve a satisfactory standard of fire safety in some buildings. Fire safety engineering is also suitable for solving problems concerning the design of the building, which although meeting the requirements of the Regulations is still problematic.

Factors that should be taken into account include:

- ability of a structure to resist the spread of fire and smoke;
- anticipated probability of a fire occurring;
- anticipated fire severity;
- consequential danger to people in and around the building.

Many processes are available for consideration such as:

- are facilities available that will assist fire and rescue services?
- are smoke controls provided and are they adequate?
- are there appropriate existing methods for controlling and extinguishing fires?
- are these methods regularly reviewed and where necessary repaired/replaced?, how adequate is the current means of preventing fire?
- how appropriate is the designed means of escape?
- is the fire separation between buildings and/or parts of buildings appropriate?
- is the rate of growth of the fire controllable?
- is the structure capable of resisting the effects of fire?
- is there an automatic early fire warning system in place?
- to what degree can the fire be contained?
- is there an existing process for training staff in fire safety and fire routines?
- do Top Management endorse the requirement for fire safety?

Risk assessment

The assessment and design of means of escape shall take into account: B1.ii

- the nature of the building structure;
- the use of the building;
- the potential of fire spread through the building; and
- the standard of fire safety management proposed.

Fire risk analysis

Part B now includes a requirement for the responsible person (i.e. the person carrying out work to a building) to make available to the owner (other than houses occupied as single private dwellings) 'fire safety information' concerning the design and construction of the building or extension plus details of the services, fittings and equipment that have been provided in order that they (when required under the new Regulatory Reform (Fire Safety) Order 2005 – Statutory Instrument 2005 No. 1541) may complete a fire risk analysis.

Although these requirements are applicable to premises whilst in operation, it would be useful for the designers of a building to carry out a preliminary fire risk assessment as part of the design process. If a preliminary risk assessment is produced, it can be used as part of the Building Regulations submission and can assist the fire safety enforcing authority in providing advice at an early stage as to what, if any, additional provisions may be necessary when the building is first occupied.

6.27.1 The requirement

Means of escape

- *There shall be an early warning fire alarm system for persons in the building.*
- *There shall be sufficient escape routes that are suitably located to enable persons to evacuate the building in the event of a fire.*
- *Safety routes shall be protected from the effects of fire.*
- *In an emergency, the occupants of any part of the building shall be able to escape without any external assistance.*

(Approved Document B1)

Internal fire spread (linings)

- *The spread of flame over the internal linings of the building shall be restricted.*
- *The heat released from the internal linings shall be restricted.*

(Approved Document B2)

Internal fire spread (structure)

Dependent on the use of the building, its size and the location of the element of construction:

- *loadbearing elements of a building structure shall be capable of withstanding the effects of fire for an appropriate period without loss of stability;*
- *the building shall be sub-divided by elements of fire-resisting construction into compartments;*
- *all openings in fire-separating elements shall be suitably protected in order to maintain the integrity of the element (i.e. the continuity of the fire separation);*
- *any hidden voids in the construction shall be sealed and sub-divided to inhibit the unseen spread of fire and products of combustion.*

(Approved Document B3)

External fire spread

- *External walls shall be constructed so as to have a low rate of heat release and thereby be capable of reducing the risk of ignition from an external source and the spread of fire over their surfaces.*

- *The amount of unprotected area in the sides of the building shall be restricted so as to limit the amount of thermal radiation that can pass through the wall.*
- *The roof shall be constructed so that the risk of spread of flame and/or fire penetration from an external fire source is restricted.*

(Approved Document B4)

Access facilities for the fire service

- *There shall be sufficient means of external access to enable fire appliances to be brought near to the building for effective use.*
- *There shall be sufficient means of access into and within the building for firefighting personnel to affect search and rescue and fight fire.*
- *The building shall be provided with sufficient internal fire mains and other facilities to assist firefighters in their tasks.*
- *The building shall be provided with adequate means for venting heat and smoke from a fire in a basement.*

(Approved Document B4)

6.27.2 Meeting the requirements

Fire resistance

An element of construction shall provide:	B3
• resistance to collapse;	
• resistance to fire penetration;	
• resistance to the transfer of excessive heat.	

The purpose in providing the structure with fire resistance is:

- to minimize the risk to the occupants;
- to reduce the risk to fire-fighters;
- to reduce the danger to people in the vicinity of the building.

Fire resistance standard

Elements such as structural frames, beams, columns, internal and external loadbearing walls, floor structures and gallery structures should have at least the fire resistance shown in Part B, Appendix A, Table Al.	B3 4.2 (V1) B3 7.2 (V2)
The fire resistance of an element of structure that supports or gives stability to another element of structure shall not be less than the other element.	B3 4.3 (V1) B3 7.3 (V2)

Loft conversions

The floor(s), both old and new, shall have the full 30 minute standard of fire resistance shown in Part B Appendix A, Table Al unless: • only one storey is being added; • the new storey contains no more than 2 habitable rooms; and • the total area of the new storey is less than 50 m^2.	B3 4.7 (V1)
In those places where the floor only separates rooms (and not circulation spaces), a modified 30 minute standard of fire resistance may be applied.	B3 4.7 (V1)
Where the conversion of an existing roof space (such as a loft conversion to a two-storey house) means that a new storey is going to be added, then the stairway will need to be protected with fire-resisting doors and partitions.	B1 2.20b

Flats

If the existing building has timber floors and these are to be retained, the requirements for fire resistance may be difficult to meet. In these cases, provided that the means of escape conforms to Part B Section 3 and are adequately protected. Then:

• those parts of the building that are **no** more than three storeys high shall need to have a 30 minute standard of fire resistance;	
• if the altered building has four or more storeys, then the full standard of fire resistance (as described in Appendix A to Part B) would be required.	B2 7.10 (V2)

The amount of fire resistance provided by the building structure and other elements of construction is determined by reference to either:

- BS 476 (National classification);
- Commission Decision 2000/367/EC of 3 May 2000 implementing Council Directive 89/1061EEC (European classification);
- BS EN 13501-2:2003;
- BS EN 13501-3:2005;
- BS EN 13501-4:2007.

See also Appendix A to Part B for Tables setting out minimum periods of fire resistance etc.

Escape routes

Planning escape routes

Common corridors that connect two or more storey exits should be sub-divided by a self-closing fire door with, if necessary, an associated fire-resisting screen. B1 2.28 (V2)

Note: Self-closing fire doors should be positioned so that smoke will not affect access to more than one stairway.

Protection of escape routes

Generally, a 30-minute standard is sufficient for the protection of means of escape. (Details of fire resistance test criteria and standards of performance are contained in Appendix A to Part B). B1 5.2 (V2)

External escape stairs

External escape stairs should meet the following provisions:

where an external escape stair is provided: B1 2.15 (a, b and c)

- all doors giving access to the stair should be fire-resisting;
- any part of the external envelope of the building within 1800 mm of (and 9 m vertically below) the flights and landings of an external escape stair should be of fire resisting construction (see Figure 6.216);
- any part of the building (including doors) within 1800 mm of the escape route shall be protected by fire resisting construction.

External escape stairs greater than 6 m in vertical extent shall be protected from the effects of adverse weather conditions. B1 2.15d

Note: Glazing in any fire-resisting construction should be fire-resisting and fixed shut.

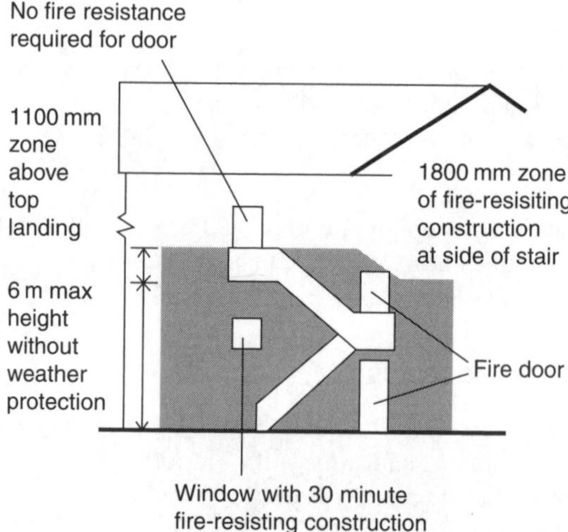

Figure 6.216 Fire resistance of areas adjacent to external stairs

Dwelling houses with one floor more than 4.5 m above ground level

The dwelling house may either have a protected stairway which: B1 2.6 (V1)

- extends to the final exit (see Figure 6.217(a)); or
- gives access to at least two escape routes at ground level, each delivering to final exits and separated from each other by fire-resisting construction and fire doors (see Figure 6.217(b));

or the top floor can be separated from the lower storeys by fire-resisting construction and be provided with its own alternative escape route leading to its own final exit (see Figure 6.217(c)).

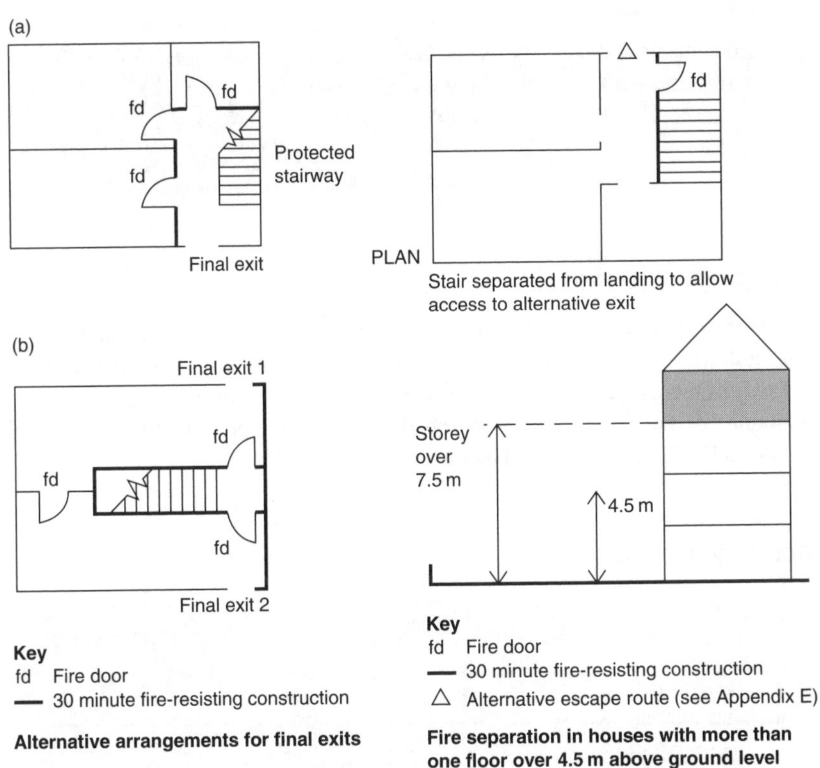

Figure 6.217 Final exits and fire separation

Flats with a floor more than 4.5 m above ground level

Where any flat has an alternative exit and the habitable rooms do not have direct access to the entrance hall (see Figure 6.217): • the bedrooms should be separated from the living accommodation by fire-resisting construction and fire doors); and • the alternative exit should be located in the part of the flat containing the bedroom(s).	B1 2.14 (V2)

Residential care homes

Generally, in care homes for the elderly it is reasonable to assume that at least a proportion of the residents will need some assistance to evacuate.

Bedrooms should be enclosed in fire-resisting construction with fire resisting doors and every corridor serving bedrooms should be a protected corridor. B1 3.48 (V2)

Shopping complexes

Part B is primarily intended to cover shops that are contained in a single separate building. If a shop forms part of a complex, such as a covered mall then the requirements for fire resistance, walls separating shop units, surfaces and boundary distances will probably be different (see Sections 5 and 6 of BS 5588-10:1991 for further guidance).

Ancillary accommodation

Ancillary accommodation such as: B1 3.50 (V2)

- day rooms;
- chemical stores;
- cleaners' rooms;
- clothes storage;
- disposal rooms;
- kitchens;
- laundry rooms;
- linen stores;
- plant rooms;
- smoking rooms (soon to be non-existent in view of future Government legislation);
- staff changing and locker rooms; and
- store rooms;

should be enclosed by fire-resisting construction.

Portal frames

Where a portal framed building is near a relevant boundary, the external wall near the boundary may need fire resistance to restrict the spread of fire between buildings. B4 12.4 (V2)

Access lobbies and corridors

Any such protected exit passageway should have the same standard of fire resistance and lobby protection as the stairway it serves.

Balconies and flat roofs

Any flat roof that forms part of a means of escape should: B1 2.10 (V1)

- be part of the same building from which escape is being made; B1 2.7 (V2)
- lead to a storey exit or external escape route; and
- provide 30 minutes fire resistance (see Appendix A, Table Al of Approved Document B for fire resistance figures for elements of structure etc.).

Note: If a balcony or flat roof is provided for escape purposes guarding may be required (see Approved Document K Protection). B1 2.11 (V1)

Ceilings

The need for cavity barriers in some concealed floor or roof spaces can be reduced by using a fire resisting ceiling below the cavity. B2 3.6 (V1) B2 6.6 (V2)

Suspended ceilings

A suspended, fire protected ceiling should meet the requirements of Table 1 and Table 6.89.

Table 6.89 Limitations on fire-protecting suspended ceilings

Height of building or separated part (m)	Type of floor	Provision for fire resistance of floor (minutes)
Less than 18	Not compartment	60 or less
	Compartment	Less than 60
18 or more	Any	60 or less
No limit	Any	More than 60

for further details see Part B, Appendix A, Table A3.

Cavity barriers

Cavity barriers should be provided at the junction between an internal cavity wall and any assembly which forms a fire resisting barrier.	B3 9.3 (V2)
Every cavity barrier should be constructed to provide at least 30 minutes fire resistance.	B3 6.5 (V1) B3 9.13 (V2)
Openings in a cavity barrier should be limited to those for: • doors which have at least 30 minutes fire resistance; • the passage of pipes which meet the provisions in Part P Section 7; • the passage of cables or conduits containing one or more cables; • openings fitted with a suitably mounted automatic fire damper; • ducts which are fire-resisting or are fitted with a suitably mounted automatic fire damper where they pass through the cavity barrier.	B3 6.8 (V1) B3 9.13 (V2)

Compartment walls and compartment floors

Every compartment wall and compartment floor should:	
• form a complete barrier to fire between the compartments they separate; and	B2 8.20a (V2) B3 5.6 (V1)
• have the appropriate fire resistance as indicated in Appendix A, Tables Al and A2.	B2 8.20b (V2) B3 5.6 (V1)
Junctions between a compartment floor and an external wall that has no fire resistance (such as a curtain wall) should be restrained at floor level to reduce the movement of the wall away from the floor when exposed to fire.	B2 8.26 (V2)

Junction of compartment wall or compartment floor with other walls

At the junction of a compartment floor with an external wall that has no fire resistance (such as a curtain wall) the external wall should be restrained at floor level to reduce the movement of the wall away from the floor when exposed to fire. B3 5.10 (V1)

Junction of compartment wall with roof

A compartment wall should be:

- taken up to meet the underside of the roof covering or deck, with fire-stopping (where necessary) at the wall/roof junction to maintain the continuity of fire resistance; B3 5.11 (V1)
- continued across any eaves. B3 8.28 (V2)

Openings in compartment walls separating buildings or occupancies

Any openings in a compartment wall which is common to two or more buildings should be limited to those for: B3 5.13 (V1) B3 8.32 (V2)

- a door which is providing 'means of access' in case of fire (and which has the same fire resistance as that required for the wall);
- the passage of a pipe.

All other openings in compartment walls or compartment floors should be limited to those for:

- doors which have the appropriate fire resistance; B3 8.34 (V2)
- the passage of pipes, ventilation ducts, service cables, chimneys, appliance ventilation ducts or ducts encasing one or more flue pipes;
- refuse chutes of non-combustible construction;
- atria designed in accordance with BS 5588-7:1997; and
- protected shafts (see B3 8.35 V2 for details of the relevant requirements).

External walls

External walls of the building should have sufficient fire resistance to prevent fire spread across the relevant boundary.	B4 8.1 (V1) B4 12.1 (V2)
The external surfaces of walls of dwellings within 1000 mm of the relevant boundary should meet Class 0 (National Class) or Class B-s3,d2 or better (European class). For all buildings other than dwellings, they should meet the relevant European requirements as shown in Diagram 40 (page 95) of Part B Volume 2.	B4 8.4 (V1) B4 12.6 (V2)

Note: Any part of an external wall which has less fire resistance than that shown in Part B, Volume 1, Appendix A, Table A2, is considered to be an unprotected area.

Fire doors

Two fire doors may be fitted in the same opening so that the total fire resistance is the sum of their individual fire resistances, provided that each door is capable of closing the opening.	App B 4 (V2)

Table 6.90 Provision for fire doors (dwellings)

Position of door	Minimum fire resistance of door in terms of integrity (minutes) when tested to BS 476-22:1987	Minimum fire resistance of door in terms of integrity (minutes) when tested to the relevant BSEN 1634 European standard
Any door		
With a cavity barrier	FD 30	E30
Between a dwelling house and a garage	FD 30s	E30Sa
Forming part of the enclosure to a protected stairway in a single family dwelling house	FD 20	E20
Within any other fire resisting construction in a dwelling house not described elsewhere in the table	FD 20	E20

Notes:

(1) Minimum fire resistance for doors in buildings other than dwellings is given in Table B1 (page 134) of Part B Volume 2

(2) BS 8214:1990 gives recommendations for the specification, design, construction, installation and maintenance of fire doors constructed with non-metallic door leaves.

Guidance on timber fire-resisting doorsets may be found in *'Timber fire-resisting doorsets: maintaining performance under he new European test standard'* published by TRADA.

Galleries

All galleries shall be provided with an alternative exit or, where the gallery floor is not more than 4.5 m above ground level, an emergency egress window (which complies with paragraph 2.8). B1 2.12 (V1) B1 2.8 (V2)

If the gallery floor is not provided with an alternative exit or escape window:

- the gallery should overlook at least 50% of the room below (see Figure 6.218);
- the distance between the foot of the access stair to the gallery and the door to the room containing the gallery should not exceed 3 m;
- the distance from the head of the access stair to any point on the gallery should not exceed 7.5 m; and
- any cooking facilities within a room containing a gallery should either:
  - be enclosed with fire-resisting construction; or
  - be remote from the stair to the gallery.

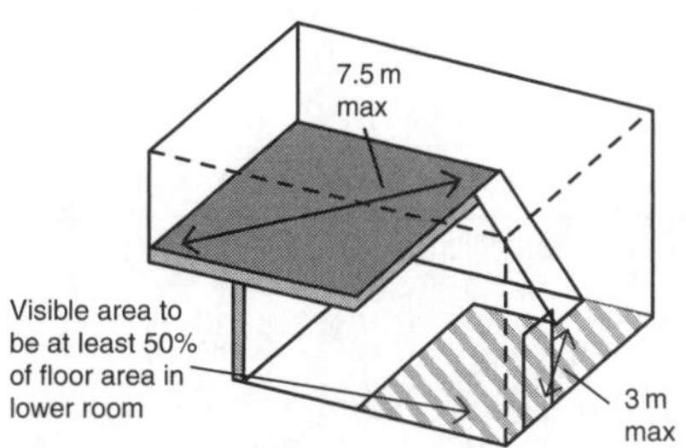

Notes:

1 This diagram does not apply where the gallery is
  i. provided with an alternative escape route; or
  ii. provided with an emergency egress window (where the gallery floor is not more than 4.5 m above gound level).

2 Any cooking facilities within a room containing a gallery should either:
  i. be enclosed with fire-resisting construction; or
  ii. be remote from the stair to the gallery and positioned such that they do not prejudice the escape from the gallery.

Figure 6.218 Gallery floors with no alternative exit

Inner rooms

Store rooms should be enclosed with fire-resisting construction.	B1 3.35 (V2)

Openings and fire-stopping

Note: Detailed guidance on door openings and fire doors is given in Part B Appendix B.

Every joint, imperfection or opening of a fire-separating element should be protected by sealing or fire-stopping so that the fire resistance of the element is not weakened.	B3 7.2 (V1) B3 10.2 (V2)

Openings for pipes

Unless the pipe is in a protected shaft, all pipes which pass through fire-separating elements should conform to one the three alternatives shown below.

	Type	Requirements
Alternative A	Proprietary seals (any pipe diameter)	Provide a proprietary sealing system which has been shown by test to maintain the fire resistance of the wall, floor or cavity barrier
Alternative B	Pipes with a restricted diameter	Where a proprietary sealing system is not used, fire stopping may be used around the pipe, keeping the opening as small as possible (see Table 3 of Part B Volume 1 for dimension details)
Alternative C	Sleeving	A pipe of lead, aluminium, aluminium alloy, fibre-cement or uPVC, with a maximum nominal internal diameter of 160 mm, may be used with a sleeving of non-combustible pipe as shown in Figure 6.219

Ventilation ducts and flues etc.

If a flue or duct passes through a compartment wall or compartment floor, or is built into a compartment wall, each wall of the flue or duct should have a fire resistance of at least half that of the wall or floor (see Figure 6.220).	B1 7.11 (V1)

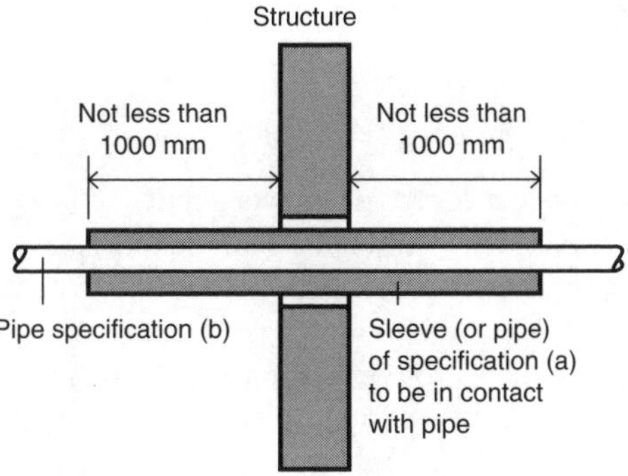

Notes:

1 Make the opening in the structure as small as possible and provide fire-stopping between pipe and structure.

Figure 6.219 Pipes penetrating structure

b. Flue built into compartment wall

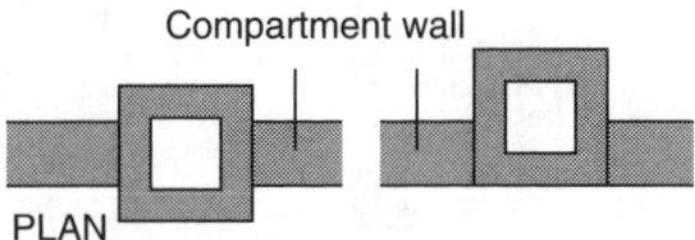

In each case flue walls should have a fire resistance at least one half of that required for the compartment wall and be of non-combustible construction.

Figure 6.220 Flues penetrating compartment walls or floors

If a flue (or a duct containing flues and/or ventilation duct(s)), passes through a compartment wall or compartment floor, or is built into a compartment wall, each wall of the flue or duct should have a fire resistance of at least half that of the wall or floor.	B3 10.16 (V2)

Passenger lifts

Passenger lifts in dwelling houses which serve any floor more than 4.5 m above ground level should either be located in the enclosure to the protected stairway or be contained in a fire resisting lift shaft.	B1 2.18

Protected shafts

Protected shaf (see Figure 6.221) should: B2 8.37 (V2)

- form a complete barrier to fire between the different compartments which the shaft connects;
- have the appropriate fire resistance given in Appendix A, Table A1; and
- meet the requirements for ventilation and the treatment of openings (see Part B Sections 8.41 and 8.42).

Protected shafts provide for the movement of people (e.g. stairs, lifts), or for passage of goods, air or services such as pipes or cables between different compartments. The elements enclosing the shaft (unless formed by adjacent external walls) are compartment walls and floors. Figure 6.221 shows three common examples which illustrate the principles.

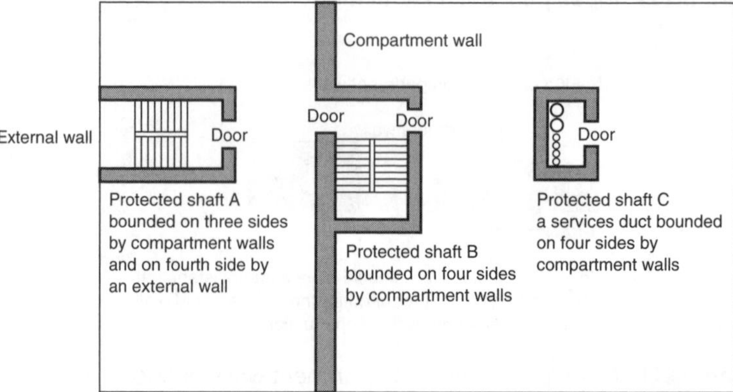

Figure 6.221 Protected shafts

An uninsulated glazed screen may be incorporated in the enclosure to a protected shaft between a stair and a lobby or corridor which is entered from the stair provided that: B2 8.38 (V2)

- the fire resistance for the stair enclosure is not more than 60 minutes;
- the glazed screen has at least 30 minutes fire resistance; and
- the lobby or corridor is enclosed to at least a 30 minute standard (see Figure 6.222).

Generally speaking, an external wall of a protected shaft does not need to have fire resistance (but see BS 5588-5:2004 for fire resistance of external walls of firefighting shafts).

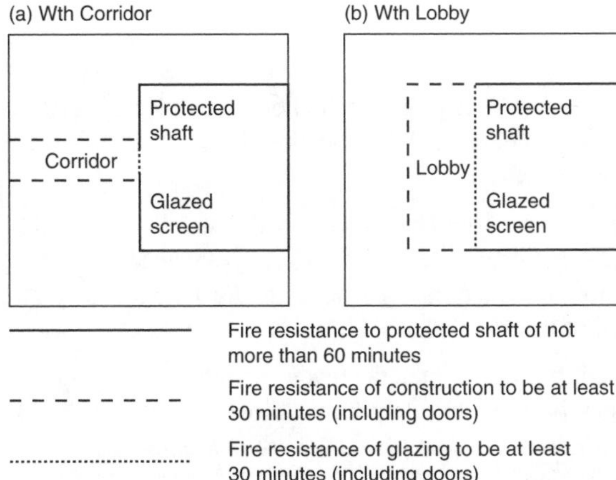

Figure 6.222 Uninsulated glazed screen separating protected shaft from lobby or corridor

Roof coverings

In thatched roofs:	B4 10.9 (V1)
• the rafters should be overdrawn with construction having not less than 30 minutes fire resistance;	B4 14.9 (V2)
• a smoke alarm should be installed in the roof space.	

Note: See Part B, Volume 1, Table 5 for limitations on roof coverings and Tables 6 and 7 for the limitations on using plastic rooflights and thermoplastic materials.

Stairs

Where an external escape stair is provided in accordance with paragraph 4.44, it should meet the following provisions: • all doors giving access to the stair should be fire-resisting and self-closing; • any part of the external envelope of the building within 1800 mm of (and 9 m vertically below) the flights and landings of an external escape stair should be fire resisting; • there is protection by fire-resisting construction for any part of the building within 1800 mm of the escape route from the stair to a place of safety; • glazing should also be fire resistant and fixed shut.	B1 5.25 (V2)

Common stairs

Any stair used as a firefighting stair should be at least 1100 mm wide (see Part B V2, Appendix C for measurement of width).	B1 2.33 (V2)
All common stairs should be situated within a fire-resisting enclosure (i.e. it should be a protected stairway), to reduce the risk of smoke and heat making use of the stair hazardous.	B1 2.36 (V2)
In single stair buildings, meters located within the stairway should be enclosed within a secure cupboard which is separated from the escape route with fire-resisting construction.	B1 2.40 (V2)

Fire detection and fire alarm systems

General

All new dwelling houses should be provided with a BS 5839-6:2004 Grade D Category LD3 fire detection and fire alarm system.	B1 1.3
There should be at least one smoke alarm on every storey of a dwelling house.	B1 1.12
An installation and commissioning certificate should be provided when a fire alarm system is installed.	B1 1.23
Occupants should be provided with information on the use of the equipment and on its maintenance (or guidance on suitable maintenance contractors).	B1 1.24
The rapid spread of smoke and fumes shall be limited.	B1.iv
The design and installation of fire detection and fire alarm systems in dwelling houses shall be in accordance with BS 5839-6:2004.	B1 1.10

Large houses

Large dwelling houses of 2 storeys (excluding basement storeys) should be provided with a BS 5839-6:2004 Grade B Category LD3fire detection and fire alarm system.	B1 1.6
Large dwelling houses of 3 or more storeys (excluding basement storeys) should be provided with a	B1 1.7

BS 5839-6:2004 Grade A Category LD3fire detection and fire alarm system in accordance with BS 5839-6:2004 Grade D Category LD3.	
Fire detectors used in large dwelling houses of 3 or more storeys (excluding basement storeys) should be sited in accordance with BS 5839-6:2004 Category L2.	B1 1.7

Material alteration

New habitable rooms that are the result of a material alteration and which are above ground floor level (or at ground floor level where no final exit has been provided) shall be equipped with: • a fire detection and fire alarm system; • smoke alarms (in accordance with paragraphs 1.10 to 1.18) in the circulation spaces.	B1 1.8

Sheltered housing

Fire detection equipment used in sheltered housing which are overseen by a warden or supervisor, shall be linked to a central monitoring point or alarm receiving centre.	B1 1.9

Smoke alarms

The provision of smoke alarms shall be based on an assessment of the risk to the occupants in the event of fire.	B1.ii
Smoke alarms should: • be mains-operated and conform to BS 5446-1:2000; • have a standby power supply such as a rechargeable (or non rechargeable) battery;	B1 1.4
• be positioned in the circulation spaces between sleeping spaces and places where fires are most like to start (e.g. in kitchens and living rooms).	B1 1.11
The design and installation of smoke alarms shall be in accordance with BS 5839-6:2004.	B1 1.10

There should be at least one smoke alarm on every storey of a dwelling house.	B1 1.12
Where the kitchen area is not separated from the stairway or circulation space by a door, there should be a compatible interlinked heat detector or heat alarm in the kitchen, in addition to whatever smoke alarms are needed in the circulation space(s).	B1 1.13
If more than one alarm is installed they should be linked so that the detection of smoke or heat by one unit operates the alarm signal in all of them.	B1 1.14

Smoke alarms/detectors should be sited so that:	B1 1.15a
• there is a smoke alarm in the circulation space within 7.5 m of the door to every habitable room;	
• they are ceiling-mounted and at least 300 mm from walls and light fittings;	B1 1.15b
• the sensor in ceiling-mounted devices is between 25 mm and 600 mm below the ceiling (25–150 mm in the case of heat detectors or heat alarms).	B1 1.15c

Smoke alarms should **not** be fixed:	B1 1.16
• over a stair or any other opening between floors;	
• next to or directly above heaters or air-conditioning outlets;	B1 1.17
• in bathrooms, showers, cooking areas or garages;	B1 1.17
• in any place where steam, condensation or fumes could give false alarms;	B1 1.17
• in places that get very hot (such as a boiler room);	B1 1.18
• in places that get very cold (such as an unheated porch);	B1 1.18
• to surfaces which are normally much warmer or colder than the rest of the space.	B1 1.18

Power supplies

The power supply for a smoke alarm system:	B1 1.19
• should be derived from the dwelling house's mains electricity supply;	

- should comprise a single independent circuit at the dwelling house's main distribution board (consumer unit or a single regularly used local lighting circuit.

It should be possible to isolate the power to the smoke alarms without isolating the lighting.	B1 1.19
The electrical installation should comply with Approved Document P (Electrical safety).	B1 1.20
Any cable suitable for domestic wiring may be used for the power supply and interconnection to smoke alarm systems (except in large buildings where the cable needs to be fire resistant (see BS 5839-6:2004),	B1 1.21
Conductors use to interconnect alarms (e.g. signalling) should be colour coded so as to distinguish them from those supplying mains power.	B1 1.21
Mains powered smoke alarms may be interconnected using radio-links, provided that this does not reduce the lifetime or duration of any standby power supply below 72 hours.	B1 1.21

Heat alarms

Heat alarms should: B1 1.4

- be mains-operated and conform to BS 5446-2:2003;
- have a standby power supply such as a rechargeable (or non rechargeable) battery;
- be designed and installed in accordance with BS 5839-6:2004. B1 1.10

Heat detectors and heat alarms should be sited so that the sensor in ceiling-mounted devices is between 25 mm and 150 mm below the ceiling. B1 1.15c

6.28 Means of escape

6.28.1 The requirement

Subject to Section 30(3) of the Fire Precautions Act 1971, if a building (or proposed building) exceeds two storeys in height and the floor of any upper storey

is more than 20ft above the surface of the street or ground on any side of the building and is:

- *let out as flats or tenement dwellings;*
- *used as an inn, hotel, boarding-house, hospital, nursing home, boarding school, children's home or similar institution; or is*
- *used as a restaurant, shop, store or warehouse and has on an upper floor sleeping accommodation for persons employed on the premises.*

then it must be equipped with adequate means of escape in case of fire, from each storey.

(Building Act 1984 Section 72)

The building shall be designed and constructed so that there are appropriate provisions for the early warning of fire, and appropriate means of escape in case of fire from the building to a place of safety outside the building capable of being safely and effectively used at all material times.

(Approved Document B1)

For a typical one- or two-storey dwelling, the requirement is limited to the provision of smoke alarms and to the provision of openable windows for emergency exit (see B1.i).

There shall be sufficient escape routes that are suitably located to enable persons to evacuate the building in the event of a fire.	B1
Safety routes shall be protected from the effects of fire.	B1
In an emergency, the occupants of any part of the building shall be able to escape without any external assistance.	B1 i
There should be alternative means of escape from '*most situations*'.	B1 v (a)
The design of means of escape shall be based on an assessment of the risk to the occupants in the event of fire.	B1.ii
If direct escape to a place of safety is impracticable, it should be possible to reach a place of relative safety such as a protected stairway within a reasonable travel distance.	B1 v (b)
Unprotected escape routes should not require people to have to travel excessive distances while exposed to the immediate danger of fire and smoke.	B1 vii
People should be able to turn their backs on a fire wherever it occurs and travel away from it to a final exit or protected escape route leading to a place of safety.	B1 vii

The following are **not** considered acceptable as a means of escape: B1

- lifts (except suitably designed and installed evacuation lifts);
- portable ladders;
- throw-out ladders;
- fold-down ladders and chutes;
- escalators (although it is recognized that they are likely to be used by people who are escaping).

These facilities may, however, be used as an additional feature.

Note: Mechanized walkways could be accepted and their capacity assessed on the basis of their use as a walking route, while in the static mode.

Risk assessment

A risk assessment shall be carried out and the design of means of escape shall take into account: B1.ii

- the nature of the building structure;
- the use of the building;
- the potential of fire spread through the building; and
- the proposed standard of fire safety management.

Dwelling houses

One or two storey dwelling houses shall be provided with: B1 2.1 (V1)

- an early warning system in the event of fire;
- suitable means for emergency egress from each storey via windows or doors.

Floors more than 7.5 m above ground shall be provided with an alternative escape route. B1 2.1 (V1)

Ground floor dwelling houses and flats

Except for kitchens, all habitable rooms on the ground floor should: B1 2.3 (V1)

- either open directly onto a hall leading to the entrance or other suitable exit; or B1 2.11 (V2)
- be provided with an emergency window (or door).

Any inner room that is a kitchen, laundry or utility room, dressing room, bathroom, WC or shower or situated not more than 4.5 m above ground level and whose only escape route is through another room, shall be provided with an emergency egress window.	B1 2.9 (V1) B1 2.5 (V2)

Note: The means of escape from a flat with a floor not more than 4.5 m above ground level is relatively simple to provide. Few provisions are specified in the 2006 edition of Part B beyond ensuring that means shall be provided for giving early warning in the event of fire and suitable means are provided for emergency egress from these storeys. With increasing height, however, the situation becomes more complex because emergency egress through upper windows will become increasingly hazardous.

Upper floors not more than 4.5 m above ground level

Except for kitchens, all habitable rooms in the upper storey(s) of a dwelling house that are served by only one stair, should be provided with: • a window (or external door); or have • direct access to a protected escape route.	B1 2.4 (V1) B1 2.12 (V2)

Dwelling houses with one floor more than 4.5 m above ground level

The dwelling house may either have a protected stairway which: • extends to the final exit (see Figure 6.222); or • gives access to at least two escape routes at ground level, each delivering to final exits and separated from each other by fire-resisting construction and fire doors (see Figure 6.222); or the top floor can be separated from the lower storeys by fire-resisting construction and be provided with its own alternative escape route leading to its own final exit (see Figure 6.222).	B1 2.6 (V1)

Dwelling houses with more than one floor more than 4.5 m above ground level

Dwelling houses with floors more than 4.5 m above ground level shall (in addition to meeting requirement B1 2.6) have: B12.7 (V1)

- an alternative escape route for each story or level that is more than 7.5 m above ground level; or
- a sprinkler system designed and installed in accordance with BS 9251:2005.

Note: The access to the alternative escape route should either be:

- via a protected stairway to an upper storey; or
- a landing within the protected stairway enclosure to an alternative escape route on the same storey; or
- the protected stairway that is at (or about) 7.5 m above ground level and which is separated from the lower storeys or levels by fire-resisting construction.

Buildings other than flats

Where the means of escape is based on phased evacuation, then a staged alarm system should be used. B1 1.25 (V2)

Where the means of escape is based on simultaneous evacuation, operation of a manual call point or fire detector should give an almost instantaneous warning from all the fire alarm sounders. B1 1.25 (V2)

Note: Automatic sprinkler systems can be used to operate a fire alarm system.

Sheltered housing

Whilst many of the provisions made in Part B 2007 for means of escape from flats are applicable to sheltered housing, the nature of the occupancy may necessitate some additional fire protection measures.

Institutional buildings

Special considerations may apply to some institutional buildings if residents need the assistance of staff to evacuate the building.

Basements

Owing to the risk that a single stairway may be blocked by smoke from a fire in the basement or ground storey:

basement storeys in a dwelling house that contain a habitable room shall be provided with either: • an external door or window suitable for egress from the basement: or • a protected stairway leading from the basement to a final exit.	B1 2.13 (V1) B1 2.6 (V2)

Galleries

All galleries shall be provided with an alternative exit or, where the gallery floor is not more than 4.5 m above ground level, an emergency egress window. If the gallery floor is not provided with an alternative exit or escape window: • the gallery should overlook at least 50% of the room below (see Figure 6.223); • the distance between the foot of the access stair to the gallery and the door to the room containing the gallery should not exceed 3 m; • the distance from the head of the access stair to any point on the gallery should not exceed 7.5 m; and • any cooking facilities within a room containing a gallery should either: – be enclosed with fire-resisting construction; or – be remote from the stair to the gallery.	B1 2.12 (V1) B1 2.8 (V2)

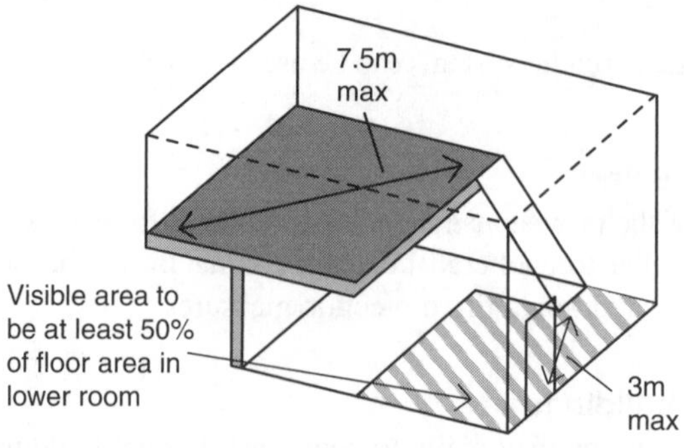

Figure 6.223 Gallery floors with no alternative exit

Balconies and flat roofs

Any flat roof that forms part of a means of escape should:	B1 2.10 (V1)
• be part of the same building from which escape is being made;	B1 2.7 (V2)
• lead to a storey exit or external escape route: and • provide 30 minutes fire resistance (see Appendix A, Table Al of Approved Document B for fire resistance figures for elements of structure etc.)	
Note: If a balcony or flat roof is provided for escape purposes, guarding may be required (see Approved Document K – Protection from falling, collision and impact).	B1 2.11 (V1)

Fire protected stairways

Fire protected stairways that, as far as reasonably possible: • exclude all flames, smoke and gases shall be designed to provide effective 'fire sterile' areas that lead to places of safety outside the building; • consist of fire-resistant material and fire resistant doors and have an appropriate form of smoke control system.	B1 1.viii

External escape stairs

Where an external escape stair is provided: • all doors giving access to the stair should be fire-resisting; • any part of the external envelope of the building within 1800 mm of (and 9 m vertically below) the flights and landings of an external escape stair should be of fire resisting construction (see Figure 6.223); • any part of the building (including doors) within 1800 mm of the escape route shall be protected by fire resisting construction.	B1 2.15 (a, b and c)
External escape stairs greater than 6 m in vertical extent shall be protected from the effects of adverse weather, conditions.	B1 2.15d

Note: Glazing in any fire-resisting construction should be fire-resisting and fixed shut.

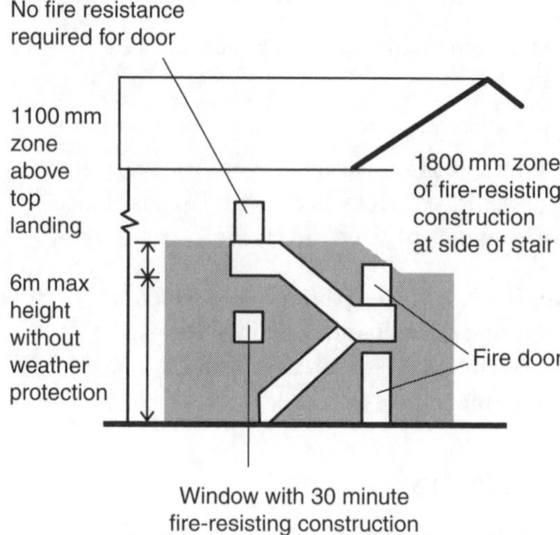

Figure 6.224 Fire resistance of areas adjacent to external stairs

Emergency egress windows and external doors

The window should be at least 450 mm high and 450 mm wide and have an unobstructed openable area of at least 0.33 m².	B1 2.8 (V1) B1 2.9 (V2)
The bottom of the openable area should be not more than 1100 m above the floor.	
The window or door should enable the person escaping to reach a place free from danger of fire (e.g. a courtyard or back garden which is at least as deep as the dwelling house is high – see Figure 6.225).	

Notes:

1. Approved Document K (Protection from falling, collision and impact) specifies a minimum guarding height of 800 mm, except in the case of a window in a roof where the bottom of the opening may be 600 mm above the floor.
2. Locks (with or without removable keys) and stays may be fitted to egress windows, provided that the stay is fitted with a child resistant elease catch.
3. Windows should be designed so that they remain in the open position without needing to be held open by the person making their escape.

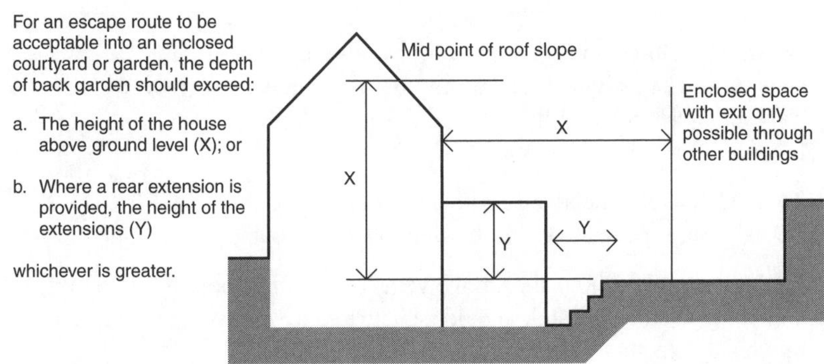

For an escape route to be acceptable into an enclosed courtyard or garden, the depth of back garden should exceed:
a. The height of the house above ground level (X); or
b. Where a rear extension is provided, the height of the extensions (Y)
whichever is greater.
Mid point of roof slope
X
X
Y
Y
Enclosed space with exit only possible through other buildings

Figure 6.225 Ground or basement storey exit into an enclosed space

Means of escape from the common parts of flats

The following requirements are primarily concerned with means of escape from the entrance doors of flats to a final exit.

Every flat should have access to alternative escape routes (but see Part B V2 paragraphs 2.20 to 2.22 for variations to this rule).	B1 2.20 (V2)
Escape routes in the common areas should comply with the limitations on travel distance shown in Table 6.91.	B1 2.23 (V2)

Table 6.91 Limitations on distance of travel in common areas of blocks of flats

Maximum distance of travel (m) from flat entrance door to common stair, or to stair lobby	
Escape in one direction only	Escape in more than one direction
7.5 m	30 m

Escape routes should be planned so that people do not have to pass through one stairway enclosure to reach another.	B1 2.23 (V2)
Common corridors should be protected corridors.	B1 2.24 (V2)
The wall between each flat and the corridor should be a compartment wall (see Section 8).	B1 2.24 (V2)
Means of ventilating common corridors/lobbies (i.e. to control smoke and so protect the common stairs) should be available.	B1 2.25 (V2)

In large buildings, the corridor or lobby adjoining the stair should be provided with a vent that is located as high as practicable and with its top edge at least as high as the top of the door to the stair. B1 2.26 (V2)

There should also be a vent, with a free area of at least $1.0\,m^2$ from the top storey of the stairway to the outside. B1 2.26 (V2)

In single stair buildings the smoke vents on the fire floor and at the head of the stair should be actuated by means of smoke detectors in the common access space providing access to the flats. B1 2.26 (V2)

In buildings with more than one stair the smoke vents may be actuated manually. B1 2.26 (V2)

Vents should either: B1 2.26 (V2)

- be located on an external wall with minimum free area of $1.5\,m^2$;
- discharge into a vertical smoke shaft that is closed at the base (and meets the criteria listed in Part b V2 Paragraph 2.26b).

Smoke control of common escape routes by mechanical ventilation is permitted provided that it meets the requirements of BS EN 12101-6:2005. B1 2.27 (V2)

Common corridors that connect two or more storey exits should be sub-divided by a self-closing fire door with, if necessary, an associated fire-resisting screen (see Figure 6.226). B1 2.28 (V2)

Note: Self-closing fire doors should be positioned so that smoke will not affect access to more than one stairway.

The dead-end portion of any common corridor should be separated from the rest of the corridor by a self-closing fire door (see Figure 6.226). B1 2.29 (V2)

Stores and other ancillary accommodation should not be located within, or entered from, any protected lobby or protected corridor forming part of the only common escape route from a flat on the same storey as that ancillary accommodation. B1 2.30 (V2)

If more than one escape route is available from a storey, or part of a building, one of those routes may be by way of a flat roof. B1 2.31 (V2)

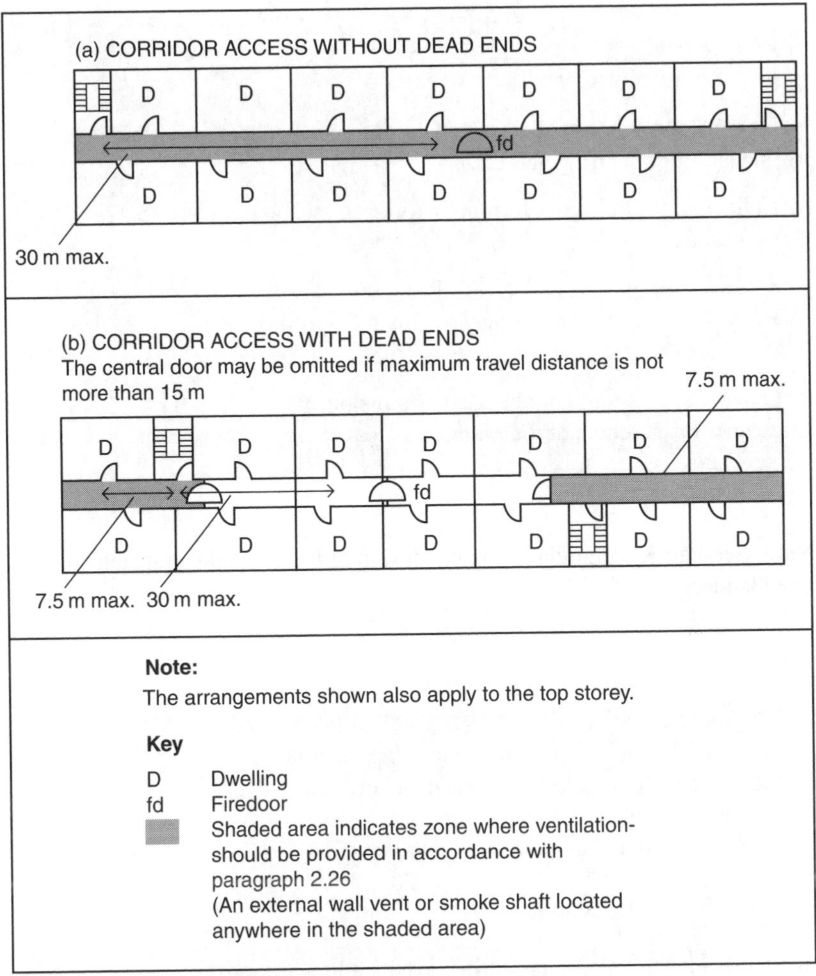

Figure 6.226 Flats served by more than one common stair

Stairs

The flights and landings of every escape stair should be constructed using materials of limited combustibility particularly if it is: B1 5.19 (V2)

- the only stair serving the building;
- within a basement storey;
- serves any storey having a floor level more than 18 m above ground or access level;

- external;
- a firefighting stair.

Single steps may only be used on escape routes where they are prominently marked. B1 5.21 (V2)

Helical and spiral stairs forming part of an escape route should be: B1 5.22a (V2)

- designed in accordance with BS 5395-2:1984;
- type B (Public stair) if they are intended to serve members of the public.

Fixed ladders should not be used as a means of escape for members of the public. B1 5.22b (V2)

Note: See Part K for guidance on the design of helical and spiral stairs and fixed ladders.

If a protected stairway projects beyond, or is recessed from, or is in an internal angle adjoining external wall of the building, then the distance between any unprotected area in the external enclosures to the building and any unprotected area in the enclosure to the stairway should be at least 1800 mm (see Figure 6.227). B1 5.24 (V2)

Where an external escape stair is provided in addition to another type of escape route (see paragraph 4.44) it should meet the following provisions: B1 5.25 (V2)

- all doors giving access to the stair should be fire-resisting and self-closing;
- any part of the external envelope of the building within 1800 mm of (and 9 m vertically below) the flights and landings of an external escape stair should be fire resisting;
- there is protection by fire-resisting construction for any part of the building within 1800 mm of the escape route from the stair to a place of safety;
- glazing should also be fire resistant and fixed shut.

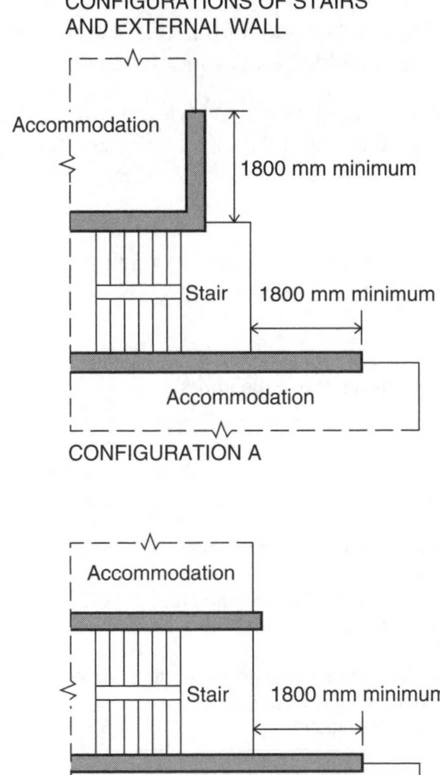

Figure 6.227 External protection to protected stairways

Width of escape stairs

The width of escape stairs should: B1 4.15 (V2)

- not be less than the width of any exit(s);
- not be less than the minimum widths given in Table 6.92;
- not exceed 1400 mm if their vertical extent is more than 30 m, **unless** it is provided with a central handrail;
- not reduce in width at any point on the way to a final exit.

In public buildings, if the width of the stair is more than 1800 mm, the stair should have a central handrail.	B1 4.16 (V2)
Every escape stair should be wide enough to accommodate the number of persons needing to use it in an emergency.	B1 4.18 (V2)

Note: For further guidance and worked examples see Appendix C to Part B and Sections 4.18 (V2) to 4.25 (V2).

Table 6.92 Minimum widths of escape stairs

	Stair situation	Maximum number of people served	Minimum stair width
1a	In an Institutional building (unless the stair is only used by staff)	150	1000 mm
1b	In an Assembly building and serving an area used for assembly purposes (unless the area is less than 100 m^2)	220	1100 mm
1c	In any other building and serving an area with an occupancy of more than 50 people	Over 2200	1000–1800 mm*
2	Any stair not described above	50	800 mm

*Depending on whether the stairs are used for simultaneous evacuation or phased evacuation (see Table 7 of Part B1 (V2).

Protection of escape stairs

Escape stairs need to have a satisfactory standard of fire protection.	B1 4.31 (V2)
Internal escape stairs should be a protected stairway within a fire-resisting enclosure.	B1 4.32 (V2)
Except for bars and restaurants, stairs may be open provided that: • it does not connect more than two storeys and reaches the ground storey not more than 3 m from the final exit; and • the storey is also served by a protected stairway; or • it is a single stair in a small premises with the floor area in any storey not exceeding 90 m^2.	B1 4.33 (V2)

Basement stairs

Because of their situation, basement stairways are more likely to be filled with smoke and heat than stairs in ground and upper storeys. Special measures are therefore needed in order to prevent a basement fire endangering upper storeys.

If an escape stair forms part of the only escape route from an upper storey of a building it should **not** be continued down to serve any basement storey (i.e. the basement should be served by a separate stair). B1 4.42 (V2)

If there is more than one escape stair from an upper storey of a building only one of the stairs serving the upper storeys of the building need be terminated at ground level. B1 4.43 (V2)

External escape stairs

An external escape stair may be used, provided that: B1 4.44 (V2)

- there is at least one internal escape stair from every part of each storey (excluding plant areas);
- in the case of an Assembly and Recreation building, the route is not intended for use by members of the public; or
- in the case of an Institutional building, the route serves only office or residential staff accommodation.

Note: External escape stairs should meet the following provisions:

- all doors giving access to the stair should be fire-resisting and self-closing;
- any part of the external envelope of the building within 1800 mm of (and 9 m vertically below) the flights and landings of an external escape stair should be fire resisting;
- there is protection by fire-resisting construction for any part of the building within 1800 mm of the escape route from the stair to a place of safety;
- glazing should also be fire resistant and fixed shut.

Common stairs

Normally a single common stair can be acceptable. In some cases, however, there should be access to more than one common stair for escape purposes.

Any stair used as a firefighting stair should be at least 1100 mm wide (see Part B V2, Appendix C for measurement of width). B1 2.33 (V2)

All common stairs should be situated within a fire-resisting enclosure (i.e. it should be a protected stairway), to reduce the risk of smoke and heat making use of the stair hazardous. B1 2.36 (V2)

Protected stairways should discharge: B1 2.38 (V2)

- directly to a final exit; or
- via a protected exit passageway to a final exit.

Where two protected stairways (or exit passageways leading to different final exits) are adjacent, they should be separated by an imperforate enclosure. B1 2.39 (V2)

A protected stairway needs to be relatively free of potential sources of fire. B1 2.40 (V2)

If an escape stair forms part of the only escape route from an upper storey of a large building it should not be continued down to serve any basement storey. B1 2.44 (V2)

The basement should be served by a separate stair.

If there is more than one escape stair from an upper storey of a building, only one of the stairs serving the upper storeys of the building need be terminated at ground level. B1 2.45 (V2)

Note: Other stairs may connect with the basement storey(s) if there is a protected lobby or a protected corridor between the stair(s) and accommodation at each basement level.

Where a common stair forms part of the only escape route from a flat, it should **not** also serve any covered car park, boiler room, fuel storage space or other ancillary accommodation of similar fire risk. B1 2.46 (V2)

Common stairs which do not form part of the only escape route from a flat may also serve ancillary accommodation if they are separated from the ancillary accommodation by a protected lobby or a protected corridor. B1 2.47 (V2)

If the stair serves an enclosed (non open-sided) car park, or place of special fire hazard, the lobby or corridor B1 2.47 (V2) should have not less than 0.4 m^2 permanent ventilation or be protected by a mechanical smoke control system.

In single stair buildings, meters located within the stairway should be enclosed within a secure cupboard which is separated from the escape route with fire-resisting construction. B1 2.40 (V2)

Gas service and installation pipes or associated meters should not be incorporated within a protected stairway unless the gas installation is in accordance with the requirements for installation and connection set out in the Pipelines Safety Regulations 1996, SI 1996 No 825 and the Gas Safety (installation and use) Regulations 1998 SI 1998 No 2451. B1 2.42 (V2)

External escape stairs

If the building (or part of the building) is served by a single access stair, that stair may be external if it serves a floor not more than 6 m above the ground level. B1 2.48 (V2)

If there is more than one escape route available from a storey (or part of a building) an external escape stair may be used, provided that there is at least one internal escape stair from every part of each storey (excluding plant areas) and the external stair(s). B1 2.49 (V2)

Flats in mixed use buildings

The stairs of buildings which are **no more than** three storeys above the ground storey, may serve both flats and other occupancies, **provided** that the stairs are separated from each occupancy by protected lobbies at all levels. B1 2.50 (V2)

The stairs of buildings which are more than three storeys above the ground storey, may serve both flats and other occupancies, **provided** that: B1 2.51 (V2)

- the flat is ancillary to the main use of the building and is provided with an independent alternative escape route;

- the stair is separated from any other occupancies on the lower storeys by protected lobbies (at those storey levels);
- any automatic fire detection and alarm system with which the main part of the building is fitted also covers the flat;
- any security measures should not prevent escape at all material times.

Live/work units

If a flat is used as a workplace, the following **additional** fire precautions will be necessary: B1 2.52 (V2)

- the maximum travel distance to the flat entrance door or an alternative means of escape (not a window) from any part of the working area should not exceed 18 m; and
- all windowless accommodation should have escape lighting (in accordance with BS 526,6-1:2005) which illuminates the route.

Design for horizontal escape buildings other than flats

Exits in the central core of a building should be remote from one another. B1 3.11 (V2)

An escape route should not be within 4.5 m of an opening between floors (i.e. such as an escalator) unless: B1 3.12 (V2)

- the direction of travel is away from the opening; or
- there is an alternative escape route which does not pass within 4.5 m of the open connection.

Any storey which has more than one escape stair should be planned so that it is **not** necessary to pass through one stairway to reach another. B1 3.13 (V2)

Storeys containing areas for the consumption of food and/or drink (and which are in addition to the main use of the building) shall have not less than two escape routes from each such area which lead directly to a storey exit without entering any kitchen or similar area of high fire hazard. B1 3.15 (V2)

The means of escape from storeys that are divided into separate occupancies: B1 3.16 (V2)

- shall ensure that each occupancy does not have to pass through other occupancy; and (if the means of escape includes a common corridor or circulation space)
- should either be a protected corridor, or be equipped with an automatic fire detection and alarm system throughout the storey.

Except doorways, all escape routes should have a clear headroom of not less than 2 m. B1 3.17 (V2)

Although the width of escape routes and exits depends on the number of persons needing to use them, they should not be less than that shown in Table 6.93. B1 3.18 (V2)

Table 6.93 Widths of escape routes and exits

Maximum number of persons	Minimum width
60	750 mm
110	850 mm
220	1050 mm
More than 220	5 per person

Protected corridors shall be installed for: B1 3.24 (V2)

- corridors serving bedrooms;
- dead-end corridors (excluding recesses not exceeding 2 m deep);
- any corridor that is common to more than one different occupancies.

If, instead of a protected corridor, the means of escape is enclosed by partitions, those partitions shall: B1 3.25 (V2)

- be carried up to the main soffit of the floor above or to a suspended ceiling:
- be fitted with doors into all openings into rooms off that corridor.

Every corridor more than 12 m long and which connects two or more storey exits, should be sub-divided by self-closing fire doors positioned approximately mid-way between the two storey exits. B1 3.26 (V2)

Unless escape stairways and corridors are protected by a pressurization system complying with BS EN 12101-6:2005, every dead-end corridor exceeding 4.5 m in length should be separated by self-closing fire doors (together with any necessary associated screens) from any part of the corridor which: B1 3.27 (V2)

- provides two directions of escape;
- continues past one storey exit to another.

If an external escape route is beside an external wall of the building, that part of the external wall that is within 1800 mm of the escape route should be fire resistant up to a height of 1100 mm above the paving level of the route. B1 3.30 (V2)

An escape over flat roofs is permissible if:

- the route does not serve an Institutional building:
- is not part of a building intended for use by members of the public. B1 3.31 (V2)

In small premises:

- floor areas should be generally undivided (except for kitchens, ancillary offices and stores) to ensure that exits are clearly visible from all parts of the floor areas; B1 3.34 (V2)
- store rooms should be enclosed with fire-resisting construction; B1 3.35 (V2)
- clear glazed areas should be provided in any partitioning separating a kitchen or ancillary office from the open floor area to enable any person within the kitchen or office to obtain early visual warning of an outbreak of fire. B1 3.36 (V2)

Escape routes

The number of escape routes and exits to be provided will depend on the number of occupants in the room, tier or storey in question and the travel distance to the nearest exit (see Table 2 of B1 (V2)). B1 3.2 (V2)

In mixed-use buildings, separate means of escape should be provided from any storeys (or parts of storeys) used for Residential or Assembly and Recreation purposes. B1 3.4 (V2)

There should be alternative escape routes from all parts of the building unless the travel distance is within set limits (see Table 6.92 of B1 (V2)).	B1 3.5 (V2)
Access control measures incorporated into the design of a building should not adversely affect fire safety provisions.	B1 3.7 (V2)
The minimum number of escape routes and exits from a room or storey shall be in accordance with the number of occupants (see Table 6.93 above).	B1 3.2 (V2)

Note: Further guidance concerning the number of occupants and exits is contained in Appendix C of Part B.

Inner rooms

If the only escape route from an inner room is through another room then: — B1 3.10 (V2)

- the occupant capacity of the inner room should not exceed 60;
- the inner room should not be a bedroom;
- the inner room should be entered directly off the access room (but not via a corridor);
- the escape route from the inner room should not pass through more than one access room;
- the travel distance from any point in the inner room to the exit(s) from the access room should not exceed the distances given in Table 6.92 of B1 (V2));
- the access room should not be a place of (i.e. potentially with) a special fire hazard;
- the access room should be in the control of the same occupier; and
- one of the following arrangements should be made;
  - the enclosures (walls or partitions) of the inner room should be stopped at least 500 mm below the ceiling; or
  - a vision panel not less $0.1\,m^2$ should be located in the door or walls of the inner room;
  - the access room should be fitted with an automatic fire detection and alarm system.

Residential care homes

Note: Generally speaking, in care homes for the elderly it is reasonable to assume that at least a proportion of the residents will need some assistance to evacuate.

Buildings should be designed for Progressive Horizontal Evacuation (PHE). — B1 3.39 (V2)

Areas used for the care of residents shall be subdivided into protected areas separated by compartment walls and compartment floors. — B1 3.41 (V2)

Note: This is to allow horizontal escape into an adjoining protected area.

Each storey used for the care of residents should be: — B1 3.42 (V2)

- divided into at least three protected areas by compartment wall; and
- all floors should be compartment floors.

Protected areas should be provided with at least two exits to adjoining, but separate, protected areas. — B1 3.43 (V2)

The maximum travel distances within a protected area to these exits should:

- not exceed those given in Table 6.92; — B1 3.43 (V2)
- not be more than 64 m to a storey exit or a final exit. — B1 3.43 (V2)

A fire in a protected area should not prevent the occupants of any other area from reaching a final exit. — B1 3.44 (V2)

Escape routes should not pass through ancillary accommodation (also see section 3.50 of Part B). — B1 3.44 (V2)

The number of residents beds in protected areas should not exceed 10. — B1 3.45 (V2)

A fire detection and alarm system should be provided to an Ll standard in accordance with BS 5839-1:2002. — B1 3.47 (V2)

Bedrooms should be enclosed in fire-resisting construction with fire resisting doors and every corridor serving bedrooms should be a protected corridor. — B1 3.48 (V2)

Bedrooms should not contain more than one bed (this includes a double bed). — B1 3.49 (V2)

Design for vertical escape

An important aspect of means of escape in multi-storey buildings is the availability of a sufficient number of adequately sized and protected escape stairs.

The number of escape stairs needed in a building (or part of a building) will be determined by: B1 4.2 (V2)

- the constraints imposed by the design of horizontal escape routes;
- whether independent stairs are required in mixed occupancy buildings;
- whether a single stair is acceptable; and
- the width for escape and the possibility that a stair may have to be discounted because of fire or smoke.

Provided that independent escape routes are not necessary from areas in different purpose groups, single escape stairs may be used from: B1 4.6 (V2)

- small premises (other than bars or restaurants);
- office buildings comprising not more than five storeys above the ground storey;
- factories comprising not more than one storey above the ground storey if the building is of normal risk (two storeys if the building is of low risk); or
- process plant buildings with an occupant capacity of not more than 10 people.

Note: In mixed use buildings (i.e. where a building contains storeys (or parts of storeys) in different purpose groups) it is important to consider the effect of one risk on another – for example, a fire in a shop, or unattended office – could have serious consequences on a residential use in the same building. It is, therefore, important to consider whether completely separate routes of escape should be provided from each different use within the building or whether other effective means to protect common escape routes can be provided.

General requirements

All escape routes should have a clear headroom of not less than 2 m with no projection below this height (except for door frames). B1 5.26 (V2)

The floors of all escape routes (including the treads of steps and surfaces of ramps and landings) should be chosen to minimize their slipperiness when wet.	B1 5.27 (V2)
Any sloping floor or tier should be constructed with a pitch of not more than 35° to the horizontal.	B1 5.28 (V2)
Where a ramp forms part of an escape route it shall meet the requirements of Part M Access to and Use of buildings (see also Part K).	B1 5.28 (V2)
Final exits should:	
• not be less in width than the minimum width required for the escape route(s) they serve;	B1 5.30 (V2)
• be sited to ensure rapid dispersal of persons from the vicinity of the building;	B1 5.31 (V2)
• not present an obstacle to wheelchair users and other people with disabilities;	B1 5.32 (V2)
• be immediately apparent to persons who may need to use them;	B1 5.33 (V2)
• be sited so that they are clear of any risk from fire or smoke in a basement, or from openings to transformer chambers.	B1 5.34 (V2)
If an escape route is over a flat roof:	B1 5.35 (V2)
• the roof should be part of the same building from which escape is being made;	
• the route across the roof should lead to a storey exit or external escape route;	
• the part of the roof forming the escape route and its supporting structure, together with any opening within 3 m of the escape route, should be fire-resisting;	
• the route should be adequately defined and guarded by walls and/or protective barriers (which meet the provisions in Approved Document K).	
All escape routes should have adequate artificial lighting.	B1 5.36 (V2)
Routes and areas listed in Table 9 of B1 (V2) should also have escape lighting which illuminates the route if the main supply fails.	B1 5.35 (V2)
Lighting to escape stairs should be on a separate circuit from that supplying any other part of the escape route.	B1 5.36 (V2)

The installation of an escape lighting system shall be in accordance with BS 5266-1:2005.

Escape routes(other than those in ordinary use and/or within a flat) should be marked by emergency exit sign(s) in accordance with BS 5499-1:2002. B1 5.37 (V2)

Note: Suitable signs should also be provided for refuges (see paragraph 4.10).

Where it is critical for electrical circuits to be able to continue to function during a fire, protected circuits (meeting the requirements of BS EN 50200:2006) are needed. B1 5.38 (V2)

Access lobbies and corridors

Escape stairs shall have a protected lobby or protected corridor at all levels (except the top storey, all basement levels and when the stair is a firefighting stair) if: B1 4.34 (V2)

- the stair is the only one serving a building which has more than one storey above or below the ground storey;
- where the stair serves any storey at a height greater than 1 8 m; or
- where the building is designed for phased evacuation.

Protected lobbies (with not more than 0.4 m^2 permanent ventilation) should be provided between an escape stairway and a place of special fire hazard. B1 4.35 (V2)

Protected stairways should discharge: B1 4.36 (V2)

- directly to a final exit; or
- by way of a protected exit passageway to a final exit.

Note: Any such protected exit passageway should have the same standard of fire resistance and lobby protection as the stairway it serves.

If two protected stairways are adjacent, they (and any protected exit passageways linking them to final exits) shall be separated by an imperforate enclosure. B1 4.37 (V2)

Protected stairways shall be free of potential sources of fire. B1 4.38 (V2)

Cavity barriers

Cavity barriers should be provided above the enclosures to a protected stairway in a dwelling house with a floor more than 4.5 m above ground level (see Figure 6.228).	B1 2.14 (V1)

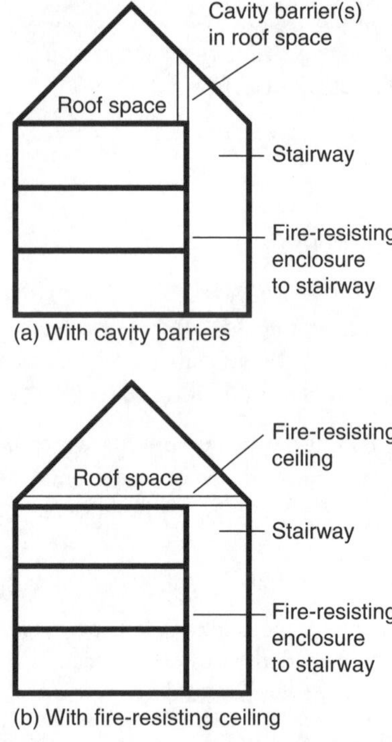

Figure 6.228 Alternative cavity barrier arrangements in roof space over protected stairway in a house with a floor more than 4.5 m above ground.

Conversion to flats

If the existing building has timber floors and these are to be retained, the requirements for fire resistance may be difficult to meet. In these cases, provided that the means of escape conforms to Part B Section 3 and are adequately protected:

doors on escape routes (both within and from the building) should be readily openable;	B1 5.10 (V2)
doors on escape routes (whether or not the doors are fire doors), should either not be fitted with lock, latch or	B1 5.11 (V2)

bolt fastenings, or they should only be fitted with simple fastenings that can be readily operated from the side approached by people making an escape, without the use of a key and without having to manipulate more than one mechanism.

Where a secure door is operated by a code, combination, swipe or proximity card, biometric data or similar means, it should also be capable of being overridden from the side approached by people making their escape. B1 5.11 (V2)

Electrically powered locks should return to the unlocked position: B1 5.11 (V2)

- on operation of the fire alarm system;
- on loss of power or system error;
- on activation of a manual door release unit.

In assembly places, shops and commercial buildings, doors on escape routes from rooms with an occupant capacity of more than 60 should either not be fitted with lock, latch or bolt fastenings (or be fitted with panic fastenings in accordance with BS EN 1125:1997). B1 5.12 (V2)

 see also Appendix B for guidance about door closing and 'hold open' devices for fire doors.

The door of any doorway or exit should be hung to open in the direction of escape. B1 5.14 (V2)

All doors on escape routes should be hung to open not less than 90 degrees. B1 5.15 (V2)

A door that opens towards a corridor or a stairway should be sufficiently recessed to prevent its swing from encroaching on the effective width of the stairway or corridor. B1 5.16 (V2)

Vision panels shall be provided where doors on escape routes sub-divide corridors, or where any doors are hung to swing both ways. B1 5.17 (V2)

 see also Parts M and N.

Revolving doors, automatic doors and turnstiles should not be placed across escape routes. B1 5.18 (V2)

Fire doors

Roller shutters across a means of escape should only be released by a heat sensor, such as a fusible link or electric heat detector, in the immediate vicinity of the door.	App B 6 (V2)

Protection of escape routes

Generally, a 30-minute standard is sufficient for the protection of means of escape. (Details of fire resistance test criteria and standards of performance are contained in Appendix A to Part B).	B1 5.2 (V2)
Walls, partitions and other enclosures that need to be fire-resisting (including roofs that form part of a means of escape), should have the appropriate performance given in Tables Al and A2 of Appendix A to Part B.	B1 5.3 (V2)
All doors that need to be fire-resisting should meet the requirements given in Table B1 of Appendix B to Part B.	B1 5.6 (V2)
The use of glazed elements in fire-resisting enclosures and/or doors depends on whether that element forms part of a protected shaft (see Appendix A, Table A4 and also Part N).	B1 5.7 (V2)

Raised storage areas

Raised free-standing floors in single storey industrial and storage buildings which are effectively galleries or a floor forming an additional storey, in certain circumstances, might not be able to meet the requirements of Appendix A, Table Al. For the purposes of fire safety, they are, however, deemed acceptable provided that:

B2 7.7 (V2)

- the structure has only one tier and is used for storage purposes only;
- the number of persons likely to be on the floor at any one time is low and does not include members of the public;
- the floor is not more than 10 m wide or long and does not exceed one half of the floor area of the space in which it is situated;
- the floor is open above and below to the room or space in which it is situated; and
- the means of escape from the floor meets the relevant requirements of Part B (particularly Sections 3).

Table 6.94 Limitations on the use of uninsulated glazed elements on escape routes

Position of glazed element	Maximum total glazed area in parts of the building with access to			
	A single stairway		More than one stairway	
	Walls	Door leaf	Walls	Door leaf
Single family dwelling houses				
1 (a) Within the enclosures of (i) protected stairway	Fixed fanlights only	Unlimited	Fixed fanlights only	Unlimited
(ii) existing stair	Unlimited	Unlimited	Unlimited	Unlimited
(b) Within fire resisting separation	100 mm from floor	100 mm from floor	100 mm from floor	100 mm from floor
(c) Existing window between an attached/ integral garage and the house	Unlimited	N/A	Unlimited	N/A
Flats and maisonettes				
2 Within the enclosures of a protected entrance hall or protected landing	Fixed fanlights only	1100 mm from floor	Fixed fanlights only	1100 mm from floor

6.29 Bathrooms

6.29.1 The requirement

All dwellings (whether they are a house, flat or maisonette) should have at least one bathroom with a fixed bath or shower, and the bath or shower should be equipped with hot **and** cold water. This ruling applies to all plans for:

- new houses;
- new buildings, part of which are going to be used as a dwelling;
- existing buildings that are going to be converted, or partially converted into dwellings.

(Building Act 1984 Section 27)

Fire safety

- *There shall be sufficient escape routes that are suitably located to enable persons to evacuate the building in the event of a fire.*
- *Safety routes shall be protected from the effects of fire.*

(Approved Document B1)

Ventilation

There shall be adequate means of ventilation provided for people in the building.

(Approved Document F)

Sanitary conveniences

***All** dwellings (houses, flats or maisonettes) should have at least one closet and one washbasin which should:*

- *be separated by a door from any space used for food preparation or where washing-up is done in washbasins;*
- *ideally, be located in the room containing the closet;*
- *have smooth, non-absorbent surfaces and be capable of being easily cleaned;*
- *be capable of being flushed effectively;*
- *only be connected to a flush pipe or discharge pipe;*
- *washbasins should have a supply of hot and cold water.*

(Approved Document G1)

All dwellings (house, flat or maisonette) should have at least one bathroom with a fixed bath or shower.

(Approved Document G3)

6.29.2 Meeting the requirement

All dwellings (houses, flats or maisonettes) should have at least one bathroom with a fixed bath or shower and the bath or shower should:

• have a supply of hot and cold water;	G2
• discharge through a grating, a trap and branch discharge pipe to a discharge stack or (if on a ground floor) discharge into a gully or directly to a foul drain;	G2 G2 H1
• be connected to a macerator and pump (of an approved type) if there is no suitable water supply or means of disposing foul water.	G2

Requirement G2 only applies to dwellings.

Bathrooms shall have either a fixed bath or shower bath that is provided with hot and cold water and connected to a foul water drainage system.

Fire safety

Bathrooms, WCs or showers that are situated less than 4.5 m above ground level whose only escape route is through another room shall be provided with an emergency egress window.	B1 2.9 (V1) B1 2.5 (V2)

Smoke alarms should not be fixed in bathrooms, showers, cooking areas or garages.	B1 1.17 (V1) B1 1.16 (V2)
A visual and audible fire alarm signal should be provided in buildings where it is anticipated that one or more persons with impaired hearing may be in relative isolation (e.g. hotel bedrooms and sanitary accommodation).	B1 1.34 (V2)
Although the use of protected shafts is normally restricted to stairs, lifts, escalators, chutes, ducts and pipes, sanitary accommodation and washrooms may also be included in them.	B2 8.36 (V2)

Ventilation

Extract ventilation concerns the removal of air directly from a space or spaces to outside. Extract ventilation may be by natural means (e.g. by passive stack ventilation, PSV) or by mechanical means (e.g. by an extract fan or central system).

All bathrooms shall be provided with extract ventilation to the outside which is capable of operating either: • intermittently at a minimum extract rate of 15 l/s or • continuously with a minimum extract rate of 8 l/s.	F 1.5
Common outlet terminals and/or branched ducts shall **not** be used for wet rooms such as a bathroom.	F App D
PSV devices shall have a minimum internal duct diameter of 100 mm and a minimum cross-sectional area of 8000 mm^2.	F
In bathrooms (without a WC) where there was no previous ventilation opening, or where the size of the original ventilation opening is not known, replacement window(s) shall have an equivalent area greater than 2500 mm^2.	F 3.6
Wall/ceiling mounted centrifugal fans (which are fitted with a 100 mm diameter flexible duct or rectangular duct and which are designed to achieve 15 l/s for kitchens) should not be ducted further than 6 metres and should have no more than one 90° bend.	F App E

Washbasins

Washbasins should have a supply of hot and cold water.	G1
Washbasins should discharge through a grating, a trap and a branch discharge pipe to a discharge stack or (if it is a ground floor location) into a gully or directly into a drain.	G1

6.30 Loft conversions

6.30.1 The requirements

Fire safety

- *There shall be an early warning fire alarm system for persons in the building.*
- *There shall be sufficient escape routes that are suitably located to enable persons to evacuate the building in the event of a fire.*
- *Safety routes shall be protected from the effects of fire.*
- *In an emergency, the occupants of any part of the building shall be able to escape without any external assistance.*

(Approved Document B1)

- *The spread of flame over the internal linings of the building shall be restricted.*
- *The heat released from the internal linings shall be restricted.*

(Approved Document B2)

Ventilation

There shall be adequate means of ventilation provided for people in the building.
(Approved Document F)

Stairs, ladders and ramps

All stairs, steps and ladders shall provide reasonable safety between levels in a building.

(Approved Document K1)

Protection from falling

Pedestrian guarding should be provided for any part of a floor (including the edge below an opening window) gallery, balcony, roof (including rooflight and other openings), any other place to which people have access and any light well, basement area or similar sunken area next to a building.

(Approved Document K2)

Requirement K2 (a) applies only to stairs and ramps that form part of the building.

6.30.2 Meeting the requirements

Fire safety

Where the conversion of an existing roof space (such as a loft conversion to a two-storey house) means that a new storey is going to be added, then the stairway will need to be protected with fire-resisting doors and partitions.	B1 2.20b
The floor(s), both old and new, shall have the full 30 minute standard of fire resistance shown in Part B Appendix A, Table Al unless: • only one storey is being added; • the new storey contains no more than 2 habitable rooms; and • the total area of the new storey is less than $50\,m^2$;	B3 4.7 (V1)
In those places where the floor only separates rooms (and not circulation spaces) a modified 30 minute standard of fire resistance may be applied.	B3 4.7 (V1)
New habitable rooms that are the result of a material alteration and which are above ground floor level (or at ground floor level where no final exit has been provided) shall be equipped with: • a fire detection and fire alarm system; • smoke alarms in accordance with BS 5839-6.	B1 2.20a (V1) B1 1.8 (V1)

Loft conversions

Fans and/or ducting placed in or passing through an unheated void or loft space should be insulated to reduce the possibility of condensation forming.	F App E
The inner radius of any bend should be greater or equal to the diameter of the ducting being used (see Figure 6.229).	F App E
Vertical duct rises may need to be fitted with a condensation trap in order to prevent the backflow of any moisture.	F App E
The circular profile of a flexible duct should be maintained throughout the full length of the duct run (see Figure 6.229).	F App E
If a back-draught device is used it may be incorporated into the fan itself.	F App E
Flexible ducting should be installed without any peaks or troughs (see Figure 6.230).	F App E

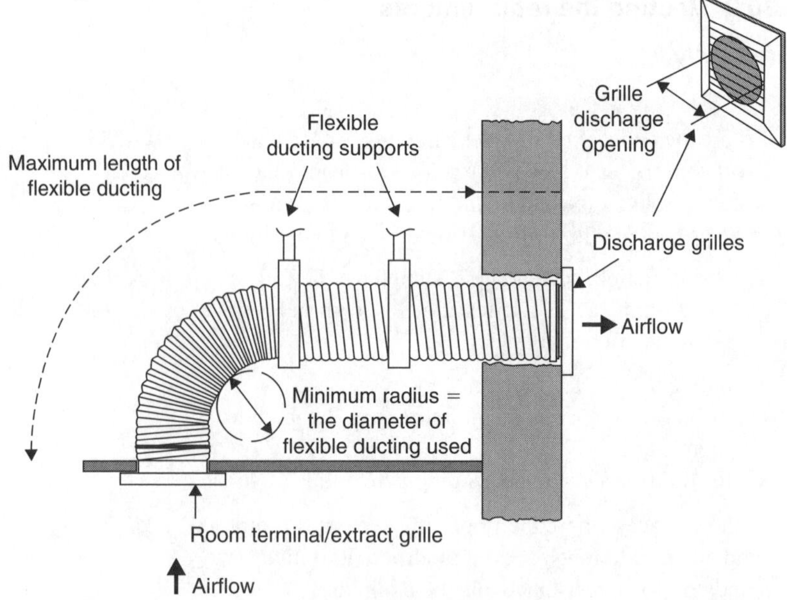

Figure 6.229 Correct installation of ducting

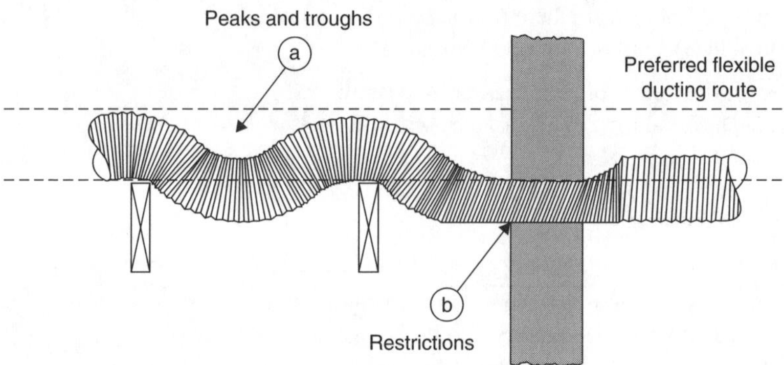

Figure 6.230 Incorrect installation of ducting

Stairs, ladders and ramps

The rise of a stair shall be between 155 mm and 220 mm with any going between 245 mm and 260 mm and a maximum pitch of 42°. The normal relationship between the dimensions of the rise and going is that twice the rise plus the going (2R + G) should be between 550 mm and 700 mm.	K1 (1.1–1.4)

Stairs with open risers that are likely to be used by children under five years should be constructed so that a 100 mm diameter sphere cannot pass through the open risers.	K1 (1.9)
Stairs that have more than 36 risers in consecutive flights should make at least one change of direction, between flights, of at least 30°.	K1 (1.14)
If a stair has straight and tapered treads, then the going of the tapered treads should not be less than the going of the straight tread.	K1 (1.20)
The going of tapered treads should measure at least 50 mm at the narrow end.	K1 (1.18)
The going should be uniform for consecutive tapered treads.	K1 (1.19) K1 (1.22–1.24)
Stairs should have a handrail on both sides if they are wider than 1 m and on at least one side if they are less than 1 m wide.	K1 (1.27)
Handrail heights should be between 900 mm and 1000 mm measured to the top of the handrail from the pitch line or floor.	K1 (1.27)
Spiral and helical stairs should be designed in accordance with BS 5395.	K1 (1.21)

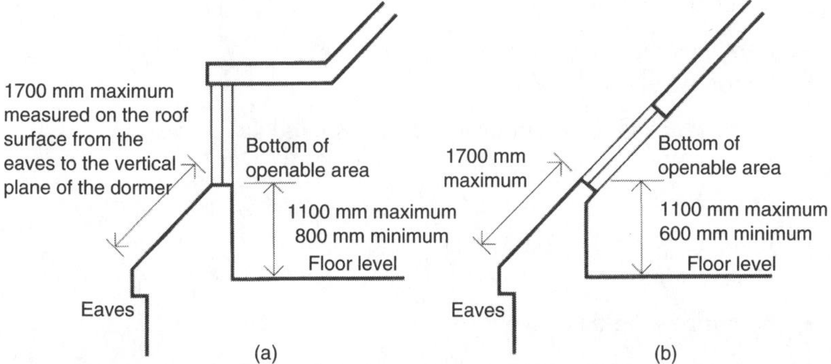

Note 1: The window or rooflight should have a clear opening which complies with paragraph 2.11a of Approved Document B and quoted on page 523.

Note 2: It is not considered necessary for the window in (b) to be provided with safety glazing.

Figure 6.231 Position of dormer window or rooflight that is suitable for emergency purposes from a loft conversion of a two-storey dwelling house. (a) Dormer window (the window may be in the end wall of the house, instead of the roof as shown). (b) Rooflight or roof window

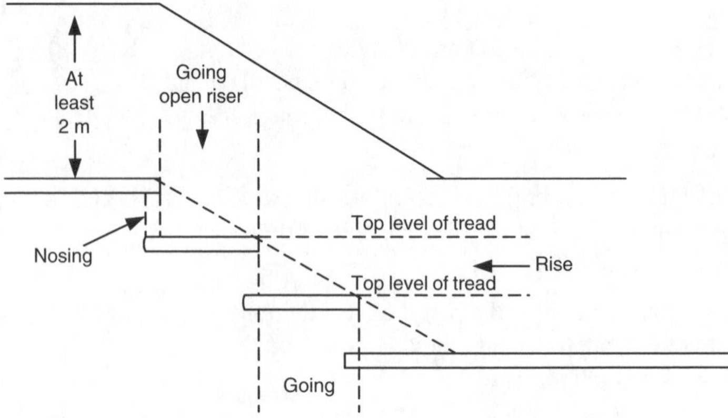

Figure 6.232 Rise and going plus headroom

Steps

• Steps should have level treads.	K1 (1.8)
• Steps may have open risers, but treads should then overlap each other by at least 16 mm.	K1 (1.8)
Steps should be uniform with parallel nosings, the stair should have handrails on both sides and the treads should have slip-resistant surfaces.	
• The headroom on the access between levels should be no less than 2 m.	K1 (1.10)
• Landings should be provided at the top and bottom of every flight.	K1 (1.15)
• The width and length of every landing should be the same (or greater than) the smallest width of the flight.	K1 (1.15)
• Landings should be clear of any permanent obstruction.	K1 (1.16)
• Landings should be level.	K1 (1.17)
• Any door (entrance, cupboard or duct) that swings across a landing at the top or bottom of a flight of stairs must leave a clear space of at least 400 mm across the full width of the flight.	K1 (1.16)
• Flights and landings should be guarded at the sides when there is a drop of more than 600 mm.	K1 (1.28–1.29)

For stairs that are likely to be used by children under five years the construction of the guarding shall be such that a 100 mm sphere cannot pass through any openings in the guarding and children will not easily climb the guarding.

For loft conversions, a fixed ladder should have fixed handrails on both sides.	K1 (1.25)

Whilst there are no recommendations for minimum stair widths, designers should bear in mind the requirements of Approved Documents B (means of escape) and M (access for disabled people).

Protection from falling

All stairs, landings, ramps and edges of internal floors shall have a wall, parapet, balustrade or similar guard at least 900 mm high.	K3 (3.2)
All guarding should be capable of resisting at least the horizontal force given in BS 6399: Part 1: 1996.	K3 (3.2)
If glazing is used as (or part of) the pedestrian guarding, see Approved Document N: Glazing – safety in relation to impact, opening and cleaning.	N
If a building is likely to be used by children under five years, the guarding should not have horizontal rails, should stop children from easily climbing it, and the construction should prevent a 100 mm sphere being able to pass through any opening of that guarding.	K3 (3.3)
All external balconies and edges of roofs shall have a wall, parapet, balustrade or similar guard at least 1100 mm high.	K3 (3.2)

Requirement K2 (a) applies only to stairs and ramps that form part of the building.

6.31 Entrance and access

6.31.1 The requirement (Building Act 1984 Sections 24 and 71)

The Building Act is very specific about exits, passageways and gangways and local authorities are required to consult with the fire authority to ensure that proposed methods of ingress and egress are deemed satisfactory (depending on the type of building). The purpose for which the building is going to be used needs to be considered as each case can be different. In particular, is the building going to be:

- a theatre, hall or other public building that is used as a place of public resort;
- a restaurant, shop, store or warehouse to which members of the public are likely to be admitted;
- a club;
- a school;
- a church, chapel or other place of worship (erected or used after the Public Health Acts Amendment Acts 1890 came into force).

At all times, the means of ingress and egress and the passages and gangways, while persons are assembled in the building, are to be kept free and unobstructed.

You are required by the Building Act 1984 to ensure that all courts, yards and passageways giving access to a house, industrial or commercial building (not maintained at the public expense) are capable of allowing satisfactory drainage of its surface or subsoil to a proper outfall.

The local authority can require the owner of any of the buildings to complete such works as may be necessary to remedy the defect.

All entrances to courts and yards must allow the free circulation of air and are not allowed to be closed, narrowed, reduced in height or in any other way altered so that it can impede the free circulation of air through the entrance.

(Building Act 1984 Section 85)

Private houses are not restricted by the actual Building Act of 1984 provided that members of the public are only admitted occasionally or exceptionally.

Fire safety

- *There shall be sufficient means of external access to enable fire appliances to be brought near to the building for effective use.*
- *There shall be sufficient means of access into, and within, the building for firefighting personnel to affect search and rescue and fight fire.*
- *The building shall be provided with sufficient internal fire mains and other facilities to assist firefighters in their tasks.*
- *The building shall be provided with adequate means for venting heat and smoke from a fire in a basement.*

(Approved Document B5)

Vehicle barriers and loading bays

- *Vehicle barriers should be provided that are capable of resisting or deflecting the impact of vehicles.*

- *Loading bays shall be provided with an adequate number of exits (or refuges) to enable people to avoid being crushed by vehicles.*

(Approved Document K3)

Disabled people

During 2002/3, Approved Document M was thoroughly overhauled and restructured in order to meet the changed requirements of the Disability Discrimination Act 1995 (which are enforced by the *Discrimination Act 1995 (Amendment) Regulations 2003* (SI 2003/1673)). Part M now covers:

- the use of a building to disabled people (redefined to include parents with children, elderly people and people with all types of disabilities – such as mobility, sight and hearing etc.) whether as residents, visitors, spectators, customers or employees, or participants in sports events, performances and conferences.

Note: See Annex A for further guidance on access and facilities for disabled people.

Access and facilities for disabled people

In addition to the requirements of the Disability Discrimination Act 1995 precautions need to be taken to ensure that:

- *new non-domestic buildings and/or dwellings (e.g. houses and flats used for student living accommodation etc.);*
- *extensions to existing non-domestic buildings;*
- *non-domestic buildings that have been subject to a material change of use (e.g. so that they become a hotel, boarding house, institution, public building or shop);*

are capable of allowing people, regardless of their disability, age or gender, to:

- *gain access to buildings;*
- *gain access within buildings;*
- *be able to use the facilities of the buildings (both as visitors and as people who live or work in them);*
- *use sanitary conveniences in the principal storey of any new dwelling.*

(Approved Document M)

Note: See Annex A for guidance on access and facilities for disabled people.

Access Statements

To assist building control bodies it is recommended that an 'Access Statement' is also provided when plans are deposited, a building notice is given, or details of a project are provided to an approved inspector.

Note: A building control file should also be prepared for all new buildings, changes of use and where extensive alterations are being made to existing buildings.

In its simplest form, an Access Statement should show where an applicant wishes to deviate from the guidance in Approved Document M, either to:

- make use of new technologies (e.g. infrared activated controls);
- provide a more convenient solution; or
- address the constraints of an existing building.

The Access Statement should include:

- the reasons for departing from the guidance;
- the rationale for the design approach adopted;
- constraints imposed by the existing structure and its immediate environment (why it is not practicable to adjust the existing entrance or provide a suitable new entrance);
- convincing arguments that an alternative solution will achieve the same, a better, or a more convenient outcome (e.g. why a fully compliant independent access is considered impracticable);
- evidence (e.g. current validated research) to support the design approach;
- the identification of buildings (or particular parts of buildings) where access needs to be restricted (e.g. processes that are carried out which might create hazards for children, disabled people or frail, elderly people).

Note: Further guidance on Access Statements is available on the Disability Rights Commission's website at http://www.drc-gb.org.

6.31.2 Meeting the requirement

Access and facilities for the fire service

Guidance

- There should be sufficient means of external access to enable fire appliances to be brought near to the building for effective use.
- There should be sufficient means of access into, and within, the building for firefighting personnel to effect search and rescue and fight fire.
- The building should be provided with sufficient internal fire mains and other facilities to assist firefighters in their tasks.
- The building should be provided with adequate means for venting heat and smoke from a fire in a basement.

Note: For dwelling houses and small buildings, it is usually only necessary to ensure that the building is sufficiently close to a point accessible to fire and rescue service vehicles.

Vehicle access

To enable high reach appliances, such as turntable ladders and hydraulic platforms, to be used and to enable pumping appliances to supply water and equipment for firefighting, search and rescue activities, access to the building is required.

There should be vehicle access for a pump appliance to within 45 m of all points within the dwelling house.	B5 11.2 (V2)
Every elevation to which vehicle access is provided should have a suitable door(s), not less than 750 mm wide, giving access to the interior of the building.	B5 11.3 (V2)

Design of access routes and hard standings (dwelling houses)

A vehicle access route may be a road or other route which, including any inspection covers and the like, meets the standards in Table 6.95.	B5 11.4 (V2) B5 16.8 (V2)

Table 6.95 Typical fire and rescue service vehicle access route specification

Appliance type	Minimum width of road between kerbs (m)	Minimum width of gateways (m)	Minimum turning circle between kerbs (m)	Minimum turning circle between walls (m)	Minimum clearance height (m)	Minimum carrying capacity (tonnes)
Pump	3.7	3.1	16.8	19.2	3.7	12.5
High reach	3.7	3.1	26.0	29.0	4.0	17.0

Turning facilities should be provided in any dead end access route that is more than 20 m long (see Figure 6.233).	B5 11.5 (V2) B5 16.9 (V2)

Buildings not fitted with fire mains

There should be vehicle access for a pump appliances to small buildings (those of up to 2000 m^2 with a top storey up to 11 m above ground level) to either: — B5 16.2 (V2)

- 15% of the perimeter; or
- within 45 m of every point on the projected plan area.

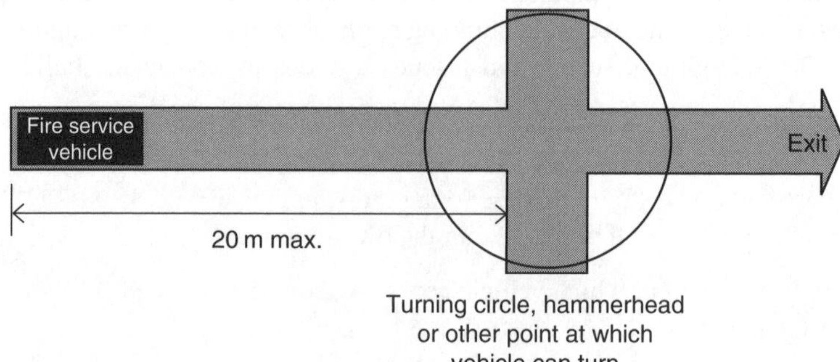
Fire and rescue service vehicles should not have to reverse more than 20 m from the end of an access road
Fire service vehicle
Exit
20 m max.
Turning circle, hammerhead or other point at which vehicle can turn

Figure 6.233 Turning facilities

There should be vehicle access for a pump appliance to blocks of flats to within 45 m of all points within each dwelling.	B5 16.3 (V2)
Every elevation to which vehicle access is provided should have a suitable door(s), not less than 750 mm wide, giving access to the interior of the building.	B5 16.5 (V2)
Door(s) should be provided such that there is no more than 60 m between each door and/or the end of that elevation (e.g. a 150 m elevation would need at least 2 doors).	B5 16.5 (V2)

Buildings fitted with fire mains

If a building is fitted with dry fire mains:

- there should be access for a pumping appliance to within 18 m of each fire main inlet connection point (typically on the face of the building); — B5 16.6 (V2)
- the inlet should be visible from the appliance.

If a building is fitted with wet mains, the pumping appliance access:

- should be to within 18 m and within sight of a suitable entrance; and
- in sight of the inlet for the emergency replenishment of the suction tank for the main. — B5 16.7 (V2)

Vehicular access

If vehicles have access to a floor or roof edge, barriers of at least 375 mm should be provided to any edges that are level with (or above) the adjoining floor or ground.	K3 (3.7)
If vehicles have access to a ramp edge, barriers of at least 610 mm should be provided to any edges that are level with (or above) the adjoining floor or ground.	K3 (3.7)
Any wall, parapet, balustrade or similar obstruction may serve as a barrier.	K3 (3.8)
All barriers should be capable of resisting forces set out in BS 6399.	K3 (3.8)
Loading bays should be provided with at least one exit point from the lower level (preferably near the centre of the rear wall).	K3 (3.9)

Access and facilities for disabled people

Access (i.e. approach, entry or exit) to a building is frequently a problem for wheelchair users, people who need to use walking aids, people with impaired sight and mothers with prams etc. In designing the approach to a building (and routes between buildings within a complex) the following should, therefore, always be taken into consideration:

- changes in level between the entrance storey and the site entry point should be minimized;
- access routes should be wide enough to let people pass each other;
- potential hazards (e.g. windows from adjacent buildings opening onto access routes) should be avoided.

Note: See also '*Mobility: A Guide to Best Practice on Access to Pedestrian and Transport Infrastructure*'.

General

The primary aim should be to make it reasonably possible for a disabled person to approach and gain access into the dwelling from the entrance point at the boundary of the site (and from any car parking that is provided on the site) to the building. It is also important that routes between buildings within a complex are also accessible.

Approach to a building

Access from the boundary of the site (and/or from car parking designated for disabled people) to the principal entrance should be level.	M (1.2, 1.4, 1.6 and 1.13) M (6.2)
If a difference in level is unavoidable (i.e. due to site constraints) the approach can have a gentle gradient (provided that it is over a long distance) or can include a number of shorter parts (at steeper gradients) as long as level landings are provided as rest points.	M (1.7)
The principal entrance (entrances, main staff entrance and any lobbies) should be accessible to disabled people and mothers pushing prams etc.	M (2.1)
If this is not possible, an alternative accessible entrance should be provided.	M (2.2)
Risks to people when entering the building should be minimal.	M (2.3)
Access routes should be wide enough to let people pass each other.	M (1.11)
Note: A surface width of 1800 mm is ideal but this can be reduced on restricted sites to 1200 mm, provided that a case is made in the Access Statement.	
The route to the principal entrance (or alternative accessible entrance) should be clearly identified and well lit.	M (1.13g)
A separate pedestrian route should be provided.	M (1.13h)
Uncontrolled vehicular crossing points should be identified by a buff coloured blister surface (see Figure 6.234).	M (1.13h)

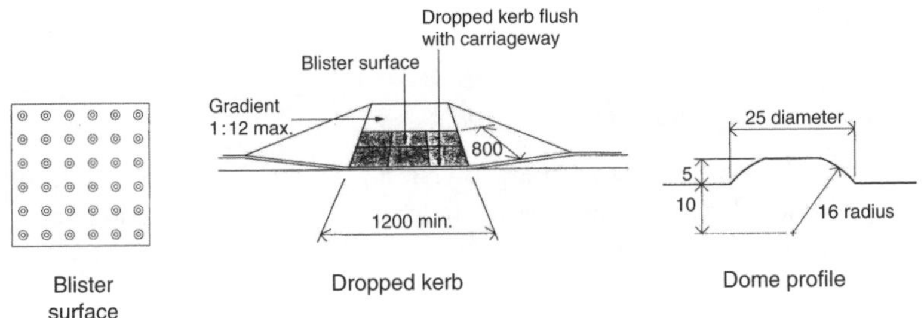

Figure 6.234 An example of tactile paving used at an uncontrolled crossing

Gradients

Approach gradients:

- should ideally be no steeper than 1:60 along their whole length; M (1.13c)
- if steeper than 1:20, should be designed as a ramped access; M (1.8)
- if less than 1:20, should be provided with level landings for each 500 mm rise of the access. M (1.13c)

Cross-fall gradients should be no steeper than 1:40. M (1.13c)

Surface

The surface of all access routes should:

- allow people to travel along them easily, without excessive effort and without the risk of tripping or falling; M (1.9)
- be at least 1.5 m wide; M (1.13a)
- be firm, durable and slip resistant; M (1.13d)
- have undulations not exceeding 3 mm (under a 1 m straight edge); M (1.13d)
- be made of the same material and similar frictional characteristics (loose sand or gravel should **not** be used). M (1.13d and e)

Building perimeters

The perimeter of the building should be well lit. M (1.12)

Passing places

Passing places should be:

- free of obstructions to a height of 2.1 m; M (1.13a)
- at least 1.8 m wide and at least 2 m long. M (1.13b)

Joints

Joints should be:

- filled flush or (if recessed) no deeper than 5 mm; M (1.13f)
- no wider than 10 mm or (if unfilled) no wider than 5 mm. M (1.13f)

The difference in level at joints between paving units should be no greater than 5 mm. M (1.13f)

On-site car parking and setting down – parking bays

At least one parking bay designated for disabled people should be provided as close as possible to the principal entrance of the building. M (1.18a)

The dimensions of the designated parking bays should be as per Figure 6.235 (with a 1200 mm accessibility zone between and a 1200 mm safety zone on the vehicular side of the parking bays and with a dropped kerb when there is a pedestrian route at the other side of the parking bay). M (1.18b)

A clearly signposted setting down point should be located on firm, level ground as close as possible to the principal (or alternative) entrance. M (1.18e)

The surface of the accessibility zone should be firm, durable and slip resistant, with undulations not exceeding 3 mm (under a 1 m straight edge). M (1.18c)

The surface of a parking bay designated for disabled people (in particular the area surrounding the bay) should allow the safe transfer of a passenger or driver M (1.15) to a wheelchair and transfer from the parking bay to the access route to the building without undue effort.

Ticket machines should: M (1.16 and 1.18d)

- have their controls between 750 mm and 1200 mm above ground level;
- be located near parking bays designated for disabled people;
- be located so that a person in a wheelchair (or a person of short stature) is able to reach the controls.

The plinth of the ticket machine should not project in front of the face of the machine. M (1.18d)

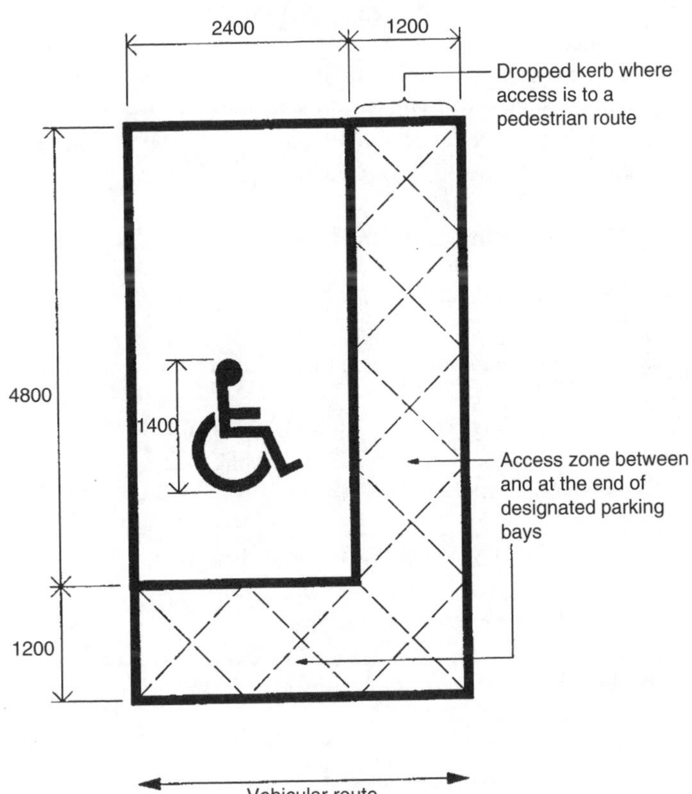

Figure 6.235 Parking bay designated for disabled people

Note: See also BS 8300 for guidance on:

- provision of parking bays designated for disabled people;
- ticket dispensing machines;
- vehicular control barriers; and
- multi-storey car parks.

Ramped access

If the constraints of the site mean that there is an approach gradient of 1 in 20 or steeper, then a ramped access should be provided as they are beneficial not only for wheelchair users but also people pushing prams and bicycles.

Where a ramped access is provided:

• the approach should be clearly signposted;	M (1.26a)
• the going should be no greater than 10 m;	M (1.26c)
• the rise should be no more than 500 mm;	M (1.26c)
• if the total rise is greater than 2 m then an alternative means of access (e.g. a lift) should be provided for wheelchair users;	M (1.26d)
• the surface width should be at least 1.5 m;	M (1.26e)
• gradients should be as shallow as practicable;	M (1.20)
• the ramp surface should be slip resistant;	M (1.26f)
• the ramp surface should be of a contrasting colour to that of the landings;	M (1.26f)
• frictional characteristics of ramp and landing surfaces should be similar;	M (1.26g)
• landings at the foot and head of a ramp should be at least 1.2 m long and clear of any obstructions;	M (1.26h)
• intermediate landings should be at least 1.5 m long and clear of obstructions;	M (1.26i)
• intermediate landings (at least 1800 mm wide and 1800 mm long) should be provided at passing places;	M (1.26j)
• all landings should: – be level; – have a maximum gradient of 1:60 along their length; – have a maximum cross fall gradient of 1:40;	M (1.26k)
• there should be a handrail on both sides;	M (1.26l)
• in addition to the guarding requirements of Part K, there should be a visually contrasting kerb on the open side of the ramp (or landing) at least 100 mm high;	M (1.26m)
• when the rise of the ramp is greater than two 150 mm steps, signposted steps should be provided;	M (1.26n)

• the gradient of a ramp flight and its going between landings should be in accordance with Table 6.96 and Figure 6.236.	M (1.26b)

Note: Approved Document K (*Protection from falling, collision and impact*) contains general guidance on stair and ramp design. The guidance in Approved Document M reflects more recent ergonomic research conducted to support BS 8300 and takes precedence over Approved Document K in conflicting areas.

Possible future amendment	Further research on stairs is currently being undertaken and will be reflected in future revisions of Approved Document K.

Table 6.96 Limits for ramp gradients

Going of a flight (m)	Maximum gradient	Maximum rise (mm)
10	1:20	500
5	1:15	333
2	1:12	166

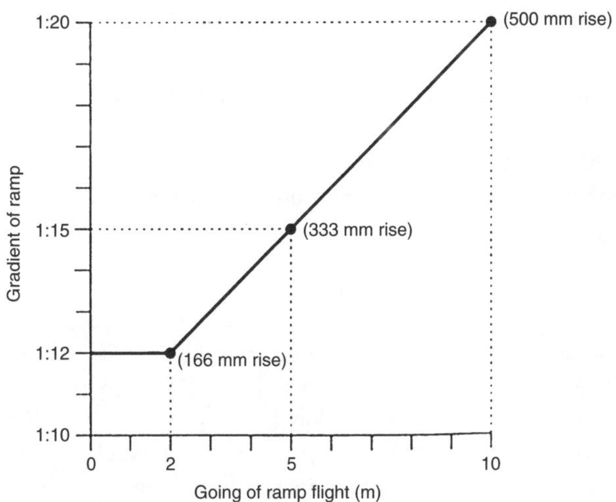

Figure 6.236 Relationship for ramp gradients to the going of a flight

Stepped access

People with impaired sight risk tripping or losing their balance if there is no warning that there are steps that provide a change in level. The risk is most hazardous at the head of a flight of stairs when a person is descending.

A stepped access should:

- have a level landing at the top and bottom of each flight; M (1.33a)
- be 1200 mm long each landing and unobstructed; M (1.33b)
- have a corduroy hazard warning surface at top and bottom landings of a series of flights so as to give advance warning of a change in level (see Figure 6.237); M (1.33c)

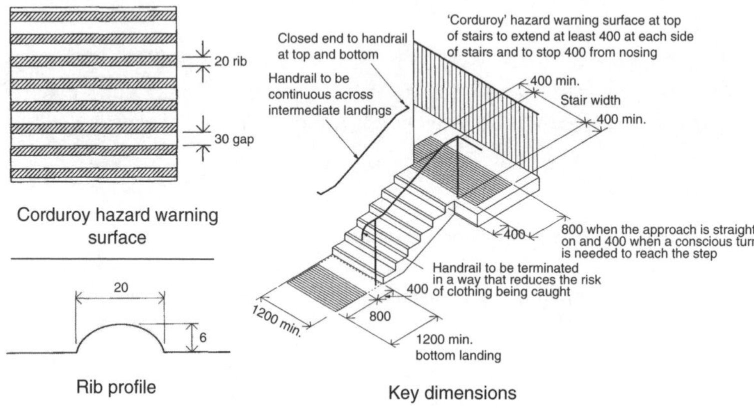

Figure 6.237 Stepped access – key dimensions and use of hazard warning surfaces

Note: Approved Document K (*Protection from falling, collision and impact*) contains general guidance on stair and ramp design. The guidance in Approved Document M reflects more recent ergonomic research conducted to support BS 8300 and takes precedence over Approved Document K.

In addition:

- side accesses onto intermediate landings should have a 400 mm deep corduroy hazard warning surface; M (1.33d)
- doors should not swing across landings; M (1.33e)
- the surface width of flights between enclosing walls, strings or upstands, should not be less than 1.2 m; M (1.33f)
- there should be no single steps; M (1.33g)
- the rise of a flight between landings should contain no more than: M (1.33h)
  - 12 risers for a going of **less** than 350 mm
  - 18 risers for a going of 350 mm or **greater** (see Figure 6.238).

Figure 6.238 External steps and stairs – key dimensions

Note: For school buildings, the preferred dimensions are a rise of 150 mm and a going of 280 mm.

Nosings for the tread and the riser should be 55 mm wide and of a contrasting material.	M (1.33i)
Step nosings should not project over the tread below by more than 25 mm (see Figure 6.239).	M (1.33j)

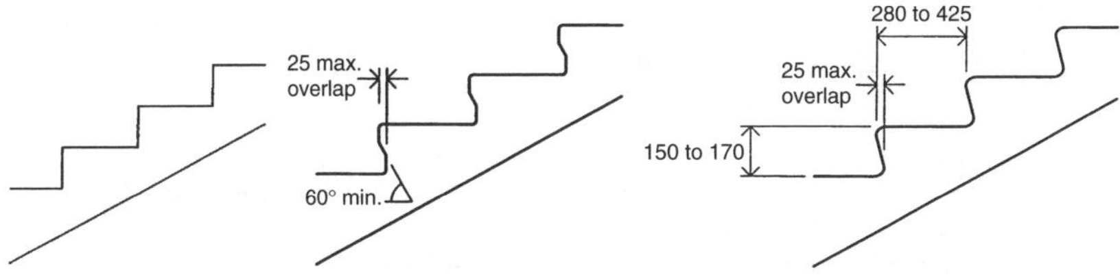

The rise and going dimensions apply to all step profiles

Figure 6.239 Examples of step profiles and key dimensions for external stairs

The rise and going of each step should be consistent throughout a flight.	M (1.33k)
The rise of each step should be between 150 mm and 170 mm.	M (1.33l)
Rises should not be open.	M (1.33n)
The going of each step should be between 280 mm and 425 mm.	M (1.33m)
There should be a continuous handrail on each side of a flight and landings.	M (1.33o)
If additional handrails are used to divide the flight into channels, then they should not be less than 1 m wide or more than 1.8 m wide.	M (1.33p)

Warnings should be placed sufficiently in advance of a hazard to allow time to stop. M (1.28)

Warnings should not be so narrow that they could be missed in a single stride. M (1.28)

Materials for treads should not present a slip hazard. M (1.29)

Handrails to external stepped and for ramped access

Handrails to external stepped or ramped access should be positioned as per Figure 6.240. M (1.37a)

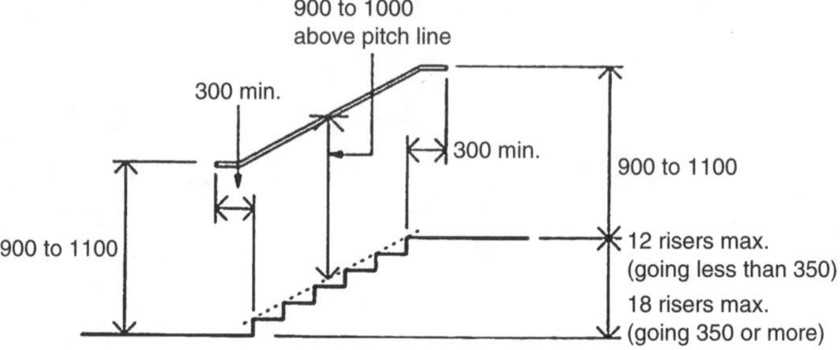

Figure 6.240 Handrails to external stepped and ramped access – key dimensions

Handrails to external stepped or ramped access should:

- be continuous across flights and landings; M (1.37c)
- extend at least 300 mm horizontally beyond the top and bottom of a ramped access; M (1.37d)
- not project into an access route; M (1.37d)
- contrast visually with the background; M (1.37e)
- have a slip resistant surface which is **not** cold to the touch; M (1.37f)
- terminate in such a way that reduces the risk of clothing being caught; M (1.37g)
- either be circular (with a diameter of between 40 and 45 mm) or oval with a width of 50 mm (see Figure 6.241). M (1.37h)

In addition, handrails should:

- not protrude more than 100 mm into the surface width of the ramped or stepped access M (1.37i) where this would impinge on the stair width requirement of Part B1;

- have a clearance of between 60 and 75 mm between the handrail and any adjacent wall surface; M (1.37j)
- have a clearance of at least 50 mm between a cranked support and the underside of the handrail; M (1.37k)
- ensure that its inner face is located no more than 50 mm beyond the surface width of the ramped or stepped access; M (1.37l)
- be spaced away from the wall and rigidly supported in a way that avoids impeding finger grip; M (1.35)
- be set at heights that are convenient for all users of the building. M (1.36)

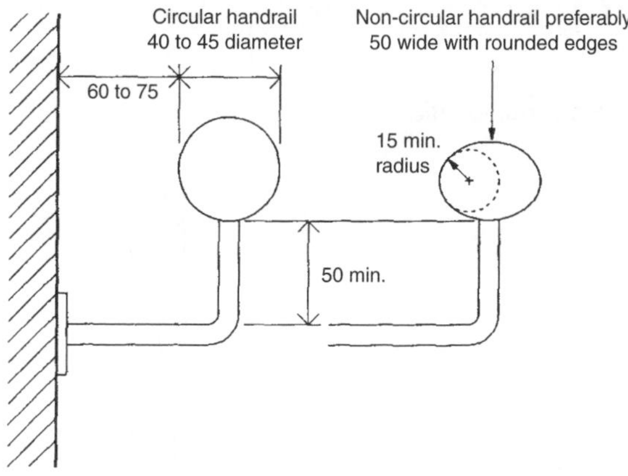

Figure 6.241 Handrail designs

Hazards on access routes

Features of a building (e.g. windows and doors) that can occasionally obstruct an access route should not present a hazard. M (1.38)

Areas below stairs or ramps with a soffit less than 2.1 m above ground level should be protected by guarding and low level cane detection. M (1.39b)

Any feature projecting more than 100 mm an access route should be protected by guarding that includes a kerb (or other solid barrier) that can be detected using a cane (see Figure 6.243). M (1.39b)

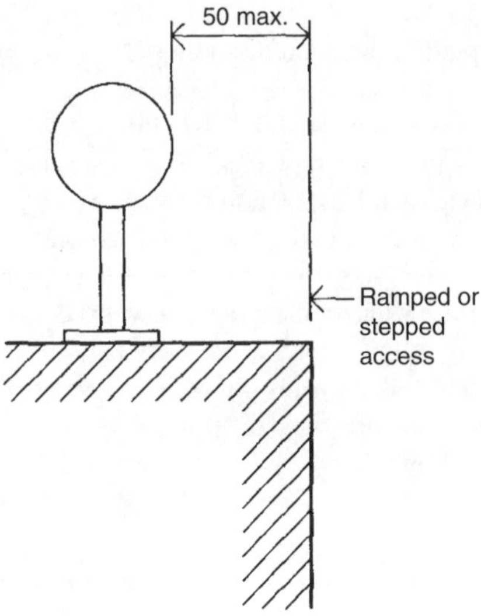

Figure 6.242 Handrail location

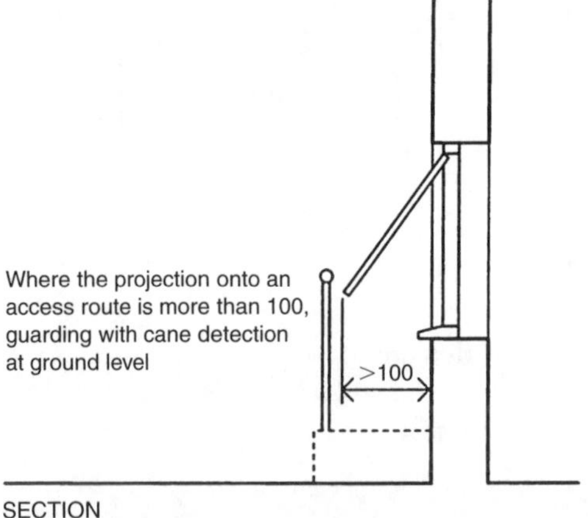

Figure 6.243 Avoiding hazards on access routes

Accessible entrances

Accessible entrances should be clearly signposted (e.g. with the International Symbol of Access) and easily recognized.	M (2.5 and 2.7a)
Accessible entrances should also:	
• be easily identifiable (e.g. by lighting and/or visual contrast);	M (2.7b)

• have a level landing at least 1500 × 1500 mm, clear of any door swings, immediately in front of the entrance;	M (2.7d)
• avoid raised thresholds (if unavoidable, then the total height should not be more than 15 mm, with a minimum number of upstands and slopes);	M (2.7e)
• ensure that all door entry systems are accessible to deaf and hard of hearing people, plus people who cannot speak;	M (2.7f)
• not have internal floor surface material (e.g. coir matting) adjacent to the threshold that could impede the movement of wheelchairs;	M (2.7h)
• not have changes in floor materials that could create a potential trip hazard;	M (2.7h)
• if mat wells are provided, have the surface of the mat level with the surface of the adjacent floor finish;	M (2.7i)
• have the route from the exterior across the threshold weather protected;	M (2.6 and 2.7a)
• not present a hazard for visually impaired people (e.g. have structural elements such as canopy supports).	M (2.5 and 2.7c)

The premises have no more than three steps

The premises are fully accessible for wheelchair users

Wheelchair assistance required

Figure 6.244 Typical access signs for disabled people

Note: See BS 8300 for further guidance on signposting.

Doors to accessible entrances

Doors to the principal (or alternative accessible) entrance should be accessible to all, particularly for wheelchair users and people with limited physical dexterity.	M (2.8)
Entrance doors should be capable of being held closed when not in use.	M (2.8)

A power operated door opening and closing system should be used if a force greater than 20 N is required to open or shut a door. M (2.13a)

Once open, all doors to accessible entrances should be wide enough to allow unrestricted passage for a variety of users, including wheelchair users, people carrying luggage, people with assistance dogs, and parents with pushchairs and small children.

People should be able to see other people approaching from the opposite direction. M (2.12)

The effective clear width through a single leaf door (or one leaf of a double leaf door) should be in accordance with Table 6.97. M (2.13b)

Table 6.97 Minimum effective clear widths of doors

Direction and width of approach	New buildings (mm)	Existing buildings (mm)
Straight-on (without a turn or oblique approach)	800	750
At right angles to an access route at least 1500 mm wide	800	750
At right angles to an access route at least 1200 mm wide	825	775
External doors to buildings used by the general public	1000	775

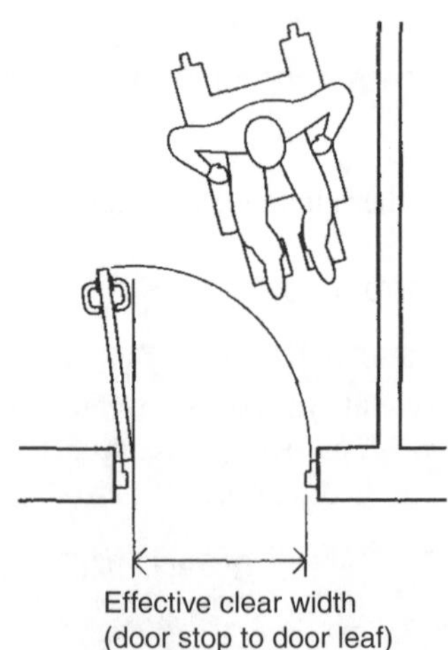

Figure 6.245 Effective clear width and visibility requirements of doors

Door leaves and side panels wider than 450 mm should incorporate vision panels: M (2.13c)

- towards the leading edge of the door;
- between 500 mm and 1500 mm from the floor (see Figure 6.245).

Manually operated non-powered entrance doors

Self-closing devices on manually operated (i.e. non-powered) doors can be a great disadvantage to people who have limited upper body strength, or people who are pushing prams or carrying heavy objects. To rectify this matter:

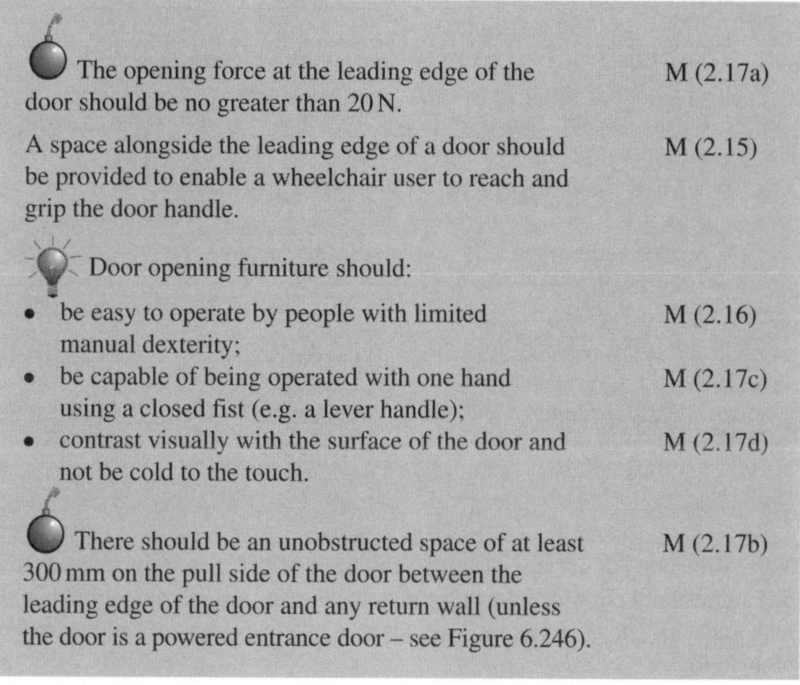

The opening force at the leading edge of the door should be no greater than 20 N. M (2.17a)

A space alongside the leading edge of a door should be provided to enable a wheelchair user to reach and grip the door handle. M (2.15)

Door opening furniture should:

- be easy to operate by people with limited manual dexterity; M (2.16)
- be capable of being operated with one hand using a closed fist (e.g. a lever handle); M (2.17c)
- contrast visually with the surface of the door and not be cold to the touch. M (2.17d)

There should be an unobstructed space of at least 300 mm on the pull side of the door between the leading edge of the door and any return wall (unless the door is a powered entrance door – see Figure 6.246). M (2.17b)

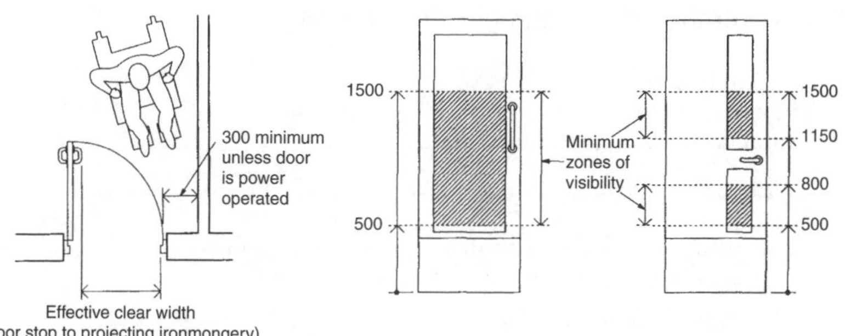

Figure 6.246 Effective clear width and visibility requirements of doors

Powered operated entrance doors

Power operated entrance doors should have a sliding, swinging or folding action controlled manually (by a push pad, card swipe, coded entry, or remote control) or be automatically controlled by a motion sensor or proximity sensor such as a contact mat.

Power operated entrance doors should:	
• open towards people approaching the doors;	M (2.21a)
• provide visual and audible warnings that they are operating (or about to operate);	M (2.21c)
• incorporate automatic sensors to ensure that they open early enough (and stay open long enough) to permit safe entry and exit;	M (2.21c)
• incorporate a safety stop that is activated if the doors begin to close when a person is passing through;	M (2.21b)
• revert to manual control (or fail safe) in the open position in the event of a power failure;	M (2.21d)
• when open, not project into any adjacent access route;	M (2.21e)
• ensure that its manual controls: – are located between 750 mm and 1000 mm above floor level; – are operable with a closed fist;	M (2.21f)
• be set back 1400 mm from the leading edge of the door when fully open;	M (2.21g)
• be clearly distinguishable against the background;	M (2.21g)
• contrast visually with the background.	M (2.19 and 2.21g)

Note: Revolving doors are not considered 'accessible' as they create particular difficulties (and possible injury) for people who are visually impaired, people with assistance dogs or mobility problems and for parents with children and/or pushchairs.

Glass entrance doors and glazed screens

The presence of the door should be apparent not only when it is shut but also when it is open.	M (2.23)
Glass entrance doors and glazed screens should:	
• be clearly marked (i.e. with a logo or sign) on the glass at two levels, 850 to 1000 mm and 1400 to 1600 mm above the floor;	M (2.24a)

Note: The logo or sign should be at least 150 mm high (repeated if on a glazed screen), or a decorative feature such as broken lines or continuous bands, at least 50 mm high.

- when adjacent to, or forming part of, a glazed screen, be provided with a high contrast strip at the top, and on both sides; M (2.24c)
- ensure that glass entrance doors (if capable of being held open) are protected by guarding to prevent the leading edge from becoming a possible hazard. M (2.24d)

For dwellings:

An external door providing access for disabled people should have a minimum clear opening width (taken from the face of the door stop on the latch side to the face of the door when open at 90°) of 775 mm. M (6.23)

The door opening width should be sufficient to enable a wheelchair user to manoeuvre into the dwelling. M (6.22)

Possible future amendment	Approved Document N (Glazing – safety in relation to impact, opening and cleaning) contains guidance on the use of symbols and markings on glazed doors and screens. The guidance now given in Approved Document M is as a result of more recent experience of ‘door manifestation’ and takes precedence over the guidance currently provided in Approved Document N in conflicting areas until such time as Approved Document N is revised.

Entrance (and internal) lobbies

Lobbies should be: M (2.27 and 3.15)

- large enough to allow a wheelchair user or a person pushing a pram to move clear of one door before opening the second door;

- capable of accommodating a companion helping a wheelchair user to open doors and guide the wheelchair through.

The minimum length of the lobby is related to the chosen door size, the swing of each door, the projection of the door into the lobby and the size of an occupied wheelchair with a companion pushing.

Within the lobby:

- glazing should not create distracting reflections; M (2.29d and 3.16d)
- floor surface materials should do not impede the movement of wheelchairs etc.; M (2.29e)
- changes in floor materials should not create a potential trip hazard; M (2.29e and 3.16e)
- the floor surface should assist in removing rainwater from shoes and wheelchairs; M (2.29f)
- any columns and ducting etc. that project into the lobby by more than 100 mm should be protected by a visually contrasting guard rail. M (2.29h and 3.16a–c)

The length and width of an entrance and/or an internal lobby should be as per Table 6.98. M (2.29a, b and c)

Table 6.98 Entrance lobbies – dimensions

	Length	Width
Entrance/internal lobby with single swing door	as per Figure 6.202	at least 1200 mm (or the width of the two doors plus 300 mm whichever is the greater)
Entrance/internal lobby with double swing doors	at least the size (i.e. width) of the two doors plus 1570 mm	at least 1800 mm

Entrance hall and reception area

If there is a reception point:

- it should be easily accessible and convenient to use; M (3.2)
- it should be located away from the principal entrance; M (3.6a)
- it should be easily identifiable from the entrance doors or lobby; M (3.6b)

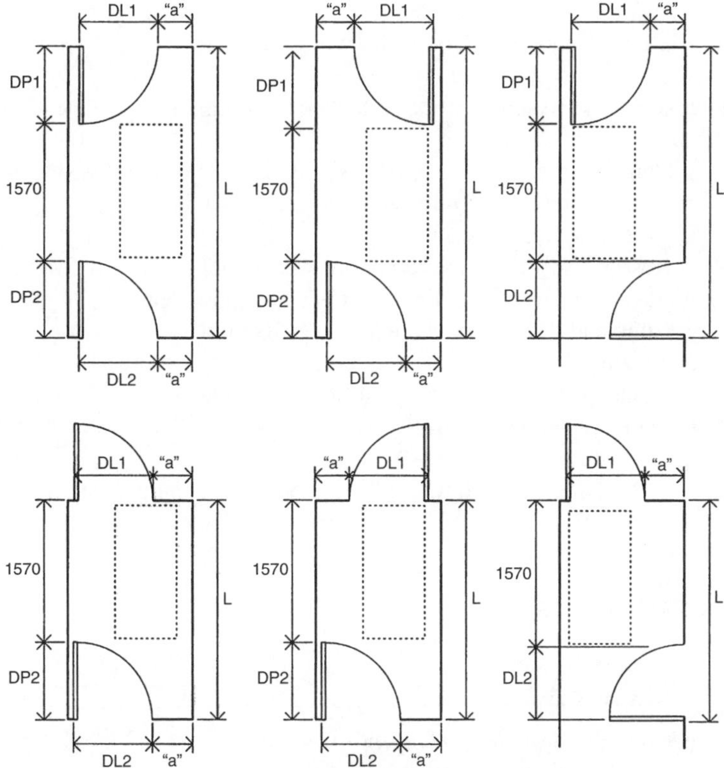

DL1 and DL2 = door leaf dimensions of the doors to the lobby
DP1 and DP2 = door projection into the lobby (normally door leaf size)
L = minimum length of lobby, or length up to door leaf for side entry lobby
"a" = at least 300 mm wheelchair access space (can be increased to reduce L)
1570 = length of occupied wheelchair with a companion pushing (or a large scooter)
NB: For every 100 mm increase above 300 mm in the dimension "a" (which gives a greater overlap of the wheelchair footprint over the door swing), there can be a corresponding reduction of 100 mm in the dimension L, up to a maximum of 600 mm reduction.

Figure 6.247 Key dimensions – lobbies and entrance doors

• relevant information about the building should be clearly available from notice boards and signs;	M (3.5)
• the floor surface should be slip resistant;	M (3.6)
• the approach to it should be direct and free from obstructions;	M (3.6b)
• the design of the approach should allow space for wheelchair users to gain access;	M (3.6c)
• there should be a clear manoeuvring space in front of any reception desk at least 1200 mm deep and 1800 mm wide.	M (3.6d)

If there is a reception desk or counter:

- it should be designed to accommodate both standing and seated visitors; M (3.6e)
- if there is a knee recess, then this should be at least 500 mm deep (or 1400 mm deep and 2200 mm wide if there is no knee recess); M (3.6d)
- at least one section of the reception desk should be no less than 1500 mm wide, its surface no higher than 760 mm and a knee recess not less than 700 mm, above floor level; M (3.6e)
- it should be provided with a hearing enhancement system (e.g. an induction loop). M (3.6f)

Note: See BS 8300 for guidance on aids to communication.

Domestic buildings

If there is a reception point:

- it should be easily accessible and convenient to use; M (3.2 to 3.5)
- information about the building should be clearly available from notice boards and signs;
- the floor surface should be slip resistant.

Approach to a dwelling

On plots which are reasonably level, wheelchair users should normally be able to approach the principal entrance. M (6.2)

Wheelchair users (having approached the entrance) should be able to gain access into the dwelling-house and/or entrance level of flats. M (6.3)

A suitable approach should be provided from the point of access to the entrance of the dwelling. M (6.11)

The whole, or part, of the approach may be a driveway. M (6.12)

The approach should:

- not have crossfalls greater than 1 in 40; M (6.11)

- be safe and convenient for disabled people as is reasonable possible; M (6.6)
- ideally be level or ramped. M (6.8)

Note: On steeply sloping plots, a stepped approach is permissible.

If a stepped approach to the dwelling is unavoidable, the aim should be for the steps to be designed to suit the needs of ambulant disabled people (see paragraph 6.19). M (6.7)

The surface of the wheelchair user's approach should: M (6.9)

- be firm enough to support the weight of the user and his or her wheelchair;
- be smooth enough to permit easy manoeuvre;
- not be made up of loose laid materials (such as gravel and shingle);
- take account of the needs of stick and crutch users.

For steeply sloping plots, it would be reasonable to provide for stick or crutch users.

Level approach

A 'level' approach should: M (6.13)

- be no steeper than 1 in 20;
- have a firm and even surface;
- have a width not less than 900 mm.

Note: The width of the approach, excluding space for parked vehicle, should take account of the needs of a wheelchair user, or a stick or crutch user.

Ramped approach

If a plot gradient exceeds 1 in 20, a ramped approach may be provided in which case: M (6.14)

- the surface should be firm and even; M (6.15a)
- the flights should have unobstructed widths of at least 900 mm; M (6.15b)

- individual flights should be no longer than 10 m (for gradients not steeper than 1 in 15) or 5 m for gradients not steeper than 1 in 12; M (6.15c)
- it should have top and bottom landings (and intermediate landings if necessary) not less than 1.2 m in length – exclusive of the swing of any door or gate which opens onto it. M (6.15c)

Stepped approach

A stepped approach should be used if the plot gradient is greater than 1 in 15. M (6.16)

A stepped approach should:

- have flights with an unobstructed width of at least 900 mm; M (6.17a)
- have a flight rise of not more than 1.8 m; M (6.17b)
- have a top and bottom (and if necessary intermediate) landing not less than 900 mm in length; M (6.17c)
- have steps: M (6.17d)
  - with suitable tread nosing profiles (see Figure 6.248);
  - with a uniform rise between 75 mm and 150 mm; M (6.17e)
- ensure that the going of each step is not less than 280 mm; M (6.17f)
- comprise three or more risers;
- have a suitable continuous handrail on one side of the flight between 850 mm and 1000 mm above the pitch line of the flight; and extend 300 mm beyond the top and bottom nosings.

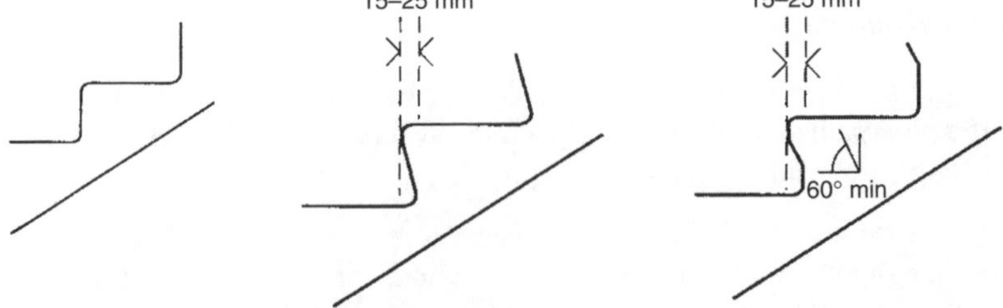

Figure 6.248 External step profiles

Approach using a driveway

Where a driveway provides a means of approach towards the entrance, the approach past any parked cars must be in accordance with requirements M6.11–6.27 above.	M (6.18)

Access into the dwelling

The point of access should be reasonably level.	M (6.11)
Where the approach to the entrance consists of a level or ramped approach, an accessible threshold at the entrance should be provided.	M (6.19)
Note: An accessible threshold into entrance level flats should also be provided.	
If a stepped approach is provided into the dwelling, the rise should be no more than 150 mm.	M (6.20)
An accessible threshold should be provided into the entrance.	M (6.21)

Note: The design of an accessible threshold should also satisfy the requirements of Part C2: 'Dangerous and offensive substances' and Part C4: 'Resistance to weather and ground moisture'.

6.32 Extensions and additions to buildings

6.32.1 The requirement

Ventilation

There shall be adequate means of ventilation provided for people in the building.
(Approved Document F)

Conservation of fuel and power

Reasonable provision shall be made for the conservation of fuel and power in buildings by:

(a) limiting heat gains and losses:
- *(i) through thermal elements and other parts of the building fabric; and*
- *(ii) from pipes, ducts and vessels used for space heating, space cooling and hot water services;*

(b) *providing and commissioning energy-efficient fixed building services with effective controls; and*
(c) *providing to the owner sufficient information about the building, the fixed building services and their maintenance requirements so that the building can be operated in such a manner as to use no more fuel and power than is reasonable in the circumstances.*

(Approved Document L1)

Responsibility for achieving compliance with the requirements of Part L rests with the person carrying out the work. That person may be, for example, a developer, a main (or sub-) contractor, or a specialist firm directly engaged by a private client.

The person responsible for achieving compliance should either themselves provide a certificate, or obtain a certificate from the sub-contractor, that commissioning has been successfully carried out. The certificate should be made available to the client and the building control body.

Fire risk analysis

Part B: 2007 now includes a requirement for the responsible person (i.e. the person carrying out work to a building) to make available to the owner (other than houses occupied as single private dwellings) 'fire safety information' concerning the design and construction of the building or extension plus details of the services, fittings and equipment that have been provided in order that they (when required under the new Regulatory Reform (Fire Safety) Order 2005 – Statutory Instrument 2005 No. 1541) may complete a fire risk analysis.

The sort of information required must include basic advice on the proper use and maintenance of systems provided in the building such as,

- emergency egress windows;
- fire doors;
- smoke alarms;
- sprinklers etc.

6.32.2 Meeting the requirement

Ventilation

The addition of a habitable room to an existing building

If the additional room is connected to an existing habitable room which now has no windows opening to outside, then the ventilation opening (or openings) shall be greater than 8000 mm^2 equivalent area.	F 3.8a(i)

If the additional room is connected to an existing habitable room which still has windows opening to outside, but with a total background ventilator equivalent area less than 5000 mm^2 equivalent area, then the ventilation opening (or openings) shall be greater than 8000 mm^2 equivalent area. F 3.8a(ii)

If the additional room is connected to an existing habitable room which still has windows opening to outside, but with a total background ventilator equivalent area of at least 5000 mm^2 equivalent area, then there should be: F 3.8a(iii)

- background ventilators of at least 8000 mm^2 equivalent area between the two rooms and
- background ventilators of at least 8000 mm^2 equivalent area between the additional room and outside.

The addition of a wet room to an existing building

Internal doors between the wet room and the existing building should have an undercut of at least minimum area 7600 mm^2 (equivalent to an undercut of 10 mm above the floor finish for a standard 760 mm width door). F 3.13

Whole building and extract ventilation can be provided by: F 3.12

- an intermittent extract and a background ventilator of at least 2500 mm^2 equivalent area; or
- a single room heat recovery ventilator; or
- a passive stack ventilator; or
- a continuous extract fan.

The addition of a conservatory to an existing building

The general ventilation rate for conservatories with a floor area greater than 30 m^2 conservatory (and adjoining rooms) can be achieved by the use of background ventilators. F 3.18

Dwellings

The area of windows, roof windows and doors in extensions should not exceed the sum of: L1B 15

- 25% of the floor area of the extension **plus**
- the area of any windows or doors, which, as a result of the extension works, no longer exist or are no longer exposed.

If the total floor area of the proposed extension exceeds these limits, then the work should be regarded as a **new** building and the requirements of Approved Document L2A and L2B should be used.

When a building is not a dwelling the area of windows and rooflights in the extension should not exceed the values given in Table 6.99. L2B 27

Table 6.99 Opening areas in an extension

Building type	Windows and personnel doors as a %age of the exposed wall	Rooflights as a %age of the roof area
Residential buildings (where people temporarily or permanently reside)	30	20
Offices, shops and places where people assemble	40	20
Industrial and storage buildings	15	20
Vehicle access doors and display windows and similar glazing	As required	N/A
Smoke vents	N/A	As required

If the extension of an existing building is a conservatory, then:

thermal and opaque elements should have U-values that are no worse than that shown in Table 6.100. L1B 22c L2B 32c

Table 6.100 Standards for thermal units (W/m^{2}K) for an existing dwelling and/or building

Existing building element	Existing dwelling	Existing building
Wall	0.35	0.70
Pitched roof – insulation at ceiling level	0.16	
Pitched roof – insulation between rafters	0.20	
Flat roof or roof with integral insulation	0.25	
Roof		0.35
Windows, roof windows, rooflights & doors		3.3
Floors	0.25	0.70

Newly constructed thermal elements that are part of an extension should meet the requirements shown in Table 6.101. L1B 50 L2B 70a

Table 6.101 Standards for new thermal elements in an extension (W/m^{2}K)

Element	Type of building	Replacement fittings (W/m^{2}K)
Walls (cavity and other)	Existing dwelling Existing building	0.30
Pitched roof – insulation at ceiling level	Existing dwelling Existing building	0.16
Pitched roof – insulation between rafters	Existing dwelling Existing building	0.20
Flat roof or roof with integral insulation	Existing dwelling Existing building	0.20
Floors	Existing dwelling Existing building	0.22

All buildings should be pressure tested except large extensions which cannot be sealed off from the existing building. L2A 74

Controlled fittings (non-domestic buildings)

The area-weighted average performance of draught-proofed units for new fittings in extensions and replacement fittings in an existing dwelling shall be no worse than given in Table 6.102. L1B 32 L2B 75

Table 6.102 Standards for controlled fittings

Element	New fittings in an extension (W/m^2K)	Replacement fittings in an existing dwelling (W/m^2K)
Windows, roof windows and glazed rooflights	1.8	2.2 (whole unit) 1.2 (centre pane)
Pedestrian doors having more than 50% of their internal face area glazed	2.2	2.2
High usage entrance doors	6.0	6.0
Vehicle access and large doors	1.5	1.5
Roof ventilators (including smoke extract ventilators)	6.0	6.0

Building services in an extension (non-domestic buildings)

The area-weighted U-value for each element type shall be $0.5\,W/m^2K$ and $0.70\,W/m^2K$ for any individual element.	L2B 29b
Automatic meter reading and data collection should be provided in all buildings with a total useful floor area that is greater than $1000\,m^2$.	L2A 43b L2B 69b
Buildings with less than $500\,m^2$ floor area need not be pressure tested.	L2A 74

Note: Conservatories with a floor area no greater than $30\,m^2$ are exempt from the Building Regulations.

Consequential improvements (non-domestic buildings)

If a building has a total useful floor area greater than $1000\,m^3$ and the proposed building work includes an extension, or the initial provision of any fixed building service, or an increase to the installed capacity of any fixed building services, then consequential improvements should be made to improve the energy efficiency of the whole building and:

- thermal units with high U-values should be upgraded;
- existing windows (but not display windows), roof windows, rooflights and doors (excluding high usage entrance doors) within the area served by the fixed building service with an increased capacity should be replaced;
- heating systems, cooling systems and air handling systems that are more than fifteen years old should either be replaced or be equipped with improved controls;
- any general lighting system serving an area greater than $100\,m^2$ which has an average lamp efficacy of less than 40 lamp-lumens per circuit watt, should be upgraded with new luminaires or improved controls;

- energy metering should be installed if less than 10% of the building's energy demand is provided by a low or zero carbon (LZC) energy system; and
- the building should be upgraded with an additional low or zero carbon energy system, **provided** that that system would achieve a simple payback within seven years or less.

6.33 External balconies

6.33.1 The requirements

Protection from falling

Pedestrian guarding should be provided for any part of a floor (including the edge below an opening window) gallery, balcony, roof (including rooflight and other openings), any other place to which people have access and any light well, basement area or similar sunken area next to a building.

(Approved Document K2)

Requirement K2 (a) applies only to stairs and ramps that form part of the building.

6.33.2 Meeting the requirements

All external balconies and edges of roofs shall have a wall, parapet, balustrade or similar guard at least 1100 mm high.	K3 (3.2)

If a balcony or flat roof is provided for escape purposes guarding may be required.	B1 2.11(V1)

6.34 Garages

6.34.1 The requirements

Electrical work

Reasonable provision shall be made in the design and installation, operation, maintenance or alteration of electrical installations in order to protect persons from fire or injury.

(Approved Document P1)

Fire precautions

As a fire precaution, all materials used for internal linings of a building should have a low rate of surface flame spread and (in some cases) a low rate of heat release.

(Approved Document B2)

6.34.2 Meeting the requirements

Cables to an outside building (e.g. garage or shed) if run underground, should be routed and positioned so as to give protection against electric shock and fire as a result of mechanical damage to a cable.	P AppA 2d
Cables concealed in floors and walls (in certain circumstances) are required to have an earthed metal covering, be enclosed in steel conduit, or have additional mechanical protection (see BS 7671 for more information).	P AppA 2d

Wall and ceiling linings

In general terms, (but see paragraphs 3.2 to 3.14 of Part B V1 and 6.2 to 6.14 of Part B V2 for more details) the surface linings of walls and ceilings should meet the following classifications:

Table 6.103 Classification of linings

Location	National Class	European Class
Domestic garages less than 40 m^2	3	D-s3, d2
Other rooms (including garages)	1	C-s3, d2

For the purpose of this requirement for surface lining, walls and ceiling are defined as follows:

	Includes	Does not include
Walls	the surface of glazing except glazing in doors	doors and door frames
	any part of a ceiling which slopes at an angle of more than 70 to the horizontal	window frames and frames in which glazing is fitted
		architraves, cover moulds, picture rails, skirtings and similar narrow members
		fireplace surrounds, mantle shelves and fitted furniture
Ceiling	the surface of glazing	trap doors and their frames
	any part of a wall which slopes at an angle of 70° or less to the horizontal	the frames of windows or rooflights and frames in which glazing is fitted
	the underside of a gallery	architraves, cover moulds, picture rails
	the underside of a roof exposed to the room below	exposed beams and similar narrow members

Compartmentation

Compartment walls and compartment floors should be provided if a domestic garage is attached to (or forms an integral part of) a dwelling house, then the garage should be separated from the rest of the dwelling house, as shown in Figure 6.249. B3 5.4 (V1)

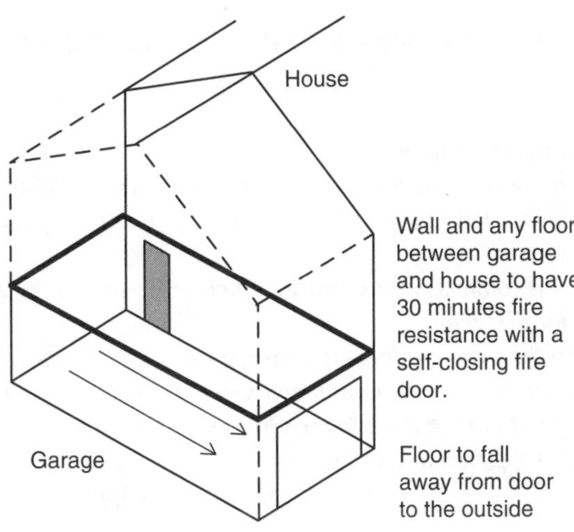
House
Wall and any floor between garage and house to have 30 minutes fire resistance with a self-closing fire door.
Garage
Floor to fall away from door to the outside

Figure 6.249 Separation between garage and dwelling house

Note: The wall and any floor between the garage and the house shall have a 30 minute standard of fire resistance. Any opening in the wall should be at least 100 mm above the garage floor level with an FD30 door.

If a door is provided between a dwelling house and the garage, the floor of the garage should: B3 5.5 (V1)

- be laid so as to allow fuel spills to flow away from the door to the outside; or
- the door opening should be positioned at least 100 mm above garage floor level.

Self-closing devices

Fire doors now **only** need to be provided with self closing devices, **if** they are between a dwelling house and an integral garage.

6.35 Conservatories

6.35.1 Requirements

Ventilation

There shall be adequate means of ventilation provided for people in the building.

(Approved Document F)

Conservation of fuel and power

Reasonable provision shall be made for the conservation of fuel and power in buildings by:

(a) limiting heat gains and losses:
  (i) through thermal elements and other parts of the building fabric; and
  (ii) from pipes, ducts and vessels used for space heating, space cooling and hot water services;

(b) providing and commissioning energy-efficient fixed building services with effective controls; and

(c) providing to the owner sufficient information about the building, the fixed building services and their maintenance requirements so that the building can be operated in such a manner as to use no more fuel and power than is reasonable in the circumstances.

(Approved Document L1)

6.35.2 Meeting the requirement

Ventilation

Ventilation through a conservatory

Habitable rooms without an openable window may be ventilated through a conservatory (see Figure 6.250) provided that that conservatory has: • purge ventilation and • an 8000 mm^2 background ventilator and • there is a closable opening between the room and the conservatory that is equipped with: – purge ventilation and – an 8000 mm^2 background ventilator.	F 1.14

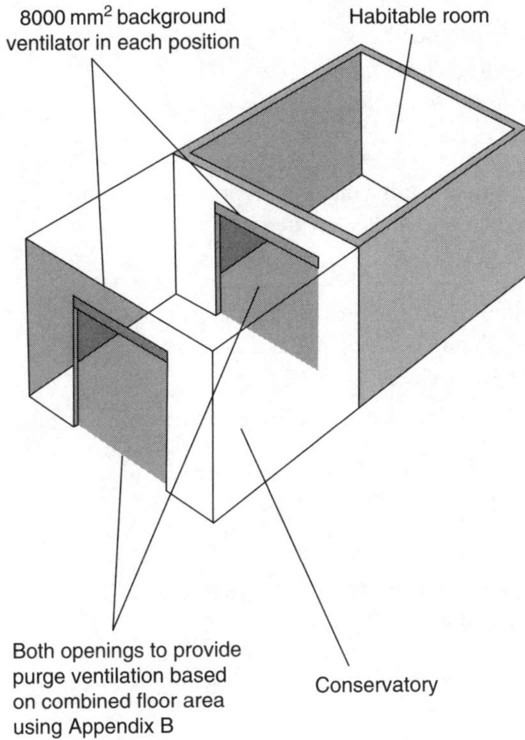

Figure 6.250 A habitable room ventilated through a conservatory

The addition of a conservatory to an existing building

The general ventilation rate for conservatories with a floor area greater than 30 m^2 conservatory (and adjoining rooms) can be achieved by the use of background ventilators.	F 3.18

Conservation of fuel and power

If a conservatory is built as part of the new dwelling, then the performance of the dwelling should be assessed as if the conservatory were **not** there and approved Document L1B should be followed in respect of the construction of the conservatory itself.

Conservatories built at ground level and with a floor area less than 30 m^2 are currently (i.e. as at July 2006) exempt from the Building Regulations.

The thermal separation between a dwelling and its conservatory must '*be constructed to a standard comparable to the rest of the external envelope of the dwelling*'.	L1A 15
There shall be an effective thermal separation between the conservatory and the heated area in the existing building.	L1B 22a L2B 32a
The walls, doors and windows between the building and the extension should be insulated and weather-stripped to at least the same extent as in the existing building.	L1B 22a L2B 32a
Independent temperature and on/off controls to all heating appliances shall be provided.	L1B 22b L2B 32b
The heating system should use heat-raising appliances with an efficiency not less than that recommended in the Heating Compliance Guided ICCM.	L1B 22b L2B 32b
The heating system should have independent temperature and on/off controls.	L1B 22b
Glazed elements shall comply with the standards given in Table 6.104.	L1B 22c L2B 32c

The U-value of any individual element should be no worse than 0.70 W/m^2K and with an area-weighted U-value of 0.25 W/m^2K.	L1B 22c L2B 32c
Newly constructed thermal elements that are part of an extension should be no worse than 0.22 W/m^2K.	L1B 50 L2B 70a
Thermal elements constructed as replacements for existing elements, or elements that are being renovated, should be no worse than 0.25 W/m^2K.	L1B 51 and 54 L2B 86 and 88
Retained thermal elements whose U-value is worse than the 0.70 W/m^2K threshold value shall be upgraded to achieve the improved U-value of 0.25 W/m^2K.	L1B 57

Table 6.104 Standards for glazed elements in conservatories

Element	Type of building	Replacement fittings (W/m^{2}K)
Windows, roof windows, rooflights and doors	Existing dwelling	2.2 (whole unit)
	Existing building	1.2 (centre pane)
Doors with more than 50% of their internal face glazed existing building	Existing dwelling	2.2 (whole unit)
	Existing building	1.2 (centre pane)
Doors with less than 50% of their internal face glazed	Existing dwelling	3.0
	Existing building	
High-usage entrance doors	Existing building	6.0
Vehicle access and large doors	Existing building	1.5
Roof ventilators	Existing building	6.0

6.36 Rooms for residential purposes

6.36.1 The requirement

'Rooms for residential purposes' are defined in Regulation 2 of the Building Regulations 2000 (as amended) and need to conform with the applicable Approved Documents (see Section 5.23) and meet the requirements for airborne and impact sound insulation shown in Table 6.105.

Table 6.105 Dwelling-houses and flats – performance standards for separating walls, separating floors and stairs that have a separating function

	Airborne sound insulation $D_{nT,w} + C_{tr}$ dB (minimum values)	Impact sound insulation $L'_{nT,w}$ dB (maximum values)
Purpose built rooms for residential purposes		
Walls	43	–
Floors and stairs	45	62

6.36.2 Meeting the requirement

Separating walls in new buildings containing rooms for residential purposes

Separating wall types 1 and 3 are considered to be the most suitable for use in new buildings containing rooms for residential purposes.

Note: Wall types 2 and 4 can be used **provided** that care is taken to maintain isolation between the leaves. Specialist advice may be needed.

Wall type 1 (solid masonry)

When using a solid masonry wall, the resistance to airborne sound depends mainly on the mass per unit area of the wall. As shown below, there are three different categories of solid masonry walls:

Table 6.106 Wall type 1 – categories

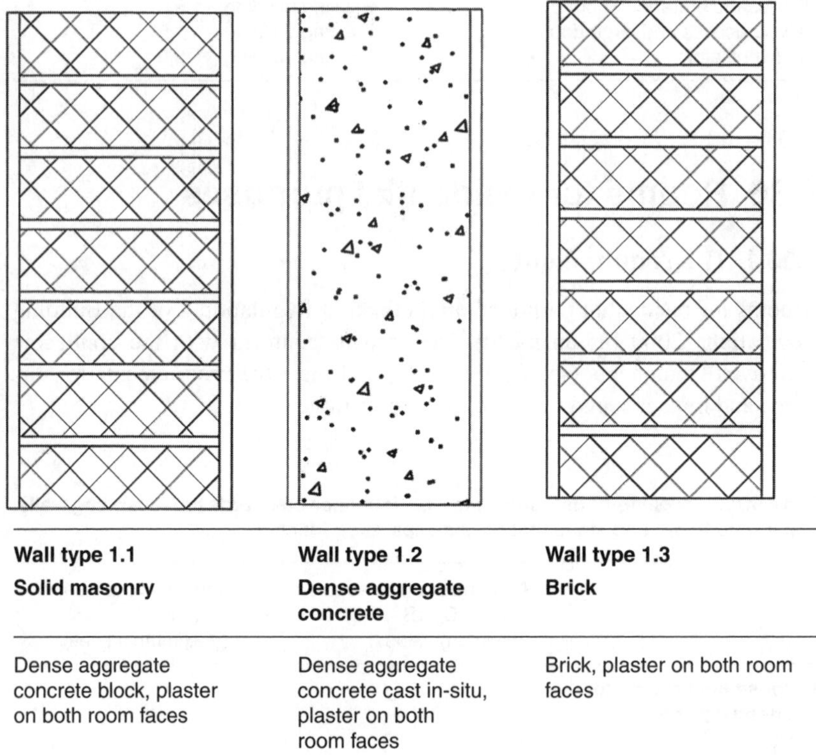

Wall type 1.1 **Solid masonry**	**Wall type 1.2** **Dense aggregate concrete**	**Wall type 1.3** **Brick**
Dense aggregate concrete block, plaster on both room faces	Dense aggregate concrete cast in-situ, plaster on both room faces	Brick, plaster on both room faces

Note: Plasterboard may be used as an alternative wall finish, provided a sheet of minimum mass per unit area 10 kg/m^2 is used on each room face.

Wall type 3 (masonry between independent panels)

Wall types 3.1 and 3.2 provide a high resistance to the transmission of both airborne sound and impact sound on the wall. Their resistance to sound depends partly on the type (and mass) of the core and partly on the isolation and mass of the panels.

Table 6.107 Wall type 3 – categories

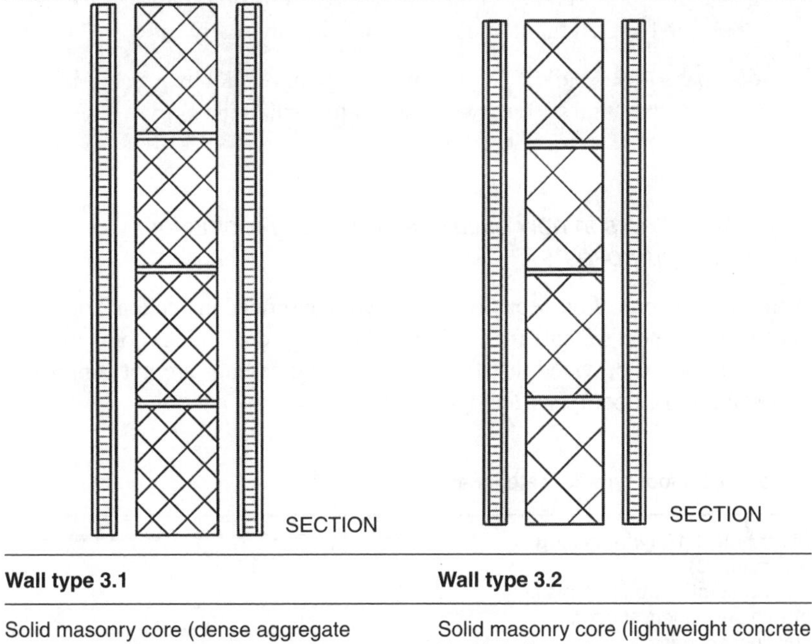

Wall type 3.1	**Wall type 3.2**
Solid masonry core (dense aggregate concrete block), independent panels on both room faces	Solid masonry core (lightweight concrete block), independent panels on both room faces

Corridor walls and doors in new buildings containing rooms for residential purposes

Separating walls as described in Table 6.107 should be used between rooms for residential purposes and corridors in order to control flanking transmission and to provide the required sound insulation between the dwelling and the corridor. Sound insulation will be reduced by the presence of a door.

All corridor doors shall have a good perimeter sealing (including the threshold where practical).	E6.6
All corridor doors shall have a minimum mass per unit area of 25 kg/m^2.	E6.6
All corridor doors shall have a minimum sound reduction index of 29 dB Rw (measured according to BS EN ISO 140-3:1995 and rated according to BS EN ISO 717-1:1997).	E6.6

All corridor doors shall meet the requirements for fire safety as described in Building Regulations Part B – Fire safety.	E6.6
Noisy parts of the building should preferably have a lobby, double door or high performance doorset to contain the noise.	E6.7

Separating floors in new buildings containing rooms for residential purposes

Although only one of the separating floor types described in Section 3 of Part E is considered most suitable for use in new buildings containing rooms for residential purposes, provided that floating floors and ceilings are not continuous between rooms, floor types 2 and 3 can also be used.

Table 6.108 Floor type 1 – categories

Floor type 1.1C (with ceiling treatment C)
Solid concrete slab (cast in-situ or with permanent shuttering), soft floor covering

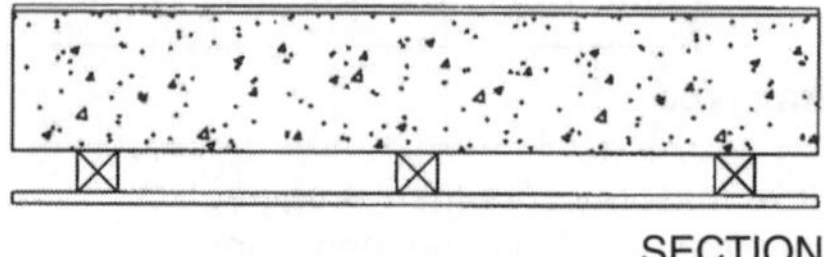

Floor type 1.2 (with ceiling treatment B)
Concrete planks (solid or hollow), soft floor covering

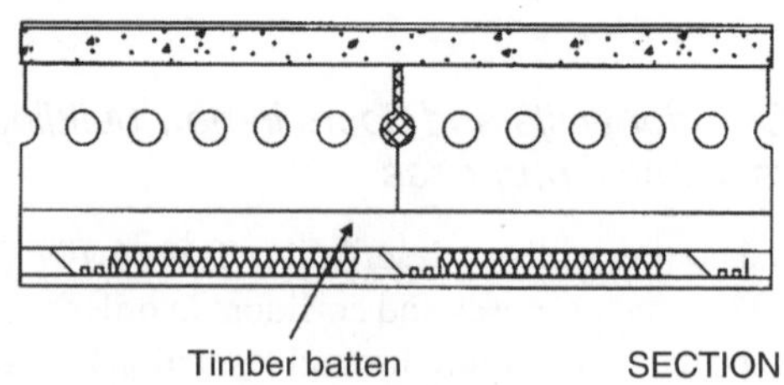

 **Note:** Specialist advice may be needed.

6.37 Rooms for residential purposes resulting from a material change of use

6.37.1 The requirement

'Rooms for residential purposes' formed by material change of use need to conform with the applicable Approved Documents (see Section 2.3) and meet the requirements for airborne and impact sound insulation shown in Table 6.109 below.

Table 6.109 Dwelling-houses and flats – performance standards for separating walls, separating floors and stairs that have a separating function

	Airborne sound insulation $D_{nT,w} + C_{tr}$ dB (minimum values)	Impact sound insulation $L'_{nT,w}$ dB (maximum values)
Rooms for residential purposes formed by material change of use		
Walls	43	–
Floors and stairs	43	64

6.37.2 Meeting the requirement

In some cases it may be that an existing wall, floor or stair in a building will achieve these performance standards without the need for remedial work, for example if the existing construction was already compliant. If this is not the case then the building work should be in compliance with the regulations concerning walls and floors as described previously in this book.

Rooms for residential purposes

Junction details

If there is a junction between a solid masonry separating wall type 1 and the ceiling void and roof space the solid wall need not be continuous to the underside of the structural floor or roof provided that:

- there is a ceiling consisting of two or more layers of plasterboard, of minimum total mass per unit area 20 kg/m^2; E6.14a
- there is a layer of mineral wool (minimum thickness 200 mm, minimum density 10 kg/m^3) in the roof void; E6.14b
- the ceiling is not perforated. E6.14c

As shown in Figure 6.251, the ceiling joists and plasterboard sheets should not be continuous between rooms for residential purposes.

Note: The ceiling void and roof space detail can only be used where the Requirements of Building Regulation Part B – Fire safety can also be satisfied. The Requirements of Building Regulation Part L – Conservation of fuel and power should also be satisfied.

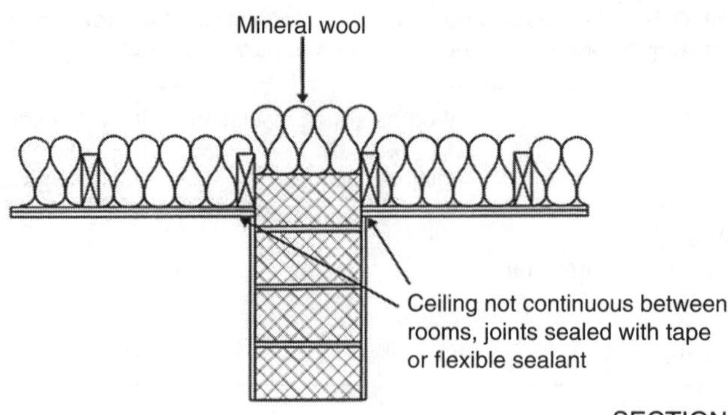

Figure 6.251 Ceiling void and roof space (only applicable to rooms for residential purposes)

Room layout and building services – design considerations

As internal noise levels are affected by room layout, building services and sound insulation, the actual layout of rooms should be considered at the design stage particularly to avoid placing noise sensitive rooms next to rooms in which noise is generated.

Note: See also:

- BS 8233:1999 Sound insulation and noise;
- Reduction for Buildings – Code of practice and sound control for homes.

6.38 Reverberation in the common internal parts of buildings containing flats or rooms for residential purposes

6.38.1 The requirement

Requirement E3 requires that '*Domestic buildings shall be designed and constructed so as to restrict the transmission of echoes.*'

6.38.2 Meeting the requirement

The guidance notes provided in Part E cover two Methods. These Methods assist in determining the amount of additional absorption to be used in corridors, hallways, stairwells and entrance halls that give access to flats and rooms for residential purposes.

	Entrance halls	Corridors	Hallways	Stairwells
Method A	Yes	Yes	Yes	Yes
Method B	Yes	Yes	Yes	No

Method A

Cover the ceiling area with the additional absorption.	E7.10
Cover the underside of intermediate landings, the underside of the other landings, and the ceiling area on the top floor.	E7.11
The absorptive material should be equally distributed between all floor levels.	E7.12
For stairwells (or a stair enclosure), calculate the combined area of the stair treads, the upper surface of the intermediate landings, the upper surface of the landings (excluding ground floor) and the ceiling area on the top floor. Either cover an area equal to this calculated area with a Class D absorber, or cover an area equal to at least 50% of this calculated area with a Class C absorber or better.	E7.11

Note: Method A can generally be satisfied by the use of proprietary acoustic ceilings.

Method B

In comparison to Method A, Method B takes account of the existing absorption provided by all of the surfaces. Section 7 of Part E provides details of how to calculate the total absorption area based on the material's absorption coefficient (α).

This can become a fairly specialist area and it would probably be advisable to seek professional advice and/or peruse BS EN 20354:1993 Acoustics – Measurement of sound absorption in a reverberation room or BS EN ISO 11654:1997 Acoustics – Sound absorbers for use in buildings – Rating of sound absorption for detailed information.

Absorption areas should be calculated for each octave band (in square metres) using the formula

$$A_T = \alpha_1 S_1 + \alpha_2 S_2 + \ldots \alpha_n S_n$$

Where:
A_T = total absorption area in sq. m
$\alpha_1 S_1$ = absorption coefficient for material 1
$\alpha_2 S_2$ = absorption coefficient for material 2
$\alpha_n S_n$ = absorption coefficient for the last type of material n

Table 6.110 Absorption coefficient data for common materials in buildings

Material	Sound absorption coefficient (α) in octave frequency bands (Hz)				
	250	500	1000	2000	4000
Fair-faced concrete or plastered masonry	0.01	0.01	0.02	0.02	0.03
Fair-faced brick	0.02	0.03	0.04	0.05	0.07
Painted concrete brick	0.05	0.06	0.07	0.09	0.08
Windows glass facade	0.08	0.05	0.04	0.03	0.02
Doors (timber)	0.10	0.08	0.08	0.08	0.08
Glazed tile/marble	0.01	0.01	0.01	0.02	0.02
Hard floor coverings (e.g. lino, parquet) on concrete floor	0.03	0.04	0.05	0.05	0.06
Soft floor coverings (e.g. carpet) on concrete floor	0.03	0.06	0.15	0.30	0.40
Suspended plaster or plasterboard ceiling (with large airspace behind)	0.15	0.10	0.05	0.05	0.05

Requirement E3 will be satisfied when:

For entrance halls, provide a minimum of 0.20 m^2 total absorption area per cubic metre of the volume. E7.17

For corridors or hallways, provide a minimum of 0.25 m^2 total absorption area per cubic metre of the volume. E7.18

For corridors or hallways, provide a minimum of 0.25 m^2 total absorption area per cubic metre of the volume.

6.39 Internal walls and floors (new buildings)

6.39.1 The requirement

Dwellings shall be designed so that the noise from domestic activity in an adjoining dwelling (or other parts of the building) is kept to a level that:

- *does not affect the health of the occupants of the dwelling;*
- *will allow them to sleep, rest and engage in their normal activities in satisfactory conditions.*

(Approved Document E1)

Dwellings shall be designed so that any domestic noise that is generated internally does not interfere with the occupants' ability to sleep, rest and engage in their normal activities in satisfactory conditions.

(Approved Document E2)

Domestic buildings shall be designed and constructed so as to restrict the transmission of echoes.

(Approved Document E3)

Table 6.111 Dwelling-houses and flats – performance standards for separating walls, separating floors and stairs that have a separating function

	Airborne sound insulation $D_{nT,w} + C_{tr}$ dB (minimum values)	Impact sound insulation $L'_{nT,w}$ dB (maximum values)
Purpose built rooms for residential purposes		
Walls	43	–
Floors and stairs	45	62
Rooms for residential purposes formed by material change of use		
Walls	43	–
Floors and stairs	43	64

6.39.2 Meeting the requirement

Note: To avoid air paths between rooms, all gaps around internal walls and floors should be filled.

There are four main types of internal wall and three types of internal floor as detailed below.

Internal wall type A or B *Timber or metal frame*	The resistance to airborne sound depends on the mass per unit area of the leaves, the cavity width, frame material and the absorption in the cavity between the leaves.
Internal wall type C or D *Concrete or aircrete block*	The resistance to airborne sound depends mainly on the mass per unit area of the wall.
Internal floor type A or B *Concrete planks or concrete beams with infilling blocks*	The resistance to airborne sound depends on the mass per unit area of the concrete base or concrete beams and infilling blocks. A soft covering will reduce impact sound at source.
Internal floor type C *Timber or metal joist*	The resistance to airborne sound depends on the structural floor base, the ceiling and the absorbent material. A soft covering will reduce impact sound at source.

Internal wall type A
Timber or metal frames with plasterboard linings on each side of frame

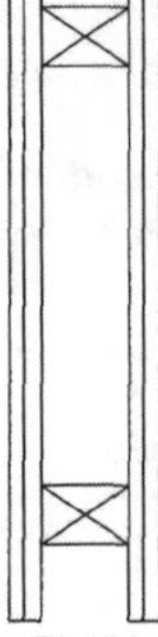

- each lining to be two or more layers of plasterboard,
- each sheet of minimum mass per unit area 10 kg/m^2
- linings fixed to timber frame with a minimum distance between linings of 75 mm (or metal frame with a minimum distance between linings of 45 mm)
- all joints are to be sealed

Internal wall type B
Timber or metal frames with plasterboard linings on each side of frame and absorbent material

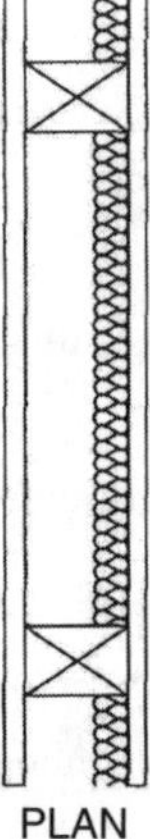

- single layer of plasterboard of minimum mass per unit area 10 kg/m^2
- linings fixed to timber frame with a minimum distance between linings of 75 mm (or metal frame with a minimum distance between linings of 45 mm)
- an absorbent layer of unfaced mineral wool batts or quilt (minimum thickness 25 mm, minimum density 10 kg/m^3) suspended in the cavity (may be wire reinforced)
- all joints well sealed

Internal wall type C
Concrete block wall, plaster or plasterboard finish on both sides

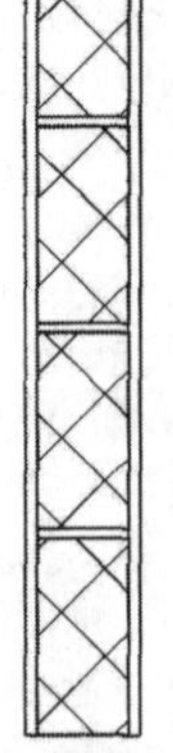

- minimum mass per unit area, excluding finish 120 kg/m^2
- all joints well sealed
- plaster or plasterboard finish on both sides

Internal wall type D
Aircrete block wall, plaster or plasterboard finish on both sides

Note: Internal wall type D should only be used with the separating walls described in section Ei (i.e. where there is no minimum mass requirement on the internal masonry walls) and it should not be used as a load-bearing wall or be rigidly connected to the separating floors.

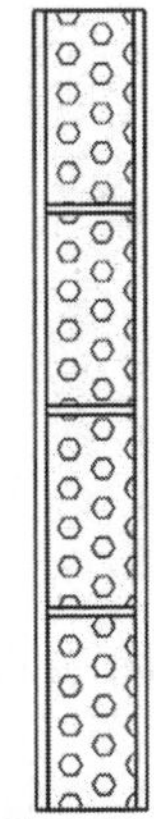
SECTION

- for plaster finish, minimum mass per unit area, including finish 90 kg/m^2
- for plasterboard finish, minimum mass per unit area, including finish 75 kg/m^2
- all joints well sealed

Internal floor type A
Concrete planks

Note: Insulation against impact sounds can be improved by adding a soft covering (e.g. carpet).

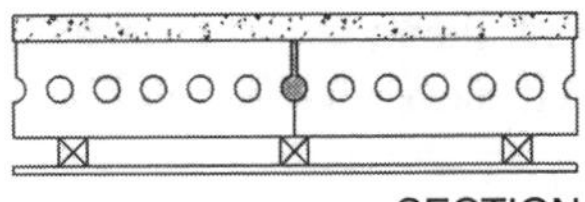
SECTION

- minimum mass per unit area 180 kg/m^2
- regulating screed optional
- ceiling finish optional

Internal floor type B
Concrete beams with infilling blocks, bonded screed and ceiling

Note: Insulation against impact sounds can be improved by adding a soft covering (e.g. carpet).

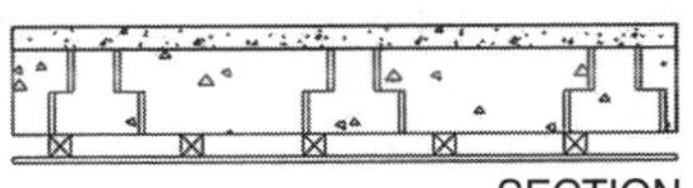
SECTION

- minimum mass per unit area of beams and blocks 220 kg/m^2
- bonded sand cement screeds with a minimum thickness of 40 mm
- ceiling finish required

Internal floor type C
Timber or metal joist, with wood-based board and plasterboard ceiling, and absorbent material

Note: Insulation against impact sounds can be improved by adding a soft covering (e.g. carpet).

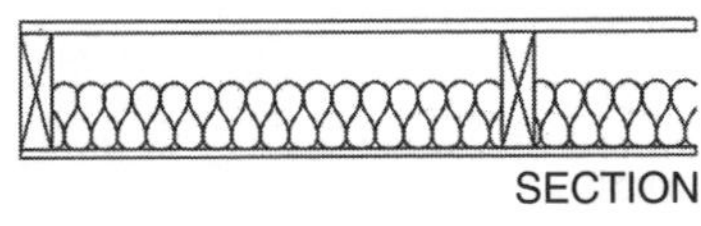
SECTION

- floor of timber or wood-based board, minimum mass per unit area 15 kg/m^2
- ceiling treatment of single layer of plasterboard, minimum mass per unit area 10 kg/m^2, fixed using any normal fixing method

	• an absorbent layer of mineral wool (minimum thickness 100 mm, minimum density 10 kg/m²) laid in the cavity

6.39.3 Other considerations

Doors

Lightweight doors with poor perimeter sealing provide a lower standard of sound insulation than walls.

This will reduce the effective sound insulation of the internal wall. Ways of improving sound insulation include ensuring that there is good perimeter sealing or by using a doorset.

Stairs

If the stair is not enclosed, then the potential sound insulation of the internal floor will not be achieved; nevertheless, the internal floor should still satisfy Requirement E2.

Noise reduction

It is good practice to consider the layout of rooms at the design stage to avoid placing noise sensitive rooms next to rooms in which noise is generated. Guidance on layout is provided in BS 8233:1999 Sound insulation and noise reduction for buildings – Code of Practice.

Electrical cables

Electrical cables give off heat when in use and special precautions may be required when they are covered by thermally insulating materials. See BRE BR 262, Thermal Insulation: avoiding risks, Section 2.3.

6.40 Regulation 7 – Materials and workmanship

6.40.1 The requirement

Building work shall be carried out –

(*a*) *with adequate and proper materials which –*
  (*i*) *are appropriate for the circumstances in which they are used;*
  (*ii*) *are adequately mixed or prepared; and*

(iii) *are applied, used or fixed so as adequately to perform the functions for which they are designed; and*

(b) *in a workmanlike manner.*

(Regulation 7)

Parts A to K and N of Schedule 1 to these regulations do not require anything to be done except for the purpose of securing reasonable standards of health and safety for persons in or about buildings (and any others who may be affected by buildings or matters connected with buildings).

6.40.2 Meeting the requirements

Materials (including products, components, fittings, naturally occurring materials, e.g. stone, timber and thatch, items of equipment, and backfilling for excavations in connection with building work) shall be:

- of a suitable nature and quality in relation to the purposes and conditions of their use;
- adequately mixed or prepared (where relevant); and
- applied, used or fixed so as to perform adequately the functions for which they are intended.

Environmental impact of building work

The environmental impact of building work can be minimized by careful choice of materials and the use of recycled and recyclable materials.	Reg. 7 (0.2)
The use of such materials must not have any adverse implications for the health and safety standards of the building work.	Reg. 7 (0.2)

Limitations

Reasonable standards of health or safety for persons in or about the building shall be assured.	Reg. 7 (0.3)
Materials and workmanship shall ensure that fuel and power is conserved.	Reg. 7 (0.3)
Access and facilities for disabled people shall be provided.	Reg. 7 (0.3)

There are no provisions under the Building Regulations for continuing control over the use of materials following the completion of building work.

Fitness of materials

The suitability of material used for a specific purpose shall be assessed using: Reg. 7 (1.2 to 1.7)

- British Standards
- European Standards (ENs)
- other national (and international) technical specifications (from other European Community member states)
- technical approvals (covered by a national or European certificate issued by a European technical approvals issuing body)

CE Marking (which provides a presumption of conformity with the stated minimum legal requirements)

- independent (UK) product certification schemes (such as those accredited by UKAS)
- tests and calculations (for example, those in conformance with UKAS's Accreditation Scheme for Testing Laboratories)
- past experience (that the material can be shown by experience, to be capable of performing the function for which it is intended)
- sampling (of materials by local authorities).

Short-lived materials and materials susceptible to changes in their properties should only be used in works where these changes do not adversely affect their performance.

Resistance to moisture

Any material that is likely to be adversely affected by condensation, moisture from the ground, rain or snow shall either resist the passage of moisture to the material or be treated or otherwise protected from moisture. Reg. 7 (1.8)

Resistance to substances in the subsoil

Any material in contact with the ground (or in the foundations) shall be capable of resisting attacks by harmful materials in the subsoil such as sulphates. Reg. 7 (1.9)

Establishing the adequacy of workmanship

The adequacy (and competence) of workmanship will normally be established by using:

- British Standard code of practice;
- workmanship specified and covered by a national or European certificate issued by a European technical approvals issuing body;
- the recommendations of an integrated management system (such as ISO 9001:2000);
- past experience (of workmanship that is capable of performing the function for which it is intended);
- tests (for example, the local authority has the power to test sewers and drains in or in connection with buildings).

Workmanship on building sites

In the main, local authorities will use the codes of practice as contained and detailed in BS 8000 (Workmanship on Building Sites) to monitor and inspect all building work. These codes of practice consist of:

- Part 1: 1989 Code of practice for excavation and filling
- Part 2: Code of practice for concrete work
  Section 2.1: 1990 Mixing and transporting concrete
  Section 2.2: 1990 Sitework with in situ and precast concrete
- Part 3: 1989 Code of practice for masonry
- Part 4: 1989 Code of practice for waterproofing
- Part 5: 1990 Code of practice for carpentry, joinery and general fixings
- Part 6: 1990 Code of practice for slating and tiling of roofs and claddings
- Part 7: 1990 Code of practice for glazing
- Part 8: 1994 Code of practice for plasterboard partitions and dry linings
- Part 9: 1989 Code of practice for cement/sand floor screeds and concrete floor toppings
- Part 10: 1995 Code of practice for plastering and rendering
- Part 11: Code of practice for wall and floor tiling
  Section 11.1: 1989 Ceramic tiles, Terrazzo tiles and mosaics (confirmed 1995)
  Section 11.2: 1990 Natural stone tiles
- Part 12: 1989 Code of practice for decorative wall coverings and painting
- Part 13: 1989 Code of practice for above ground drainage and sanitary appliances
- Part 14: 1989 Code of practice for below ground drainage
- Part 15: 1990 Code of practice for hot and cold water services (domestic scale)
- Part 16: 1997 Code of practice for sealing joints in buildings using sealants.

6.41 Work on existing constructions

If an existing building is going to be converted into dwelling houses (and/or flats) via a 'material change of use' then a certain amount of remedial work to the existing construction will first have to be completed. In the paragraphs below I have listed the most important areas that should be addressed, but if you are contemplating carrying out this sort of work, then it would be best to have a chat with your local authorities as they might have specific requirements and rulings for your own particular area.

6.41.1 Walls

In any type of building, the amount of sound resistance provided by a wall depends on:

- its construction;
- the type of independent panel(s) it uses (if any);
- the isolation of these panel(s);
- the type of absorbent material that has been used.

If the existing wall is masonry (and has a thickness of at least 100 mm and is plastered on both faces) then the following wall treatment (commonly referred to as *Wall treatment 1: Independent panel(s) with absorbent material*) is recommended.

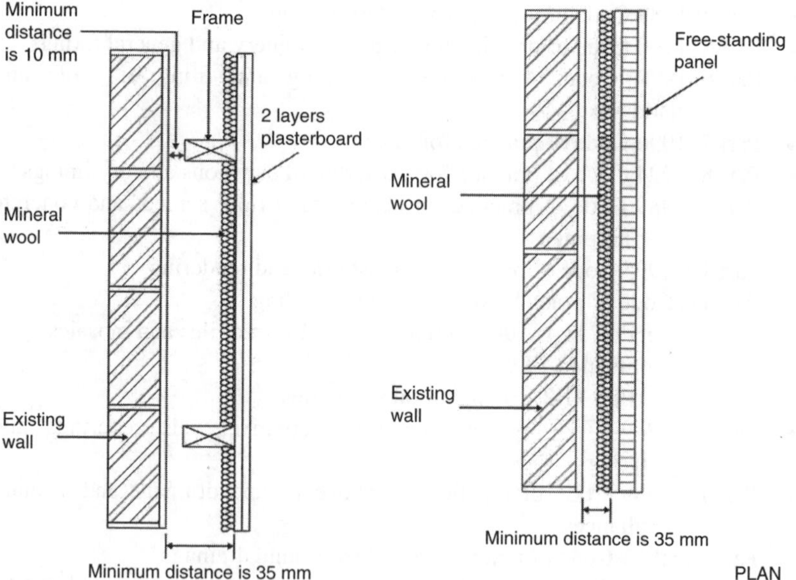

Figure 6.251 Wall treatment 1: Independent panel(s) with absorbent material

In particular:

The independent panel and its supporting frame should not be in contact with the existing wall.	E4.25a
The perimeter of the independent ceiling should be sealed with tape or sealant.	E4.25b
The absorbent material should not be tightly compressed as this may bridge the cavity.	E4.25

In addition:

- the minimum mass per unit area of panel (excluding any supporting framework) should be 20 kg/m^2;
- each panel should consist of at least 2 layers of plasterboard with staggered joints;
- if the panels are free-standing they should be at least 25 mm from masonry core;
- if the panels are supported on a frame, there should be a gap of at least 10 mm between the frame and the face of the existing wall;
- mineral wool (minimum density 10 kg/m^3 and minimum thickness 35 mm) should be used in the cavity between the panel and the existing wall.
- if a wall is common to two or more buildings, then it must be designed and constructed so that it adequately resists the spread of fire between those buildings;
- external walls must be constructed so as to have a low rate of heat release and thereby be capable of reducing the risk of ignition from an external source and the spread of fire over their surfaces;
- the amount of unprotected area in the sides of the building shall be restricted so as to limit the amount of thermal radiation that can pass through the wall.

(Approved Document B and B4)

Note: Wall linings may be required to reduce flanking transmission.

6.41.2 Floors

In buildings, the amount of resistance to airborne and impact sound will depend on:

- the combined mass of the existing floor and the independent ceiling;
- the amount of absorbent material;
- the isolation of the independent ceiling;
- the airtightness of the whole construction.

Two types of floor treatment are recommended dependent on the type of construction and material that has been used for the existing floor. The following requirements are common to both treatments:

- if the existing floor is timber, then gaps in the floor boarding should be sealed by overlaying with hardboard (alternatively, the gap should be filled with sealant);
- if floor boards are going to be replaced, boarding (minimum thickness 12 mm) and mineral wool (minimum thickness 100 mm, minimum density 10 kg/m^3) should be laid between the joists in the floor cavity;
- if the existing floor is concrete (and the mass per unit area of the concrete floor is less than 300 kg/m^2) then the mass of the floor should be increased to at least 300 kg/m^2;
- any air gaps through the concrete floor should be sealed.

Floor treatment 1: Independent ceiling with absorbent material

The resistance to airborne and impact sound from Floor treatment 1 depends on:

- the combined mass of the existing floor;
- the independent ceiling;
- the absorbent material;
- the isolation of the independent ceiling;
- the airtightness of the whole construction.

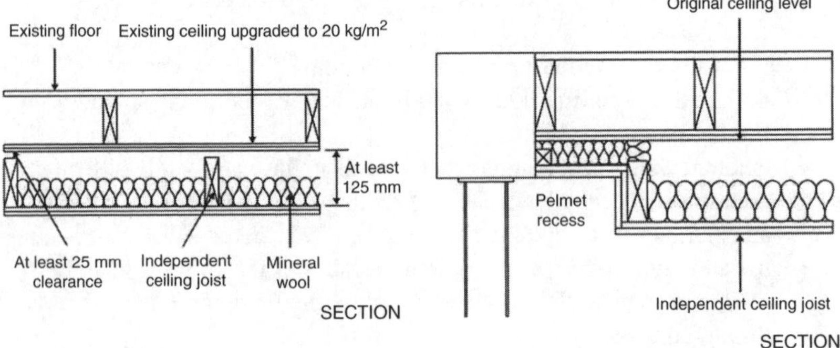

Figure 6.253 Floor treatment 1: Independent ceiling with absorbent material

Specifically:

The ceiling should have: E4.27

- at least 2 layers of plasterboard with staggered joints;
- a minimum total mass per unit area 20 kg/m^2;
- an absorbent layer of mineral wool laid on the ceiling; (minimum thickness 100 mm, minimum density 10 kg/m^3).

The ceiling should be supported by either: E2.47

- independent joists fixed only to the surrounding walls;

Note: A clearance of at least 25 mm should be left between the top of the independent ceiling joists and the underside of the existing floor construction

or

- independent joists fixed to the surrounding walls with additional support provided by resilient hangers attached directly to the existing floor base.

Where a window head is near to the existing ceiling, the new independent ceiling may be raised to form a pelmet recess. E4.29

The perimeter of the independent ceiling should be sealed with tape or sealant. E4.30

A rigid or direct connection should not be created between the independent ceiling and the floor base. E4.30

The absorbent material should not be tightly compressed as this may bridge the cavity. E4.30

Floor treatment 2: Platform floor with absorbent material

With Floor treatment 2, the resistance to airborne and impact sound depends on:

- the total mass of the floor;
- the effectiveness of the resilient layer;
- the absorbent material.

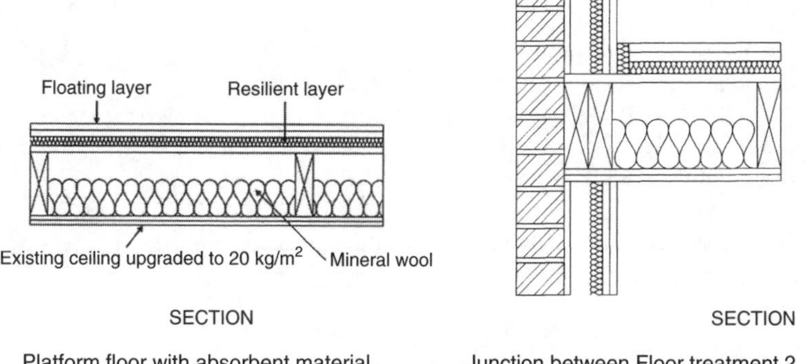

Figure 6.254 Floor treatment 2: Platform floor with absorbent material

Specifically:

Where this treatment is used to improve an existing timber floor, a layer of mineral wool (100 mm, minimum density 10 kg/m^3) should be laid between the joists in the floor cavity.	E4.32
The floating layer should be a: • minimum of two layers of board material; • minimum total mass per unit area 25 kg/m^2.	E4.32
Each layer should be: • a minimum thickness of 8 mm; • fixed together (e.g. spot bonded or glued/screwed) with joints staggered; • laid loose on a resilient layer.	E4.32
The resilient layer should be: • mineral wool (minimum thickness 25 mm, density 60–100 kg/m^3); • paper faced on the underside.	E4.32
The correct density of resilient layer (to carry the anticipated load) should be assured.	E4.32
The probable movement of materials after laying (e.g. expansion of chipboard) should be taken into account.	E4.32
The resilient layer should be carried up at all room edges to isolate the floating layer from the wall surface.	E4.32
A small 5 mm gap should be left between the skirting and floating layer and filled with a flexible sealant.	E4.32
Resilient materials should be laid in sheets with joints tightly butted and taped.	E4.32
The perimeter of any new ceiling should be sealed with tape or sealant.	E4.32
The floating layer and the base or surrounding walls should not be bridged with services or fixings that penetrate the resilient layer.	E4.32

Loadbearing elements of structure

When altering an existing two-storey, single-family dwelling house to provide additional storeys, the floor(s), both old and new, shall have the full 30 minute standard of fire resistance.	B3 4.7 (V1)
All loadbearing elements of a structure shall have a minimum standard of fire resistance.	B3 4.1 (V1) B3 7.1 (V2)
Structural frames, beams, floor structures and gallery structures should have at least the fire resistance given in Appendix A of Approved Document B.	B3 4.2 (V1) B3 7.2 (V2)

Fire resistance – compartmentation

To prevent the spread of fire within a building, whenever possible, the building should be sub-divided into compartments separated from one another by walls and/or floors of fire-resisting construction.	B3 5.1 (V1) B3 8.1 (V2)
Parts of a building that are occupied mainly for different purposes should be separated from one another by compartment walls and/or compartment floors.	B3 5.3 (V1) B3 8.11 (V2)
The wall and any floor between the garage and the house shall have a 30 minute fire resistance. Any opening in the wall to be at least 100 mm above the garage floor level with an FD30 door.	B3
In buildings containing flats or maisonettes compartment walls or compartment floors shall be constructed between: • every floor (unless it is within a maisonette); • one storey and another within one dwelling; • every wall separating a flat or maisonette from any other part of the building; • every wall enclosing a refuse storage chamber.	B3 8.13 (V2)
Every compartment floor should:	B3 5.6 (V1)
• form a complete barrier to fire between the compartments they separate; and • have the appropriate fire resistance as indicated in Appendix A of Approved Document B, Tables A1 and A2.	B3 8.20 (V2)

Where a compartment wall or compartment floor meets another compartment wall, or an external wall, the junction should maintain the fire resistance of the compartmentation.	B3 5.9 (V1) B3 8.25 (V2)
Junctions between a compartment floor and an external wall that has no fire resistance (such as a curtain wall) should be restrained at floor level to reduce the movement of the wall away from the floor when exposed to fire.	B2 8.26 (V2)
Compartment walls should be able to accommodate the predicted deflection of the floor above by either: • having a suitable head detail between the wall and the floor, that can deform but maintain integrity when exposed to a fire; or • the wall may be designed to resist the additional vertical load from the floor above as it sags under fire conditions and thus maintain integrity.	B2 8.27 (V2)

Under floor voids

Extensive cavities in floor voids should be subdivided with cavity barriers.	B3

6.41.3 Ceilings

If there is an existing lath and plaster ceiling it should be retained as long as it satisfies Building Regulation Part B – Fire safety.

If the existing ceiling is not lath and plaster, it should be upgraded to provide at least two layers of plasterboard with joints staggered, total mass per unit area 20 kg/m^2.

Note: Care should be taken at the design stage to ensure that adequate ceiling height is available in all rooms to be treated.

Ceiling linings

To inhibit the spread of fire within the building, ceiling internal linings shall:

- adequately resist the spread of flame over their surfaces; and
- have, if ignited, a rate of heat release or a rate of fire growth which is reasonable in the circumstances.

Note: Flame spread over wall or ceiling surfaces is controlled by ensuring that the lining materials or products meet given performance levels which are measured in terms of performance with reference to Tables A1 and A3 of Part B.

6.41.4 Corridor walls and doors

A separating wall should be used between the new dwellings (and/or flats) and the adjacent corridor in order to control flanking transmission and provide the required amount of sound insulation.

Note: It is likely that the sound insulation will be reduced by the presence of a door.

In particular, measures should be taken to ensure that any door:

- has good perimeter sealing (including the threshold where practical); E4.20
- has a minimum mass per unit area of 25 kg/m^2 or a minimum sound reduction index of 29 dB R_W (measured according to BS EN ISO 1403: 1995 and rated according to BS EN ISO 7171: 1997);
- satisfies the Requirements of Building Regulation Part B – Fire safety. E4.20

Noisy parts of the building should preferably have a lobby, double door or high performance doorset to contain the noise. E4.21

Note: These facilities should also comply with Building Regulation Part M – Access and facilities for disabled people.

6.41.5 Stairs

If stairs form a separating function then they are subject to the same sound insulation requirements as floors. In all cases, the resistance to airborne sound depends mainly on:

- the mass of the stair;
- the mass and isolation of any independent ceiling;
- the airtightness of any cupboard or enclosure under the stairs;
- the staircovering (which reduces impact sound at source).

The following wall treatment (commonly referred to as *Stair treatment 1: Stair covering and independent ceiling with absorbent material*) is recommended.

It should be noted:

The soft covering should be: E4.37

- at least 6 mm thickness;
- laid over the stair treads;

- be securely fixed (e.g. glued) so it does not become a safety hazard.

If there is a cupboard under all, or part, of the stair: E4.37

- the underside of the stair within the cupboard should be lined with plasterboard (minimum mass per unit area 10 kg/m^2) together with an absorbent layer of mineral wool (minimum density 10 kg/m^3);
- the cupboard walls should be built from two layers of plasterboard (or equivalent), each sheet with a minimum mass per unit area 10 kg/m^2;
- a small, heavy, well fitted door should be fitted to the cupboard.

If there is no cupboard under the stair, an independent ceiling should be constructed below the stair (see Floor treatment 1). E4.37

Where a staircase performs a separating function it shall conform to Building Regulation Part B – Fire safety. E4.38

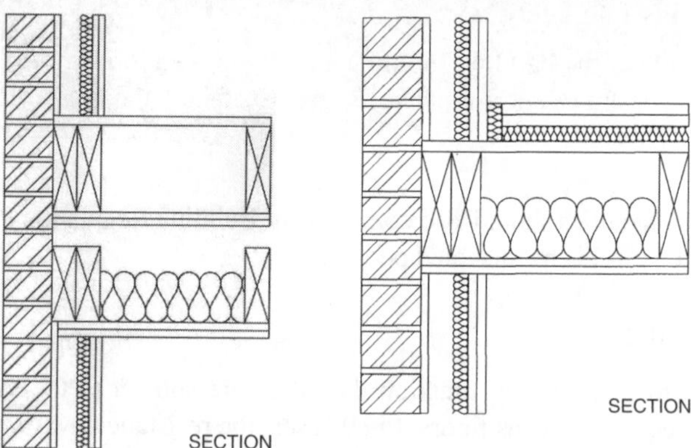

Figure 6.255 Stair treatment 1: Stair covering and independent ceiling with absorbent material

6.41.6 Junction requirements for material change of use

There are three recommended types of junctions that can be made:

Junctions with abutting construction

The perimeter of any new ceiling should be sealed with tape or caulked with sealant. E4.41

For floating floors:

- the resilient layer shall be carried up at all room edges to isolate the floating layer from the wall surface; E4.39
- a small 5 mm gap, filled with flexible sealant, should be left between the skirting and floating layer. E4.40

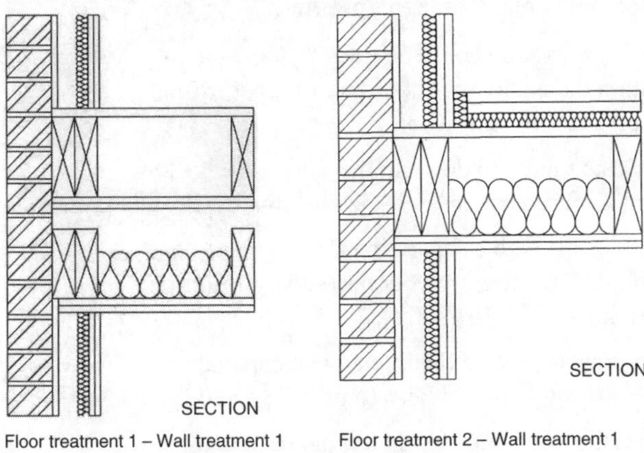

Figure 6.256 Junctions with abutting construction

Junctions with external or load bearing walls

If the adjoining masonry wall has a mass per unit area less than 375 kg/m^2, then the walls should be lined with either:

- an independent layer of plasterboard, or E4.43a
- a laminate of plasterboard and mineral wool. E4.43b

Note: Specialist advice may be needed on the diagnosis and control of flanking transmission.

Junctions with floor penetrations

Piped services (excluding gas pipes) and ducts which pass through separating floors in conversions should be surrounded with sound absorbent material for their full height and enclosed in a duct above and below the floor. E4.45

The joint between casings and ceiling should be sealed with tape or E4.46

A nominal 5 mm gap, sealed with sealant, should be left between the casing and any floating layer.	E4.46
Pipes and ducts that penetrate a floor separating habitable rooms in different flats should be enclosed for their full height in each flat.	E4.46
The enclosure should be constructed of material having a mass per unit area of at least 15 kg/m^2.	E4.47
Either the enclosure should be lined or the duct or pipe within the enclosure should be wrapped with 25 mm unfaced mineral wool.	E4.48
The enclosure may go down to the floor base if Floor treatment 2 is used but ensure isolation from the floating layer.	E4.49
Penetrations through a separating floor by ducts and pipes should have fire protection to satisfy Building Regulation Part B – Fire safety.	E4.50
Fire stopping should be flexible and be capable of preventing a rigid contact between the pipe and floor.	E4.50
Gas pipes may be contained in a separate (ventilated) duct or can remain unenclosed.	E3.120
If a gas service is installed it shall comply with the Gas Safety (Installation and Use) Regulations 1998, SI 1998 No 2451.	E3.120

Note: All of these facilities should also comply with Building Regulation Part M – Access and facilities for disabled people.

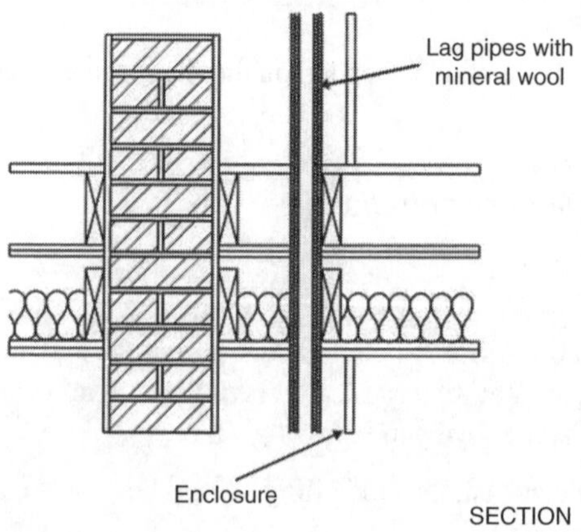

Figure 6.257 Floor penetrations

Bibliography

Standards referred to

British Standards

Compliance with a British Standard does not of itself confer immunity from legal obligations. British Standards can, however, provide a useful source of information which could be used to supplement or provide an alternative to the guidance given in this Approved Document.

When an Approved Document makes reference to a named standard, the relevant version of the standard is the one listed at the end of the publication. However, if this version of the standard has been revised or updated by the issuing standards body, the new version may be used as a source of guidance provided it continues to address the relevant requirements of the Regulations.

Drafts for Development (DDs) are not British Standards. They are issued in the DD series of publications and are of a provisional nature. They are intended to be applied on a provisional basis so that information and experience of their practical application may be obtained and the document developed. Where the recommendations of a DD are adopted then care should be taken to ensure that the requirements of the Building Regulations are adequately met. Any observations that a user may have in relation to any aspect of a DD should be passed on to BSI.

Title	Standard
Acoustics – Measurement of sound absorption in a reverberation room	BS EN 20354:1993
Acoustics – Measurement of sound insulation in buildings and of building elements Part 3: Laboratory measurement of airborne sound insulation of building elements	BS EN ISO 140-3:1995
Acoustics – Measurement of sound insulation in buildings and of building elements Part 4: Field measurements of airborne sound insulation between rooms	BS EN ISO 140-4:1998
Acoustics – Measurement of sound insulation in buildings and of building elements Part 8: Laboratory measurements of the reduction of transmitted impact noise by floor coverings on a heavyweight standard floor	BS EN ISO 140-8:1998

Title	Standard
Acoustics – Measurement of sound insulation in buildings and of building elements Part 6: Laboratory measurements of impact sound insulation of floors	EN ISO 140-6:1998
Acoustics – Measurement of sound insulation in buildings and of building elements Part 7: Field measurements of impact sound insulation of floors	BS EN ISO 140-7:1998
Acoustics – Method for the determination of dynamic stiffness Part 1: Materials used under floating floors in dwellings.	BS EN 29052-1:1992
Acoustics – Rating of sound insulation in buildings and of building elements Part 1: Airborne sound insulation	BS EN ISO 717-1:1997
Acoustics – Rating of sound insulation in buildings and of building elements Part 2: Impact sound insulation	BS EN ISO 717-2:1997
Acoustics – Sound/absorbers for use in buildings – Rating of sound absorption	BS EN ISO 11654:1997
Aggregates for concrete	BS EN 12620:2002
Application of fire safety engineering principles to the design of buildings. Code of practice	BS 7974:2001
Building components and building elements – Thermal resistance and thermal transmittance – Calculation method	BS EN ISO 6946:1997
Building hardware. Panic exit devices operated by a horizontal bar. Requirements and test methods	BS EN 1125:1997
Building materials and products – Hygrothermal properties – Tabulated design values	BS EN 12524:2000
Buildings and Structures for Agriculture. Various relevant parts including – Part 33: 1991 Guide to the control of odour pollution and Part 52: 1991 Code of practice for design of alarm systems, emergency ventilation and smoke ventilation for livestock housing	BS 5502
Capillary and compression tube fittings of copper and copper alloy Part 2: 1983 Specification for capillary and compression fittings for copper tubes	BS 864
Cast iron pipes and fittings, their joints and accessories for the evacuation of water from buildings Requirements, test methods and quality assurance	BS EN 877:1999
Cement Part 1: 2000: Composition, specifications and conformity criteria for common elements Part 2: 2000: Conformity evaluation	BS EN 197

Chimneys. Clay/ceramic flue blocks for single wall chimneys. Requirements and test methods	BS EN 1806: 2000
Chimneys. Clay/ceramic flue liners. Requirements and test methods	BS EN 1457: 1999
Chimneys. General requirements	BS EN 1443: 1999
Chimneys. Metal chimneys. Test methods	BS EN 1859:2000
Code of practice for accommodation of building services in ducts	BS 8313: 1997
Code of practice for assessing exposure of walls to wind-driven rain	BS 8104:1992
Code of practice for building drainage	BS 8301: 1985
Code of practice for daylighting	BS 8206 Part 2
Code of practice for design and installation of damp-proof courses in masonry construction	BS 8215: 1991
Code of practice for design and installation of natural stone cladding and lining	BS 8298:1994
Code of practice for design and installation of non-loadbearing precast concrete cladding	BS 8297:2000
Code of practice for design and installation of small sewage treatment works and cesspools	BS 6297: 1983
Code of practice for design of non-loadbearing external vertical enclosures of buildings	BS 8200: 1985
Code of practice for drainage of roofs and paved areas	BS 6367: 1983
Code of practice for earth retaining structures	BS 8002: 1994
Code of practice for external renderings	BS 5262:1991
Code of practice for fire door assemblies with non-metallic leaves	BS 8214:1990
Code of practice for flues and flue structures in buildings	BS 5854: 1980 (1996)
Code of practice for foundations	BS 8004: 1986
Code of practice for mechanical ventilation and air-conditioning in buildings	BS 5720:1979
Code of practice for non-automatic firefighting systems in buildings	BS 9990:2006
Code of practice for oil firing Part 1: 1977 Installations up to 44 kW output capacity for space heating and hot water supply purposes Part 2: 1978 Installations of 44 kW or above output capacity for space heating, Hot water and steam supply purposes	BS 5410
Code of practice for powered lifting platforms for use by disabled persons (amendment due 1999)	BS 6440: 1983
Code of practice for protection of structures against water from the ground	BS 8102:1990

Title	Standard
Code of practice for protective barriers in and about buildings	BS 6180: 1995
Code of practice for sanitary pipework	BS 5572: 1978
Code of practice for sheet roof and wall coverings	BS CP 143
Code of practice for sheet roof and wall coverings: Part 14: 1975 Corrugated asbestos-cement	BS 5247
Code of practice for site investigations	BS 5930:1999
Code of practice for stone masonry	BS 5390: 1976 (1984)
Code of practice for the audio-frequency induction-loop systems (AFILS)	BS 7594:1993
Code of practice for the control of condensation in buildings	BS 5250:2002
Code of practice for the operation of fire protection measures. Electrical actuation of gaseous total flooding extinguishing systems	BS 7273-1:2006
Code of practice for the operation of fire protection measures. Electrical actuation of pre-action sprinkler systems	BS 7273-3:2000
Code of practice for the operation of fire protection measures. Mechanical actuation of gaseous total flooding and local application extinguishing systems	BS 7273-2:1992
Code of practice for the storage and on-site treatment of solid waste from buildings	BS 5906: 1980 (1987)
Code of practice for thermal insulation of cavity walls (with masonry or concrete inner and outer leaves) by filling with urea-formaldehyde (UF) foam systems	BS 5618: 1985
Code of practice for use of masonry Part 1: 1992 Structural use of un-reinforced masonry Part 2: 2000 Structural use of reinforced and prestressed masonry Part 3: 2001 Materials and components, design and workmanship	BS 5628
Code of practice for ventilation principles and designing for natural ventilation	BS 5925: 1991
Components for residential sprinkler systems. Specification and test methods for residential sprinklers	DD 252:2002
Components for smoke and heat control systems: Part 2: 1990 Specification for powered smoke and heat exhaust ventilators Part 6: 2005 Components for smoke and heat control systems. Specifications for cable systems	BS 7346

Part 7: 2006 Components for smoke and heat control systems. Code of practice on functional recommendations and calculation methods for smoke and heat control systems for covered car parks

Concrete — BS 8500
Part 1: 2002 Method of specifying and guidance for the specifier
Part 2: 2002 Specification for constituents materials and concrete

Concrete — BS 5328
Part 1: 1990 Guide to specifying concrete
Part 2: 1990 Method for specifying concrete mixes
Part 3: 1990 Specification for the procedures to be used in producing and transporting concrete
Part 4: 1990 Specification for the procedures to be used in sampling, testing and assessing compliance of concrete

Construction and testing of drains and sewers — BS EN 1610:1998

Copper and copper alloys. Plumbing fittings — BS EN 1254:1998
Part 1: Fittings with ends for capillary soldering or capillary brazing to copper tubes
Part 2: Fittings with compression ends for use with copper tubes
Part 3: Fittings with compression ends for use with plastics pipes
Part 4: Fittings combining other end connections with capillary or compression ends
Part 5: Fittings with short ends for capillary brazing to copper tubes

Copper and copper alloys. Seamless, round copper tubes for water and gas in sanitary and heating applications — BS EN 1057:1996

Copper indirect cylinders for domestic purposes
Specification for double feed indirect cylinders — BS 1566-1:1984

Design of buildings and their approaches to meet the needs of disabled people – Code of practice — BS 8300:2001

Discharge and ventilating pipes and fittings, sand-cast or spun in cast iron — BS 416
Part 1: 1990 Specification for spigot and socket systems
Part 2: 1990 Specification for socket less systems

Drain and sewer systems outside buildings — BS EN 752
Part 1: 1996 Generalities and definitions
Part 2: 1997 Performance requirements
Part 3: 1997 Planning
Part 4: 1997 Hydraulic design and environmental aspects

Title	Standard
Part 5: 1997 Rehabilitation Part 6: 1998 Pumping installations Part 7: 1998 Maintenance and operations	
Ductile iron pipes, fittings, accessories and their joints for sewerage applications. Requirements and test methods	BS EN 598:1995
Emergency lighting. Code of practice for the emergency lighting of premises	BS 5266-1:2005
Emergency lighting: Part 1: 1988 Code of practice for the emergency lighting of premises other than cinemas and certain other specified premises used for entertainment	BS 5266
Factory-made insulated chimneys Part 1: 1990 (1996) Methods of Test, AMD 8379 Part 2: 1990 (1996) Specification for chimneys with stainless steel flue linings for use with solid fuel fired appliances Part 3: 1990 (1996) Specification for chimneys with stainless steel flue lining for use with oil fired appliances	BS 4543
Fibre cement flue pipes, fittings and terminals Part 1: 1991 (1998) Specification for light quality fibre cement flue pipes, fittings and terminals. Specifications for heavy quality cement flue pipes, fittings and terminals	BS 7435
Part 2: 1991	
Fire detection and alarm systems for buildings Part 1: 2002 Code of practice for system design, installation and servicing Part 2: 1983 Specification for manual call points Part 6: 1995 Code of practice for the design and installation of fire detection and alarm systems in dwellings Part 6: 2004 Code of practice for the design, installation and maintenance of fire detection and fire alarm systems in dwellings Part 8: 1998 Code of practice for the design, installation and servicing of voice alarm systems Part 9: Code of practice for the design, installation, commissioning and maintenance of emergency voice communication systems	BS 5839
Fire detection and fire alarm devices for dwellings. Specification for smoke alarms	BS 5446-1:2000
Fire detection and fire alarm devices for dwellings. Specification for heat alarms	BS 5446-2:2003

Fire detection and fire alarm systems. Manual call points BS EN 54-11:2001

Fire extinguishing installations and equipment on premises BS 5306
Part 1: 1976 (1988) Hydrant systems, hose reels and foam inlets
Part 2: 1990 Specification for sprinkler systems

Fire performance of external cladding systems BS 8414-2:2005
Test method for non-loadbearing external cladding systems fixed to and supported by a structural steel frame

Fire performance of external cladding systems BS 8414-1:2002
Test methods for non-loadbearing external cladding systems applied to the face of a building

Fire precautions in the design, construction and use of buildings BS 5588
Part 0: 1996 Guide to fire safety codes of practice for particular premises
Part 1: 1990 Code of practice for residential buildings
Part 4: 1998 Code of practice for smoke control using pressure differentials
Part 5: 1991 Code of practice for firefighting stairs and lifts
Part 6: 1991 Code of practice for places of assembly
Part 7: 1997 Code of practice for the incorporation of atria in buildings
Part 8: 1999 Code of practice for means of escape for disabled people
Part 9: 1989 Code of practice for ventilation and air conditioning ductwork
Part 10: 1991 Code of practice for shopping complexes
Part 11: 1997 Code of practice for shops, offices, industrial, storage and other similar buildings
Part 12: 2004 Code of practice for construction and use of buildings. Managing fire safety

Fire resistance tests for door and shutter assemblies BS EN 1634-2:2007
Part 2: Fire door hardware

Fire resistance tests for door and shutter assemblies BS EN 1634-1:2000
Fire doors and shutters

Fire resistance tests for door and shutter assemblies BS EN 1634-3:2001
Smoke control doors and shutters

Fire resistance tests for loadbearing elements BS EN 1365-1:1999
Part 1: 1999. Walls

Fire resistance tests for non-loadbearing elements BS EN 1364
Part 1: 1999 Walls
Part 2: 1999 Ceilings
Part 3: 2006 Curtain walling. Full configuration (complete assembly)

Fire resistance tests for service installations BS EN 1366

Title	Standard
Part 1: 1992 Ducts Part 2: 1999 Fire dampers Part 3: 2004 Penetration seals Part 4: 2006 Linear joint seals Part 5: 2003 Service ducts and shafts Part 6: 2004 Raised access and hollow core floors	
Fire safety signs, notices and graphic symbols: Part 1: 1990 Specification for fire safety signs	BS 5449
Fire tests on building materials and structures. Classification and method of test for external fire exposure to roofs	BS 476-3:2004
Fire tests on building materials and structures. Method for assessing the heat emission from building materials	BS 476-11:1982
Fire tests on building materials and structures. Method for determination of the fire resistance of elements of construction (general principles)	BS 476-20:1987
Fire tests on building materials and structures. Method of test for fire propagation for products	BS 476-6:1989
Fire tests on building materials and structures. Method of test to determine the classification of the surface spread of flame of products	BS 476-7:1997
Fire tests on building materials and structures. Methods for determination of the fire resistance of loadbearing elements of construction	BS 476-21:1987
Fire tests on building materials and structures. Methods for determination of the fire resistance of non-loadbearing elements of construction	BS 476-22:1987
Fire tests on building materials and structures. Methods for determination of the contribution of components to the fire resistance of a structure	BS 476-23:1987
Fire tests on building materials and structures. Non-combustibility test for materials	BS 476-4:1970
Fire tests on building materials and structures. Test methods and criteria for the fire resistance of elements of building construction (withdrawn)	BS 476-8:1972
Fire tests on building materials and structures. Method for determination of the fire resistance of ventilation ducts	BS 476-24:1987
Fire tests on building materials and structures: Part 3: 1958 External fire exposure roof tests Part 4: 1970 (1984) Non-combustibility test for materials Part 6: 1981 Method of test for fire propagation for products	BS 476

Part 6: 1989 Method of test for fire propagation for products
Part 7: 1971 Surface spread of flame tests for materials
Part 7: 1987 Method for classification of the surface spread of flame of products
Part 7: 1997 Method of test to determine the classification of the surface spread of flame of products. BS 476: Fire tests on building materials and structures
Part 8: 1972 Test methods and criteria for the fire resistance of elements of building construction
Part 11: 1982 (1988) Method for assessing the heat emission from building materials
Part 20: 1987 Method for determination of the fire resistance of elements of construction (general principles)
Part 21: 1987 Methods for determination of the fire resistance of loadbearing elements of construction
Part 22: 1987 Methods for determination of the fire resistance of non-loadbearing elements of construction
Part 23: 1987 Methods for determination of the contribution of components to the fire resistance of a structure
Part 24: 1987 Method for determination of the fire resistance of ventilation ducts
Part 31: Methods for measuring smoke penetration through doorsets and shutter assemblies
Section 31.1: 1983 Measurement under ambient temperature conditions

Fire-resistance tests. Fire dampers for air distribution systems. Classification, criteria and field of application of test results — BS ISO 10294-2:1999

Fire-resistance tests. Fire dampers for air distribution systems. Intumescent fire dampers — BS ISO 10294-5:2005

Fixed firefighting systems. Automatic sprinkler systems. Design, installation and maintenance — BS EN 12845:2004

Flat roofs with continuously supported coverings – Code of practice — BS 6229:2003

Flue blocks and masonry terminals for gas appliances — BS 1289-1: 1986
Part 1: 1986 Specification for precast concrete flue blocks and terminals
Part 2: 1989 Specification for clay flue blocks and terminals

Fuel oils for non-marine use — BS 2869: 1998
Part 2: 1988 Specification for fuel oil for agricultural and industrial engines and burners (Classes A2, Cl, C2, D, E, F, G and H)

Title	Standard
Glossary of terms relating to solid fuel burning equipment Part 1: 1994 Domestic appliances	BS 1846
Graphical symbols and signs. Safety signs, including fire safety signs. Specification for geometric shapes, colours and layout	BS 5499-1:2002
Gravity drainage systems inside buildings Part 1: Scope, definitions, general and performance requirements Part 2: Wastewater systems, layout and calculation Part 3: Roof drainage layout and calculation Part 4: Effluent lifting plants, layout and calculation Part 5: Installation, maintenance and user instructions	BS EN 12056:2000
Guide for design, construction and maintenance of single-skin air supported structures	BS 6661: 1986
Guide to assessment of suitability of external cavity walls for filling with thermal insulants Part 1: 1985 Existing traditional cavity construction	BS 8208
Guide to development and presentation of fire tests and their use in hazard assessment	BS 6336: 1998
Guide to the principles of the conservation of historic buildings	BS 7913: 1998
Heating Boilers. Heating boilers with forced draught burners. Terminology, general requirements, testing and marketing	BS EN 303-1: 1999
Hygrothermal performance of building components and building elements. Internal surface temperature to avoid critical surface humidity and interstitial condensation. Calculation methods	BS EN ISO 13788:2001
Installation and Maintenance of Flues and Ventilation for Gas Appliances of Rated Input not exceeding 70 Kw net Part 1: 2000 Specification for installation and maintenance of flues Part 2: 2000 Specification for installation of	BS 5440
chimneys and flues for domestic appliances burning solid fuel (including wood and peat) Part 1: 1984 (1998) Code of practice for masonry chimneys and flue pipes	BS 6461
Installation of domestic heating and cooking appliances burning solid mineral fuels Part 1: 1994 Specification for the design of installations Part 2: 1994 Specification for installing and commissioning on site	BS 8303

Part 3: 1994 Recommendations for design and on site installation

Installation of factory-made chimneys to BS 4543 for domestic appliances — BS 7566
Part 1: 1992 (1998) Method of specifying installation design information
Part 2: 1992 (1998) Specification for installation design
Part 3: 1992 Specification for site installation
Part 4: 1992 (1998) Recommendations for installation design and installation

Installations for separation of light liquids (e.g. petrol or oil) — BS EN 858:2001
Part 1: Principles of design, performance and testing, marking and quality control

Internal and external wood doorsets, door leaves and frames — BS 4787
Part 1: 1980 (1985) Specification for dimensional requirements

Investigation of potentially contaminated land. Code of practice — BS 10175:2001

Lifts and service lifts — BS 5655
Part 1: 1986 Safety rules for the construction and installation of electric lifts
(Part 1 to be replaced by BS EN 81-1, when published)
Part 2: 1988 Safety rules for the construction and installation of hydraulic lifts
(Part 2 to be replaced by BS EN 81-2, when published)
Part 5: 1989 Specifications for dimensions for standard lift arrangements
Part 7: 1983 Specification for manual control devices, indicators and additional fittings amendment slip

Lighting for buildings. Code of practice for daylighting — BS 8206-2:1992

Loading for buildings — BS 6399
Part 1: 1996 Code of practice for dead and imposed loads
Part 2: 1997 Code of practice for wind loads
Part 3 :1988 Code of practice for imposed roof loads

Measurement of sound insulation in buildings and of building elements — BS 2750
Part 1: 1980 Recommendations for laboratories
Part 3: 1980 Laboratory measurement of airborne sound insulation of building elements
Part 4: 1980 Field measurement of airborne sound insulation between rooms

Title	Standard
Part 6: 1980 Laboratory measurement of impact sound insulation of floors Part 7: 1980 Field measurements of impact sound insulation of floors	
Method for specifying thermal insulating materials for pipes, tanks, vessels, ductwork and equipment operating within the temperature range −40°C to +70°C	BS 5422:2001
Method of test for ignitability of fabrics used in the construction of large tented structures	BS 7157: 1989
Method of test for resistance to fire of unprotected small cables for use in emergency circuits	BS EN 50200:2006
Methods for rating the sound insulation in building elements Part 1: 1984 Method for rating the airborne sound insulation in buildings and interior building elements Part 2: 1984 Method for rating the impact sound insulation	BS 5821
Methods of test for flammability of textile fabrics when subjected to a small igniting flame applied to the face or bottom edge of vertically oriented specimens, Test 2	BS 5438: 1989
Methods of testing plastics Part 1: Thermal properties: Methods 120A to 120E: 1990 Determination of the Vicat softening temperature of thermoplastics	BS 2782
Methods of testing. Plastics. Introduction	BS 2782-0:2004
Oil Burning Equipment	
Part 5: 1987 Specification for Oil Storage Tanks	BS 799
Part 1: 2002 Fire classification of construction products and building elements. Classification using test data from reaction to fire tests Part 2: 2003 Fire classification of construction products and building elements. Classification using data from fire resistance tests, excluding ventilation services Part 3: 2005 Fire classification of construction products and building elements. Classification using data from fire resistance tests on products and elements used in building service installations: fire resisting ducts and fire dampers Part 4: 2005 Fire classification of construction products and building elements Part 4: Classification using data from fire resistance tests on smoke control systems	BS EN 13501

Part 5: 2005 Fire classification of construction products and building elements. Classification using data from external fire exposure to roof tests	
Part 2: 2000 Floors and roofs Part 3: 2000 Beams Part 4: 1999 Columns	BS EN 1365
Particleboards. Specifications. Requirements for load-bearing boards for use in humid conditions	BS EN 312-5:1997
Plastics piping systems for non-pressure underground drainage and sewerage. Unplasticized poly (vinylchloride) (PVC-U). Specifications for pipes, fittings and the system	BS EN 1401-1:1998
Plastics piping systems for soil and waste (low and high temperature) within the building structure. Acrylonitrilebutadiene-styrene (ABS). Specifications for pipes, fittings and the system	BS EN 1455-1:2000
Plastics piping systems for soil and waste discharge (low and high temperature) within the building structure. Unplasticized polyvinyl chloride (PVC-U). Specifications for pipes, fittings and the system	BS EN 1329-1:2000
Plastics piping systems for soil and waste discharge (low and high temperature) within the building structure. Polypropylene (PP). Specifications for pipes, fittings and the system	BS EN 1451-1:2000
Plastics piping systems for soil and waste discharge (low and high temperature) within the building structure. Polyethylene (PE). Specifications for pipes, fittings and the system	BS EN 1519-1:2000
Plastics piping systems for soil and waste discharge (low and high temperature) within the building structure	BS EN 1565-1:2000
Plastics piping systems for soil and waste discharge (low and high temperature) within the building structure. Chlorinated polyvinyl chloride) (PVC-C). Specification for pipes, fittings and the system	BS EN 1566-1:2000
Plastics. Thermoplastic materials. Determination of Vicat softening temperature (VST)	BS EN 150 306:2004
Plastics. Thermoplastic materials. Determination of Vicat softening temperature (VST)	BS EN ISO 306:2004
Precast concrete masonry units Part 1: 1981 – Specification for precast concrete masonry units	BS 6073
Precast concrete pipes fittings and ancillary products. Part 2: 1982 Specification for inspection chambers and street gullies Part 100: 1988 Specification for un-reinforced and reinforced pipes and fittings with flexible joints	BS 5911

Title	Standard
Part 101: 1988 Specification for glass composite concrete (GCC) pipes and fittings with flexible joints	
Part 120: 1989 Specification for reinforced jacking pipes with flexible joints	
Part 200: 1989 Specification for un-reinforced and reinforced manholes and soakaways of circular cross section	
Pressure sewerage systems outside buildings	BS EN 1671:1997
Profiled fibre cement. Code of practice	BS 8219:2001
Protection of buildings against water from the ground	BS CP 102:1973
Reaction to fire tests for building products. Building products excluding footings exposed to thermal attack by a single burning item	BS EN 13823:2002
Reaction to fire tests for building products. Conditioning procedures and general rules for selection of substrates	BS EN 13238:2001
Reaction to fire tests for building products. Determination of the heat of combustion	BS EN ISO 1716:2002
Reaction to fire tests for building products. Non-combustibility test	BS EN 150 1182:2002
Reaction to fire tests for building products. Non-combustibility test	BS EN ISO 1182:2002
Reaction to fire tests. Ignitability of building products subjected to direct impingement of flame. Single-flame source test	BS EN ISO 11925-2:2002
Recommendations for the storage and exhibition of archival documents	BS 5454: 2000
Refrigerating systems and heat pumps – safety and environmental requirements: Installation site and personal protection	BS EN 378 Part 3: 2000
Requirements for electrical installations (IEE Wiring Regulations 16th Edition)	BS 7671: 2001 (incorporating Amendments No l: 2002 and No 2: 2004)
Safety and control devices for use in hot water systems	BS 6283
Part 2: 1991 Specification for temperature relief valves for pressures from 1 bar to 10 bar	
Part 3: 1991 Specification for combined temperature and pressure relief valves for pressures from 1 bar to 10 bar	
Safety rules for the construction and installation of lifts – Particular applications for passenger and	BS EN 81-70:2003

good lifts – Accessibility to lifts for persons including persons with disability

Safety rules for the construction and installation of lifts. Electric lifts — BS EN 81-1:1998

Safety rules for the construction and installation of lifts. Hydraulic lifts — BS EN 81-2:1998

Safety rules for the construction and installation of lifts. Particular applications for passenger and goods passenger lifts. Firefighters' lifts — BS EN 81-72:2003

Sanitary installations — BS 6465
Part 1: 1984 Code of practice for scale of provision, selection and installation of sanitary appliances

Sanitary tapware. Waste fittings for basins, bidets and baths. General technical specifications — BS EN 274:1993

Small wastewater treatment plants less than 50 PE — BS EN 12566-1:2000

Smoke and heat control systems. Specification for powered smoke and heat exhaust ventilators — BS EN 12101-3:2002

Smoke and heat control systems. Specification for pressure differential systems. Kits — BS EN 12101-6:2005

Sound insulation and noise reduction for buildings – code of practice — BS 8233:1999

Specification for aggregates from natural sources for concrete — BS 882: 1983

Specification for ancillary components for masonry — BS EN 845
Part 1: 2001 Ties, tension straps, hangers and brackets
Part 2: 2001 Lintels
Part 3: 2001 Bed joint reinforcement of steel meshwork

Specification for asbestos-cement pipes, joints and fittings for sewerage and drainage — BS 3656: 1981 (1990)

Specification for calcium silicate (sandlime and flintlime) bricks — BS 187: 1978

Specification for cast iron spigot and socket drain pipes and fittings — BS 437: 1978

Specification for cast iron spigot and socket flue or smoke pipes and fittings — BS 41: 1973 (1981)

Specification for clay and calcium silicate modular bricks — BS 6649: 1985

Specification for clay bricks — BS 3921: 1985

Specification for clay flue linings and flue terminals — BS 1181: 1999

Specification for copper and copper alloys. Tubes — BS 2871
Part 1: 1971 Copper tubes for water, gas and sanitation

Specification for copper direct cylinders for domestic purposes — BS 699:1984

Title	Standard
Specification for copper hot water storage combination units for domestic purposes	BS 3198:1981
Specification for dedicated liquified petroleum gas appliances. Domestic flueless space heaters (including diffusive catalytic combustion heaters)	BS EN 449: 1997
Specification for design and construction of fully supported lead sheet roof and wall coverings	BS 6915:2001
Specification for design, installation, testing and maintenance of services supplying water for domestic use within buildings and their curtilages	BS 6700: 1987
Specification for direct surfaced wood chipboard based on thermosetting resins	BS 7331:1990
Specification for electrical controls for household and similar general purposes	BS 3955: 1986
Specification for fabrics for curtains and drapes Part 2: 1980 Flammability requirements	BS 5867
Specification for fibre boards	BS 1142: 1989
Specification for flexible joints for grey or ductile cast iron drain pipes and fittings (BS 437) and for discharge and ventilating pipes and fittings (BS 416)	BS 6087: 1990
Specification for impact performance requirements for flat safety glass and safety plastics for use in buildings	BS 6206: 1981
Specification for installation in domestic premises of gas-fired ducted-air heaters of rated input not exceeding 60 kW	BS 5864: 1989
Specification for installation of domestic gas cooking appliances (lst, 2nd and 3rd family gases)	BS 6172: 1990
Specification for installation of gas fired catering appliances for use in all types of catering establishments (lst, 2nd and 3rd family gases)	BS 6173: 2001
Specification for installation of gas fires, convector heaters, fire/back boilers and decorative fuel effect gas appliances Part 1: 2001 Gas fires, convector heaters and fire/back boilers and heating stoves (lst, 2nd and 3rd family gases) Part 2: 2001 Inset live fuel effect gas fires of heat input not exceeding 15 kW (2nd and 3rd family gases) Part 3: 2001 Decorative fuel effect gas appliances of heat input not exceeding 20 kW (2nd and 3rd family gases) AMD 7033	BS 5871

Specification for installation of gas-fired hot water boilers of rated input not exceeding 60 kW	BS 6798:2000
Specification for installation of hot water supplies for domestic purposes, using gas fired appliances of rated input not exceeding 70 kW	BS 5546: 2000
Specification for ladders for permanent access to chimneys, other high structures, silos and bins	BS 4211: 1987
Specification for low-voltage switchgear and controlgear assemblies. Particular requirements or low-volt age switchgear and control assemblies intended to be installed in places where unskilled persons have access to their use	BS EN 60439-3:1991
Specification for masonry units Part 1: 2003 Clay masonry units Part 2: 2001 Calcium silicate masonry units Part 3: Aggregate concrete masonry units Part 4: 2001 Autoclaved aerated concrete masonry units Part 5: Manufactured stone masonry units Part 6: 2001 Natural stone masonry units	BS EN 771
Specification for masonry units. Clay masonry units	BS EN 771-1:2003
Specification for metal flue pipes, fittings, terminals and accessories for gas-fired appliances with a rated input not exceeding 60 kW, AMD 8413	BS 715: 1993
Specification for metal ties for cavity wall construction	BS 1243: 1978
Specification for modular co-ordination in building	BS 6750: 1986
Specification for mortar for masonry Part 2: 2002: Masonry mortar	BS EN 998
Specification for open fireplace components	BS 1251: 1987
Specification for performance requirements for cables required to maintain circuit integrity under fire conditions	BS 6387: 1994
Specification for performance requirements for domestic flued oil burning appliances (including test procedures)	BS 4876: 1984
Specification for plastics inspection chambers for drains	BS 7158:2001
Specification for plastics waste traps	BS 3943: 1 979 (1988)
Specification for Portland cements	BS 12: 1989
Specification for powered stairlifts	BS 5776: 1996
Specification for prefabricated drainage stack units in galvanized steel	BS 3868:1995
Specification for quality of vitreous china sanitary appliances	BS 3402:1969

Title	Standard
Stairs, ladders and walkways. Code of practice for the design, construction and maintenance of straight stairs and winders	BS 5395-1:2000
Specification for safety aspects in the design, construction and installation of refrigerating appliances and systems	BS 4434: 1989
Specification for sizes of sawn and processed softwood	BS 4471: 1987
Specification for softwood grades for structural use	BS 4978: 1988
Specification for the use of structural steel in building Part 2: 1969 – Metric units	BS 449
Specification for thermoplastics waste pipe and fittings	BS 5255: 1989
Specification for thermostats for gas-burning appliances	BS 4201: 1979 (1984)
Specification for tongued and grooved softwood flooring	BS 1297: 1987
Specification for topsoil	BS 3882:1994
Specification for un-plasticized polyvinyl chloride (PVC- U) pipes and plastics fittings of nominal sizes 110 and 160 for below ground drainage and sewerage	BS 4660: 1989
Specification for unplasticized PVC pipe and fittings for gravity sewers	BS 5481: 1977 (1989)
Specification for unvented hot water storage units and packages	BS 7206: 1990
Specification for urea-formaldehyde (UF) foam systems suitable for thermal insulation of cavity walls with masonry or concrete inner and outer leaves	BS 5617: 1985
Specification for vitreous-enamelled low-carbon-steel fluepipes, other components and accessories for solid-fuel-burning appliances with a maximum rated output of 45 kW	BS 6999. 1989 (1996)
Specification for vitrified clay pipes, fittings and ducts, also flexible mechanical joints for use solely with surface water pipes and fittings	BS 65: 1991
Specification. Indicator plates for fire hydrants and emergency water supplies	BS 3251:1976
Sprinkler systems for residential and domestic occupancies. Code of practice	BS 9251:2005
Stainless steels. List of stainless steels	BS EN 10 088-1: 1995
Stairs, ladders and walkways Part 1: 1977 Code of practice for stairs Part 2: 1984 Code of practice for the design of helical and spiral stairs	BS 5395

Part 3: 1985 Code of practice for the design of industrial type stairs, permanent ladders and walkways	
Steel plate, sheet and strip, Part 2: 1983 Specification for stainless and heat-resisting steel plate, sheet and strip	BS 1449
Steel plate, sheet and strip. Carbon and carbon manganese plate, sheet and strip. General specifications	BS 1449-1: 1991
Structural design of buried pipelines under various conditions of loading. General requirements	BS EN 1295-1:1998
Structural design of low-rise buildings Part 1: 1995 Code of practice for stability, site investigation, foundations and ground floor slabs for housing Part 2: 1996 Code of practice for masonry walls for housing Part 3: 1996 Code of practice for timber floors and roofs for housing Part 4: 1995 Code of practice for suspended concrete floors for housing	BS 8103
Structural fixings in concrete and masonry Part 1: 1993 Method of test for tensile loading	BS 5080: 1993
Structural use of aluminium Part 1: 1991 Code of practice for design amendment slip Part 2: 1991 Specification for materials, workmanship and protection	BS 8118
Structural use of concrete Part 1: 1997 Code of practice for design and construction Part 2: 1985 Code of practice for special circumstances Part 3: 1995 Design charts for single reinforced beams, doubly reinforced beams and rectangular columns	BS 8110
Structural use of steelwork in building Part 1: 2000 Code of practice for design Part 2: 2001 Specification for materials, fabrication and erection Part 3: 1990 Design in composite construction Part 4: 1994 Code of practice for design of composite slabs with profiled steel sheeting Part 5: 1998 Code of practice for design of cold formed thin gauge sections	BS 5950
Structural use of timber Part 2: 2002 Code of practice for permissible stress design, materials and workmanship Part 3: 1998 Code of practice for trussed rafter roofs	BS 5268

Title	Standard
Part 6: Code of practice for timber framed walls Part 6.1: 1988 Dwellings not exceeding three storeys	
Test methods for external fire exposure to roofs	DD ENV 1187-.2002
Test methods for external fire exposure to roofs	ENV 1187:2002, test 4
The principles that should be applied when proposing work on historic buildings	BS 7913: 1998
Thermal bridges in building construction – calculation of heat flows and surface temperatures Part 1: General methods	BS EN ISO 10211-1:1996
Thermal bridges in building construction – calculation of heat flows and surface temperatures Part 2: Linear thermal bridges	BS EN ISO 10211-2:2001
Thermal insulation – determination of steady-state thermal transmission properties – calibrated and guarded hot box	BS EN ISO 8990:1996
Thermal insulation for use in pitched roof spaces in dwellings Part 5: Specification for installation of man-made mineral fibre and cellulose fibre insulation	BS 5803-5:1985 (as amended 1999)
Thermal insulation of cavity walls by filling with blown man-made mineral fibre Part 1: 1982 Specification for the performance of installation systems Part 2: 1982 Code of practice for installation of blown man-made mineral fibre in cavity walls with masonry and/or concrete leaves	BS 6232
Thermal performance of building materials and products – determination of thermal resistance by means of guarded hot plate and heat flow meter methods – dry and moist products of low and medium thermal resistance	BS EN 12664:2001
Thermal performance of building materials and products – determination of thermal resistance by means of guarded hot plate and heat flow meter methods – products of high and medium thermal resistance	BS EN 12667:2000
Thermal performance of building materials and products – determination of thermal resistance by means of guarded hot plate and heat flow meter methods – thick products of high and medium thermal resistance	BS EN 12939:2001
Thermal performance of buildings – heat transfer via the ground – calculation methods	BS EN ISO 13370:1998

Thermal performance of windows and doors – determination of thermal transmittance by hot box method Part 1: Complete windows and doors	BS EN ISO 12567-1:2000
Thermal performance of windows, doors and shutters – calculation of thermal transmittance Part 1: Simplified methods	BS EN ISO 10077-1:2000
Thermoplastics piping systems for non-pressure underground drainage and sewerage – structure walled piping systems of unplasticized poly (vinyl chloride) (PVC-U), Polypropylene (PP) and Polyethylene (PE) Part 1: Specification for pipes, fittings and the system Part 1: 1997 Guide to specifying concrete Part 2: 1997 Methods for specifying concrete mixes Part 3: 1990 Specification for the procedures to be used in producing and transporting concrete Part 4: 1990 Specification for the procedures to be used in sampling, testing and assessing compliance of concrete	BS EN 13476-1:2001
Unplasticized PVC soil and ventilating pipes of 82.4 mm minimum mean outside diameter, and fittings and accessories of 82.4 mm and of other sizes. Specification	BS 4514:2001
Vacuum drainage systems inside buildings.	BS EN 12109:1999
Vacuum sewerage systems outside buildings.	BS EN 1091:1997
Ventilation for buildings – performance testing of components/products for residential ventilation Part l: Externally and internally mounted air transfer devices	BS EN 13141-1: 2004
Ventilation for buildings – performance testing of components/products for residential ventilation Part 3: Range hoods for residential use devices	BS EN 13141-3: 2004
Ventilation for buildings – performance testing of components/products for residential ventilation Part 4: Fans used in residential ventilation systems	BS EN 13141-4: 2004
Ventilation for buildings – Performance testing of components/products for residential ventilation Part 6: Exhaust ventilation system packages used in a single dwelling	BS EN 13141-6: 2004
Ventilation for buildings – performance testing of components/products for residential ventilation Part 7: Performance testing of mechanical supply and exhaust ventilation units (including heat recovery) for mechanical ventilation systems intended for single family dwellings	BS EN 13141-7: 2003.

Title	Standard
Ventilation for buildings – performance testing of components/products for residential ventilation Part 8: Performance testing of unducted mechanical supply and exhaust ventilation units (including heat recovery) for mechanical ventilation systems intended for a single room	PrEN 13141-8: 2004
Ventilation for buildings – performance testing of components/products for residential ventilation Part 9: Humidity controlled external air inlet	PrEN 13141-9: 2004
Ventilation for buildings – performance testing of components/products for residential ventilation Part 10: Performance testing of unducted mechanical supply and exhaust ventilation units (including heat recovery) for mechanical ventilation systems intended for a single room	prEN 13141-10: 2004
Vitreous china washdown WC pans with horizontal outlet. Specification for WC pans with horizontal outlet for use with 7.5 l maximum flush capacity cisterns	BS 5503-3:1990
Vitrified clay pipes and fittings and pipe joints for drains and sewers Part 1: 1991 Test requirements Part 2: 1991 Quality control and sampling Part 3: 1991 Test methods	BS 295:
Vitrified clay pipes and fittings and pipe joints for drains and sewers Part 1: 1991 Test requirements Part 2: 1991 Quality control and sampling Part 3: 1991 Test methods Part 6: 1996 Requirements for vitrified clay manholes	BS EN 295:
Wall hung WC pen, specification for wall hung WC pans for Specification for WC pans with horizontal outlet for use with 7.5 l maximum flush capacity cisterns	BS 5504-4:1990
Wastewater lifting plants for buildings and sites – principles of construction and testing Part 1: Lifting plants for wastewater containing faecal matter Part 2: Lifting plants for faecal-free wastewater Part 3: Lifting plants for wastewater containing faecal matter for limited application	BS EN 12050:2001
Windows, doors and rooflights Part 1: Code of practice for safety in use and during cleaning of windows and doors (including guidance on cleaning materials and methods)	BS 8213: Part 1: 1991
Wood preservatives. Guidance on choice, use and application	BS 1282:1999

Wood stairs Part 1: 1989 Specification for stairs with closed risers for domestic use, including straight and winder flights and quarter and half landings	BS 585
Wood-based panels for use in construction – Characteristics, evaluation of conformity and marking	BS EN 13986: 2002
Workmanship on building sites Part 6: 1990 Code of practice for slating and tiling of roofs and claddings Part 13: 1989 Code of practice for above ground drainage and sanitary appliances Part 14:1989 Code of practice for below ground drainage	BS 8000

Note: Copies of all British Standards are available from: BSI, PO Box 16206, Chiswick, London W4 4ZL. Website: www.bsonline.techindex.co.uk

Other publications

Air Tightness Testing and Measurement Association (ATTMA) www.attma.org

Measuring Air Permeability of Building Envelopes, 2006.

Association for Specialist Fire Protection (ASFP) www.asfp.org.uk

- ASFP Red book – *Fire stopping and penetration seals for the construction industry*, 2nd edition, ISBN 1 87040 923 X
- ASFP Yellow book – *Fire protection for structural steel in buildings*, 4th edition, ISBN 1 87040 925 6
- ASFP Grey book – *Fire and smoke resisting dampers*, ISBN 1 87040 924 8
- ASFP Blue book – *Fire resisting ductwork*, 2nd edition, ISBN 1 87040 926 4.

The British Automatic Sprinkler Association (BAFSA) www.bafsa.org.uk

- Sprinklers for Safety. *Use and Benefits of Incorporating Sprinklers in Buildings and Structures*, 2006. ISBN 0 95526 280 1.

Building Research Establishment Ltd (BRE) www.bre.co.uk

- BR 262 *Thermal insulation: avoiding risks, 2001.* ISBN 1 86081 515 4
- BR 364 *Solar Shading of Buildings, 1999* ISBN 1 86081 2759

- BRE Digest 498 *Selecting lighting controls, 2006.* ISBN 1 86081 905 2
- BR 443 *Conventions for U-value calculations, 2006* (available at www.bre.co.uk/uvalues)
- Information Paper 1P1/06 *Assessing the effects of thermal bridging at junctions and around openings in the external elements of buildings,* 2006. ISBN 1 86081 904 4
- *Delivered energy emission factors for 2003* (available at www.bre.co.uk/filelibrary/2003emissionfactorupdate.pdf)
- *CO_2 emission figures for policy analysis,* July 2005 (available at www.bre.co.uk/filelibrary/co2emissionfigures2001.pdf)
- *Simplified Building Energy Model (SBEM) User Manual and Calculation Tool* (available at www.odpm.gov.uk).
- BR 128 *Guidelines for the construction of fire resisting elements*, 1988. ISBN 0 85125 293 1
- BR 135 *Fire performance of external thermal insulation for walls of multistorey buildings,* 2003. ISBN 978 1 86081 622 2
- BR 187 *External fire spread. Building separation and boundary distances*, 1991. ISBN 978 1 86081 465 5
- BR 208 *Increasing the fire resistance of existing timber floors*, 1988. ISBN 978 1 86081 359 7
- BR 274 *Fire safety of PFTE based materials used in buildings*, 1994. ISBN 978 1 86081 653 6
- BR 369 *Design methodologies for smoke and exhaust ventilation*, 1999. ISBN 978 1 86081 289 7
- BR 498 *Selecting lighting controls*, 2006. ISBN 1 86081 905 2
- BR 454 *Multi-storey timber frame buildings – a design guide*, 2003. ISBN 1 86081 605 3.

Builders Hardware Industry Federation
www.firecode.org.uk

- *Hardware for Fire and Escape Doors*, 2006. ISBN 0 95216 422 1.

Centre for Window and Cladding Technology
www.cwct.co.uk

- *Thermal assessment of window assemblies, curtain walling and non-traditional building envelopes,* 2006. ISBN 1 87400 338 6.

CIBSE
www.cibse.org

- CIBSE Commissioning Code M *Commissioning Management,* 2003. ISBN 1 90328 733 2
- CIBSE Guide A *Environmental Design, 2006.* ISBN 1 90328 766 8

- AM 10 *Natural ventilation in non-domestic buildings,* 2005. ISBN 1 90328 756 1
- TM 31 *Building Log Book Toolkit,* CIBSE, 2006. ISBN 1 90328 771 5
- TM 36 *Climate change and the indoor environment. 28 750 2 impacts and adaptation,* 2005. ISBN 1 903 28750 2
- TM 37 *Design for improved solar shading control,* 2006. ISBN 1 90328 757 X
- TM 39 *Building energy metering, 2006.* ISBN 1 90328 707 7.

Department for Communities and Local Government (DCLG) www.communities.gov.uk

- Regulatory Reform (Fire Safety) Order *Fire safety in adult placements: a code of practice*, 2005. ISBN 0 11 072 945 5.

Department for Education and Skills (DFES)

- Building Bulletin 1 *01 Ventilation of School Buildings, School Building and Design Unit, 2005* (download from www.teachernet.gov.uk/iaq).

Department of the Environment, Food and Rural, Affairs (Defra) www.defra.gov.uk

- The Government's Standard Assessment Procedure for energy rating of dwellings, SAP 2005 (available at www.bre.co.uk/sap2005).

Department of Health (DH) www.dh.gov.uk

- HTM 05 - 02 *Guidance in support of functional provisions for healthcare premises.*

Department of Transport, Local Government and the Regions (DTLR)

- *Limiting thermal bridging and air leakage: Robust construction details for dwellings and similar buildings,* Amendment 1. Published by TSO, 2002. ISBN 0 1 1 753 631 8 (available to download from Energy Saving Trust (EST) website on http://portal.est.org.uk/housingbuildings/calculators/robustdetails/).

Door and Shutter Manufacturers' Association (DSMA) www.dhfonline.org.uk

- *Code of practice for fire-resisting metal doorsets*, 1999.

Electrical Contractors' Association (ECA) and National Inspection Council for Electrical Installation Contracting (NICEIC)

- *ECA comprehensive guide to harmonised cable colours, BS 7671:2001 Amendment No 2,* ECA, March 2004.
- *Electrical Installers' Guide to the Building Regulations,* NICEIC and ECA, August 2004. Available from www.niceic.org.uk and www.eca.co.uk.
- *New fixed wiring colours – A practical guide,* NICEIC, Spring 2004.

Energy Saving Trust (EST) www.est.org.uk

- **CE66** *Windows for new and existing housing, 2006*
- CE129 *Reducing overheating – a designer's guide,* 2006.
- GPG268 *Energy efficient ventilation in dwellings a guide for specifiers,* 2006
- GIL20 *Low energy domestic lighting, 2006.*

English Heritage www.english-heritage.org.uk

- *Building Regulations and Historic Buildings,* 2002 (revised 2004).

Environment Agency www.environment-agency.gov.uk

- Pollution Prevention Guidelines (PPG18) *Managing Fire, Water and Major Spillages.*

Football Licensing Authority www.flaweb.org.uk/home.php

- *Concourses,* ISBN 0 95462 932 9.

Fire Protection Association (FPA) www.thefpa.co.uk

- *Design guide.*

Glass and Glazing Federation (GGF) www.ggf.org.uk

- *A guide to best practice in the specification and use of fire resistant glazed systems.*

Health and Safety Executive (HSE) www.hse.gov.uk

- *L24 Workplace Health, Safety and Welfare: Workplace (Health, Safety and Welfare) Regulations 1992, Approved Code of Practice and Guidance, The Health and Safety Commission, 1992.* ISBN 0 71760 413 6.

Heating and Ventilating Contractors Association

- DW/143 *A practical guide to ductwork leakage testing,* 2060. ISBN 0 90378 330 4
- DW/144 *Specification for sheet metal ductwork,* 1998. ISBN 0 90378 327 4.

Institution of Engineering Technology www.theiet.org

- *Electrician's guide to the Building Regulations, 2005.* ISBN 0 86341 463 X available from www.iee.org
- *IEE Guidance Note 1: Selection and erection of equipment,* 4th edition, 2002. ISBN 0 85296 989 9
- *IEE Guidance Note 2: Isolation and switching,* 4th edition, 2002. ISBN 0 85296 990 2
- *IEE Guidance Note 3: Inspection and testing,* 4th edition, 2002. ISBN 0 85296 991 0
- *IEE Guidance Note 4: Protection against fire,* 4th edition, 2003. ISBN 0 85296 992 9
- *IEE Guidance Note 5: Protection against electric shock,* 4th edition, 2003. ISBN 0 85296 993 7
- *IEE Guidance Note 6: Protection against overcurrent,* 4th edition, 2003. ISBN 0 85296 994 5
- *IEE Guidance Note 7: Special locations,* 2nd edition (incorporating the 1st and 2nd amendments), 2003. ISBN 0 85296 995 3
- *IEE On-Site Guide* (BS 7671 IEE Wiring Regulations, 16th edition), 2002. ISBN 0 85296 987 2
- *New wiring colours,* 2004. Leaflet available to download at www.iee.org/cablecolours.

International Association of Cold Storage Contractors (IACSC) www.iarw.org/iacsc/european_division

- *Design, construction, specification and fire management of insulated envelopes for temperature controlled environments*, 1999.

Metal Cladding and Roofing Manufacturers Association www.mcrma.co.uk

- *Guidance for design of metal cladding and roofing to comply with Approved Document L2.*

Modular and Portable Buildings Association (MPBA)
www.mpba.biz

- *Energy performance standards for modular and portable buildings, 2006.*

National Association of Rooflight Manufacturers
www.narm.org.uk

- *Use of rooflights to satisfy the 2002 Building Regulations for the Conservation of Fuel and Power.*

NBS (on behalf of ODPM)
www.thebuildingregs.com

- *Domestic Heating Compliance Guide,* 2006. ISBN 1 85946 225 1
- *Low or Zero Carbon Energy Sources: Strategic Guide,* 2006. ISBN 1 85946 224 3

Passive Fire Protection Federation (PFPF)
www.pfpf.org

- *Ensuring best practice for passive fire protection in buildings*, ISBN 1 87040 919 1.

Steel Construction Institute (SCI)
www.steel-sci.org

- SCI P197 *Designing for structural safety: A handbook for architects and engineers*, 1999. ISBN 1 85942 074 5
- SCI Publication 288 *Fire safe design. A new approach to multi-storey steel-framed buildings*, 2nd edition, 2000. ISBN 1 85942 169 5
- SCI Publication P313 *Single, storey steel framed buildings in fire boundary conditions*, 2002. ISBN 1 85942 135 0.

Thermal Insulation Manufacturers and Suppliers Association (TIMSA)
www.timsa.org.uk

- *HVAC Guidance for Achieving Compliance with Part L of the Building Regulations, 2006.*

Timber Research and Development Associations (TRADA)
www.trada.co.uk

Timber Fire-Resisting Doorsets; maintaining performance under the new European test standard, ISBN 1 90051 035 9.

Legislation

- SI 1991/1620 Construction Products Regulations 1991
- SI 1992/2372 Electromagnetic Compatibility Regulations 1992
- SI 1994/3051 Construction Products (Amendment) Regulations 1994
- SI 1994/3080 Electromagnetic Compatibility (Amendment) Regulations 1994
- SI 1994/3260 Electrical Equipment (Safety) Regulations 1994.
- SI 2001/3335 Building (Amendment) Regulations 2001
- SI 2005/1726 Energy Information (Household Air Conditioners) (No. 2) Regulations 2005
- SI 2006/652 Building And Approved Inspectors (Amendment) Regulations 2006.

Other

- 94/61 1 /EC implementing Article 20 of the Council Directive 89/1 06, IEEC on construction products
- Commission Decision 2000/147/EC of 8th February 2000 implementing Council Directive 89/106/EEC
- Commission Decision 2000/55&EC of 6th September 2000 implementing Council Directive 89/106/EEC
- Commission Decision 20001367/EC of 3rd May 2000 implementing Council Directive 89/106/EEC
- Commission Decision 2001/671/EC of 21 August 2001 implementing Council Directive 89/106/EC as regards the classification of the external fire performance of roofs and roof coverings
- Commission Decision 2005/823iEC of 22 November 2005 amending Decision 2001/671/EC regarding the classification of the external fire performance of roofs and roof coverings
- Commission Decision 961603/EC of 4th October 1996
- Construction Product (Amendment) Regulations 1 994 (SI 1 994 No 3051)
- Construction Products Regulations 1 991 (SI 1 991 No 1620)
- Disability Discrimination Act 1995.
- Education Act 1996
- Electrical Equipment (Safety) Regulations 1994 (SI 1994 No 3260)
- Electromagnetic Compatibility (Amendment) Regulations 1994 (SI 1994 No 3080)
- Electromagnetic Compatibility Regulations 1992 (ISI 1992 No 2372)
- European tests: Commission Decision 20001367,1 @C of 3rd May 2000 implementing Council Directive 89/106/EEC
- Health and Safety (safety signs and signals) Regulations 1996
- Pipelines Safety Regulations 1996, SI 1996 No 825 and the Gas Safety (Installation and Use) Regulations 1998 SI 1998 No 2451
- The Workplace (Health, Safety and Welfare) Regulations 1992

Acronyms

ach	air changes per hour
AHIPP	The Association of Home Information Pack Providers
ATTMA	Air Tightness Testing and Measurement Association
BAFSA	British Automatic Sprinkler Association
BER	Building (CO_2) Emission Rate
BHIF	British Hardware Industry Federation
BRE	Building Research Establishment
BS	British Standard
CIBSE	Chartered Institution of Building Services Engineers
CIRIA	Construction Industry Research and Information Association
CoPSO	Council of Property Search Organisations
CP	Competent Persons self-certification scheme
DCER	Dwellings Carbon Emission Rate
DCLG	Department for Communities and Local Government
DEA	Domestic Energy Assessors
DER	Dwellings (CO_2) Emission Rate
DFES	Department For Education and Skills
DH	Department of Health
DSMA	Door and Shutter Manufacturers Association
EPC	Energy Performance Certificate
EST	Energy Saving Trust
FPA	Fire Protection Association
GGF	Glass and Glazing Federation
HIP	Home Information Pack
HSE	Health and Safety Executive
HVAC	Heating Ventilation And Cooling
IACSC	International Association of Cold Storage Contractors
LZC	Low and Zero Carbon
MEV	Mechanical Extract Ventilation
MVHR	Continuous mechanical supply and extract with heat recovery
PCCB	Property Codes Compliance Board
PSV	Passive Stack Ventilation
RVA	Residential Ventilation Association
SAP	Standard Assessment Procedure
SBEM	Simplified Building Energy Model
SCI	Steel Construction Institute
SRHRV	Single Room Heat Recovery Ventilator
TEHVA	The Electric Heating and Ventilation Association
TER	Target Emissions Rate
TRADA	Timber Research and Development Association
TVOC	Total Volatile Organic Compound

Books by the same author

Title	Details	Publisher
WIRING REGULATIONS IN BRIEF	This essential reference will prove an invaluable guide for anyone based in the electrical industry working on electrical systems (design, installation, inspection and testing), who requires a comprehensive source of information on the specific requirements of the IEE Wiring Regulations (published by the IET), without having to trawl through the lengthy, complicated coverage of the Regulations themselves	Elsevier ISBN-13: 978 0 7506 6851 4
ISO 9001:2000 for Small Businesses (3rd edn) 	A guide to cost-effective compliance with the requirements of ISO 9001:2000	Butterworth-Heinemann ISBN: 0 7506 6617 X
ISO 9001:2000 in Brief (2nd edn) 	A 'hands-on' book providing practical information on how to cost-effectively set up an ISO 9001:2000 Quality Management System	Butterworth-Heinemann ISBN: 0 7506 6616 1

Title	Details	Publisher
ISO 9001:2000 Audit Procedures (2nd edn)	A complete set of audit check sheets and explanations to assist authors in completing internal, external and third part audits of *newly implemented, existing* and *transitional* QMSs	Butterworth-Heinemann ISBN: 0 7506 6615 3
Quality Management System for ISO 9001:2000  Quality Management System for ISO 9001:2000	'Quality Management System for ISO 9001: 2000' and accompanying CD is probably the most comprehensive set of ISO 9001:2000 compliant documents available world-wide. Fully customisable, it can be used as a basic template for any organisation wishing to work in compliance with, or gain registration to, ISO 9001:2000	ISBN: 0 9548 6474 3
Optoelectronic and Fiber Optic Technology 	An introduction to the fascinating technology of fiber optics	Butterworth-Heinemann ISBN: 0 7506 5370 1

Title	Details	Publisher
CE Conformity Marking	Essential information for any manufacturer or distributor wishing to trade in the European Union. Practical and easy to understand	Butterworth-Heinemann ISBN: 0 7506 4813 9
Environmental Requirements for Electromechanical and Electronic Equipment	Definitive reference containing all the background guidance, ranges, test specifications, case studies and regulations worldwide	Butterworth-Heinemann ISBN: 0 7506 3902 4
MDD Compliance using Quality Management Techniques	Easy to follow guide to MDD, enabling the purchaser to customize the Quality Management System to suit his own business	Butterworth-Heinemann ISBN: 0 7506 4441 9
Quality and Standards in Electronics 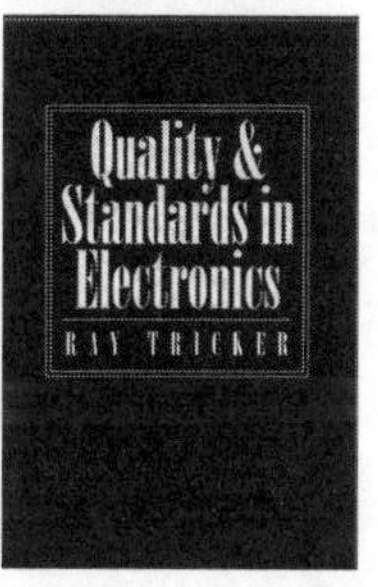	Ensures that manufacturers are aware of all the UK, European and international necessities, and know the current status of these regulations and standards, and where to obtain them	Butterworth-Heinemann ISBN: 0 7506 2531 7

CDs by the same author

Title	Details	ISBN
ISO 9001:2000 Quality Manual & Audit Checksheets	A CD containing a soft copy of the generic Quality Management System featured in *ISO 9001:2000 for Small Businesses* (3rd edn) plus a soft copy of all the checksheets and example audit forms contained in *ISO 9001:2000 Audit Procedures* (2nd edn)	0 9548647 2 7
ISO 9001:2000 Audit Checklists  	A CD containing all of the major audit checksheets and forms required to conduct either: • a simple internal review of a particular process; • an external (e.g. third party) assessment of an organization against the formal requirements of ISO 9001:2000; • an assessment of an organization against the formal requirements of other associated management standards such as **ISO 140001** and **OHSAS etc**	0 9548647 1 9
Aide Mémoire for ISO 9001:2000  	An electronic book (in CD format) that cuts out all the lengthy explanations, hypothesis and discussions about ISO 9001:2000 and just provides the reader with the bare details of the standard (e.g. What does the standard actually say? What needs to be checked? Where is a particular requirement covered in the standard? What will an auditor be looking for?, etc.). Also includes additional auditors checksheets and example forms etc. from *ISO 9001:2000 Audit Procedures* (2nd edn) and *ISO 9001:2000 for Small Businesses* (3rd edn)	0 9548647 0 0

Useful contact names and addresses

The following professional body is willing to provide general and informal advice about the Act. **However**, any advice given should **not** be seen as being endorsed by the Office of the Deputy Prime Minister!

The Royal Institution of Chartered Surveyors (RICS)
Technical Services Unit
12 Great George Street
London, SW1P 3AD
Tel: 020 7222 7000 (extension 492)
Fax: 020 7222 9430

The following bodies hold lists of their members who may be willing to provide professional advice or act as a 'surveyor' under the Act – again with the proviso that any advice given should **not** be seen as being endorsed by the Office of the Deputy Prime Minister!

Architecture and Surveying Institute
Register of Party Wall Surveyors
St Mary House
15 St Mary Street
Chippenham
Wiltshire, SN15 3WD
Tel: 01249 444505
Fax: 01249 443602

The Association of Building Engineers (ABE)
Private Practice Register
Lutyens House
Billing Brook Road
Weston Favell
Northampton, NN3 8NW
Tel: 01604 404121
Fax: 01604 784220

The Pyramus & Thisbe Club
Florence House
53 Acton Lane
London, NW10 8UX
Tel: 020 8961 3311
Fax: 020 8963 1689

The Royal Institute of British Architects (RIBA)
Clients Advisory Service
66 Portland Place
London, W1N 4AD
Tel: 020 7307 3700
Fax: 020 7436 9112

The Royal Institution of Chartered Surveyors (RICS)
Information Centre
12 Great George Street
London, SW1P 3AD
Tel: 020 7222 7000
Fax: 020 7222 9430

Professional contacts

Asbestos specialists

Asbestos Information Centre Ltd
PO Box 69
Widnes
Cheshire

WA8 9GW
0151 420 5866

Concrete specialist

British Ready Mixed Concrete Association
The Bury
Church Street
Chesham
Buckinghamshire
HP5 1JE
01494 791050

Damp, rot, infestation

British Wood Preserving & Damp Proofing Association
Building No. 6
The Office Village
4 Romford Road
Stratford
London
E15 4EA
020 8519 2588

English Nature
Northminster House
Northminster Road
Peterborough
PE1 1VA
01733 340345

Countryside Council for Wales
Plaspenrhos
Penrhos Road
Bangor
Gwynedd
LL57 2LG
01248 370444

Scottish Natural Heritage
12 Hope Terrace
Edinburgh
EH9 2AS
0131 447 4784

Local Department of Environmental Health
Refer to your local directory

Decorators

British Decorators Association
32 Coton Road
Nuneaton
Warwickshire
CV11 5TW
01203 353776

Scottish Decorators Federation
41A York Place
Edinburgh
EH1 3HT
0131 557 9345

Electricians

National Inspection Council for Electrical Installation Contracting
37 Albert Embankment
London
SE1 7UJ
020 7582 7746

Fencing erectors

Fencing Contractors Association
Warren Rd
Trellech
Monmouthshire
NP25 4PQ
07000 560722

Glazing specialist

Glass and Glazing Federation
44–48 Borough High Street
London
SE1 1XB
020 7403 7177

Heating installers

British Gas Regional Office
Refer to your local directory

Electricity Supply Company
Refer to your local directory

British Coal Corporation
Hobart House
Grosvenor Place

London
SW1X 7AE
020 7235 2020

Heating and Ventilating Contractors Association
Esca House
34 Palace Court
London
W2 4JG
020 7229 2488

National Association of Plumbing, Heating and Mechanical Services Contractors
Ensign House
Ensign Business Centre
Westwood Way
Coventry
CV4 8JA
01203 470626

Home security

Local Crime Prevention Officer
Refer to your local directory

Local Fire Prevention Officer
Refer to your local directory

National Approval Council for Security Systems
Queensgate House
14 Cookham Road
Maidenhead
S16 8AJ
01628 37512

Master Locksmiths Association
Units 4–5
Woodford Halse Business Park
Great Central Way
Woodford Halse
Daventry
NN1 6PZ
01327 62255

British Security Industry Association
Security House
Barbourne Road
Worcester
WR1 1RS
01905 21464

Insulation installers

Draught Proofing Advisory Association Ltd,
External Wall Insulation Association,
National Cavity Insulation Association,
National Association of Loft Insulation Contractors
PO Box 12
Haslemere
Surrey
GU27 3AH
01428 654011

Plasterers

Federation of Master Builders
14 Great James Street
London
WC1N 3DP
020 7242 7583

Plumbers

National Association of Plumbing, Heating and Mechanical Services Contractors
Ensign House
Ensign Business Centre
Westwood Way
Coventry
CV4 8JA
01203 470626

Roofers

Builders' Merchants Federation
15 Soho Square
London
W1V 5FB
020 7439 1753

National Federation of Roofing Contractors
24 Weymouth Street

London
W1G 7LX
020 7436 0387

Ventilation

Heating and Ventilating Contractors Association
Esca House
34 Palace Court
London
W2 4JG
020 7229 2488

Other useful contacts

British Board of Agrément (BBA)
PO Box 195
Bucknalls Lane
Garston
Watford
WD2 7NG
Tel: 01923 665300
Fax: 01923 665301
E-mail: bba@btinternet.com
Internet: Http://www.bbacerts.co.uk

BSI
British Standards Institution
389 Chiswick High Road
London
W4 4AL
Tel: 0181 996 9001
Fax: 0181 996 7001
E-mail: info@bsi.org.uk
Internet: www.bsi.org.uk

Fensa Ltd
Fenestration Self-Assessment Scheme
44–48 Borough High Street
London SE1 1XB
Tel: 020 7207 5874
Fax: 020 7357 7458

HETAS Ltd
HETAS Ltd
12 Kestrel Walk
Letchworth
Hertfordshire
SG6 2TB
Tel: 01462 634721
Fax: 01462 674329

Note: Registration scheme for companies and engineers involved in the installation and maintenance of domestic solid fuel fired equipment.

HMSO
The Stationery Office
The Publications Centre
PO Box 29
Norwich
NR3 1GN
Telephone orders/General enquiries
0870 600 5522
Fax orders 0870 600 5533
www.thestationeryoffice.com

Institute of Plumbing
64 Station Lane
Hornchurch
Essex
RM12 6NH
Tel: 01708 472791
Fax: 01708 448987

Note: Approved Contractor Person Scheme (Building Regulations).

OFTEC
Oil Firing Registration Scheme
Century House
100 High Street
Banstead
Surrey
SM7 2NN
Tel: 01737 373311
Fax: 01737 373553

Robust Details Limited
PO Box 7289
Milton Keynes
MK14 6ZQ
Business Line: 0870 240 8210
Technical Support Line: 0870 240 8209
Fax: 0870 240 8203

UKAS
United Kingdom Accreditation Service
21–47 High Street
Feltham, Middlesex
TW3 4UN
Tel: 0208 917 8400
Fax: 0208 917 8500

WIMLAS
WIMLAS Limited
St Peter's House
6–8 High Street
Aver, Buckinghamshire
SL0 9NG
Tel: 01753 737744
Fax: 01753 792321
E-mail: wimlas@compuserve.com

Air Tightness Testing and Measurement Association (ATTMA)	www.attma.org
BRE	www.bre.co.uk
Centre for Window and Cladding Technology (CWCT)	www.cwct.co.uk
Chartered Institution of Building Services Engineers (CIBSE)	www.cibse.org
Department for Education and Skills (DFES)	www.dfes.gov.uk
Department of the Environment, Food and Rural Affairs (Defra)	www.defra.gov.uk
Department of Transport, Local Government and the Regions (DTLR)	www.dtlr.gov.uk
Energy Saving Trust (EST)	www.est.org.uk
English Heritage	www.english-heritage.org.uk
Health and Safety Executive (HSE)	www.hse.gov.uk
Heating and Ventilating Contractors Association (HVCA)	www.hvca.org.uk
Metal Cladding and Roofing Manufacturers Association (MCRMA)	www.mcrma.co.uk
Modular and Portable Buildings Association (MPBA)	www.mpba.biz
National Association of Rooflight Manufacturers (NARM)	www.narm.org.uk
NBS (on behalf of ODPM)	www.thebuildingregs.com
The Planning Portal	www.planningportal.gov.uk/england/genpub/en/
Thermal Insulation Manufacturers and Suppliers Association (TIMSA)	www.timsa.org.uk
TrustMark	www.**trustmark**.org.uk

Useful websites

The Building Act and Building Regulations www.communities.gov.uk (the new DCLG site)

Approved Documents
www.planningportal.gov.uk/england/professionals/en11531410382.html

Assessable thresholds	www.tso.co.uk
British Standards	www.bsonline.techindex.co.uk
Building near trees	www.nhbc.co.uk
Building Research	www.bre.co.uk
Carbon dioxide from natural sources and mining areas	www.bgs.ac.uk www.tso.co.uk www.defra.gov.uk www.ciria.org.uk
Cladding	www.mcrma.co.uk
Concrete in aggressive ground	www.bre.co.uk
Contaminated land	www.defra.gov.uk www.ciria.org.uk www.hse.co.uk
Contamination in disused coal mines	www.tso.co.uk
Demolition	www.ciria.org.uk
Electrical safety	www.theiet.org
Environmental aspects	www.arup.com www.environment-agency.gov.uk
Excavation and disposal	www.defra.gov.uk www.ciria.org.uk
Flood protection	www.ciria.org.uk www.safety.odpm.gov.uk
Flooding from sewers	www.defra.gov.uk www.ciria.org.uk
Foundations	www.bre.co.uk
Gas contaminated land	www.defra.gov.uk www.bre.co.uk www.ciria.org.uk
Geoenvironmental and geotechnical investigations	www.ags.org.uk
Glass and glazing	www.ggf.org.uk
Hardcore	www.bre.co.uk
Health and safety	www.hse.co.uk www.defra.gov.uk
Land quality	www.environment-agency.gov.uk
Landfill gas	www.ciwm.co.uk www.defra.gov.uk www.gassim.co.uk
Laying water pipelines in contaminated ground	www.fwr.org
Low-rise buildings	www.bre.co.uk
Materials and workmanship	www.tso.co.uk
Methane	www.bgs.ac.uk www.tso.co.uk www.defra.gov.uk www.ciria.org.uk
Oil seeps from natural sources and mining areas	www.bgs.ac.uk www.tso.co.uk www.defra.gov.uk www.ciria.org.uk

Petroleum retail sites	www.petroleum.co.uk
Pollution control	www.odpm.gov.uk
Protection of ancient buildings	www.spab.org.uk
Radon	www.bre.co.uk
Robust construction details	www.tso.co.uk
Roofing design	www.mcrma.co.uk
Shrinkable clay soils	www.bre.co.uk
Soil sampling	www.defra.gov.uk
Soils, sludge and sediment	www.ciria.org.uk
Subsidence	www.bre.co.uk
Thermal bridging	www.bre.co.uk www.tso.co.uk
Thermal insulation	www.bre.co.uk
Timbers	www.bre.co.uk

Index